PROPERTY OF
THE NATIONAL INSTITUTES OF HEALTH

Laboratory of Neurosciences
NIA, NIH

MOLECULAR GENETIC NEUROSCIENCE

Molecular Genetic Neuroscience

Editors

Francis O. Schmitt, Ph.D. Stephanie J. Bird, Ph.D.

Floyd E. Bloom, M.D.

Raven Press ■ New York

Raven Press, 1140 Avenue of the Americas, New York, New York 10036

© 1982 by Raven Press Books, Ltd. All rights reserved. This book is protected by copyright. No part of it may be reproduced, stored in a retrieval system, or transmitted, in any form or by any means, electronic, mechanical, photocopying, recording, or otherwise, without the prior written permission of the publisher.

Made in the United States of America

Library of Congress Cataloging in Publication Data
Main entry under title:

Molecular genetic neuroscience.

Includes bibliographical references and index.
1. Nervous system—Diseases—Genetic aspects.
2. Molecular genetics. 3. Gene expression. I. Schmitt, Francis Otto, 1903– . II. Bird, Stephanie J.
III. Bloom, Floyd E.; 1936– . [DNLM: 1. Genetics, Biochemical—Congresses. 2. Neurochemistry—Congresses. WL 104 M718 1981]
RC346.M64   599.01'88   82-7699
ISBN 0-89004-744-8   AACR2

Great care has been taken to maintain the accuracy of the information contained in the volume. However, Raven Press cannot be held responsible for errors or for any consequences arising from the use of the information contained herein.

Preface

Exciting developments that may profoundly alter the life sciences are currently being made in two fields: molecular genetics and neuroscience. This volume was developed to stimulate further advances by encouraging the application of current research concepts and techniques in molecular biology to the nervous system.

In the last two decades, discoveries in molecular genetics have revealed much about the very basic mechanisms on which life depends. More recently, advances have been made in recombinant DNA technology, including genetic engineering, that may introduce dramatically new and productive methodologies which, in their effects, could rival or exceed the advances produced by the Industrial Revolution when power tools and assembly lines replaced hand tools and hand crafting in the factory. In the genetic engineering revolution, now in an early stage of its development, microorganisms and macromolecular complexes, including specifically sequenced polynucleotides ("artificial genes"), may be used in synthesizing valuable new products as well as variations of others that are now available only in small quantities. Even more important are the clues that may be generated by the molecular genetic reevaluation concerning functional aspects at the cellular and the system (e.g., brain) levels. These developments are expected to have a great impact on science, medicine, agriculture, industry, and human welfare in general.

In the neurosciences, a true community has been formed "... by joining together specialists from a number of previously separate communities in the shared pursuit of penetrating the complexities of the human brain and behavior," as the science historian Judith Swazey has aptly put it. Neuroscience is now at the cutting edge of the new biology.

If the explosive developments in molecular biology and neuroscience could be coupled, the impact on basic science and on clinical medicine would be very powerful indeed. This might also nucleate the formation of a single, strong community of molecular genetic neuroscientists.

To initiate such a coupling, it was decided in the summer of 1980 that the staff of the Neurosciences Research Program should begin planning a conference that would include a judicious combination of tutorial lectures on various aspects of molecular genetics, and lectures by innovative neuroscientists who had already made substantial applications of molecular genetics in their research programs. From this there emerged the week-long conference, held at Woods Hole, Massachusetts, on which this volume is based.

Francis O. Schmitt
Stephanie J. Bird
Floyd E. Bloom

Acknowledgments

The primary goal of the Neurosciences Research Program, an interdisciplinary research organization of the Massachusetts Institute of Technology and Rockefeller University, is to facilitate the investigation of the nervous system and behavior. To this end, the NRP conducts scientific meetings to explore crucial problems in the neurosciences and to issue publications that result from them.

Grateful recognition is given to those who assisted the editors in planning the conference on which this volume is based: H. M. Goodman, J. F. Habener, W. E. Hahn, J. B. Martin, B. S. McEwen, A. Rich, and E. M. Shooter, together with NRP staff members, G. Adelman, L. K. Gerbrandt, F. E. Samson, and F. G. Worden.

Sincere gratitude is due Frederic G. Worden for invaluable administrative support of this project from its inception; to Katheryn Cusick for assistance at the administrative level and for valuable advice and assistance with all problems that arose, great or small; to Lauren K. Gerbrandt for assistance in the early planning; to Frederick E. Samson for valuable planning and editorial assistance; to George Adelman for assistance and advice concerning editorial and publication matters; to Joyce Taylor Snow for expert assistance in the final stages of editing; and to the secretarial staff for typing and proofreading manuscripts.

Sponsorship and Support

Grateful acknowledgment for the direct support of the conference on which this volume is based is made to the Camille and Henry Dreyfus Foundation, Inc. The Neurosciences Research Program is supported in part through the Massachusetts Institute of Technology by the National Institute of Neurological and Communicative Disorders and Stroke (Grant No. NS 15690) and the National Institute of Mental Health (Grant No. MH 23132) and through the Neurosciences Research Foundation, Inc., by the Lita Annenberg Hazen Charitable Trust, the Beverly and Harvey Karp Foundation, the John D. and Catherine T. MacArthur Foundation, the van Ameringen Foundation, Inc., the G. Unger Vettlesen Foundation, and the Vollmer Foundation, Inc.

Contents

I. Introductory Essay

1   A Protocol for Molecular Genetic Neuroscience   1
    Francis O. Schmitt

II. Molecular Organization of DNA in Relation to Gene Function

2   Left-handed DNA   13
    Alexander Rich

III. Control of Gene Expression

3   Regulatory Mechanisms of Gene Expression in Higher Eukaryotes   25
    James E. Darnell, Jr., and Michael C. Wilson

4   Components of an Efficient Molecular Switch   37
    Mark Ptashne

5   Regulation of Gene Expression   47
    Phillip A. Sharp and Michael C. Wilson

IV. Intercellular Gene Transfer

6   Applications of Somatic Cell Genetics and Gene Transfer Techniques for the Analysis of the Genetic Control of Developments   63
    Frank H. Ruddle

V. Structure and Diversification of Mammalian Genes

7   Antibody Genes: Arrangements and Rearrangements   75
    Leroy E. Hood

8   Molecular Genetics of Human Globin Gene Expression   87
    Thomas Maniatis, Pamela Mellon, Vann Parker, Nicholas Proudfoot, and Brian Seed

9   Structure, Evolution, and Expression of Mammalian Insulin Genes   103
    Howard M. Goodman, Paul Berg, Steve Clark, Barbara Cordell, Don Diamond, Chi Nguyen-Huu, Yuet W. Kan, and Roger V. Lebo

VI. Antibodies: Synthesis and Use as Tools

10 Antibodies to Chemically Synthesized Peptides from DNA Sequences as Probes of Gene Expression 119
    Richard A. Lerner

11 Applications of Monoclonal Antibodies in Biological Research 127
    Edgar Haber

12 Analysis of Retina and Other Neural Tissues Using Cell-Specific Antibodies 137
    Colin J. Barnstable

VII. Posttranslational Processing

13 Proteolytic Processing of Secretory Proteins 149
    Donald F. Steiner

14 The Processing of Cell-Surface Glycoproteins 161
    Phillips W. Robbins

15 Precursor Processing and the Neurosecretory Vesicle 171
    Harold Gainer

16 Neuropeptide Degradation 189
    Jeffrey F. McKelvy, James E. Krause, and J. P. Advis

VIII. Molecular Genetics of Opioid Peptides

17 Endorphins and Enkephalins: Historical Aspects of Opioid Peptides 203
    Dorothy T. Krieger

18 Regulation of Expression of Pro-Opiomelanocortin and Related Genes in Various Tissues: Use of Cell-Free Systems and Hybridization Probes 219
    Edward Herbert, Olivier Civelli, Neal Birnberg, Patricia Rosa, and Michael Uhler

19 Structure and Expression of Hormone Genes 231
    John D. Baxter, Peter L. Whitfeld, Peter H. Seeburg, Andrea Barta, Nancy E. Cooke, Norman L. Eberhardt, Robert I. Richards, Guy Cathala, Maurice Wegnez, Joseph A. Martial, and John Shine

20 A Multivalent Proenkephalin and Its Processing 239
    Sidney Udenfriend

21 Dynorphin and the Dynorphin Receptor: Some Implications of Gene Duplication of the Opioid Message 249
    Avram Goldstein

IX. Steroid Hormones and Gene Expression

22 Steroid Hormone Action in the Brain: Cellular and Behavioral Effects   265
   Bruce S. McEwen

23 Steroid-Controlled Gene Expression in a *Drosophila* Cell Line   277
   Peter Cherbas, Charalambos Savakis, Lucy Cherbas, and
     M. Macy D. Koehler

24 Putative Steroid Receptors: Genetics and Development   289
   Thomas O. Fox, Kathie L. Olsen, Christine C. Vito, and Steven J. Wieland

25 Regulation of Hormonally Controlled mRNA Synthesis and Gene
      Expression in Variant Somatic Cells and Cell Hybrids   307
   E. Brad Thompson and Jeanine S. Strobl

X. Genetic Expression in the Nervous System

26 Overview of the Molecular Genetics of Mouse Brain   323
   William E. Hahn, Jeffrey Van Ness, and Nirupa Chaudhari

27 The Regulation of Gene Expression During Terminal Neurogenesis   335
   François Gros, Bernard Croizat, Marie-Madeleine Portier,
     Frances Berthelot, and Armando Felsani

28 Somatostatin, Glucagon, and Calcitonin: A Molecular Approach to
      Biosynthesis in Peripheral and Neural Tissues   349
   Joel F. Habener, Richard H. Goodman, John W. Jacobs, and P. Kay Lund

29 Somatostatin in the Nervous System   359
   Seymour Reichlin

30 Early Translational Events in the Synthesis of Acetylcholine Receptor   373
   David J. Anderson, Peter Walter, and Günter Blobel

XI. Genetically Determined Disorders of the Nervous System

31 Mutations Affecting the Central Nervous System in the Mouse   389
   Richard L. Sidman

32 Genetic Disorders of the Human Nervous System: A Commentary   401
   Victor A. McKusick

33 Huntington's Disease: Genetically Determined Cell Death in the Mature
      Human Nervous System   407
   Joseph B. Martin

34 Molecular Genetic Approaches to Neural Degenerative Disorders   415
   David E. Housman and James F. Gusella

XII. Trophic and Instructive Factors

35 Modulation of Neuronal Function by Nerve Growth Factor — 425
Eric M. Shooter, Peter Frey, Peter W. Gunning, Gary E. Landreth, Arne Sutter, and Bruce A. Yankner

36 Cellular and Hormonal Interactions in the Development of Sympathetic Neurons — 437
Paul H. Patterson

37 A Possible Role for Cytoplasmic Calcium in the Regulation of Gene Expression — 445
Anthony N. Martonosi

XIII. Molecular Genetic Approach to Neurobiological Problems

38 Strategies for Application of Molecular Genetics to Neurobiology — 459
Hans Thoenen

XIV. Selected Summary

39 Prospects from Retrospect — 467
Floyd E. Bloom

List of Abbreviations — 477

Amino Acid Symbols and Genetic Code — 479

Subject Index — 481

Contributing Fellows

To facilitate rapid publication of the conference proceedings, 10 Contributing Fellows, listed below and included in the list of participants, were appointed to provide rough drafts for use by lecturers in preparing their manuscripts for publication.

Xandra O. Breakefield	Jiri Novotny
Louis J. De Gennaro	Petro E. Petrides
Thomas H. Fraser	Charalambos Savakis
James F. Gusella	J. Gregor Sutcliffe
Robert J. Milner	Michael C. Wilson

Contributors

J. P. Advis
Department of Neurobiology and Behavior
State University of New York
Stony Brook, New York 11794

David J. Anderson
Laboratory of Cell Biology
The Rockefeller University
New York, New York 10021

Andrea Barta
Endocrine Research Division
University of California
San Francisco, California 94143

Colin J. Barnstable
Department of Neurobiology
Harvard Medical School
Boston, Massachusetts 02115

John D. Baxter
Endocrine Research Division
University of California
San Francisco, California 94143

Paul Berg
Department of Biochemistry and Biophysics
University of California School of Medicine
San Francisco, California 94143

Frances Berthelot
Institut Pasteur
75724 Paris, France

Stephanie J. Bird
Neurosciences Research Program
Jamaica Plain, Massachusetts 02130

Neal Birnberg
Department of Chemistry
University of Oregon
Eugene, Oregon 97403

Günter Blobel
Laboratory of Cell Biology
The Rockefeller University
New York, New York 10021

Floyd E. Bloom
Arthur V. Davis Center for Behavioral
  Neurobiology
The Salk Institute for Biological Studies
San Diego, California 92138

Xandra O. Breakefield
Department of Human Genetics
Yale University School of Medicine
New Haven, Connecticut 06510

Guy Cathala
Endocrine Research Division
University of California
San Francisco, California 94143

Nirupa Chaudhari
Department of Anatomy
School of Medicine
University of Colorado
Denver, Colorado 80262

Lucy Cherbas
Biological Laboratories
Harvard University
Cambridge, Massachusetts 02138

Peter Cherbas
Biological Laboratories
Harvard University
Cambridge, Massachusetts 02138

Olivier Civelli
Department of Chemistry
University of Oregon
Eugene, Oregon 97403

Steve Clark
Department of Biochemistry and Biophysics
University of California School of Medicine
San Francisco, California 94143

Nancy E. Cooke
Endocrine Research Division
University of California
San Francisco, California 94143

CONTRIBUTORS

Barbara Cordell
Department of Biochemistry and Biophysics
University of California School of Medicine
San Francisco, California 94143

Francis H. C. Crick
The Salk Institute for Biological Studies
San Diego, California 92138

Bernard Croizat
College de France
Biochemie Cellulaire
75231 Paris, France

James E. Darnell
The Rockefeller University
New York, New York 10021

Louis J. De Gennaro
Department of Pharmacology
Yale University School of Medicine
New Haven, Connecticut 06510

Don Diamond
Department of Biochemistry and Biophysics
University of California School of Medicine
San Francisco, California 94143

Norman L. Eberhardt
Endocrine Research Division
University of California
San Francisco, California 94143

Gerald M. Edelman
Department of Biochemistry
The Rockefeller University
New York, New York 10021

Armando Felsani
Institut Pasteur
75724 Paris, France

Thomas O. Fox
Department of Neuroscience
Children's Hospital Medical Center
Boston, Massachusetts 02115

Thomas H. Fraser
The Upjohn Company
Kalamazoo, Michigan 49001

Peter Frey
Department of Neurobiology
Stanford University School of Medicine
Stanford, California 94305

Harold Gainer
Functional Neurochemistry
Laboratory of Developmental Neurobiology
National Institute of Child Health and Human Development
National Institutes of Health
Bethesda, Maryland 20205

Avram Goldstein
Addiction Research Foundation
Palo Alto, California 94304

Howard M. Goodman
Department of Biochemistry and Biophysics
University of California School of Medicine
San Francisco, California 94143

Richard H. Goodman
Laboratory of Molecular Endocrinology
Massachusetts General Hospital
Boston, Massachusetts 02114

François Gros
Institut Pasteur
75724 Paris, France

Peter W. Gunning
Department of Neurobiology
Stanford University School of Medicine
Stanford, California 94305

James F. Gusella
Neurology and Children's Services
Massachusetts General Hospital
Boston, Massachusetts 02114

Joel F. Habener
Laboratory of Molecular Endocrinology
Massachusetts General Hospital
Boston, Massachusetts 02114

Edgar Haber
Molecular and Cellular Research Laboratory
Massachusetts General Hospital
Boston, Massachusetts 02114

William E. Hahn
Department of Anatomy
School of Medicine
University of Colorado
Denver, Colorado 80262

Edward Herbert
Department of Chemistry
University of Oregon
Eugene, Oregon 97403

CONTRIBUTORS

Leroy E. Hood
Division of Biology
California Institute of Technology
Pasadena, California 91125

David E. Housman
Department of Biology and Center for Cancer Research
Massachusetts Institute of Technology
Cambridge, Massachusetts 02139

Leslie L. Iversen
MRC Neurochemical Pharmacology Unit
Medical Research Council Centre
Medical School
Cambridge CB2 2QH, England

John W. Jacobs
Laboratory of Molecular Endocrinology
Massachusetts General Hospital
Boston, Massachusetts 02114

Yuet W. Kan
Department of Biochemistry and Biophysics
University of California School of Medicine
San Francisco, California 94143

M. Macy D. Koehler
Biological Laboratories
Harvard University
Cambridge, Massachusetts 02138

Masakazu Konishi
Division of Biology
California Institute of Technology
Pasadena, California 91125

James E. Krause
Department of Neurobiology and Behavior
State University of New York
Stony Brook, New York 11794

Dorothy T. Krieger
Division of Endocrinology
Mount Sinai Medical Center
New York, New York 10029

Gary E. Landreth
Department of Neurobiology
Stanford University School of Medicine
Stanford, California 94305

Roger V. Lebo
Department of Biochemistry and Biophysics
University of California School of Medicine
San Francisco, California 94143

Richard A. Lerner
Department of Cellular and Developmental Immunology
Research Institute of Scripps Clinic
La Jolla, California 92037

Rodolfo R. Llinas
Department of Physiology and Biophysics
New York University Medical Center
New York, New York 10016

P. Kay Lund
Laboratory of Molecular Endocrinology
Massachusetts General Hospital
Boston, Massachusetts 02114

Thomas Maniatis
Department of Biochemistry and Molecular Biology
Harvard University
Cambridge, Massachusetts 02138

Joseph A. Martial
Endocrine Research Division
University of California
San Francisco, California 94143

Joseph B. Martin
Department of Neurology
Massachusetts General Hospital
Boston, Massachusetts 02114

Anthony N. Martonosi
Department of Biochemistry
State University of New York
Upstate Medical Center
Syracuse, New York 13210

Bruce S. McEwen
Department of Neurobiology
The Rockefeller University
New York, New York 10021

Jeffrey F. McKelvy
Department of Neurobiology and Behavior
State University of New York
Stony Brook, New York 11794

Victor A. McKusick
Department of Medicine
The Johns Hopkins Hospital
Baltimore, Maryland 21205

Pamela Mellon
Department of Biochemistry and Molecular Biology
Harvard University
Cambridge, Massachusetts 02138

Robert J. Milner
Arthur V. Davis Center for Behavioral
  Neurobiology
The Salk Institute for Biological Studies
San Diego, California 92138

Robert Y. Moore
Department of Neurology
State University of New York
Stony Brook, New York 11794

Paul Mueller
Department of Biochemistry and Biophysics
University of Pennsylvania
School of Medicine
Philadelphia, Pennsylvania 19174

Chi Nguyen-Huu
Department of Biochemistry and Biophysics
University of California School of Medicine
San Francisco, California 94143

Jiri Novotny
Molecular and Cellular Research Laboratory
Massachusetts General Hospital
Boston, Massachusetts 02114

Kathie L. Olsen
Department of Neuroscience
Children's Hospital Medical Center
Boston, Massachusetts 02115

Sanford L. Palay
Department of Anatomy
Harvard Medical School
Boston, Massachusetts 02115

Vann Parker
Department of Biochemistry and Molecular
  Biology
Harvard University
Cambridge, Massachusetts 02138

Paul H. Patterson
Department of Neurobiology
Harvard Medical School
Boston, Massachusetts 02115

Petro E. Petrides
Department of Neurobiology
Stanford University Medical School
Stanford, California 94305

Marie-Madeleine Portier
Institut Pasteur
75724 Paris, France

Nicholas Proudfoot
Department of Biochemistry and Molecular
  Biology
Harvard University
Cambridge, Massachusetts 02138

Mark Ptashne
The Biological Laboratories
Department of Biochemistry
Harvard University
Cambridge, Massachusetts 02138

Seymour Reichlin
Department of Medicine
New England Medical Center Hospital
Boston, Massachusetts 02111

Alexander Rich
Department of Biology
Massachusetts Institute of Technology
Cambridge, Massachusetts 02139

Robert I. Richards
Endocrine Research Division
University of California
San Francisco, California 94143

Phillips W. Robbins
Center for Cancer Research
Massachusetts Institute of Technology
Cambridge, Massachusetts 02139

Patricia Rosa
Department of Chemistry
University of Oregon
Eugene, Oregon 97403

Frank H. Ruddle
Kline Biology Tower
Yale University
New Haven, Connecticut 06511

Fred E. Samson
R. Smith Center for Mental Retardation
University of Kansas Medical Center
Kansas City, Kansas 66103

Charalambos Savakis
Biological Laboratories
Harvard University
Cambridge, Massachusetts 02138

Francis O. Schmitt
Neurosciences Research Program
Jamaica Plain, Massachusetts 02130

CONTRIBUTORS

Peter H. Seeburg
Endocrine Research Division
University of California
San Francisco, California 94143

Brian Seed
Department of Biochemistry and Molecular Biology
Harvard University
Cambridge, Massachusetts 02138

Phillip A. Sharp
Department of Biology
Massachusetts Institute of Technology
Cambridge, Massachusetts 02139

John Shine
Endocrine Research Division
University of California
San Francisco, California 94143

Eric M. Shooter
Department of Neurobiology
Stanford University School of Medicine
Stanford, California 94305

Richard L. Sidman
Department of Neurosciences
Children's Hospital
Boston, Massachusetts 02115

Donald F. Steiner
Department of Biochemistry
University of Chicago
Chicago, Illinois 60637

Charles F. Stevens
Department of Physiology
Yale University School of Medicine
New Haven, Connecticut 06510

Jeanine S. Strobl
Laboratory of Biochemistry
National Cancer Institute
National Institutes of Health
Bethesda, Maryland 20205

J. Gregor Sutcliffe
Department of Immunopathology
Research Institute of Scripps Clinic
La Jolla, California 92037

Arne Sutter
Department of Neurobiology
Stanford University School of Medicine
Stanford, California 94305

Hans Thoenen
Department of Neurochemistry
Max Planck Institute for Psychiatry
8033 Martinsried-Munich
West Germany

E. Bradbridge Thompson
Laboratory of Biochemistry
National Cancer Institute
National Institutes of Health
Bethesda, Maryland 20205

Sidney Udenfriend
Roche Institute of Molecular Biology
Nutley, New Jersey 07110

Michael Uhler
Department of Chemistry
University of Oregon
Eugene, Oregon 97403

Jeffrey Van Ness
Department of Anatomy
School of Medicine
University of Colorado
Denver, Colorado 80262

Christine C. Vito
Department of Neuroscience
Children's Hospital Medical Center
Boston, Massachusetts 02115

Peter Walter
Laboratory of Cell Biology
The Rockefeller University
New York, New York 10021

Maurice Wegnez
Endocrine Research Division
University of California
San Francisco, California 94143

Peter L. Whitfield
Endocrine Research Division
University of California
San Francisco, California 94143

Steven J. Wieland
Department of Neuroscience
Children's Hospital Medical Center
Boston, Massachusetts 02115

Michael C. Wilson
Department of Immunopathology
Research Institute of Scripps Clinic
La Jolla, California 92037

Frederic G. Worden
Neurosciences Research Program
Jamaica Plain, Massachusetts 02130

Bruce A. Yankner
Department of Neurobiology
Stanford University School of Medicine
Stanford, California 94305

A Protocol for Molecular Genetic Neuroscience

Francis O. Schmitt

ABSTRACT

There are many ways to document the view that introducing the concepts and technologies of molecular genetics and recombinant DNA into the mainstream of neuroscience thinking and experimentation will prove highly rewarding both to neuroscientists and to molecular geneticists. As an introduction, it seemed desirable to point out some of the exciting and highly challenging aspects of neuroscience that would gain enormously from a research program in which molecular geneticists and neuroscientists collaborated. It is hoped that this volume will help to stimulate such interaction.

THE IMPORTANCE OF CHEMICAL CIRCUITRY IN NEUROSCIENCE

Brain science has traditionally been dominated by concepts of circuitry: structural (cellular) bioelectrical, and biochemical (see Schmitt, 1979). Structural and bioelectrical circuitry have been fruitfully studied over many years; chemical circuitry emerged only recently and is highly relevant to the subject matter of this volume.

Within the elongate axons and dendrites—collectively called neurites for convenience (Shepherd, 1979)—and the cell body (soma), there is fast (ca. 400 mm/day, i.e., 5 μm/sec) transport of chemicals contained as "cargo" within thin-walled vesicles ca. 50 to 100 nm in diameter (Fig. 1). It is thought that the vesicles are propelled along a fibrous protein "cytoskeletal" lattice by a chemomechanical force generated much like that producing muscle contraction and contractility in living cells generally. With the light microscope one can, under suitable conditions, observe a two-way traffic of particulates, in both the anterograde and the retrograde directions. By a less clearly understood mechanism, substances traverse synaptic junctions.

Molecular fast transport plays an important role in molecular genetic neuroscience because it is the process by which neuroactive ligands capable of evoking gene expression may be transported rapidly over relatively large distances (millimeters or centimeters). Such a ligand, synthesized at ribosomal or other sites in the soma, is transported through neurites, over substantial distances to a terminal where it is released into extracellular space, or often into a synaptic cleft. Furthermore, neuroactive substances may be taken up by a target cell and transported intracellularly from the cell membrane, through the cytoplasm (via parallel, radially oriented, fibrous cytoskeletal proteins) to the cell center. Here these substances may enter the nucleoplasm where interaction with the cell's DNA can take place. Transport can also occur in the reverse direction, i.e., from the DNA through the cytoplasm to the cell membrane (Fig. 2).

The human cerebral cortex is about 3 to 5 mm thick and covers the entire brain. Its neurons are entrained mostly in local circuits, i.e., their axons do not extend for long distances (as in the spinal cord and in peripheral nerves), but synapse with nearby neurons (e.g., stellate and pyramidal cells) or with neurons in immediately adjacent brain tissue (e.g., the thalamus).

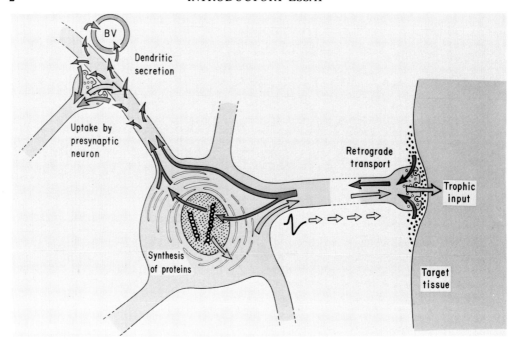

FIG. 1. Bidirectional fast transport of neuroactive materials within neurons and transsynaptically between neurons and target cells. Retrogradely transported substances *(dark arrows)* may affect gene expression and, by release from dendrites, the neuronal environment, including blood vessels (BV). *Light arrows* indicate anterograde transport. (From Schmitt, 1979.)

Thus, ligands capable of directly or indirectly altering gene function can be transported from their site of synthesis to their target neurons in the cortex in a few minutes. The rate-limiting reaction is clearly not the transport of substances from one part of the brain to another, but rather processes involved in gene expression triggered by the binding of the ligand to its receptor; these may require many minutes or hours.

NEUROACTIVE SUBSTANCES AND THEIR ROLE IN BRAIN FUNCTION

For many years the function of "classic" transmitters has been thought to be the excitation or inhibition of postsynaptic target cells by altering membrane potential, usually by changing the permeability of ionic (NA^+, K^+, Cl^-) channels. There are seven major classic transmitters: acetylcholine (ACh), norepinephrine (NE), dopamine (DA), 5-hydroxytryptamine (5HT, serotonin), γ-aminobutyric acid (GABA), glycine (Gly), and glutamic acid (Glu). In addition, there are several dozen peptides and certain purines that are candidate transmitters (see Bloom, 1981). Why so many kinds of transmitters when it would seem that two might suffice, i.e., one to excite and one to inhibit target cells?

Clearly neurotransmitters must participate in other processes in addition to the fast bioelectric function of altering membrane polarization by changing the permeability of ion channels. Bloom (1975, 1979) pointed out that bioelectric changes may not be the only, or in some cases even the primary, function of transmitters. In Bloom's view, the neuron is a living cell, not a transistor, and the effect of a transmitter on a target cell may include a set of holistic, metabolic, and trophic (i.e., life-supporting biosynthetic) changes.

Neurotransmitters are active in the concentration range of 10^{-6} to 10^{-9} M. Other substances also act on nervous tissue in this concentration range, including hormones, peptides, purines,

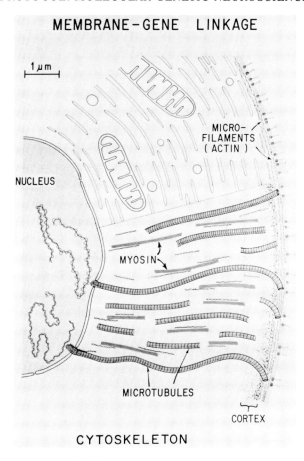

FIG. 2. Fibrous protein substrate for bidirectional transport of substances from the cell membrane to, and into, the nucleus and its genome. (From Schmitt, 1979.)

TABLE 1. *Coexistence of a classic transmitter and a peptide in individual neurons[a]*

5-HT + Substance P
5-HT + TRH
NE + Somatostatin
NE + Enkephalin
NE + Pancreatic polypeptide
DA + CCK
ACh + VIP

[a] 5-HT, 5-hydroxytryptamine; NE, norepinephrine; DA, dopamine; ACh, acetylcholine; CCK, cholecystokinin; VIP, vasoactive intestinal polypeptide.
Data taken from Hökfelt et al. (1980) and from T. Hökfelt *(personal communication)*.

and probably yet undiscovered categories of substances; these will here be collectively termed "neuroactive substances" (NAS).

For determining the transmitter or other NAS produced by particular neurons, very sensitive methods have been developed, including fluorescent microscopy (Falck et al., 1962) and highly specific immunohistochemical methods. With such methods Hökfelt and coworkers (1980 a,b), as well as several other investigators (Chan-Palay et al., 1978; Chan-Palay, 1979; Björklund et al., 1979), made the important discovery that in many neurons of the central nervous system and endocrine cells that they studied, one or more peptides coexist with a classic transmitter. Different mixes of classic transmitters and peptides coexist in different neuronal types (Table 1). Hökfelt suggested that such peptides may modulate (potentiate or reduce) the effects of classic transmitters and that this action may involve presynaptic as well as postsynaptic mechanisms of the particular classic transmitter.

The seven classic transmitters are generated in the neuronal soma or in axon terminals by the action of enzymes on small molecular precursors. In contrast, fairly large polypeptide precursors are ribosomally synthesized and, during their transport down the axon, while enclosed within vesicles, they are subjected to the action of several kinds of hydrolytic enzymes that cleave the precursor to produce the final peptides. These peptides and the classic transmitter are released, by exocytosis from the presynaptic ending (Fig. 3). If a second peptide coexists with the classic transmitter, all three NASs are thought to be released at the terminal, and possibly into intercellular space as well as into the synaptic cleft. Such a process departs widely from the concept of a unitary transmitter visualized in the process of classic neurohumoral transmission.

NEUROACTIVE SUBSTANCES RELEASED IN THE CEREBRAL CORTEX BY CENTRAL CORE RETICULAR NEURONS

The neurons of the cerebral cortex are "innervated" by several sets of ascending neurons originating from the major integrative center of the brain stem, the central core reticular

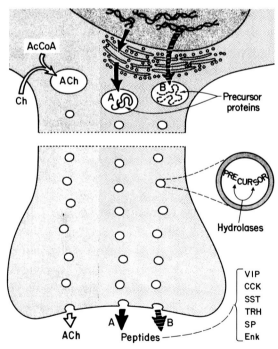

FIG. 3. Schematic representation of the coexistence within the neuroplasm of a classic transmitter (here shown as acetylcholine, ACh) and peptides. The polypeptide precursors (A and B) are ribosomally produced and cleaved by hydrolytic enzymes to form the peptides for release from the axon terminal. AcCoA, acetyl CoA; Ch, choline; VIP, vasoactive intestinal polypeptide; CCK, cholecystokinin; SST, somatostatin; TRH, thyrotropin-releasing hormone; SP, substance P; Enk, enkephalin. (From Hökfelt et al., 1980a,b.)

formation. The distribution of the three members of the monoamine system throughout the cortex is shown diagrammatically in Fig. 4. In the locus coeruleus, the axons, after leaving the cells of origin in the brain stem, branch thousands of times to form a reticular network of very thin (0.1 μm) axons, with periodically occurring small bag-like varicosities within which are the vesicles that contain the classic transmitter norepinephrine, the characteristic fluorescence of which marks the course of the axons within and on the cerebral cortex. Norepinephrine is thought to be released from these vesicles and varicosities into the intercellular, ambient fluid in which the cells of the cerebral cortex are immersed.

In similar fashion, fine, varicosity-laden axons proceed from raphe neurons and project to virtually all cells of the cerebral cortex. The transmitter in the vesicles and varicosities in these neurons is serotonin, which is also thought to be released into intercellular fluid.

A third classic transmitter, dopamine, is released from dendrites of neurons in the substantia nigra (Geffen et al., 1976). These dendrites contain no vesicles, varicosities, or other synaptic morphological specialization. Dopamine is stored in the cisterns of the smooth endoplasmic reticulum according to Cheramy and colleagues (1981) and Iversen (1979). The latter suggested that release of transmitters from dendrites may be widespread in the central nervous system, involving no morphologically defined specializations, and that classic transmitters and possibly other NASs, e.g., peptides, may be thus released in many areas of the central nervous system.

It is widely believed that virtually all of the cells of the cerebral cortex are exposed to various combinations and concentrations of NASs including not only norepinephrine, serotonin, and dopamine released from the central core reticular neurons that innervate the cortex, but also many peptides (more than 25 are now known) and possibly also purines. The functions of these NASs may include not only regulation or modulation of cortical activity, as is generally believed, but also the evocation of gene expression of various kinds, to be discussed below.

For support of the point of view developed in this chapter, it is important to establish that the various NASs do, in fact, exist in solution in the intercellular ambient milieu bathing cortical cells. Iversen pointed out that concepts of chemical neurotransmission for the visceral autonomic nervous system have historically played a key role in the interpretation of chemical transmission

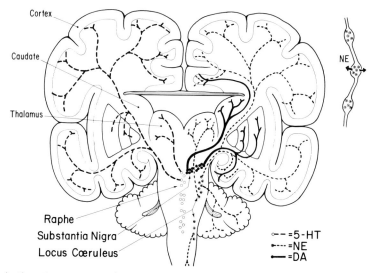

FIG. 4. Projection of monoaminergic neurons to neurons of the cerebral cortex. The projection is, in each case, bilateral but is shown as unilateral for clarity of illustration. Paracrine release into intercellular space of norepinephrine (NE) from a varicosity of a neuron of the locus coeruleus is shown in the *upper right*. 5-HT, 5-hydroxytryptamine; DA, dopamine.

in the central nervous system (Iversen et al., 1979). Because it is an accepted fact that transmitters of autonomic neurons are released from varicosities at points distant from their target cells, with no morphologically demonstrable synaptic structure, it is reasonable to believe that neurons of the central nervous system, which contain vesicle-filled varicosities, may also release transmitters and other NASs from these varicosities into the ambient surround of their target cells, i.e., of the cerebral cortex.

More direct evidence is that described in a series of papers by Glowinski (1979,1981) and Cheramy and coworkers (1981), who demonstrated the release of dopamine from dendrites of nigrostriatal neurons by the use of multiply implanted push-pull cannulae, to superfuse and monitor the release of dopamine from the nigra; he suggested that such dendritic release is the physiological event that contributes to the local transfer of information in nigral neurons. Glowinski (1981), using similar methods, also demonstrated the release of GABA from neurons of the caudate nucleus and of Met-enkephalin from the basal ganglia of cats.

That certain NASs are indeed in solution in the intercellular fluids of brain tissue has been proved, by the use of specially designed electrodes specifically sensitive to catecholamines and capable of monitoring continuously ongoing chemical changes in the central nervous system of behaving animals. The electrodes were inserted into brain tissue of anesthetized rats. Continuous recording of voltametric signals permitted determination of the nature and concentration of monoamines, including norepinephrine, serotonin, and dopamine. A detailed review of these methods and of the results obtained so far has been prepared by Adams and coworkers (1981) and Adams (1982) who report that data they obtained on dopamine release in substantia nigra agree not only qualitatively but also semiquantitatively with those of Glowinski. Adams also found substantial increase in dopamine release as a result of stimulation of the nigra.

Loullis et al. (1980) have provided an important link between the direct monitoring of amines and their metabolites and chemical analysis of the intercellular fluid. This was made possible by combining a push-pull cannula with electrochemical detection to measure release of norepinephrine, serotonin, and dopamine into perfusates collected through the cannula.

NEUROPARACRINE INTERACTION OF CENTRAL CORE RETICULAR NEURONS WITH THOSE OF THE CEREBRAL CORTEX

No morphological differentiations representing synaptic structures have been demonstrated in the cortical distribution of central core reticular monoamine neurons. It has therefore been called a "nonsynaptic" system. However, the word "paracrine" may be more appropriate. According to Vale and Brown (1979), the concept of paracrine mediation by which cells in general, through release of activating substances, modify the function of neighboring cells, was first proposed by Feyrter (1953). Applied to the nervous system, paracrine activity is intermediate between a "private" one-to-one interaction between neurons (classic synapse) and a to-whom-it-may-concern, neurosecretory type of signal carried in the bloodstream to react with any cell bearing a receptor specific to that hormonal signal (see also Scharrer, 1972, 1975). Certain peptides possess paracrine as well as endocrine action, which may be manifested by neuromodulator or neurotransmitter action in the central nervous system. Vale and Brown (1979) suggest that peptides probably serve as extracellular brain messengers, and that, as neurotransmitters or neuromodulators, they may be key participants in brain function.

The neurons of the human cerebral cortex are arrayed in a complex pattern both within laminar planes parallel to the surface and in columnar or modular units that extend vertically, i.e., radially, throughout the thickness of the cortex. The various kinds of cortical neurons have characteristic structure and synaptic connectivity with both intracortical and extracortical inputs and outputs (see Schmitt et al., 1981). Although the cortex has a considerable degree of structural and functional plasticity, the "wiring diagram" and the pattern of electrical activity appear to be reasonably stable and reproducible within a species. The neurotransmitters that mediate excitatory or inhibitory synaptic activity of the classic type in this complex circuitry are known to a considerable extent.

From the data described above, this highly specific and complex neuronal matrix of classic neuroanatomy and neurophysiology, the cerebral cortex must now be viewed as being immersed in an extracellular ambient milieu containing possibly dozens of highly potent NASs, a very bizarre arrangement indeed for that portion of the brain thought to subserve the highest human functions, which might be thought a priori to require connectivity and neuronal control of the most specific kind. This seemingly random mode of action of highly potent NASs departs widely from conventional views of neurotransmitter and neuromodulator action. Although alternative explanations may suggest themselves, the hypothesis put forward here posits that specificity and selectivity of interaction of particular NASs with cortical neurons may, as in the neurosecretory case, be determined by their ability to bind with high affinity to specific receptors presumably arrayed on neuronal surfaces that interface with the ambient intercellular milieu. The complexes formed by the receptor binding of specific neuroactive ligands may, through specific coupling with DNA, activate one or more genes in particular target neurons.[1] Thus, the process is highly relevant to the theme of this volume.

This suggestion, although hypothetical, would bring to bear molecular biological factors seldom considered by anatomists and physiologists or, for that matter, by neurochemists. It concerns not merely metabolic processes of cellular maintenance, but also the role of many gene-directed processes in brain development, such as specificity of position and orientation of neurons in the brain, and of a number of other properties, and indeed eventually even the mediation of higher brain functions.

MOLECULAR GENETICS AS A KEY FACTOR IN THE DEVELOPMENT OF NEW AND UNIFYING PRINCIPLES OF BRAIN FUNCTION

Neuroscience has in the past benefited enormously by the use of new concepts and techniques in other fields that had developed to a stage of explosive advance. Two such cases might be cited as illustrations.

First, in the second and third decades of this century, electronic theory and its fantastic applications in electronic engineering was revolutionizing many sciences, technologies, and industries. For many years neurophysiologists had realized that a major difficulty in understanding the function of the nervous system was the lack of techniques with which to record, with high temporal and electrical fidelity, action potential waves as they are propagated along the axons of different nerve types. Through the ingenuity of several leading neurophysiologists, particularly Herbert Gasser and Joseph Erlanger, newly developed techniques of electronic voltage amplification, visualization, and recording of waves by means of the cathode ray oscilloscope were applied, making it possible to record action potential waves with high fidelity. This was a major breakthrough in neuroscience, marked by the award of the Nobel Prize to Gasser and Erlanger in 1944.

---

[1]Such coupling may occur by: (1) internalization of the complex followed by migration to the DNA where gene activation occurs; (2) binding of NAS (usually hormones) to cytosolic receptors followed by transport to, and activation of a gene; (3) activation of a second messenger, e.g., Ca^{2+} or cyclic nucleotide, which interacts with specific nuclear sites and alters gene expression; and (4) diffusion of the NAS (e.g., triiodothyronine) to the nucleus where it binds to a DNA-bound receptor, thereby activating the gene.

Second, in the third and fourth decades of this century there occurred a dramatic series of discoveries and technical developments in electron optics that culminated in the construction of the electron microscope and the development of techniques for applying this device to the study of cells and tissues by means of ultra-thin sectioning. This increased the resolution obtainable from 0.2 μm (light microscopy) down to and below 1 nm (electron microscopy). This revolutionized microscopic anatomy, leading to a major breakthrough in our understanding of the molecular organization of cellular structures, including the genome.

The conference at Woods Hole in May 1981 was organized to nucleate yet another highly productive interaction of neuroscience with a field that is presently exploding dramatically, namely molecular genetics. That a close association between these two major communities of science will lead to historic new breakthroughs in neuroscience seems abundantly clear. Recombinant DNA techniques are now being greatly augmented by the development of completely automated equipment to pass from phenotype (e.g., a protein, or merely its amino acid sequence) to the gene or to the nucleotide sequence of the genotype, and in the opposite direction (see L. Hood, *this volume*). Therefore, by applying recombinant techniques we can expect in the near future to witness the discovery and synthesis via microorganismic or viral vectors of new hormones, enzymes, proteins, and conjugated proteins that are produced by the nervous system. This will doubtless yield material of great scientific and clinical value. However, even more valuable will be discoveries that may lead to new concepts of mechanisms of brain function, possibly including higher brain function.

F. Bloom *(this volume)* depicts important possibilities, general strategies, and abounding opportunities that may emerge from interdisciplinary interaction. Other kinds of discoveries yet undreamed of may be confidently expected to be made in the future, and these will equally confidently lead to the establishment of a new biology and concepts of life and its meaning to humankind.

REFERENCES

Adams, R. N. (1982): Electrochemical detection methods for monoamine measurements *in vitro* and *in vivo*. In: *Handbook of Psychopharmacology, Vol. 15: New Techniques in Psychopharmacology*, edited by L. L. Iversen, S. D. Iversen, and S. H. Snyder. Plenum Press, New York *(in press)*.

Adams, R. N., Lane, R. F., Wightman, R. M., Justice, J., Knott, P., Rebec, G., and Plotsky, P. M. (1981): In vivo electrochemistry: Principles and applications. *Soc. Neurosci. Abstr.*, 7:827.

Björklund, A. J., Emson, P. C., Gilbert, R. F., and Skagerberg, G. (1979): Further evidence for the possible coexistence of 5-hydroxytryptamine and substance P in medullary raphe neurones of rat brain. *Br. J. Pharmacol.*, 66:112P–113P.

Bloom, F. E. (1975): The role of cyclic nucleotides in central synaptic function. *Rev. Physiol. Biochem. Pharmacol.*, 74:1–103.

Bloom, F. E. (1979): Chemical integrative processes in the central nervous system. In: *The Neurosciences Fourth Study Program*, edited by F. O. Schmitt and F. G. Worden, pp. 51–58. MIT Press, Cambridge, Mass.

Bloom, F. E. (1981): Neuropeptides. *Sci. Am.*, 245:148–168.

Chan-Palay, V. (1979): Combined immunocytochemistry and autoradiography after in vivo injections of monoclonal antibody to substance P and ³H-serotonin: Coexistence of two putative transmitters in single raphe cells and fiber plexuses. *Anat. Embryol. (Berl.)*. 156:241–254.

Chan-Palay, V., Jonsson, G., and Palay, S. L. (1978): Serotonin and Substance P co-exist in neurons of the rat's central nervous system. *Proc. Natl. Acad. Sci. USA*, 75:1582–1586.

Cheramy, A., Leviel, V., and Glowinski, J. (1981): Dendritic release of dopamine in the substantia nigra. *Nature*, 289:537–542.

Falck, B., Hillarp, N. A., Thieme, G., and Torp, A. (1962): Fluorescence of catecholamines and related compounds condensed with formaldehyde. *J. Histochem. Cytochem.*, 10:349–361.

Feyrter, F. (1953): *Über die Peripheren Endokrinen (Parakrinen) Drüsen des Menschen*. Maudrich, Vienna and Dusseldorf.

Geffen, L. B., Jessell, T. M., Cuello, A. C., and Iversen, L. L. (1976): Release of dopamine from dendrites in rat substantia nigra. *Nature*, 260:258–260.

Glowinski, J. (1979): Some properties of the ascending dopaminergic pathways: Interactions of the nigrostriatal dopaminergic system with other neuronal pathways. In: *The Neurosciences Fourth Study Program*, edited by F. O. Schmitt and F. G. Worden, p. 1069. MIT Press, Cambridge, Mass.

Glowinski, J. (1981): In vivo release of transmitters in the cat basal ganglia. *Fed. Proc.*, 40:135–141.

Hökfelt, T., Johansson, O., Ljungdahl, A., Lundberg, J. M., and Schultzberg, M. (1980a): Peptidergic neurons. *Nature*, 284:515–521.

Hökfelt, T., Lundberg, J. M., Schultzberg, M., Johansson, O., Ljungdahl, A., and Rehfeld, J. (1980b): Co-existence of peptides and putative transmitters in neurons. In: *Neural Peptides and Neuronal Communication*, edited by F. Costa and M. Trabucci, pp. 7–23. Raven Press, New York.

Iversen, L. L. (1979): Neurotransmitter interactions in the substantia nigra: A model for local circuit chemical interactions. In: *The Neurosciences Fourth Study Program*, edited by F. O. Schmitt and F. G. Worden, pp. 1085–1092. MIT Press, Cambridge, Mass.

Iversen, L. L., Hökfelt, T., and Burnstock, G. (1979): Non-adrenergic, noncholinergic autonomic neurotransmission mechanisms. *Neurosci. Res. Program Bull.*, 17:482.

Loullis, C. C., Hingtgen, J. N., Shea, P. A., and Aprison, M. H. (1980): In vivo determination of endogenous biogenic amines in rat brain using HPLC and push-pull cannula. *Pharmacol. Biochem. Behav.*, 12:959–963.

Scharrer, B. (1972): Neuroendocrine communication (neurohormonal, neurohumoral and intermediate). *Prog. Brain Res.*, 38:7–18.

Scharrer, B. (1975): The concept of neurosecretion and its place in neurobiology. In: *The Neurosciences, Paths of Discovery*, edited by F. G. Worden, J. P. Swazey, and G. Adelman, p. 231. MIT Press, Cambridge, Mass.

Schmitt, F. O. (1979): The role of structural, electrical and chemical circuitry in brain function. In: *The Neurosciences Fourth Study Program*, edited by F. O. Schmitt and F. G. Worden, pp. 5–19. MIT Press, Cambridge, Mass.

Schmitt, F. O., Worden, F. G., Adelman, G., and Dennis, S. G., eds. (1981): *The Organization of the Cerebral Cortex*, MIT Press, Cambridge, Mass.

Shepherd, G. M., ed. (1979): *The Synaptic Organization of the Brain*, 2nd ed., p. 20. Oxford University Press, New York.

Vale, W., and Brown, M. (1979): Neurobiology of peptides. In: *The Neurosciences Fourth Study Program*, edited by F. O. Schmitt and F. G. Worden, pp. 1027–1041. MIT Press, Cambridge, Mass.

Section II

Molecular Organization of DNA in Relation to Gene Function

The explosive advances in molecular genetics are almost universally regarded as having originated with the use of X-ray diffraction analysis to resolve the three-dimensional structure of DNA. The insightful hypothetical structure put forward by Watson and Crick in 1953 successfully integrated hard data on the characteristic X-ray diffraction patterns of the DNA molecule with quantitative chemical data on the relative frequency of the four nucleotide bases, adenine (A), thymidine (T), guanine (G), and cytosine (C).

As is well known, in the Watson-Crick double helix two right-handed helical polynucleotide chains coil around the same central axis, such that the purine and pyrimidine bases are in parallel planes on the inside of the helix and perpendicular to the axis. The bases of each chain are precisely paired in each plane through hydrogen bonding of A to T and of G to C; only these A–T and G–C pairings can yield a regular structure with the known dimensions of DNA. The 3.4 Å major period and the 34 Å minor periods of the X-ray diffraction pattern can then be accounted for as the stacking distance between base pair planes, given that 10 base pair distances also constitute a complete turn of the helix. The diffraction patterns also revealed that, depending on the degree of hydration, two forms of DNA were possible, called A and B, of which the B form is considered to be the natural configuration. The B-DNA theory, with its concept of base-pair complementarity, has provided a wealth of insights into the molecular mechanisms underlying DNA replication, RNA transcription, and protein translation. Moreover, by positioning the hydrophobic bases on the inside of the coil and the hydrophilic sugars and phosphates on the exterior, the structure not only met the physical needs of the X-ray diffraction data and the chemical composition, but also provided a reasonable macromolecule for a living system.

A. Rich describes recent work that explores the possibility that still other conformations of DNA may exist with their own unique biological attributes. By analyzing the X-ray diffraction patterns of relatively short synthetic deoxynucleotide chain pairs composed of alternating GC bases, Rich and his colleagues have deduced a third form of DNA helix, which differs markedly from the A and B right-handed helical forms. From synthetic GC DNA, crystallographic evidence was obtained for a left-handed helical configuration of the base pair planes with the phosphate deoxyribose backbone taking an irregular "zig-zag" course around the molecule. This Z-DNA form exhibits other important structural differences, including the number of bases in a turn and their relative positions within the helix. Notably, there is a greatly enhanced accessibility to the C-5 position of the cytosine; frequently methylated in the DNA of inactive genes, this site may be involved in regulation of transcription. Although short segments of alternating GC bases are found in some DNAs, it is not yet known whether the rather extreme chemical requirements necessary to induce the Z-DNA secondary structure occur *in vivo*.

Nevertheless, the possibility that appropriate chemical or physical conditions could exist transiently to convert B-DNA to the Z form offers a novel mechanism to be explored. In their most recent studies, Rich and his colleagues have provided initial support for the natural occurrence of the Z-DNA form by immunocytochemical staining of *Drosophila* genes. The possible genetic role of Z-DNA is also discussed.

Left-Handed DNA

Alexander Rich

ABSTRACT

DNA can exist in both right-handed and left-handed forms. A particular left-handed form, Z-DNA, has been found in crystalline structure and in the interband regions of *Drosophila* polytene chromosomes. It is likely that the left-handed form plays a role in regulating gene expression.

Molecular genetics began in 1953 when Watson and Crick described their model for the structure of DNA (Watson and Crick, 1953). For the first time the structure of the genetic material could be expressed in precise chemical terms. Their proposed model for DNA was consistent with x-ray crystallographic data obtained from fibers of long DNA molecules. Although the data were sufficient to demonstrate the helical nature of DNA, the patterns were disordered, in part because the long DNA fibers did not have an ordered sequence of bases. Atoms could not be visualized from these data, and significant interpretation and assumptions were required. To define the structure of DNA at the atomic level, it is necessary to use oligonucleotides with a defined base sequence that can form highly ordered crystals. When the three-dimensional structure of the first deoxyhexanucleotide was solved to atomic resolution (<1 Å), to the surprise of most molecular biologists, it was found to have a left-handed helical structure (Wang et al., 1979), in contrast to the right-handed helix of the classic Watson-Crick DNA model.

The differences in the overall structures of the left-handed and right-handed helices are due to changes in the conformation of the functional groups that compose each helix. The structure of the left-handed DNA demonstrates that fine changes in conformation at the nucleotide level can produce considerable changes in the overall structure and probably alter the function of a biological macromolecule. These changes may have implications for genetic regulation in the nervous system and other tissues.

THE CRYSTAL STRUCTURE OF LEFT-HANDED DNA

The deoxyhexanucleotide pentaphosphate d(CpGpCpGpCpG) or d(CG)$_3$ was crystallized and its three-dimensional structure solved to a resolution of 0.9 Å (Wang et al., 1979). This allowed the positions of all atoms to be defined precisely. In addition to the nucleotide atoms, the crystal also fixed the positions of 62 water molecules, 1 hydrated magnesium ion, and 2 molecules of the polyamine spermine, which cocrystallized with each pair of hexanucleotides. Because the guanine (G) residues base-pair with the cytosine (C) residues, the hexanucleotide paired with itself and the crystal actually consisted of a six base-pair length of double-stranded DNA. In the crystal these are stacked end to end and approximate a continuous double-stranded DNA molecule. This enables the structures of left-handed DNA to be analyzed at the level of the hexamer and at the level of a polymeric DNA.

The structure of the left-handed DNA is shown in Fig. 1. The phosphate deoxyribose backbone of the DNA follows a zigzag path as it winds around the molecule; the left-handed DNA structure is called Z-DNA. The more familiar right-handed DNA helix is referred to as B-DNA. The Z-DNA helix has 12 base pairs per helical turn and a diameter of 18 Å; B-DNA has close to 10 base pairs per turn and is 20 Å wide. Furthermore, the deep concave major groove surface in B-DNA has entirely disappeared in Z-DNA and is replaced by a convex surface (Fig. 1).

The zigzag course of the phosphate deoxyribose backbone in Z-DNA is due to the fact that there are two different types of nucleotide conformations in the chain. The glycosidic bond between the sugar ring and the guanine base can exist in one of two possible positions (Fig. 2): *syn*, with the guanine above the plane of the sugar ring, and *anti*, with the base partly below the deoxyribose. The sugar ring itself can have different conformations or puckers. In Z-DNA, the guanine residues have a *syn* ring position and a C3' endo sugar pucker; the cytosine residues have the *anti* position with a C2' endo sugar pucker. Only the latter conformation is found in all residues of B-DNA. Alternate residues of left-handed Z-DNA thus have quite different conformations, and this generates the zigzag path of the phosphate deoxyribose backbone. The pyrimidines (cytosine and thymine) do not readily assume the *syn* conformation, because of steric hindrance, and therefore left-handed helices may be favored by alternating

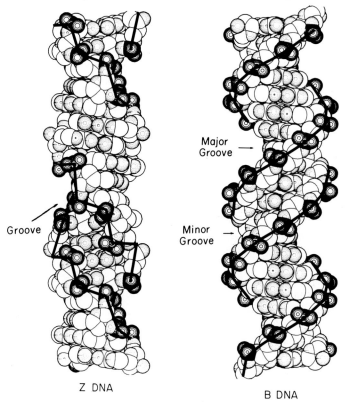

FIG. 1. Van der Waals side views of Z-DNA and B-DNA. The irregularity of the Z-DNA backbone is illustrated by the *heavy lines* that go from phosphate to phosphate residue along the chain. This includes positions where the phosphate residues are missing in the crystal structure but would be occupied in a continuous double helix. The groove in Z-DNA is quite deep, extending to the axis of the double helix. In contrast, B-DNA has a smooth line connecting the phosphate groups and two grooves, neither of which extends into the helix axis of the molecule.

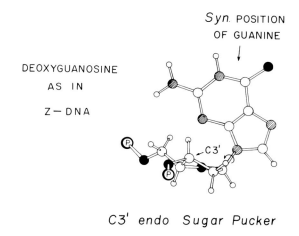

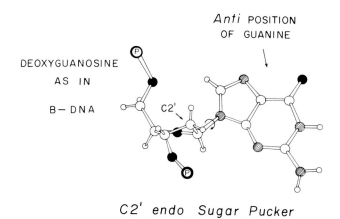

FIG. 2. Conformation of deoxyguanosine in B-DNA and in Z-DNA. The sugar is oriented so that the plane defined by C1'-O1'-C4' is horizontal. Atoms lying above this plane are in the endo conformation. The C3' is endo in Z-DNA whereas in B-DNA the C2' is endo. These two different ring puckers are associated with significant changes in the distance between the phosphorus atoms. In addition, Z-DNA has guanine in the *syn* position, in contrast to the *anti* position in B-DNA. The *curved arrow* around the glycosidic carbon-nitrogen linkage indicates the site of rotation. ◯, carbon; ○, hydrogen; ●, oxygen; ◐ nitrogen; ⓟ phosphate group.

purine-pyrimidine sequences (Wang et al., 1979; Drew et al., 1980). Because of the alternating pattern, the repeating unit of Z-DNA is a dinucleotide, in this case dCpG; in B-DNA the asymmetric unit is a mononucleotide.

The *syn* conformation of dG residues in Z-DNA also influences the base stacking between one base pair and the base pair immediately above and below it in the helical ladder. In B-DNA, each base lies almost on top of the base below it. In left-handed Z-DNA, however, the dCpG bases are sheared relative to one another: the cytosine residues lie partly over the next cytosine on the opposite stand, and guanine residues are stacked on the O1' oxygen atoms of the deoxyribose of the next nucleotide below. In Z-DNA, the base pairs are found on the outside of the helix, whereas in B-DNA the base pairs are concentrated in the core of the helix. The major groove of B-DNA is a wide concave surface, whereas in Z-DNA these atoms form a convex surface on the outside of the molecule. Part of this surface is made up of guanine residues exposed to the outside and therefore susceptible to possible chemical modification.

The Z-DNA helix can best be described as a band of material wrapped around an empty core. These differences are clearly seen in end views of the helices (Fig. 3). These views also show the sixfold symmetry of Z-DNA in contrast to the fivefold symmetry of B-DNA.

The structure of left-handed DNA has also been found in several other crystals of d(CG)$_3$ (Wang et al., 1981). In all these crystal structures there was variation in the conformation of some GpC phosphate groups, particularly those interacting with magnesium ions. For this reason, there are two different conformations, ZI and ZII, with different phosphate orientations. Left-handed DNA has also been found in crystals of the tetramer d(CG)$_2$ (Crawford et al., 1980; Drew et al., 1980) and in fibers of synthetic poly (dG-dC)·poly (dG-dC) (Arnott et al., 1980). The differences in conformation between Z-DNA and B-DNA are summarized in Fig. 4.

SOLUTION STUDIES: THE INTERCONVERSION OF Z-DNA AND B-DNA

To translate the molecular structure crystallographic results into biological systems, it is essential to show that the structure of the molecule in the crystal is the same as the structure of the molecule in solution. This has been shown for Z-DNA. Several years ago Pohl and Jovin (1972) discovered that the double-stranded DNA polymer, poly(dG-dC), changed its solution conformation in the presence of a high salt concentration. The transition between the two confirmations was freely reversible. Patel and colleagues (1979) found that the phosphorus nuclear magnetic resonance (NMR) spectrum of poly(dG-dC) in high salt has two peaks of roughly equal intensity, but only one peak in low salt. This is consistent with Z-DNA structure in high salt solutions. A study has been made of the Raman spectra of crystalline d(CG)$_3$, which has the Z-DNA structure, and the high and low salt forms of poly(dG-dC). It has been shown that the Raman spectrum of crystalline d(CG)$_3$ is identical to that of the high salt form of poly(dG-dC), but not the low salt form (Thamann et al., 1981). This demonstrates that the Z-DNA structure is found in the high salt solution.

The existence of the two forms of DNA in solution allowed for the study of the interconversion of Z-DNA and B-DNA. Pohl and Jovin (1972) used circular dichroism measurements to dis-

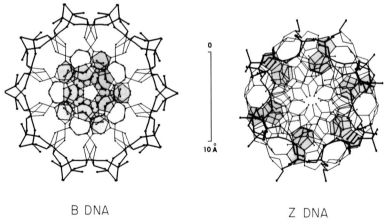

FIG. 3. End views of B-DNA and Z-DNA in which the guanine residues of one strand have been shaded. The Z-DNA figure represents a view down the complete c axis of the crystal structure encompassing two molecules. The shaded guanine residues illustrate the approximate six-fold symmetry. The imidazole part of the guanine residue forms a segment of the outer cylindrical wall of the molecule together with the phosphate residues. The B-DNA figure represents one full helix turn. In contrast to Z-DNA, the guanine residues in B-DNA are located closer to the center of the molecule and the phosphates are on the outside.

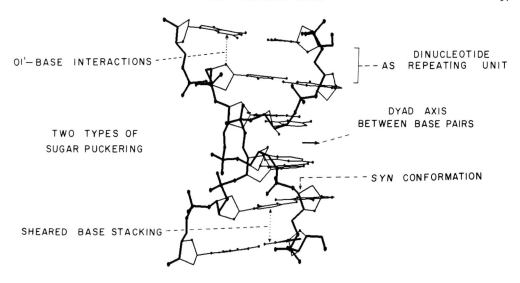

FIG. 4. Schematic diagram showing a projection of the left-handed double helix of the spermine-magnesium d(CpGpCpGpCpG) hexamer. Six structural features are found in this left-handed DNA conformation, which are different from those found in right-handed B-DNA, as indicated in the diagram.

tinguish the high and low salt conformations of poly(dG-dC). Circular dichroism (CD) is an optical technique that measures the difference in the molar extinction coefficient of left- and right-handed polarized beams after passing plane-polarized light through a solution. Left- and right-handed helical RNA molecules would be expected to have different effects on polarized light, and this is seen in the drastically different CD spectra of the high and low salt forms of poly(dG-dC) (Pohl and Jovin, 1972). CD spectra can be used to monitor the interconversion of Z-DNA and B-DNA; however, it must be done cautiously as other effects can alter the CD spectrum.

A variety of different agents have been found to affect this interconversion. The initial Pohl and Jovin studies (1972) demonstrated that the CD spectrum changed at a midpoint of 2.7 M sodium chloride or 0.7 M magnesium chloride. The high cation concentration appears to be required to neutralize the phosphate groups that are closer together in Z-DNA than in B-DNA. Ethanol can also stabilize Z-DNA; addition of the dye ethidium bromide, which can intercalate between base pairs, returned the conformation to B-DNA (Pohl et al., 1972). The DNA-binding carcinogen acetoxyacetylaminofluorine (AAF) will also cause the conversion of poly(dG-dC) to Z-DNA (Santella et al., 1981). It binds to the C-8 position of the guanine residue and probably causes the guanine to be stabilized in the *syn* conformation. The bulky AAF molecule sterically hinders the guanine from assuming the *anti* position and stabilizes the Z-DNA conformation (Santella et al., 1981). However, it is not necessary to put so bulky a group as AAF on the guanine C-8 position. Even a bromine atom on guanine C-8 will stabilize the *syn* conformation, resulting in a low salt stable form of Z-DNA (Fig. 5) (Lafer et al., 1981).

Other derivatives on the bases will also stabilize the Z-DNA conformation: these include methylation at the 7 position of guanine (Möller et al., 1981) and the 5 position of cytosine (Behe and Felsenfeld, 1981). In Z-DNA, the methyl group on 5-methyl cytosine makes an additional hydrophobic interaction with hydrophobic parts of the sugar molecule and stabilizes the Z-DNA conformation. The most common form of DNA methylation in eukaryotic systems is on the C-5 of cytosine in the sequence CpG (Razin and Riggs, 1981), and it may be significant that the methylation also stabilizes the Z-DNA conformation.

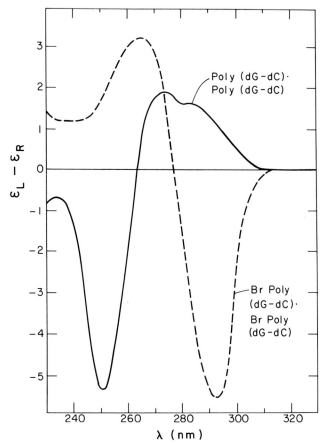

FIG. 5. Circular dichroism spectra of poly(dG-dC)·poly(dG-dC) (B form) *(solid line)* and brominated (Br) poly(dG-dC)·poly(dG-dC) (Z form) *(broken line)* in 15 mM Tris HCl, pH 7.2/150 mM NaCl/1 mM EDTA. The inversion of the brominated polymer spectrum is similar to that seen when the salt concentration is raised to 4 M NaCl for the unbrominated polymer.

Conversion of B-DNA to Z-DNA requires an inversion in the stacking of the bases. The guanine bases turn "upside down" (Fig. 6) by rotating about the glycosidic bond, whereas the entire cytidine residue, base plus sugar, rotates (Wang et al., 1979). Because of the zigzag course of the phosphate deoxyribose backbone in Z-DNA, the distance between the phosphate groups across the helix will vary. The phosphates in the GpC segments are approximately the same distance apart across the helix as they are in B-DNA, whereas the phosphate groups in CpG segments are closer together. Thus, to insert a segment of Z-DNA in a B-DNA helix, the Z must be in units of dCpG. This suggests that stretches of Z-DNA in the middle of B-DNA are likely to involve an even number of nucleotides. The dCpG sequence appears to be particularly stable in the Z-DNA conformation; however, the structure can probably accommodate any alternating sequence of purines and pyrimidines.

LEFT-HANDED DNA IN BIOLOGICAL SYSTEMS

The Z-DNA conformation can clearly exist in solution, but does it also exist in biological systems? We have approached this problem by producing an antibody that is specific for the Z-DNA conformation and can be used to detect possible Z-DNA sequences in cellular DNA.

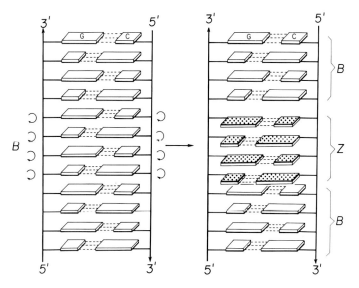

FIG. 6. A diagram illustrating the change in topological relationship if a four-base-pair segment of B-DNA were converted into Z-DNA. This conversion could be accomplished by rotation of the bases relative to those in B-DNA. This rotation is shown diagrammatically by shading one surface of the bases. All the *dark shaded areas* are at the bottom in B-DNA. In the segment of Z-DNA, however, four of them are turned upward. The turning is indicated by the *curved arrows*. Rotation of the guanine residues about the glycosidic bond produces deoxyguanosine in the *syn* conformation, whereas for dC residues both cytosine and deoxyribose are rotated. The altered position of the Z-DNA segment is drawn to indicate that these bases do not stack directly on the base pairs in the B-DNA segment.

Z-DNA can be stabilized by bromination of poly(dG-dC) in the guanine C-8 position. The brominated guanosine stays in the *syn* conformation and the polymer assumes the Z-DNA conformation (Lafer et al., 1981). When this polymer is bound to methylated bovine serum albumin and injected into rabbits, antibodies that specifically recognize Z-DNA are produced. The antibodies do not recognize B-DNA, single-stranded DNA, or brominated polynucleotides but only polymers in the Z conformation. The methylated bovine serum albumin, which is positively charged, appears to stabilize the Z-DNA conformation of poly(dG-dC), since injection of unbrominated poly(dG-dC) with it also results in the production of antibodies specific against Z-DNA, although at a reduced titer (Lafer et al., 1981).

These antibodies were used to stain the large polytene chromosomes found in the salivary gland of *Drosophila*. The polytene chromosomes have accumulated 1,000 to 2,000 copies of the normal genome, making them easy to visualize. The giant chromosomes have distinct banding patterns: the dark bands have most of the DNA, seen as particles in the electron microscope, whereas the lighter interband regions are seen to have a fibrillar structure in the nucleoprotein. When these chromosomes were stained with antibodies against Z-DNA using the indirect immunofluorescence technique, only the interband regions were found to stain (Fig. 7). Blocking of the antibodies with brominated poly(dG-dC) Z-DNA abolished all specific staining with the anti-Z-DNA antiserum, and adding poly(dG-dC) as B-DNA had no effect (Nordheim et al., 1981). These data indicate that Z-DNA does exist in nature and suggest that it is confined to specific regions of the chromosome. The intensity of interband staining is highly reproducible, but varies almost 10-fold in magnitude among different interband regions. The Z-DNA segments thus seem to be a fundamental feature of the polytene chromosome. Since the band-interband organization is believed to be found in all eukaryotic chromosomes, it is likely that Z-DNA will be found to be widely distributed.

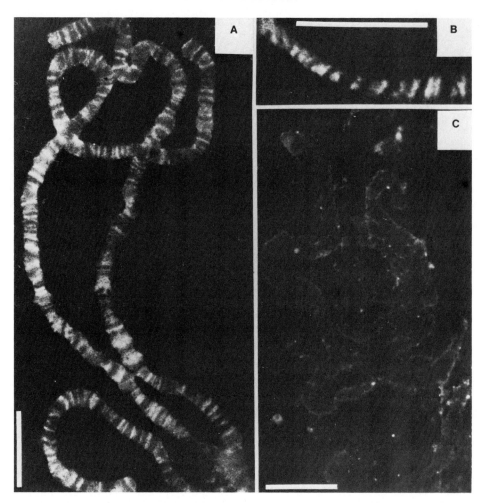

FIG. 7. Fluorescence micrographs of *Drosophila* polytene chromosomes demonstrating the specificity of the reaction between anti-Z-DNA antibody and fixed chromosomes. **A:** Chromosomes stained with anti-Z-DNA antibody. **B:** Chromosomes stained with anti-Z-DNA antibody that has been preincubated with poly(dG-dC)·poly(dG-dC) in the B conformation. The pattern of fluorescence staining is identical to that produced by the antibody without B-DNA competitor. **C:** Chromosomes stained with anti-Z-DNA antibody that has been preincubated with the Z-form brominated poly(dG-dC)·poly(dG-dC). No chromosomal fluorescence staining is detectable although the precipitated antigen-antibody complex produces an increased general background of fluorescence. Bar equals 10 μm.

An attractive hypothesis is that the Z-DNA conformation may be concerned with gene regulation: the process of gene inactivation may be associated with a change in the conformation from B-DNA to Z-DNA in part of the gene or its controlling sequences. The phosphate groups in Z-DNA are closer together than in B-DNA and thus require something to stabilize the conformation. Stabilization could be associated with (1) negative supercoiling, (2) Z-DNA binding proteins, (3) the binding of specific cations such as spermine, or (4) the methylation of dCpG sequences.

It is striking that the predominant methylated base in eukaryotic DNA is 5-methyl cytosine, usually in the sequence dCpG (Razin and Riggs, 1981). The frequency of the sequence dCpG in the DNA of most higher eukaryotes is approximately 10 to 20% of the frequency that would be predicted from the base frequencies of cytosine and guanine. Many of these dCpG sequences have a methylated cytosine that can be detected by using specific restriction enzymes (Bird,

1978). The extent of methylation at these sites varies in different tissues and with the developmental stage of the animal. For example, the methylation of cytosine at defined d(CpG) sites in rabbit or human β-globin genes has been shown to vary considerably between different tissues: in sperm DNA all sites are methylated but in somatic tissues some sites are unmethylated, particularly in tissues that are actively expressing the β-globin gene (van der Ploeg and Flavell, 1980). Patterns of DNA methylation are maintained through several cell divisions (Bird, 1978). In fact, it has recently been shown that the methylation patterns were maintained in DNA fragments, which were methylated *in vitro* and transferred into eukaryotic cell lines by microinjection (Wigler, 1981). These experiments have been interpreted as evidence for an inverse association between gene activity and the degree of DNA methylation (Razin and Riggs, 1981; Wigler, 1981). It should be pointed out, however, that although there is some relation between gene inactivity and DNA methylation, there are tissues with decreased methylation of the β-globin gene, for example, that do not synthesize β-globin (van der Ploeg and Flavell, 1980). In most eukaryotes but not necessarily in *Drosophila*, Z-DNA may be stabilized by methylation of cytosine in the C-5 position when it is found in a dCpG sequence. As mentioned above, poly(dG-m^{5}dC) exists as Z-DNA in physiological salt solutions (Behe and Felsenfeld, 1981).

The relevance for neuroscience is that fetal mammalian brain has highly methylated DNA (van der Ploeg and Flavell, 1980) and a high activity of the appropriate methyl transferase enzymes (Salas et al., 1979). Possibly, DNA methylation in the brain is involved in the fixation of gene inactivity during the development of a large number of different cell lineages. Alternatively, methylation may also reflect the cessation of DNA synthesis and cell division in the mature brain.

The discovery of left-handed DNA and its existence in nature has implications for all functions of the DNA molecule. The possibility that the conformation of the DNA molecule itself may be modifiable must now be included in considerations of the mechanisms of gene regulation. It is hoped that firm evidence will be forthcoming on the relationships between DNA conformation, methylation, and gene activity. This will have obvious importance for the nervous system, particularly during its development, which must involve a highly regulated pattern of gene switching in different cell images.

ACKNOWLEDGMENTS

This research was supported by grants from the NIH, NSF, NASA, and the American Cancer Society.

REFERENCES

Arnott, S., Chandrasekaran, S., Birdsall, D. L., Leslie, A. G. W., and Ratliff, R. L. (1980): Left-handed DNA helices. *Nature*, 283:743–745.
Behe, M., and Felsenfeld, G. (1981): Effects of methylation on a synthetic polynucleotide: The B-Z transition in poly(dG-m^{5}dC)·poly(dG-m^{5}dC). *Proc. Natl. Acad. Sci. USA*, 78:1619–1623.
Bird, A. P. (1978): Use of restriction enzymes to study eukaryotic DNA methylation. II: The symmetry of methylated sites supports semi-conservative copying of the methylation pattern. *J. Mol. Biol.*, 118:49–60.
Crawford, J. L., Kolpak, F. J., Wang, A. H.-J., Quigley, G. J., van Boom, J. H., van der Marel, G., and Rich, A. (1980): The tetramer d(CpGpCpG) crystallizes as a left-handed double helix. *Proc. Natl. Acad. Sci. USA*, 77:4016–4020.
Drew, H., Takano, T., Tanaka, S., Itakura, K., and Dickerson, R. E. (1980): High-salt d(CpGpCpG), a left-handed Z-DNA double helix. *Nature*, 286:567–570.
Lafer, E. M., Moller, A., Nordheim, A., Stollar, B. D., and Rich, A. (1981): Antibodies specific for left-handed Z-DNA. *Proc. Natl. Acad. Sci. USA*, 78:3546–3550.
Möller, A., Nordheim, A., Nichols, S. R., and Rich, A. (1981): 7-Methylguanine in poly(dG-dC)·poly(dG-dC) facilitates Z-DNA formation. *Proc. Natl. Acad. Sci. USA*, 78:4777–4781.
Nordheim, A., Pardue, M. L., Lafer, E. M., Möller, A., Stollar, B. D., and Rich, A. (1981): Antibodies to left-handed Z-DNA bind to interband regions of Drosophila polytene chromosomes. *Nature*, 294:417–422.

Patel, D. J., Canuel, L. L., and Pohl, F. M. (1979): Alternating B-DNA conformation for the oligo (dG-dC) duplex in high-salt solution. *Proc. Natl. Acad. Sci. USA*, 76:2508–2511.

Pohl, F. M., and Jovin, T. M. (1972): Salt-induced cooperative conformational change of a synthetic DNA: Equilibrium and kinetic studies with poly(dG-dC). *J. Mol. Biol.*, 67:375–396.

Pohl, F. M., Jovin, T. M., Baehr, W., and Holbrook, J. J. (1972): Ethidium bromide as a cooperative effector of a DNA structure. *Proc. Natl. Acad. Sci. USA*, 69:3805–3809.

Razin, A., and Riggs, A. D. (1981): DNA methylation and gene function. *Science*, 210:604–610.

Salas, C. E., Prohl-Leszkowicz, A., Lang, M. C., and Dirkheimer, G. (1979): Effect of modification by N-acetoxy-N-2-acetylaminofluorene on the level of DNA methylation. *Nature*, 278:71–72.

Santella, R. M., Grunberger, D., Weinstein, I. B., and Rich, A. (1981): Induction of the Z conformation in poly(dG-dC)·poly(dG-dC) by binding of N-2-acetylaminofluorene to guanine residues. *Proc. Natl. Acad. Sci. USA*, 78:1451–1455.

Thamann, T. J., Lord, R. C., Wang, A. H.-J., and Rich, A. (1981): The high salt form of poly(dG-dC)·poly(dG-dC) is left handed Z-DNA: Raman spectra of crystals and solutions. *Nucleic Acids Res.*, 9:5443–5457.

Van der Ploeg, L. H. T., and Flavell, R. A. (1980): DNA methylation in the human gamma, alpha, beta-globin locus in erythroid and nonerythroid tissues. *Cell*, 19:947–958.

Wang, A. H.-J., Quigley, G. J., Kolpak, F. J., Crawford, J. L., van Boom, J. H., van der Marel, G., and Rich, A. (1979): Molecular structure of a left-handed double helical DNA fragment at atomic resolution. *Nature*, 282:680–686.

Wang, A. H.-J., Quigley, G. J., Kolpak, F. J., van der Marel, G., van Boom, J. H., and Rich, A. (1981): Left-handed double helical DNA: Variations in the backbone conformation. *Science*, 211:171–176.

Watson, J. D., and Crick, F. H. C. (1953): A structure for deoxyribose nucleic acid. *Nature*, 171:737–738.

Wigler, M. H. (1981): The inheritance of methylation patterns in vertebrates. *Cell*, 24:285–286.

Section III

Control of Gene Expression

Although the secondary and tertiary structure of DNA remains an interesting area for exploration, the crucial structural feature that is the focal point for gene expression is the precise molecular complementarity between the primary sequence of bases in one strand of DNA with the antiparallel sequence of the second strand. This primary base pair complementarity does not "merely" provide a check-and-balance, double referential copy system to reduce errors during DNA replication; the complementarity also provides the mechanism by which DNA is transcribed into RNA and thereby eventually translated into gene products.

The questions under consideration in this section focus on the rules by which the gene segments of DNA in a prokaryotic organism or in the differentiated cells of a eukaryote are selectively expressed and regulated for optimal cell function. Put more simply, what is it that turns genes on and off and provides the appropriate amounts of gene products necessary for a given cell in its changing environment?

Several years ago, it appeared that gene expression could be studied and interpreted only in microorganisms. Intensive study of the mechanisms by which *E. coli* could adapt to adverse growth conditions in which a carbohydrate or amino acid source was missing led to important concepts that were later useful in studies of expression in eukaryotes. In particular, the "operon" concept of Jacob and Monod, established by multiple experiments, developed a multi-element array that illuminated mechanisms of responsive control of transcription. A regulatory gene product, termed the repressor, prevents transcription of "downstream" (i.e., toward the 3' terminus of the DNA) structural genes by preventing RNA polymerase from reading from its preferred site of attachment, the "promotor" gene, through an adjacent "operator" gene locus (where the repressor attached) to the structural genes, which would code for the eventual mRNA. When environmental conditions arise that demand the transcription and translation of the repressed gene products, environmentally produced cellular metabolites inhibit the expression of the repressor gene, the intracellular level of repressor decreases, and the operon is no longer repressed. RNA polymerase now attaches to the promotor zone and successfully reads through the operator gene into the structural gene domains, producing mRNA for translation to protein.

A somewhat more involved molecular transcription control mechanism for which the individual genes and gene products have now been explicitly detailed is described by M. Ptashne. He examines the means by which the genome of bacteriophage λ can be switched from a lysogenic state, in which most gene segments are repressed, to a lytic growth state, in which specific genes are efficiently and rapidly activated by relatively transient environmental signals, such as a pulse of UV light. These studies exemplify the complex arrangements that have evolved in prokaryotes for selective and efficient gene expression and its dynamic regulation. Similarly rapid expression control mechanisms might also serve useful purposes for neurons.

In eukaryotes, the total content of genetic information is several orders of magnitude larger than the genome of prokaryotes. A eukaryotic cell has a nucleus and a nuclear membrane. Within the nucleus, very large DNA molecules, containing hundreds to thousands of genes, exist as individual chromosomes to which are bound very basic peptides, termed histones. The DNA of a single chromosome, together with its protein attachments, is referred to as chromatin. Through a variety of physical manipulations that perturb the DNA molecule, it became possible to assess the rate at which the DNA strands reassociate from single to dual strands. It was established that the rate at which the perturbed DNA reassembled into dual-stranded DNA is directly related to the degree of homologous sequences that co-existed in the genome. Eukaryotic DNA was found to possess surprisingly large amounts of repetitive DNA sequence, as many as hundreds to thousands of copies per cell; these repetitive DNA sequences may code for some general regulatory products. More important for the functioning of differentiated cells is the occurrence of "single copy" DNA, which encodes specialized gene products such as hemoglobin and ovalbumin. Single copy DNA can nevertheless lead to the rapid production of its gene products by coding for many copies of a stable mRNA, which can then be translated to the product.

As the techniques of recombinant DNA technology became widely applied, it was possible to exploit the base pair complementarity property to use isolated mRNAs, or synthetic single- or double-stranded cDNAs, as probes for gene sequence structure. In one of the most unexpected developments in recent molecular genetics, it was discovered, through detailed structural analysis of the gene DNA primary sequences, that eukaryotic genes need not be linear arrays of bases transcribable directly into a colinear mRNA. Rather, in many cases, the DNA sequences contain segments (called exons) that are expressed, and others (called introns) that are not expressed in the resultant mRNA, but intervene between the exons. The primary DNA transcript of the gene, i.e., heterogeneous nuclear RNA or hnRNA, is processed to mRNA by cleavage and splicing operations that remove the introns. Specific sequences on DNA upstream (i.e., toward the 5′ terminus) from the site of transcription are also necessary to determine the attachment of the proper RNA polymerase to commence transcription of the DNA at the appropriate spot. Furthermore, before the mRNA is transported from the nucleus to the cytoplasm for translation by the ribosomes, it is processed so as to receive a special 5′ "cap" that appears necessary for proper translation, as well as a long tail composed of a poly(A) sequence attached downstream of the stop codon sequence in the mRNA.

J. Darnell and M. Wilson explicate the current status of these multiple control processes as they operate in eukaryotic cells; many of the clues to these operations are provided by cells infected by relatively simple viruses where gene segments of the virus can be followed in detail as they are transcribed and processed by the host cell.

P. Sharp and M. Wilson consider some of the implications of RNA splicing during transcription, and the molecular features of the RNA sequences that act as clues to the splicing ends of a transcription unit. Although such processes are not yet well studied in neurons, there seems to be little question that these control mechanisms will be found to operate there as well.

Regulatory Mechanisms of Gene Expression in Higher Eukaryotes

James E. Darnell, Jr., and Michael C. Wilson

ABSTRACT

A brief review of transcription unit design, including an exploration of the boundaries of the mouse β-globin transcription unit, is given. The order of steps in mRNA processing and the question of independence of the steps is discussed. Knowing the boundaries of transcription units and how primary transcripts are processed, we can assess the levels of gene control. Several types of transcriptional control are apparent during adenovirus infection, and tissue-specific mRNAs also can be readily demonstrated to be largely under transcriptional control. With proper assays, changes in the levels of several different adenovirus mRNAs can be demonstrated to be due not to transcriptional controls but to differential processing and cytoplasmic stability.

The goal of eukaryotic molecular biology is to elucidate the rules and ultimately the mechanisms that govern gene expression in higher organisms. It is evident that the regulation of the concentration of individual mRNA sequences plays an extremely important role in setting the abundance of specific proteins that in turn determines the phenotype of the cell. An investigation of the control of mRNA abundance must address three questions: What is the unit of transcription of the mRNA precursor? What is the pathway of processing of the nuclear precursor into the cytoplasmic mature mRNA? At what stage during the biogenesis of mRNA is the abundance of that mRNA regulated?

The expression of the adenovirus genome during a lytic infection of human HeLa cells has provided a valuable model system for the study of gene expression in eukaryotes. The extensive knowledge of both the structural and transcriptional map of the adenovirus type 2 (ad-2) genome (see Fig. 1) has facilitated the development of techniques to dissect the pathway of mRNA synthesis. Through these investigations, it has become apparent that control of gene expression is regulated at various stages during the biogenesis of mRNA. As these techniques are only now being applied to the study of cellular gene expression, this chapter focuses on the concepts demonstrated in the adenovirus system with inclusion of a few examples where similar evidence has been obtained for cellular genes.

DEFINITION OF THE TRANSCRIPTION UNIT AND THE PATHWAY OF mRNA FORMATION

The production of a mature, cytoplasmic mRNA begins with the synthesis of a nuclear mRNA transcript. The extent of this primary transcript defines the transcriptional unit for that mRNA. In the cell, nuclear transcripts compose a population of molecules that vary in length from 2,000 to 20,000 nucleotides (Derman et al., 1976). Although it has become clear that cyto-

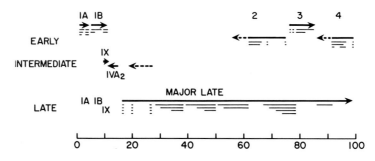

FIG. 1. Genomic map positions of early, intermediate, and late adenovirus type 2 transcription units. The viral genome is divided into 100 units (1 unit = 350 base pairs). *Arrows* indicate the position and polarity of the primary transcripts of each transcription unit. Uncertainty about the exact length of the primary transcription product is indicated by *breaks in the arrows*. Exons, the sequences spliced to form the mature mRNAs, are indicated as *thin lines* below the primary transcript. (For a detailed map of the adenovirus genome, see P. Sharp and M. Wilson, *this volume*, Fig. 1.)

plasmic mRNA molecules are derived from larger molecules within the heterogeneous nuclear RNA (hnRNA) population, it is not evident whether each hnRNA molecule gives rise to a cytoplasmic RNA.

Two experimental approaches have been successfully used to define the transcription unit of mRNA and subsequently support a precursor-product relationship between nuclear primary transcripts and mature mRNA. The first approach is the nascent chain analysis pioneered by Dintzis (1961). A pulse label of nuclear RNA, brief enough to label only the 5' ends of nascent nucleotide chains, provides a uniform label representative of the sequences transcribed before any processing can occur. The length distribution of these nascent molecules can be used to calculate the size of the primary transcripts and therefore, the transcriptional unit. In HeLa cells, Derman et al. (1976) calculated that the average hnRNA molecule was 5,000 to 6,000 nucleotides, or about three to four times the average length of cytoplasmic mRNA molecules. If the genomic DNA of a particular gene sequence is available, molecular hybridization of the nascent labeled RNA to fragments generated from the genomic DNA will distinguish the DNA sequences that are transcribed and the molarity of that transcription. Coupled with the fractionation of the nascent molecules, the molarity of transcription can be determined. The smallest labeled nuclear transcript will hybridize to the DNA fragments proximal to the RNA start site, the longest to the most distal region of the transcription unit. By this means, the boundary of the primary transcript and, therefore, of the transcription unit of a specific gene sequence can be defined.

A second approach, similar in principle, is the use of ultravoilet (UV) irradiation to place random lesions along the DNA, causing premature termination of RNA transcripts. As first shown by Sauerbier and his colleagues for prokaryote transcription units and the eukaryote ribosomal RNA transcription unit, the target size and thus the sensitivity of the production of an RNA molecule to premature termination is directly related to its distance from the transcription initiation site (for review see Sauerbier and Hercules, 1978). With this technique, Goldberg and coworkers (1977) demonstrated that the UV target size of the cytoplasmic mRNA is three to five times that of the length of the mRNA and closely fits the estimate for the length of hnRNA obtained by Derman and coworkers (1976). These two experiments together demonstrated an obligatory precursor-product relationship between hnRNA and mRNA in the eukaryote cell.

The first application of these two approaches to define a specific eukaryote mRNA transcription unit was to the adenovirus major late mRNAs. Nascent label experiments of Bachenheimer and Darnell (1975) provided the first evidence that a large nuclear transcript, found in viral-infected cells, did contain RNA sequences that were present in cytoplasmic mRNA molecules. Their data showed, in fact, that this transcript carried the sequences of several viral mRNAs, in a colinear fashion. Again using the nascent labeling technique, Weber et al. (1977) and Evans et al. (1977) established the boundaries of this major transcription unit at the viral map coordinates of 16.45 and 99, a distance of 26,000 nucleotides. (For a detailed map of the adenovirus genome, see P. Sharp and M. Wilson, *this volume*, Fig. 1.) That significant initiation within the boundaries of this transcript could not be detected suggests that the large transcript was the sole precursor to many viral mRNAs. The role of the large transcript as the obligatory precursor to the cytoplasmic mRNA came from the UV irradiation data. The sensitivity of newly synthesized cytoplasmic mRNA sequences after UV irradiation was directly related to their distance from the initiation site (at map coordinate 16.45) (Goldberg et al., 1978).

The subsequent use of these techniques has led to the definition of transcription units for other abundant adenovirus mRNAs synthesized early and late in infection from different regions of the viral genome, as well as for mRNAs of other DNA viruses such as SV40 and polyoma. These techniques, moreover, can be applied to dissect the boundaries of small overlapping transcription units. For example, a high dose of UV irradiation effectively limits the extent of transcription to within 500 to 700 nucleotides from the RNA initiation site. These newly synthesized promotor proximal sequences are therefore diagnostic for the presence of an RNA initiation site. With this technique, Wilson and colleagues (1979a) detected three transcription units within the left 3,500 nucleotides of the adenovirus genome.

The precise location of the initiation site for RNA synthesis is the landmark for the 5' end of the transcription unit. In prokaryotes this site can be identified during transcription by the initial tetraphosphate nucleotide, which is lacking in eukaryote mRNAs and their precursors (Salditt-Georgieff et al., 1980). The earliest detectable event during mRNA biogenesis is the formation of a cap structure, produced by the 5'-5' linkage of the terminal nucleotide with guanosine triphosphate (GTP) and subsequent methylation of the terminal nucleotides (Shatkin, 1976). Although this structure, which may be released from the body of the mRNA by ribonuclease treatment, is diagnostic for the 5' end of mRNA molecules, its presence initially precluded the immediate appreciation of the fact that the 5' end of mRNA represents the site of initiation within the transcription unit. The demonstration that the first nucleotide following GTP within the cap structure was, in fact, the first nucleotide transcribed came from the work of Ziff and Evans (1978). Comparing the sequence of pulse-labeled nascent RNA molecules of the adenovirus major late transcription unit with the DNA sequence surrounding the 5' end of the transcription unit, they observed that there was no detectable transcription of sequences upstream from the first nucleotide within the cap structure of mature cytoplasmic mRNA.

In a similar analysis, Weil and Colleagues (1979) used a reconstituted *in vitro* transcription system with purified eukaryote RNA polymerase II and a crude cell extract to obtain transcripts from adenovirus DNA, which had been cloned and propagated as recombinant DNA in bacteria. In this system they found the initiation of capped transcripts with precisely the same oligonucleotide sequence as the *in vivo* synthesized mRNA. Together, the *in vivo* and *in vitro* results argue strongly that the 5' capped terminus of the mature mRNA represents the site of initiation of mRNA synthesis and the 5' boundary of the transcription unit.

The precise genomic location of the end of a primary transcript and, therefore, the 3' boundary of a transcription unit has yet to be defined. At the 3' end of most mature mRNA molecules is a stretch of poly(A), which is added as a posttranscriptional modification of the primary nuclear transcript (Brawerman, 1976; Edmonds and Winters, 1976). Although it is tempting to suggest that the site of poly(A) addition is also the site of termination of transcription, two observations clearly rule this out. First, analysis of labeled nascent RNA transcripts demonstrates that for three adenovirus transcription units, equimolar transcription extends downstream beyond the site of polyadenylation, before termination within a discrete region of the genome (Fraser et al., 1979a; Nevins et al., 1980). The UV sensitivity of the transcripts of these regions requires that they originate from the upstream RNA initiation site and do not, therefore, represent the initiation of new transcripts at the site of polyadenylation (Fraser et al., 1979a; Nevins et al., 1980). The observation that complex transcription units, such as the adenovirus major late and the recently described immunoglobulin heavy chain transcription units (Early et al., 1980) contain multiple poly (A) sites, provides further evidence against the termination of the primary transcript at the site of poly(A) addition. Selection of different poly(A) sites within a single transcription unit requires that polymerase molecules be able to pass the proximal poly(A) site to reach the distal site, ruling out obligatory termination at the site of polyadenylation.

PROCESSING OF THE PRIMARY TRANSCRIPT

With the definition of a transcription unit extending from the site of RNA initiation beyond the poly(A) site, it has become evident that the maturation of cytoplasmic mRNA molecules must include cleavage of the primary nuclear transcript. The first step is the cleavage of the 3' end segment to reveal the site of polyadenylation. The addition of poly(A) is rapid, occurring within a minute after the polymerase transverses the site. It has recently become evident that the signal for the site of cleavage and polyadenylation is encoded within the primary transcript. Located about 30 nucleotides upstream from the poly(A) of all polyadenylated mRNAs is a highly conserved oligonucleotide sequence AAUAAA. Using a series of deletion mutants of SV40 constructed *in vitro*, Fitzgerald and Shenk (1981) demonstrated that polyadenylation occurs at a fixed distance from this site regardless of the sequence to which the poly(A) is directly added.

The addition of poly(A) results in the formation of a colinear nuclear molecule, which contains internal sequences that must be removed to produce functional cytoplasmic mRNA. The initial discovery of the requirement for the removal of these intervening sequences, or introns, came from the independent investigations of Sharp and collaborators (Berget et al., 1977) and the members of the Cold Spring Harbor Laboratory (Chow et al., 1977; Klessig, 1977). During the course of mapping the location of the specific adenovirus mRNAs encoding hexon and fiber proteins, these workers found that the sequences at the 5' terminus of the mRNAs did not hybridize with, and therefore were not transcribed by, the genomic DNA from which the body of the mRNA was synthesized. A search for the viral DNA sequence complementary to the 5' terminus of the mature mRNAs determined that the viral mRNAs produced from the major late transcription unit share three sequences transcribed from the site of initiation of the primary transcript. These sequences, which form the tripartite leader, are therefore joined to the protein-encoding body of the mRNA by intramolecular ligation of distant sequences, termed splicing, of the primary transcript. Subsequent examination of other viral mRNAs, including SV40 and polyoma as well as adenovirus, has shown that splicing is required not only to join the 5' terminus, or leader, sequences to a mRNA but also to form a continuous protein-coding sequence within the mRNA. Moreover, it has been demonstrated that alternate forms of mRNAs can be produced by different splicing patterns of the primary transcript. In the adenovirus major late transcription unit, selection of a specific mRNA sequence from a group of potential mRNAs occurs primarily at the 5' terminus of the body of the mRNA. Other transcription units of adenovirus as well as SV40 and polyoma use variable internal splicing to produce different mRNA species (see review by Ziff, 1980). These alternate splicing patterns change the trans-

RNA SPLICING IN EUKARYOTES

The analysis of cellular genes in many laboratories has shown that the principles observed in the structure of the adenovirus transcription unit and the processing of its primary transcript are universal in eukaryote gene expression. These studies have confirmed that most genes are composed of discontinuous coding sequences and are transcribed into large RNA precursors that must undergo polyadenylation and splicing to form mature mRNA (for a discussion of the significance of splicing, see Darnell, 1978). The recent investigation of structure of the major β-globin transcriptional unit of the mouse serves as an example of these principles for a common eukaryote gene. A fragment of DNA, isolated from the genome of the mouse by recombinant DNA cloning procedures contains the structural gene for β-globin (Fig. 2). The polyadenylated nuclear precursor spans the entire distance of 1,500 nucleotides between the genomic sequences encoding the 5' terminal capped oligonucleotide and the polyadenylation site. Exons, the sequences conserved in cytoplasmic mRNA, are interrupted by two intervening sequences of introns, which are removed by RNA splicing from the transcript during mRNA maturation (Ross and Knecht, 1978; Tilghman et al., 1978; Curtis et al., 1979). To define the boundaries of this transcription unit, Hofer and Darnell (1981) determined the molarity of transcription across this region of the mouse genome. Labeled nascent RNA transcripts were isolated from pulse-labeled cell cultures actively engaged in globin synthesis, and hybridized to fragments of DNA representing sequences both within and surrounding the 5' and 3' terminus of the polyadenylated nuclear transcript. Their results demonstrate that the primary transcript initiates at or very near the DNA region containing the 5' terminus of the mRNA but extends more than 1,000 nucleotides beyond the poly(A) site. Thus the transcription unit for this cellular gene, and probably most eukaryote genes, obeys the same rules governing the structure first described for the viral genes of adenovirus and SV40.

The early addition of poly(A) to the primary transcript before splicing is suggestive of a possible functional relationship between these two processing events. An experiment was designed by Zeevi and coworkers (1981) to test whether splicing could be achieved in the absence

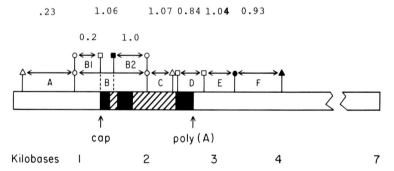

FIG. 2. Map of the mouse β-globin major transcription unit and molarity of transcription. The location of the 5' terminal cap site, the three exons *(black boxes)*, the small and large introns *(two diagonally striped boxes)*, and the poly(A) site are depicted. Arrows indicate the restriction fragments prepared and recloned from the major β-globin DNA used in this study. Restriction enzymes used were: *open triangle*, Pst I; *open circle*, Hind III; *open square*, Sau 3A; *solid triangle*, Bgl II; *solid circle*, XbA I; *solid square*, Mbo II. B1 and B2 are further restriction fragments recloned from B. The molarity of transcription was obtained from the hybridization of RNA transcripts labeled by the incorporation of [^{32}P]-triphosphates in isolated nuclei and calculated by dividing the cpm hybridized to the indicated fragments by the length of fragment. (From Hofer and Darnell.)

of poly(A) addition. The transcriptional unit chosen to explore this question was adenovirus region 2, located within the viral genome between map coordinates 75 and 59 (Fig. 3). Two mRNAs of about 2,000 nucleotides are produced from this transcription unit. The formation of the mature mRNAs from a 3,500-nucleotide polyadenylated precursor requires the splicing of two or three small segments to the main body of the mRNA. Using the drug cordycepin, 3′ deoxyadenosine, to selectively inhibit polyadenylation, these workers were able to detect the production of spliced nuclear RNAs lacking the normal 230 nucleotides of poly(A). Two lines of evidence suggest that the nuclear molecules formed were correctly spliced. First, the nuclear molecules synthesized after inhibition of poly(A) synthesis were 2,000 nucleotides, an appropriate 200 nucleotides shorter than the control molecules with poly(A). Second, the molecules produced in the presence of cordycepin were shown by hybridization to contain sequences both proximal and distal to the RNA initiation site, separated in the genome by 3,500 nucleotides. These results demonstrate that although the processing of mRNA transcripts follows a well-ordered series of events, production of the 3′ terminus by cleavage of the transcript, polyadenylation, and the internal splicing of mRNA segments are independent events.

LEVELS OF CONTROL OF mRNA ABUNDANCE

In bacterial systems, mRNA concentration is controlled by transcription. But the more complex pathway of mRNA biosynthesis in eukaryote cells allows for regulation at other levels. Beyond the level of transcription there is the opportunity to regulate the formation of mRNAs of different polypeptide coding capacity by differential splicing of the primary transcript. Control of the efficiency of transport of the mRNA to the cytoplasm and of the rate of cytoplasmic turnover of the mRNA can also dramatically affect the concentration of mRNAs. The contribution of each of these potential levels of regulation has been examined in the adenovirus system, with the conclusion that although considerable control is maintained at the level of transcription, regulation at other levels also affects the expression of various adenovirus transcription units.

The demonstration of transcriptional control requires an assay for the initiation and passage of polymerase molecules within a transcription unit. Pulse labeling of nascent RNA transcripts, either in the whole cell or in isolated nuclei, provides a measurement of the rate of transcription. The rate of initiation of transcription, on the other hand, must be determined by the quantitation of the promotor proximal RNA synthesis (see Evans et al., 1977; Wilson et al., 1979a).

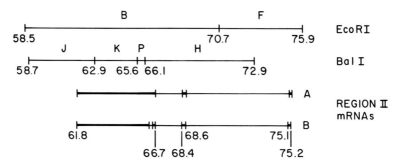

FIG. 3. The transcription unit of adenovirus 2 early region 2. Cap site location (probably initiation site) is at position 75.2 and the polarity of transcription is right to left, with the 3′ site of polyadenylation at 61.8 (1 unit = 350 base pairs). The exons are drawn in *heavy lines* and splice sites are indicated as *vertical lines*. The early region 2 mRNAs are calculated to be 2,200 *(A)* and 2,100 *(B)* nucleotides long including the 230 nucleotides for the poly(A). Sites of cleavage by restriction endonucleases Ba1 I and EcoRI in the genomic DNA are shown. Letters identify restriction fragments.

The expression of the mRNA of pIX, a capsid protein of adenovirus, provides an example of temporal regulation of transcription (see Fig. 1 for map of the adenovirus genome). At the left end of the adenovirus gene, three transcription units have been detected by the identification of UV-resistant promotor proximal transcripts (Wilson et al., 1979a). The existence of these independent RNA initiation sites has been confirmed by sequence analysis of the 5' termini of the mRNA and the corresponding genomic DNA (Baker and Ziff, 1979). Two of these transcription units, 1A and 1B, are active throughout infection, each producing at least two mRNAs by alternative splicing patterns (Spector et al., 1978). The single mRNA of the third transcription unit, encoding pIX, becomes abundant only after 6 hr of infection. Examination of the synthesis of promotor proximal RNA has determined that no initiation of transcripts from this region occurs prior to the appearance of mRNA in the cytoplasm. The expression of this transcriptional unit is therefore qualitatively controlled at the level of initiation of transcription.

During the course of an adenovirus infection, several examples of quantitative regulation of the rate of transcription are found. The most striking example is the synthesis of transcripts originating from the promotor map position at 16.45, the major late transcription unit. Measurement of the rate of transcription of promotor proximal sequences by pulse labeling of nascent RNA has shown a 50-fold increase of activity from this region during the transition from early to late infection (Fraser et al., 1979b). As the rate of synthesis was measured relative to the activity of the three transcriptional units at the left end of the viral genome (1A, 1B, and pIX), this increase reflects a selective control of the rate of transcription among viral transcription units. During the early stage of infection, Nevins and colleagues (1980) have demonstrated the independent regulation of activity of other viral transcription units. They observed that the transcription rate of region 1A reached a maximum rapidly within the first 2 hr of infection, followed closely by transcription units 3 and 4. In contrast, the maximal transcription from region 2 required 6 hr. The selective control of activity of these transcriptional units is further exhibited by their repression. Synthesis from region 4, and to some extent region 3, declines 10-fold soon after maximum transcription is reached 3 hr after infection. Similarly, the activity of region 2 is also rapidly diminished after 6 hr of infection. The transcription rate of the 1A transcription unit, however, appears to remain constant throughout this period of infection. It has now become apparent that this pattern of transcription is at least in part regulated by viral proteins. The repression of region 4 and possibly region 2 is inhibited by cycloheximide, indicating the need for a newly synthesized polypeptide. The studies of Nevins and Jensen-Winkler (1980) and Blanton and Carter (1979) employing a temperature-sensitivity mutant of adenovirus, ts^{125}, moreover, suggest that the repression of transcription of these regions is due to, or at least requires, a functional viral protein—the 72,000-dalton protein encoded by the region 2 transcription unit.

The evidence presented above describes the role played by transcriptional control in adenovirus gene expression. The extent to which transcriptional regulation governs the expression of eukaryotic cellular genes has not been carefully analyzed. In a few cases of specialized genes, most notably ovalbumin, transcriptional regulation has been indicated. Derman and colleagues (1981) have undertaken an investigation designed to document regulation of transcription for a series of mRNAs in different tissues. To provide the probes to assay transcription rates, complementary DNA (cDNA) to the entire population of mRNAs of mouse liver was generated by reverse transcriptase and DNA polymerase, joined to a bacterial plasmid vector and finally cloned in *Escherichia coli*. The cloned recombinant DNA molecules bearing the sequences of individual mRNAs therefore could be used as hybridization probes to detect labeled transcripts in experiments identical to those outlined for adenovirus.

Two families of cDNA clones were obtained; those complementary to mRNAs found exclusively in liver, and those mRNAs common to both liver and L cells, a general tissue-culture cell line displaying no phenotypic hepatic characteristics. Table 1 shows the relative cellular abundance of 16 of these mRNAs in liver, hepatoma, and whole brain as assayed by hybridization to the cDNA clones. The steady-state concentration of liver-specific mRNAs varies over more than two orders of magnitude but was extremely low or virtually undetectable in the hepatoma and brain mRNA populations. The common mRNAs, detected by hybridization to clones 12 to 16, appear in all mRNA populations at approximately the same concentration. The same collection of clones was then used to assay the rates of transcription by hybridization of nascent RNA transcripts labeled in isolated nuclei (Table 2). It is apparent that the liver-specific mRNA sequences are actively transcribed in the liver, in fact at a rate higher than that found for the five common mRNAs. In the brain little or no significant amounts of the liver-specific transcripts are found, although transcription of common mRNA is at the same level as that observed in liver nuclei. These results argue strongly that, as in the production of adenovirus mRNAs, the major component of gene regulation in eukaryotes lies at the level of transcription.

Changes in the concentration of mRNA formed by alternative splicing patterns from a single transcription unit require regulation beyond the level of transcription. Regulation could take place within the nucleus, with the selection of mRNA species, at the level of polyadenylation, splicing, or simply transport into the cytoplasm. Alternatively, the cytoplasmic stability of the different mRNAs may vary differentially to allow for a greater accumulation of an individual mRNA species. With the exception of the immunoglobin heavy chain (see L. Hood, *this volume*), the frequency of complex genes, producing more than one mRNA, is not known for eukaryotes. In viral systems, however, these types of transcription units are the rule rather than the exception. Recent evidence demonstrates regulation both of nuclear mRNA levels and of cytoplasmic stability during adenovirus gene expression.

Although the activity of the major late transcription unit is greatly enhanced during late infection, RNA transcripts can be detected from this region during the early phase of infection. Shaw and Ziff (1980) have shown that the initiation of these early transcripts is at precisely the same nucleotide as late in infection, suggesting that the same promotor of transcription is recognized at all times of infection. Electron microscopic analysis of the mRNA produced from

TABLE 1. *Percentage of poly (A) in each mRNA*

Clone input		Liver	Hepatoma	Brain
pliv-S	1	8.0	a	
	2	0.4	a	
	3	0.2	a	
	4	0.17	0.03	
	5	0.1	a	0.01
	6	0.1	0.02	0.003
	7	0.07	a	
	8	0.07	0.002	
	9	0.05	a	
	10	0.04	0.004	
	11	0.015	a	
pliv-C	12	0.25	0.22	0.32
	13	0.17	0.11	0.12
	14	0.15	0.10	0.13
	15	0.27	0.17	0.12
	16	0.35	0.30	0.15

[a] < 0.001
From Derman et al., 1981.

TABLE 2. *Hybridization of tissue-specific nascent hnRNA*

Clone input		Liver 6×10^7 cpm	Hepatoma 4×10^7 cpm	Brain 8×10^7 cpm
pliv-S	1	6,150	20	30
	2	2,800	34	49
	3	2,430	66	114
	4	1,275	22	52
	5	1,215	40	33
	6	1,520	66	44
	7	575	30	30
	8	485	16	33
	9	1,105	30	30
	10	550	25	30
	11	300	20	18
pliv-C	12	220	150	150
	13	80	200	275
	14	110	50	180
	15	185	470	270
	16	170	60	80
pBR-322		27	30	38

these transcripts hybridized to viral DNA showed that the difference between early and late mRNAs was the splicing of an additional 400-nucleotide piece of the primary transcript into the early mRNA (Chow et al., 1979). Furthermore, the studies of several laboratories demonstrated that the expression of this transcription unit early in infection was limited to a single mRNA (Akusjarvi and Persson, 1980; Thomas and Mathews, 1980; Nevins and Wilson, 1981). In contrast, late in infection, 13 or more different mRNAs are produced from this transcription unit by alternative splicing and the addition of poly(A) at five distinct sites (see Fig. 1). Examination of the nuclear RNA early in infection showed that the selection of this mRNA was a nuclear event, and not determined in the cytoplasm (Nevins and Wilson, 1981). The transcription originating at the 16.45 site did not extend beyond the third group of mRNAs, L_3, preventing the expression of the final 10 Kb of the transcription unit. There was a great preference for polyadenylation at the 3' end of the first mRNA group. The splicing of the capped 5' sequence, and presumably the extra 400-nucleotide "leader" sequence, was largely to the single mRNA to be transported to the cytoplasm. Little or none of the other mRNA produced from this region by the alternative splicing pattern used during late infection was observed.

Differential mRNA stability is another potential mechanism for regulating cytoplasmic mRNA concentration. One example for control at this level of mRNA biogenesis is the expression of the mRNAs produced from the 1A and 1B transcription units at the left end of the adenovirus genome. As shown in Fig. 4, the relative proportions of the mRNAs formed by alternative splicing from a single transcription unit change dramatically during the transition from early to late infection (Spector et al., 1978; Wilson and Darnell, 1981). The appearance of pIX mRNA in late infected cells is due entirely to the de novo initiation of its independent transcription unit located within the 3' terminus of the 1B region. An examination of the nuclear polyadenylated RNAs showed that no change in the ratio of the choice of splicing pattern of the 1A and 1B mRNAs could be detected in early and late infected cells.

The half-life of these mRNAs was determined by following the rate of cytoplasmic accumulation of newly synthesized mRNA to steady state. Previously it had been shown that the turnover of these mRNAs is extremely rapid during early infection, with a half-life of 10 min

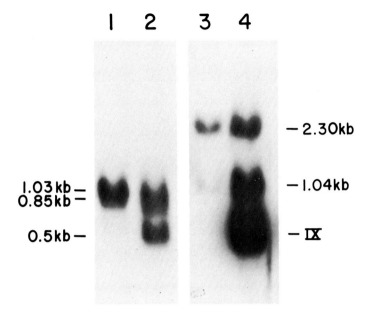

FIG. 4. RNA blot analysis of 1A and 1B mRNAs. Samples of approximately 5 μg of poly(A) cytoplasmic mRNA of HeLa cells infected with ad-2 for 3 or 15 hr were denatured in the presence of glyoxal and fractionated on a 1.5% agarose slab gel. The RNA was transferred to activated DBM-paper and hybridized with *in vitro* [^{32}P]-labeled ad-2 plasmid DNA complementary to either the 3' end of 1A (lanes 1 and 2) or of 1B (lanes 3 and 4). Early poly(A) mRNA, lanes 1 and 3; late poly(A) mRNA, lanes 2 and 4.

or less (Wilson et al., 1979b). When the stability of the mRNAs was examined late in infection, the half-life of the majority of the mRNAs was found to increase 6- to 10-fold. The sole exception to this increased stability was the 2.3 Kb 1B mRNA, which retained its rapid rate of turnover (Wilson and Darnell, 1981). The differential change in cytoplasmic stability appears, therefore, to dictate the change in ratio of these two 1B mRNAs during the course of the viral infection.

The adenovirus system has provided an initial view of the mechanisms controlling gene expression in eukaryotes. The utility of this model system is due to the availability of easily isolated fragments of genomic viral DNA that could be used to explore the pathway of mRNA biogenesis. Moreover, the rather high rate of viral mRNA transcription coupled to the synchronous progression of the infection among a population of cells has greatly facilitated the development of a number of experimental procedures that have pinpointed various levels of regulation of mRNA abundance. With the advent of mRNA and genomic DNA cloning by recombinant DNA technology, these procedures can now be used to approach the mechanisms governing the expression of cellular genes. It is clear that a delineation of the levels at which the control of gene expression is maintained is not sufficient for a full understanding of these processes. The availability of viral mutants, induced either *in vivo* or through the manipulation of isolated gene sequence (see P. Sharp and M. Wilson, *this volume*), holds the promise that these mechanisms of control will be examined in finer detail, in terms of nucleotide sequence, proteins, and cellular structures. Although our understanding of events dictating eukaryote gene expression remains quite primitive with respect to what has been achieved in prokaryote systems, it is hoped that such levels of understanding will be attained in the near future.

ACKNOWLEDGMENTS

This work was supported in part by grants from the National Cancer Institute (CA 16006-7 and CA 09256-04) and the American Cancer Society (MV 39 J). We are grateful to J. Nevins and E. Ziff for many stimulating discussions.

REFERENCES

Akusjarvi, G., and Persson, H. (1980): Controls of RNA splicing and termination in the major late adenovirus transcription unit at early times of infection. *Nature*, 292:420–425.
Bachenheimer, S., and Darnell, J. E., Jr. (1975): Adenovirus Type 2 mRNA is transcribed as part of a high molecular weight precursor RNA. *Proc. Natl. Acad. Sci. USA*, 72:4445–4449.
Baker, C., and Ziff, E. (1979): Biogenesis, structures, and sites of encoding of the 5' termini of Adenovirus-2 mRNAs. *Cold Spring Harbor Symp. Quant. Biol.*, 44:415–428.
Berget, S. M., Moore, C., and Sharp, P. A. (1977): Spliced segments at the 5' terminus of adenovirus late mRNAs. *Proc. Natl. Acad. Sci. USA*, 71:3171–3175.
Blanton, R., and Carter, T. (1979): Autoregulation of adenovirus 5 early gene expression. III. Transcription studies in isolated nuclei. *J. Virol.*, 29:458–465.
Brawerman, G. (1976): Characteristics and significance of the polyadenylate sequence in mammalian mRNA. In: *Progress in Nucleic Acid Research and Molecular Biology*, Vol. 17, edited by W. E. Cohn, pp. 149–179. Academic Press, New York.
Chow, L. T., Broker, T. R., and Lewis, J. B. (1979): Complex splicing patterns of RNAs from the early regions of adenovirus-2. *J. Mol. Biol.*, 134:265–303.
Chow, L. T., Gelinas, R., Broker, T., and Roberts, R. (1977): An amazing sequence arrangement at the 5' ends of adenovirus-2 messenger RNA. *Cell*, 12:1–8.
Curtis, P. J., Mantei, N., and Weissman, C. (1979): Characterization and kinetics of synthesis of 15S β-globin mRNA. *Cold Spring Harbor Symp. Quant. Biol.*, 42:971–984.
Darnell, J. E., Jr. (1978): Implications of RNA·RNA splicing in evolution of eukaryote cells. *Science*, 202:1257–1260.
Derman, E., Goldberg, S., and Darnell, J. E. (1976): HnRNA in hela cells: Distribution of transcript sizes estimated from nascent molecular profile. *Cell*, 9:465–472.
Derman, E., Krauter, K., Walling, L., Weinberger, C., Ray, M., and Darnell, J. E. (1981): Transcriptional control in the production of liver specific mRNAs. *Cell*, 23:731–739.
Dintzis, H. (1961): Assembly of the peptide chains of hemoglobin. *Proc. Natl. Acad. Sci. USA*, 47:247–261.
Early, P., Rogers, T., Davis, M., Calame, K., Bond, M., Wall, R., and Hood, L. (1980): Two mRNAs can be produced from a single globulin μ gene by alternative RNA splicing pathways. *Cell*, 20:313–329.
Edmonds, M., and Winters, M. A. (1976): Polyadenylate polymerases. In: *Progress in Nucleic Acid Research and Molecular Biology*, Vol. 17, edited by W. E. Cohn, pp. 149–179. Academic Press, New York.
Evans, R., Frazer, N. W., Ziff, E., Weber, J., Wilson, M., and Darnell, J. E., Jr. (1977): The initiation sites for RNA transcription in Ad-2 DNA. *Cell*, 12:733–739.
Fitzgerald, M., and Shenk, T. (1981): The sequence 5'-AAUAAA-3' forms part of the recognition site for polyadenylation of late SV40 mRNAs. *Cell*, 24:251–260.
Fraser, N. W., Nevins, J. R., Ziff, E., and Darnell, J. E., Jr. (1979a): The major late adenovirus type-2 transcription unit termination is downstream from the last polyA site. *J. Mol. Biol.*, 129:643–656.
Fraser, N. W., Sehgal, P. B., and Darnell, J. E., Jr. (1979b): Multiple discrete sites for premature RNA chain termination late in adenovirus-2 infection: Enhancement by 5,6 dichloro-β-D-ribo-furanosylbenzimidazole. *Proc. Natl. Acad. Sci. USA*, 76:2571–2575.
Goldberg, S., Nevins, J., and Darnell, J. E., Jr. (1978): Evidence from UV transcription mapping that late adenovirus type 2 mRNA is derived from a large precursor molecule. *J. Virol.*, 25:806–810.
Goldberg, S., Schwartz, H., and Darnell, J. E. (1977): Evidence from UV transcription mapping in Hela cells that heterogeneous nuclear RNA is the messenger RNA precursor. *Proc. Natl. Acad. Sci. USA*, 74:4502–4523.
Hofer, E., and Darnell, J. E. (1981): The primary transcription unit of the mouse β-major globin gene. *Cell*, 23:585–593.
Klessig, D. F. (1977): Two adenovirus mRNAs have a common 5' terminal leader sequence encoded at least 10Kb upstream from their main coding regions. *Cell*, 12:9–21.
Nevins, J. R., Blanchard, J. M., and Darnell, J. E. (1980): Transcription units of adenovirus type 2: Termination of transcription beyond the polyA addition site in early regions 2 and 4. *J. Mol. Biol.*, 144:377–386.
Nevins, J. R., and Jensen-Winkler, J. (1980): Regulation of early adenovirus transcription: A protein product of early region 2 specifically represses region 4 transcription. *Proc. Natl. Acad. Sci. USA*, 77:1893–1897.
Nevins, J. R., and Wilson, M. C. (1981): Regulation of adenovirus-2 gene expression at the level of transcriptional termination and RNA processing. *Nature*, 290:113–118.
Ross, I., and Knecht, D. A. (1978): Precursors of alpha and beta globin messenger RNAs. *J. Mol. Biol.*, 119:1–20.

Salditt-Georgieff, M., Harpold, M., Chen-Kiang, S., and Darnell, J. E., Jr. (1980): The addition of 5' cap structures occur early in hnRNA synthesis and prematurely terminated molecules are capped. *Cell*, 19:69–78.

Sauerbier, W., and Hercules, K. (1978): Gene and transcription mapping by radiation effects. *Annu. Rev. Genet.*, 12:328–363.

Shatkin, A. J. (1976): Capping of eukaryote mRNAs. *Cell*, 9:645–653.

Shaw, A. R., and Ziff, E. B. (1980): Transcripts from the adenovirus-2 major late promotor yield a single copy family of 3' coterminal mRNAs and five late families. *Cell*, 22:905–916.

Spector, D. J., McGrogan, M., and Raskas, H. J. (1978): Regulation of cytoplasmic RNA synthesis from region 1 of the adenovirus 2 genome. *J. Mol. Biol.*, 126:395–414.

Thomas, G. P., and Mathews, M. B. (1980): DNA replication and the early to late transition in adenovirus infection. *Cell*, 22:523–533.

Tilghman, S. M., Curtis, P. J., Teimeier, D. S., Leber, P., and Weissman, C. (1978): The intervening sequence of a mouse beta-globin gene is transcribed within the 15S beta-globin mRNA precursor. *Proc. Natl. Acad. Sci. USA*, 75:1309–1313.

Weber, J., Jelinek, W., and Darnell, J. E., Jr. (1977): The definition of large viral transcription unit late in Ad2 infection of Hela cells: Mapping of nascent RNA molecules labeled in isolated nuclei. *Cell*, 10:612–617.

Weil, P. A., Luse, D. S., Segall, J., and Roeder, R. G. (1979): Selective and accurate initiation of transcription of the Ad2 major late promotor in a soluble system dependent on purified RNA polymerase II and DNA. *Cell*, 18:469–484.

Wilson, M. C., and Darnell, J. E., Jr. (1981): Control of messenger RNA concentration by differential cytoplasmic half-life: Adenovirus mRNAs from transcription units 1A and 1B. *J. Mol. Biol.*, 148:231–251.

Wilson, M. C., Frasier, N. W., and Darnell, J. E., Jr. (1979a): Mapping of RNA initiation sites by high doses of UV irradiation: Evidence for three independent promotors within the left 11% of the Ad-2 genome. *Virology*, 94:175–184.

Wilson, M. C., Nevins, J. R., Blanchard, J. M., Ginsberg, H. S., and Darnell, J. E., Jr. (1979b): Metabolism of mRNA from the transforming region of Adenovirus 2. *Cold Spring Harbor Symp. Quant. Biol.*, 44:447–455.

Zeevi, M., Nevins, J. R., and Darnell, J. E., Jr. (1981): Nuclear RNA is spliced in the absence of poly A addition. *Cell*, 26:39–46.

Ziff, E. B. (1980): Transcription and RNA processing by the DNA tumor viruses. *Nature*, 287:491–499.

Ziff, E. B., and Evans, R. M. (1978): Coincidence of the promotor and capped 5' terminus of RNA from the Adenovirus 2 major late transcription unit. *Cell*, 15:1463–1475.

Molecular Genetic Neuroscience, edited by
F. O. Schmitt, S. J. Bird, and F. E. Bloom.
Raven Press, New York © 1982.

Components of an Efficient Molecular Switch

Mark Ptashne

ABSTRACT

The bacteriophage λ has two contrasting lifestyles—quiescent passive growth in the lysogenic state or active lytic growth resulting in virus production. The transition from one state to the other is rapid; λ uses two sets of genes and two alternate regulatory proteins, each of which is capable of inhibiting transcription of one set of genes but not the other. The detailed molecular interactions that occur at the operator account for the extreme sensitivity and authority of this switch.

The bacterial virus λ can infect an *E. coli* cell, turn off most of its own genes, and then take up residency inside the bacterial cell (Ptashne et al., 1976; Echols, 1980). Inside the lysogenic bacterium, the quiescent state of the viral genome is extremely stable, and the bacterial cells will grow many generations without releasing virus. If that lysogenic cell is treated transiently with any of a variety of compounds that have also been shown to be carcinogens (Witkin, 1976; Gottesman, 1981), virtually all lysogenic cells switch on the phage genes, and lytic growth of the λ virus ensues, killing the host bacterium as a burst of virus is released. How does nature maintain a system that is very stable in one state, yet can respond to a transient signal and with nearly 100% efficiency switch to another state? More specifically, what are the molecular details of such a finely tuned biological switch?

One simplifying pleasantry in dealing with a prokaryotic system is that one need not worry about several of the complexities of eukaryotic gene expression discussed elsewhere (J. Darnell and M. Wilson, P. Sharp and M. Wilson, *this volume*), namely, mRNA capping, RNA splicing, and poly(A) addition. To the best of our knowledge, all prokaryotic genes are colinear with their RNA transcripts, and these transcripts are colinear with their protein products. In part, this simplicity is a consequence of translation occurring on nascent RNA transcripts—a situation possible because bacterial cells have no segregated nucleus.

The λ virus is able to turn its genes off by making repressor, a protein that binds to specific sequences on the DNA and turns off the transcription of all lytic-cycle genes. The phage DNA maintains its attachment to the bacterial chromosome, and the only virus gene being expressed is that for repressor. When the transient inducer is presented, a series of events occur, which results in proteolytic cleavage of the repressor, and consequently its inactivation, and hence the phage lytic-cycle genes are able to be expressed. Repressor, then, is responsible for maintaining the virus in its quiescent lysogenic state, and inactivation of repressor throws the switch. But what guarantees that the repressed lysogen is stably maintained for generations, and what ensures that induction of the lytic state is all or none?

To understand the problem at a deeper level, consider the established model diagrammed in Fig. 1, which shows a small part of the λ genome. There are about 50 genes in the entire virus genome. The *c*I and *cro* genes, shown in the figure, are next to each other. The small box

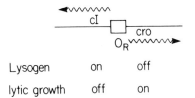

FIG. 1. Alternate physiological states of λ. In a lysogen, repressor is made but *cro* is not. During the early stages of lytic growth, *cro* protein is synthesized, and repressor is not. O_R, right operator.

between them is called an operator (O_R, right operator; O_L, left operator, is not shown). To understand how the λ molecular switch operates in controlling the lysogenic state, this is the only part of the viral genome of interest. Each of the two genes, *c*I and *cro*, encodes a repressor protein that represses the expression of the repressor produced by the other gene. For example, in a lysogen, the *c*I gene is on, actively transcribed from right to left. It directs the synthesis of a repressor that sits at the site indicated by O_R and keeps the *cro* gene turned off. After the *c*I repressor has been inactivated and lytic growth ensues, the first gene whose transcription is turned on is the *cro* gene, leading to synthesis of *cro* repressor. *Cro* also binds to O_R and prevents transcription of the *c*I gene. The two repressors operate as a biphasic switch: in a lysogen, one repressor prevents synthesis of the other. When that repressor is knocked out the system flips, and the second repressor takes over, turning off synthesis of the first (Ptashne et al., 1976).

In a lysogen, the phage can keep the *c*I gene on all the time so that the system virtually never breaks down spontaneously. And yet, with an outside inducer, the system can be flipped. One might consider that this biphasic switch puts the system in danger, because any leakage of transcription of the *cro* gene and synthesis of the *cro* protein could cause the system to flip. In fact, in the presence of *c*I repressor, *cro* repressor will cause the lytic growth cycle to be pursued for reasons discussed below. How does repressor keep *cro* synthesis shut off entirely, and then turn it on dramatically in response to a signal? The answer lies in an understanding of the molecular details of how these two proteins work.

The product of that *c*I gene is a small protein with molecular weight of about 26,000. It has been isolated in quantity by using recombinant DNA techniques to make large amounts of a protein otherwise present in very small amounts. The repressor is an extreme example of a two-domain protein. There are about 100 amino acids in each spherical domain, with a connector of about 40 amino acids (Pabo et al., 1979). There are various ways of showing that a protein has two domains. In this case it can be shown that these two domains denature independently, as though they were two separate proteins that just happened to be joined by a connector. They can also be separated by proteolytic cleavage.

The monomer does not bind to DNA with appreciable affinity, certainly not enough to do what it has to do *in vivo*. In a bacterial cell, about 80% of the repressor is in the form of dimers, in concentration equilibrium with monomers. Dimerization is mediated by contacts between the carboxyl terminal domains as shown by experiments using the separated domains (Pabo et al., 1979). The amino terminal domains function to permit repressor to bind to DNA.

As shown in Fig. 2, the dimer binds to DNA. The site on DNA that the repressor sees is 17 base pairs long and has twofold rotational symmetry. That is, the sequence on one of the DNA single strands occurs in inverted orientation on the complementary strand. This means that the dimeric repressor uses the twofold symmetry in the binding site. There is a vast amount of information from genetic and biochemical experiments that tells us precisely which of the bases are recognized by this particular repressor. It is the amino terminal domain of *c*I repressor that recognizes DNA. These isolated domains can be demonstrated by chemical experiments

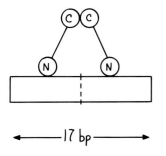

FIG. 2. A repressor dimer bound to a single operator site. Each repressor monomer contains two domains. The amino terminal domain N binds DNA and the carboxyl terminal domain C, forms important dimerization contacts. The DNA site is 17 base pairs (bp) long and is approximately twofold rotationally symmetric about the middle base pair.

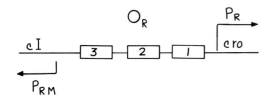

FIG. 3. The λ right operator O_R and surrounding DNA. Boxes show the three sites O_R1, O_R2, and O_R3 in O_R. Gene cl is transcribed leftward from promoter P_{RM} and gene cro is transcribed rightward from promoter P_R.

to recognize precisely the same DNA sequences and make the same contacts made by the intact repressor dimers. Since the energy of interaction between the carboxyl domains is lost, the energy goes down to what it is with monomer repressor. To recapitulate, the amino terminal domains recognize the DNA; the carboxyl terminal domains function in part to dimerize repressor monomers (Pabo et al., 1979; Sauer et al., 1979).

Long before these facts were known, we knew that the operator contained three binding sites. Figure 3 shows the cI and cro genes. The cI gene is transcribed leftward from the promotor, called P_{RM} in λ terminology. The promotor for transcription of the cro gene is called P_R, for promotor rightward. Promotors are biochemically defined as the sites that bind RNA polymerase enzyme to initiate the synthesis of RNA. Between these two transcription start points, there are three of the 17 base-pair cI repressor binding site sequences. The three sites are similar in sequence but not identical and are called, right to left, O_R1, O_R2, and O_R3. Repressor binds to them in a special way.

From the structural studies of the repressor proteins, the size and approximate shape of the two domains are known. In addition, the operator DNA sequences, the exact region of the operator that each repressor covers when bound to DNA, and the general structure of DNA have been determined. Before one can determine what repressor looks like when bound on DNA, one more important bit of information is needed. What does the protein do when it sees DNA? That is, does it wrap around it or unwind it? With a series of chemical probes developed by Gilbert and colleagues (Siebenlist et al., 1980), it has been shown that repressor binds along one face of the helix contacting bases in the major groove. For example, Fig. 4 shows the results of a particular probe experiment that delineates those phosphates that repressor contacts to bind to one of its 17 base-pair recognition sites. No other phosphates seem to be involved, or, to put it another way, if blocking groups are put on any other phosphates, they have no effect on repressor binding to operator. If the phosphates indicated in the figures are blocked, repressor will not bind to operator. These studies demonstrate the axis of twofold rotational symmetry in these phosphate contacts, each set of five flanking the major groove. Such symmetry appears to be a common feature of the binding of dimeric proteins to DNA.

Putting all the facts together, one would be tempted to draw the model shown in Fig. 5. Using each of the three binding sites within the operator, the dimeric repressor can be positioned. One subunit amino terminal domain makes contacts in the major groove, the other amino

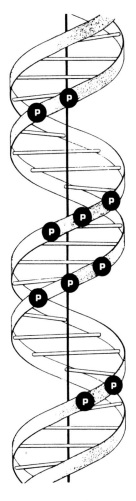

FIG. 4. Phosphates contacted at O_R1 by repressor. Indicates those phosphates that, when ethylated, hinder λ repressor binding at O_R1. These were determined using the chemical probe ethylnitrosourea as described by Siebenlist and Gilbert (1980) and A. Maxam and W. Gilbert *(unpublished data)*. O_R1 is represented here as DNA in the B form. As determined in a similar way, *cro* contacts a subset of those phosphate groups contacted by repressor; that is, the inner six but not the outer four. Most of the N^7 positions of guanines that are protected from methylation by dimethyl sulfate are exposed along the face of the helix shown.

terminal domain of the dimer makes contacts in the adjacent major groove. In a sense, the figure shows the "Newtonian" version of how repressor looks when bound to the three sites. In some ways, it is a good approximation, given the known sizes of the carboxyl terminal domains that are making the dimer contacts and the sizes of the amino terminal domains contacting the major grooves. However, this primitive model is incomplete because it neglects the important cooperative protein-protein interactions between adjacent repressor dimers. Although the binding of repressor to the operator sites is really quite tight, involving on the order of 10 to 12 kcal, the protein-protein interactions involve only a few kilocalories. Nevertheless, this protein-protein interaction can determine the entire pattern of how this gene control system works (Johnson et al., 1979). In brief, these cooperative interactions ensure that, in a lysogen, O_R1 and O_R2 are occupied by repressor but O_R3 is not.

These conclusions were shown in part by the following experiments that measure the relative strengths of the repressor-operator interactions. A restriction-enzyme-generated DNA fragment containing the binding sites for repressor is isolated and labeled with [^{32}P] at one end. If this fragment is exposed to deoxyribonuclease (DNase) for a short enough time, it is possible to control the reaction such that each DNA molecule is cut once or not at all. The reaction products are then resolved on a polyacrylamide gel and the gel autoradiographed. A ladder of bands is

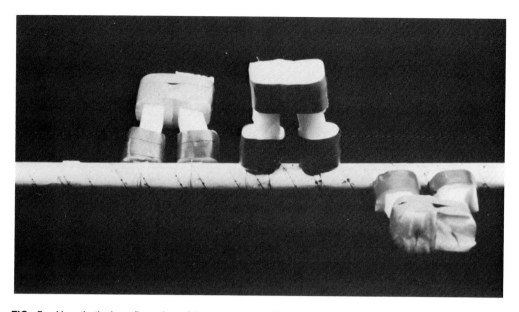

FIG. 5. Hypothetical configuration of three repressor dimers bound to O_R in the absence of interactions between adjacent dimers. Carboxyl terminal domains dimerize and amino terminal domains bind DNA. Approximately to scale, the chief uncertainty is the configuration of the 40 amino acid "connectors" between amino and carboxyl domains.

seen (Fig. 6), the rungs representing DNA containing the [^{32}P]-labeled end of increasing lengths (the shorter length fragments migrated faster on the gel than the longer fragments). If a protein, such as repressor, were present in the reaction, it might shield particular bonds in the DNA fragment from DNase cleavage, and on the autoradiograph of the gel one would see that the protein had left its "footprint." The footprint pattern can be used to determine which nucleotides are involved in the protein-DNA interaction as well as the strength of that interaction.

On a piece of DNA that contains all three binding sites, low concentrations of repressor, first shield cleavage at sites 1 and 2, and then at higher concentrations at site 3. On wild-type DNA, repressor binds to sites 1 and 2 with high affinity and simultaneously; it binds much more weakly to site 3. These data are interpreted to mean that on wild-type, operator-containing fragments, repressor binds 25 times more tightly to sites 1 and 2 than to site 3. It is not the case, however, that repressor has those relative dissociation constants for these sites ($O_R1:O_R2:O_R3::1:1:25$). If one goes one step further to ask what the intrinsic affinity for repressor for each of these sites is—for instance, by mutating site 2 and site 3 and measuring the binding to site 1—one finds that the relative intrinsic K_ds are $O_R1:O_R2:O_R3::2:25:25$. Site 1 has almost the same affinity as in the wild-type case, but the intrinsic affinity of site 2 is much lower than in wild type. This comparison means that repressor in the wild-type DNA situation must bind to sites 1 and 2 cooperatively. If repressor is absent from site 1, the affinity of site 2 is reduced. When repressor is cooperating at 1 and 2, it is having no effect on site 3. However, if, in these experiments, site 1 is mutant, sites 2 and 3 make a remarkable response. The binding to site 2 is reduced, but not to its intrinsic level, and binding to site 3 is increased ($O_R1:O_R2:O_R3::0:5:5$). Mutation at site 1 causes binding at site 3 to go up, apparently an action at a distance. The mechanism is that on a wild-type template 1 and 2 can collaborate; on a mutant template on which site 1 is vacant, 2 and 3 can collaborate. This cooperativity is called "alternate pairwise" (Johnson et al., 1979). The molecular explanation for this is shown in

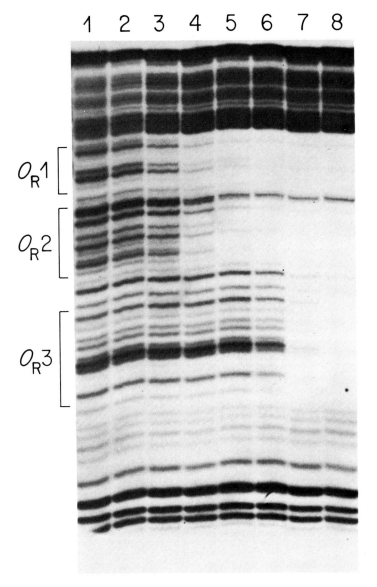

FIG. 6. Visualization of repressor binding to a wild-type template. A DNA fragment bearing wild-type O_R and labeled at one end with [^{32}P] (the *Alu/Hha* I 160 fragment of Humayun et al., 1977) was partially digested with DNase I in the presence of various concentrations of repressor. The products were visualized by autoradiography after electrophoresis through a polyacrylamide gel. The regions of the gel displaying fragments produced by cleavage within the regions O_R1, O_R2, and O_R3 are indicated. The total active repressor concentration (expressed in moles of repressor monomers per liter) in each reaction was as follows: slot 1, 0; slot 2, 3.5 nM; slot 3, 8.8 nM; slot 4, 18 nM; slot 5, 35 nM; slot 6, 88 nM; slot 7, 180 nM; and slot 8, 350 nM. The experiment provides a measure of the concentration at which each site is filled in half the molecules to an error of less than ± 30%. The concentration of repressor dimers, the active binding form, is calculated from the total repressor concentration using the dimer-monomer dissociation constant K_d = 20 nM (Chadwick et al., 1970; Sauer, 1979; Johnson et al., 1980).

Figs. 7a and b. To further this argument, if one looks just at the binding of the isolated amino terminal domains of cI repressor to the three sites, one finds the ratios of the intrinsic affinities (1:25:25). In other words, the amino terminal domains bind noncooperatively, which is why it is believed that the cooperative energy is provided by contacts between adjacent carboxyl terminal domains.

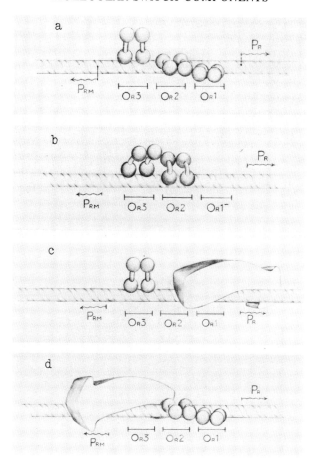

FIG. 7. Arrangements of repressor and RNA polymerase molecules bound in the vicinity of O_R. Repressor dimers are shown bound to one or more of the sites in O_R, and RNA polymerase molecules are shown bound to P_R and P_{RM}. The starting points of transcription of P_R and P_{RM}, located on different strands on opposite sides of the helix, are shown. **a:** Repressor dimers at all three sites in O_R, with those bound at O_R1 and O_R2 interacting cooperatively. **b:** O_R1 is mutant so that it cannot bind repressor, and the alternate pairwise interaction between repressor dimers at O_R2 and O_R3 is shown. **c:** Repressor at O_R3 and polymerase at P_R. Note that repressor at O_R3 only should have no effect on polymerase at P_R. **d:** Repressor at O_R1 and O_R2 and polymerase at P_{RM}. Note that an amino terminal domain of a repressor bound at O_R2 could plausibly touch a polymerase at P_{RM}. The components are drawn roughly to scale. An RNA polymerase molecule is drawn as a shape that implies its direction of transcription *(fat arrow)*. A repressor monomer is drawn to indicate its two domains and the connector that joins them. Repressor dimers are stabilized by contacts between the carboxyl terminal domains, and DNA is contacted by amino terminal domains. In **a, b,** and **d,** repressor dimers are shown interacting through additional contacts, presumed to be between carboxyl terminal domains. The DNA helix is drawn in B form with 10.4 base pairs per turn. The positioning of bound repressors was determined by chemical probe experiments such as that of Fig. 2. The positioning of bound polymerase molecules was deduced from a comparison of the P_R and P_{RM} sequences with those of promoters on which chemical probe experiments have been performed. (Based in part on Humayun et al., 1977; Johnson et al., 1979; Pabo et al., 1979; Sauer et al., 1979; Simpson, 1979; Wang, 1979; Siebenlist and Gilbert, 1980; and A. Johnson, unpublished data.).

Figure 7a shows what happens in a case that virtually never exists in nature, namely when there is enough cI repressor in the cell so that sites 1, 2, and 3 are filled. The repressor dimers at sites 1 and 2 are touching, and the dimer at site 3 is binding independently. If site 1 is unoccupied, then sites 2 and 3 can collaborate (7b). Figures 7c and d show what is physiologically important. Not as much is known about the structure of RNA polymerase (represented by the fat arrow) as about repressor. If no repressor is present at sites 1 or 2, then RNA polymerase will bind and transcribe P_R. In a lysogen, sites 1 and 2 are occupied, each by a dimer, the dimers collaborating with one another, and this binding of repressor to these two sites (O_R1 and O_R2) has two consequences. First, it prevents binding of polymerase to P_R, so P_R is turned off. Second, repressor also acts as a positive regulator of P_{RM} (Ptashne et al., 1976). Repressor bound to site 2 actually turns on transcription of its own gene. The positive control seems to result from protein contacts made between the amino terminal domain of repressor and RNA polymerase. Various probe experiments show that these two molecules virtually touch. If a repressor molecule is occupying only site 2 (achieved by mutating the other sites), positive control is elicited. The amino terminal domains are sufficient to demonstrate positive control. In fact, polymerase at P_{RM} and repressor at O_R2 bind cooperatively; one helps the other.

Three kinds of protein-protein interactions that are important for the system have been discussed: (1) the formation of dimers by cI repressor; (2) the interaction of the dimers so that they bind to sites 1 and 2; and (3) the interaction of an amino terminal domain of cI repressor with polymerase to turn on transcription of P_{RM}. That is the picture of what happens in a lysogen—the stable state. Why is it so complicated? Why bother having monomers that go to dimers, dimers that collaborate at sites 1 and 2, positive control of the gene that codes for the dimer repressor, and negative control of the lytic gene?

Recall that when cI repressor is inactivated by cleavage, the *cro* gene is activated. *Cro* then acts to turn off the synthesis of cI repressor. *Cro* is 66 amino acids long (compared to 96 for the N-terminal domain of cI) and forms a stable dimer. Despite the fact that these proteins have entirely different amino acid sequences, they bind to precisely the same three sites. The difference is that *cro* binds to site 3 (O_R3) most tightly, whereas repressor fills sites 1 and 2. In a lysogen, sites 1 and 2 are filled. In the presence of a carcinogen, repressor is inactivated and *cro* is made. *Cro* goes to site 3. When it occupies site 3, more repressor cannot be made, polymerase cannot bind to P_{RM}, and the system flips. Sites 1 and 2 are open and more *cro* is made. At high concentrations, in fact, *cro* does bind to these other two sites to turn down its own transcription. So λ has two proteins, cI and *cro*. Both are repressors and see the same three sites, but do so with opposite affinity orders.

Studies of other temperate phages reveal that these systems behave exactly the same way; their repressors are cleaved in the same way, yet their repressors (cI and *cro*) and their operators have sequences totally different from those of λ. One other case, bacteriophage P22, is shown in Fig. 8 *(top)*. Every formal rule enunciated for λ is precisely true for P22. That is to say, monomers make dimers, dimers bind cooperatively to operator sites 1 and 2, there is an alternate pairwise cooperativity, and repressor bound to sites 1 and 2 turns off the genes to the right and turns on the genes to the left. What seems astonishing is that the site at which polymerase is bound near the cI repressor bound at O_R2 has been moved over 11 base pairs to make a sandwich with polymerase (compare to Fig. 8 *(bottom)*—the λ case). However, the probe experiments suggest that the contact point with RNA polymerase is the same, the difference being in the detailed mechanism. Along this stretch of the DNA there is apparently protein on both sides of the helix: on one side it is polymerase; on the other side, repressor. It would appear that nature has gone out of the way either to recreate or to maintain the same formal control rules, despite differences in sequence and minor differences in detailed mechanisms. Therefore, there must be something fundamentally important about this particular means of genetic control. The complexity must be necessary to solve the problem of making a molecular switch. For comparison, consider a single binding site to which the protein binds very tightly, such as in the

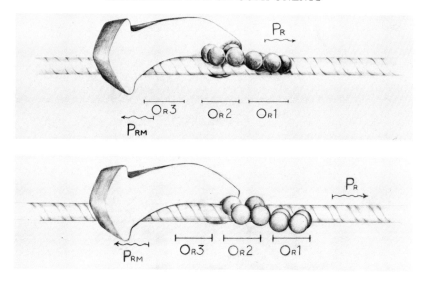

FIG. 8. Configuration of molecules bound to O_R/P_R in a P22 lysogen. Dimers of P22 **(top)** repressors are bound cooperatively to O_R1 and O_R2, repressing P_R and activating P_{RM}. Compare with the λ case **(bottom)**.

lac operon (Riggs et al., 1970). Since induction of *lac* enzyme synthesis goes up by a factor of 1,000 there is a 1,000-fold repression by a tight binding *lac* repressor. Since the rate of synthesis is inversely proportional to how much repressor is present (to first approximation), if 90% of the *lac* repressor is inactivated only 1% of maximal synthesis is achieved. Therefore, a control system based on a simple model in which only a single repressor binds very tightly to a DNA site would require that virtually all repressor molecules must be inactivated to switch from fully off to fully on. However, the difficulty in doing so may explain why systems like the *lac* operon are rarely fully induced into nature. Now just consider what happens in this case, with all the cooperativity built in. When we know the concentration of repressor, the various binding constants, the various dissociation constants, and the energy of cooperative interaction, what happens when repressor is inactivated? If the cell starts at 99.7% repression, how much synthesis occurs as repressor is inactivated? Inactivating about 70% of repressor is sufficient to flip the system because at that point P_R is activated to about 50% of its maximal level. If just that much *lac* repressor were inactivated, there would be very little synthesis from the promotor. To get the *lac* system to go, several orders of magnitude more repressor would need to be inactivated.

In general, what this system shows is that for this case nature has taken advantage of the simple proposition that in building in several layers of cooperativity (here mediated by protein–protein interactions), it can achieve a very stable state that, by rather modest changes in protein concentration, can flip quite spectacularly into another state.

ACKNOWLEDGMENTS

This work is supported by grants from NSF and NIH.

REFERENCES

Chadwick, P., Pirrotta, V., Steinberg, R., Hopkins, N., and Ptashne, M. (1970): The λ and 434 phage repressors. *Cold Spring Harbor Symp. Quant. Biol.*, 35:283–294.
Echols, H. (1980): Bacteriophage λ development. *The Molecular Genetics of Development*, edited by T. Leighton and W. F. Loomis, pp. 1–14. Academic Press, New York.
Gottesman, S. (1981): Genetic control of the SOS system in E. coli. *Cell*, 23:1–2.

Humayun, Z., Kleid, D., and Ptashne, M. (1977): Sites of contact between λ operators and λ repressor. *Nucleic Acids Res.*, 4:1595–1607.

Johnson, A. D., Meyer, B. J., and Ptashne, M. (1979): Interactions between DNA-bound repressors govern regulation by the λ phage repressor. *Proc. Natl. Acad. Sci. USA*, 76:5061–5065.

Johnson, A. D., Pabo, C. O., and Sauer, R. T. (1980): Bacteriophage λ repressor and cro protein: Interactions with operator DNA. *Methods Enzymol.*, 65:839–856.

Pabo, C. O., Sauer, R. T., Sturtevant, J. M., and Ptashne, M. (1979): The λ repressor contains two domains. *Proc. Natl. Acad. Sci. USA*, 76:1608–1612.

Ptashne, M., Backman, K., Humayun, M. Z., Jeffrey, A., Mauer, R., Meyer, B., and Sauer, R. T. (1976): Autoregulation and function of a repressor in bacteriophage lambda. *Science*, 194:156–161.

Riggs, A. D., Suzuki, H., and Bourgeois, S. (1970): *Lac* repressor-operator interaction I. Equilibrium studies. *J. Mol. Biol.*, 48:67–83.

Sauer, R. T. (1979): Molecular characterization of the λ repressor and its gene cI. Ph.D. dissertation, Harvard University.

Sauer, R. T., Pabo, C. O., Meyer, B. J., Ptashne, M., and Backman, K. C. (1979): Regulatory functions of the λ repressor reside in the aminoterminal domain. *Nature*, 279:396–400.

Siebenlist, U., and Gilbert, W. (1980): Contacts between *Escherichia coli* RNA polymerase and an early promotor of phage T7. *Proc. Natl. Acad. Sci. USA*, 77:122–126.

Siebenlist, U., Simpson, R. B., and Gilbert, W. (1980): *E. coli* RNA polymerase interacts homologously with two different promotors. *Cell*, 20:269–281.

Simpson, R. B. (1979): Contacts between *Escherichia coli* RNA polymerase and thymines in the *lac* UV5 promoter. *Proc. Natl. Acad. Sci. USA*, 76:3233–3237.

Wang, J. C. (1979): Helical repeat of DNA in solution. *Proc. Natl. Acad. Sci. USA*, 76:200–203.

Witkin, E. M. (1976): Ultraviolet mutagenesis and inducible DNA repair in *Escherichia coli*. *Bacteriol. Rev.*, 40:869–907.

Regulation of Gene Expression

Phillip A. Sharp and Michael C. Wilson

ABSTRACT

Understanding the mechanism of regulation of mammalian genes is important in studying the development of the nervous system and brain. The major stage of gene regulation in mammalian cells is the initiation of transcription. *In vitro* and *in vivo* systems have been developed to study this process. Although still preliminary, these studies have defined the DNA sequences necessary for initiating transcription. Less is known about the biochemistry of posttranscriptional processing of RNA, i.e., RNA splicing and polyadenylation. It is readily apparent that this stage of gene expression must also be complex.

The success of molecular biology is due, in large part, to the application of new methods that reduce the overwhelming complexity of the eukaryote genome to a rather simple series of nucleotide sequences. The sensitivity of this methodology has now reached the level where the detection and structural analysis of one mRNA in 10 cells are possible. The potential of a molecular approach to the neurosciences is that the questions addressed in these analyses result in definitive answers in terms of nucleic acids and proteins. Most strikingly, rapid progress in these techniques has led to the determination of the structure of eukaryote genes and an understanding of the processes required for the expression of those genes as mature polypeptide coding mRNAs. It is now possible to determine at what level regulation of gene expression is maintained—at transcription, processing, mRNA stability, or translation (see also Darnell and M. Wilson, *this volume*). The demonstration of common nucleotide sequences within the primary structure of various viral and cellular genes suggests the existence of protein recognition signals within the genome. The locations of these common sequences within the transcriptional unit—at the 5' end, near splicing junctions, and at the 3' site of the mRNA—imply their function during the synthesis and processing of mRNA transcripts. Direct evidence for their role in these processes is only now being accumulated. The molecular mechanisms involved in the regulation of gene expression through these nucleotide sequences, however, remains as yet unknown.

To identify the rules governing the epigenetic expression of the eukaryote genome, several questions must be answered. How is regulation of cell type achieved, leading to, for example, the selective production of globin in erythrocytes and not in fibroblasts? What are the sequences within the genome that receive those signals? Unlike prokaryote gene regulation, such as the molecular switch governing lysogeny in λ (see M. Ptashne, *this volume*), the interactions between proteins and perhaps RNA molecules directing mRNA biogenesis in eukaryotes remain virtually unexplored. Although an appropriate eukaryote model system has yet to be attained, it is evident that the concepts and techniques are now well enough defined to begin to answer these questions.

Another question important to the study of the nervous system is: What is the relationship between gene expression and morphological development in higher eukaryotes? The tools available to examine this question are rather primitive and will require long-term investment before answers as definitive as those obtained for the regulation of gene expression in specific cell types are achieved.

NUCLEOTIDE SEQUENCES DEFINING EUKARYOTE mRNA PROMOTERS

Earlier genetic analysis of prokaryotes defined the promoter of transcription as the initial binding site for the polymerase and the adjacent operator sequence as the site of repressor binding. The interaction between these proteins and the DNA sequence of these regions completely governs the initiation of RNA transcripts in prokaryotes. In eukaryotes, promoter is less precisely defined but, by analogy to prokaryotes, is taken to designate the nucleotide sequences just upstream from the start of mRNA synthesis. It is not clear, however, if in the intact cell the interaction of factors specifically at this site is solely, or even in part, responsible for the regulation of initiation of transcription in higher organisms.

The complex organization of DNA-containing chromatin may permit sequences some distance from an initiation site to control expression. Nuclear DNA is present in association with protein as a chromatin structure, composed of basic repeating units of 200 nucleotide lengths of DNA wound around a core of two each of the four histones: H2A, HaB, H3, and H4. Recent attempts to probe the nature of this association between DNA and protein have proved that the chromatin structure is not uniform with respect to DNA sequence. Initially, an assay of the nuclease sensitivity of DNA sequences within intact nuclei demonstrated that the chromatin structure containing actively transcribing genes is more accessible to nucleases than the structure associated with inactive sequences (Weintraub and Groudine, 1976). It has been shown, moreover, that the enhanced nuclease sensitivity of the chromatin structure extends well beyond the DNA sequences encoding the 5' terminus and the 3' site of polyadenylation of the primary mRNA transcript (Stadler et al., 1980). Within the regions of this nuclease-sensitive chromatin, further subtle modifications have been observed. In the regions surrounding and partially upstream from active genes, the interaction of histones and DNA within the nucleosomes appears to be sequence-specific, resulting in a uniform "phased" pattern of nucleosomes with respect to the structure of the transcription unit (Gottesfeld and Bloomer, 1980; Wu, 1980). In contrast, the nucleosome structure of inactive genes is randomly formed. Moreover, recent evidence has shown that the region immediately adjacent to the putative promoter site of RNA initiation is readily accessible to nuclease (Samal et al., 1981; Wu, 1980). Again these hypersensitive regions are associated only with the chromatin structure of active, or potentially active, genes of a particular cell type.

It is difficult to establish whether the complex organization of chromatin directly controls the expression of genes or, inversely, whether its complex organization is a product of local interactions resulting in active expression of a gene. With the capability to isolate specific DNA segments and the development of soluble systems that specifically transcribe RNA from these sequences, it is now possible to study some of the local interactions that lead to RNA synthesis. This has led to a number of correlates between primary nucleotide sequences and the mechanisms governing the initiation of RNA transcription.

The system used most extensively to study the regulation of transcription has been the adenovirus 2 during lytic infection of human cells (for recent reviews, see Tooze, 1980; Ziff, 1980). Although alterations in splicing patterns, termination of transcription, polyadenylation, and mRNA stability have all been observed to determine, in part, the concentration of viral mRNAs (see J. Darnell and M. Wilson, *this volume*), control at the level of initiation of transcription appears to be fundamental to viral gene expression during the course of infection

(Fig. 1). At the onset of infection, transcription of region 1A (0 to 4.4 map units) is initiated at the left end of the genome, in a rightward direction. A gene product of this region, expressed as two mRNAs by an alternated splicing pattern, appears to facilitate the subsequent transcription of the remaining viral transcription units. Experiments employing deletion mutants of this region have demonstrated, however, that the requirement for this region can be overcome by infection with a high multiplicity of viral genomes, suggesting that the region 1A protein enhances transcription but is not absolutely required for initiation (J. R. Nevins and T. Shenk, *personal communication*). During the course of early infection, initiation of transcription of the other early regions, 1B (4.4 to 11.2), 2 (11.2 to 75.0), 3 (76.0 to 84.0), and 4 (92.0 to 99.0 map units), proceeds according to a defined temporal pattern. Transcription from these regions, scattered across the 35,000-nucleotide double-stranded genome, occurs on both rightward and leftward strands and is independent of viral DNA replication. The intermediate phase of infection, commencing about 6 hr after infection, marks the initiation of synthesis at the independent transcription units producing the mRNAs encoding protein 1X and IVa$_2$. The late phase of infection, initiated by the replication of the viral genome, is dominated by the transcription from the major late promoter, located at map coordinate 16.45. Although transcription initiating from the promoter of the major late transcription unit can be detected throughout the early phase of infection, the rate of synthesis from this region is dramatically increased after viral DNA replication has begun. Moreover, the mRNAs formed from this transcript are markedly different between these times. Prior to viral DNA replication, a single major mRNA is produced from the most promoter proximal mRNA coding sequences (Chow et al., 1979; Akusjärvi and Persson, 1981; Nevins and Wilson, 1981); during late infection at least 13 different mRNAs, encoded throughout the length of the 26,000-nucleotide transcript, are formed. It is evident that the well-understood temporally coordinated expression of the nine transcription

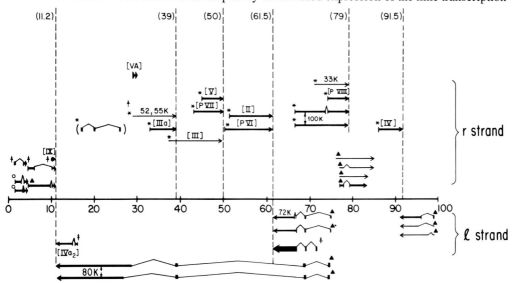

FIG. 1. Transcription map of adenovirus 2 mRNAs. The structure of the major RNAs expressed from different regions of adenovirus 2 are presented by lines drawn above or below the heavy line marked in units of 100. Sequences spliced together are joined by a *caret*. The mRNAs are divided into four groups: O, pre-early; ▲, early, ‡, intermediate; and *, late. By convention, regions of the genome transcribed into early mRNAs are referred to as early regions 1A (0-4.4), 1B (4.4-11.2), 2 (11.2-75), 3 (76-84), and 4 (92-99). All late mRNAs have a tripartite leader set from 16.5, 19.5, and 26.5 spliced to their 5'-termini and these mRNAs can be assigned to one of five families with polyadenylated termini at either 39, 50, 61.5, 79, or 91.5.

units within this viral genome, in conjunction with the recently derived nucleotide sequence of these regions, offers a unique opportunity to investigate the importance of primary nucleotide sequences in the regulation of transcription of eukaryote genes.

The analysis of the nucleotide sequences that are required to initiate transcription has been facilitated by the development of an *in vitro* cell-free system, dependent on exogenous DNA templates, and capable of initiating transcription at precisely the same site used *in vivo*. The potential of these systems is twofold. First, the dissection of the protein factors required for faithful transcription can be attained by reconstitution of the components of the extracts after separation by conventional protein fractionation techniques. Second, the templates initially isolated by recombinant DNA techniques may be modified by *in vitro* deletion and substitutions to generate mutant genes that can be assayed for transcriptional activity.

As an initial approach, several laboratories have employed an S100 cytoplasmic extract (i.e., an extract centrifuged at 1,000 g for 1 hr) to reproduce faithfully the *in vivo* transcription of isolated 5S genes of *Xenopus laevis in vitro* by polymerase III (Birkenmeyer et al., 1978; Ng et al., 1979). Supplementing the S100 cytoplasmic extract with purified eukaryote RNA polymerase II, Roeder and co-workers have demonstrated cell-free synthesis of precursors to mRNA transcripts initiating at *in vivo* start sites. The initial template used in this analysis was an isolated DNA fragment bearing the adenovirus major late transcription unit (Weil et al., 1979). After demonstrating initiation of transcription in permeable nuclei at the late promoter site of adenovirus 2, Manley and colleagues (1979, 1980) developed a simpler preparation that specifically initiated transcription by RNA polymerase II on exogenously added DNA. One advantage of this procedure is its relative ease of preparation, eliminating the often difficult requirement of isolating RNA polymerase II.

The characterization of the RNA transcript synthesized *in vitro* provides evidence that RNA polymerase II does recognize the same promoter sequences *in vitro* on purified DNA templates as observed *in vivo*. The most convenient assay of faithful transcription *in vitro* has proved to be the measurement of the length of the transcript from templates cleaved at distinct sites by restriction enzymes. An example of this "run-off assay" is shown in Fig. 2 (Hu and Manley, 1981). An isolated DNA fragment containing the initial 400 nucleotides of the major late adenovirus transcription unit was cloned into a plasmid of *Escherichia coli*. This plasmid was cleaved by the enzyme EcoR1, at a distance of 4,360 nucleotides from the *in vivo* RNA initiation site. After incubation in the soluble cell-free system in the presence of [^{32}P]triphosphates, the transcripts synthesized from this template were analyzed by gel electrophoresis. The major product of this reaction was a transcript of 4,360 nucleotides in length, as expected if the polymerase initiated transcription at the correct *in vivo* site. Further cleavage of the template with other enzymes at distances 3,610, 1,420, 650, and 370 nucleotides from the initiation site resulted in transcripts generated *in vitro* with lengths consistent with the distance between the *in vivo* start site and the site of restriction enzyme cut. These data provide strong evidence that site-specific initiation has been achieved in vitro. Analysis of the 5' terminal oligonucleotide confirms, moreover, that the first nucleotide of the transcripts generated *in vitro* is identical to the second nucleotide found *in vivo* (Ziff and Evans, 1978; Weil et al., 1979; Manley et al., 1980). As *in vivo*, the 5' terminus of the *in vitro* transcripts are modified by the addition of a guanylate residue in a 5'-5' phosphate linkage forming the "cap" structure found in all eukaryote mRNAs (Shatkin, 1976). It is important that in these *in vitro* studies transcription was not found upstream from the penultimate nucleotide found in the mature cytoplasmic mRNA. This evidence supports the earlier conclusion from *in vivo* studies that the 5' capped terminus of the mature mRNA represents the site of RNA initiation and that the 5' end of mRNA is not generated by cleavage of the primary transcript during processing. The location of the 5' terminus

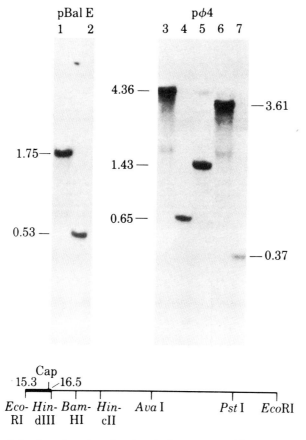

FIG. 2. *In vitro* transcription from the adenovirus 2 major late promoter. The 15.3 to 16.5 map-unit fragment of adenovirus 2 DNA containing the major late promoter was cloned into the pBR322 plasmid of *Escherichia coli*. This plasmid (pØ4) was cleaved by a series of restriction endonucleases and the DNA used as template for *in vitro* transcription reactions. For comparison, a segment of adenovirus 2 also containing the major late promoter (pBal E) was used as template in reactions resolved in lanes 1 and 2. RNAs synthesized were extracted, denatured with glyoxal, and resolved by electrophoresis in a 1.4% agarose gel. Sizes are expressed in kilobases. The top of the panel corresponds to the origin of electrophoresis. Lanes 1 and 2: pBal E DNA digested with Bam HI and Sma I, respectively. Lanes 3–7: pØ4 DNA digested with Eco RI, Hinc II, Ava I, Pst I, and Bam HI, respectively. The structure of pØ4 and the cleavage sites of the restriction enzymes used are shown at the bottom.

of the transcript, therefore, defines the position of the upstream nucleotide sequences that contain the putative promoter of the transcription unit.

With the demonstration that faithful initiation of RNA transcription is achieved *in vitro* and can be monitored by the runoff assay, it is now possible to dissect the sequences of the nucleotides surrounding the site of initiation and begin to define the structure of the eukaryote promoter. Hu and Manley (1981) have constructed a series of mutant DNA templates by deleting various regions from the adenovirus late promoter. The technique used to generate these deletions is simply the digestion of the viral template with a progressive exonuclease, Bal131, and subsequent recloning of the fragments into the bacterial plasmids as recombinant DNA. Shown in Fig. 3 is a diagram of the templates used by these workers and a summary of the results obtained from the transcription assay. The progressive deletion of viral sequences extending from − 66 to − 47 nucleotides upstream from the initiation site (position + 1) produced

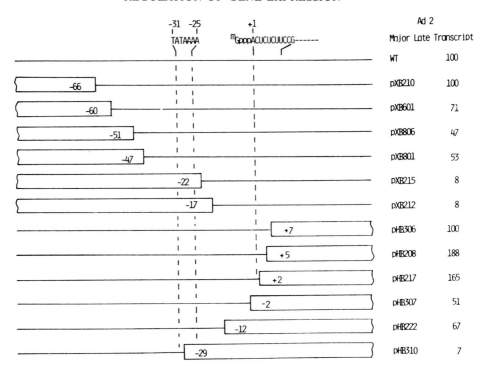

FIG. 3. Localization of the upstream and downstream boundaries of the region required for the initiation of transcription *in vitro*. The top line represents the wild-type (*WT*) adenovirus 2 (Ad 2) genome with the positions of the cap site of major late transcripts and the TATA region as indicated. Sequences removed in deletion clones are indicated by *open bars*. The *numbers* within the *open bars* indicate the positions of the deletion end point. The level of transcription from each deletion clone was estimated by first scanning autoradiograms of [^{32}P]uridine triphosphate runoff transcripts and then integrating the intensity of the band corresponding to the position of the runoff transcript. The numbers obtained were normalized to the background level in each lane and were expressed as percentages of the value obtained with the parental clone pØ4. The variability of these numbers is ≤ 10%.

templates that retained the potential to initiate transcription in vitro, although the efficiency of transcription was somewhat reduced by the loss of sequences at −51 to −47. Deletion of the sequences downstream from the initiation site did not result in the loss of initiation, indicating that the sequence to be transcribed was not recognized, at least *in vitro*, by the polymerase. Extending the deletions of template sequences through the initiation site until at least 12 nucleotides upstream from the *in vivo* start site still resulted in the transcription of specific transcripts, but at either enhanced or depressed levels. These transcripts are initiated at sites in the abutted plasmid DNA. Loss of specific initiation was only observed when the sequences between −47 to −12 were deleted from the viral template. These results demonstrate that *in vitro* the only recognition sites essential for initiation of transcription are the sequences between −47 and −12 nucleotides upstream from the actual start of the RNA transcript.

Within this region, positioned by the template deletions, lies a nucleotide sequence, TATAAAA. This sequence, first recognized by Goldberg and Hogness to be present upstream from each of the histone coding regions in *Drosophila* and hence called the Goldberg-Hogness box, or TATA region, is present approximately 30 nucleotides upstream from most, but not all, eukaryote mRNA transcription units. Several other studies have also shown the importance of the TATA consensus sequence in the *in vitro* reaction (Corden et al., 1980; Mathis and Chambon, 1981; Tsai et al., 1981). In fact, Wasylyk and co-workers (1980) showed that conversion of the third base, T, to a G in the conalbumin promoter sequence of TATAAA almost abolishes the *in vitro* reaction.

Although the *in vitro* transcription studies demonstrate the singular importance of the TATA sequence 30 nucleotides upstream from the start site of RNA transcription by RNA polymerase II *in vitro*, it is vital to determine whether this sequence alone produces the entire information required for correct initiation of transcription *in vivo*. Comparison of the nucleotide sequence surrounding several transcription units, in fact, has located another common region, in addition to the TATA consensus sequence, with the sequence GGTCAATCT approximately 70 to 80 nucleotides upstream from the initiation site of most eukaryote genes. As shown above, deletion of this region has little or no effect on transcription *in vitro* (see Fig. 3). To determine the importance of this sequence and other regions upstream from the start site of transcription, the approach of several laboratories has been to assay the transcriptional activity of exogenous templates with *in vitro* constructed deletions in an *in vivo* environment. The results obtained in three investigations are summarized in Fig. 4. Grosschell and Birnstiel (1980) injected deleted templates of sea urchin histone H2A into the nuclei of frog oocytes and observed that the deletion of the common sequences 80 nucleotides upstream from the initiation site actually resulted in the enhancement of transcription. In contrast, the removal of a region 180 nucleotides upstream led to the inhibition of transcription of the sea urchin histone gene. Similarly, the *in vivo* expression of early SV40 genes, transfected into culture cells, required the presence of sequences between 125 and 198 nucleotides upstream from the initiation site (Benoist and Chambon, 1981). An apparently conflicting result was obtained by McKnight and colleagues with the thymidine kinase gene of herpes simplex virus (HSV-tk); these workers observe that the common sequence located 80 nucleotides upstream from the start site was required for recognition of the polymerase in the frog oocyte system (S. McKnight, *personal communication*). Common to these systems, however, is the observation that deletion of upstream sequences beyond the TATA region at -30 nucleotides results in loss of *in vivo* transcriptional activity. These sequences, required for recognition by RNA polymerase *in vivo* but apparently not *in vitro*, therefore also constitute part of the promoter for eukaryote transcription units.

Deletion of the TATA sequence from the region upstream from the histone, SV40, and thymidine kinase start sites did not result in significant loss of transcription activity of templates assayed *in vivo* either in oocytes or transfected cells (see Fig. 4). The 5' terminus of the transcripts of these templates were, however, found to be heterogeneous. The *in vivo* results suggest, therefore, that the role of this region 30 nucleotides from the start site is to guide the polymerase to initiate at the proper nucleotide in the template.

The template activity of transcription units in a cell-free system is strongly dependent on the presence of the TATA sequence. The *in vitro* transcription of viral genes naturally lacking this sequence further demonstrates the specificity of this recognition. Adenovirus transcription units 2 and IVa$_2$ are clearly active *in vivo*, although producing transcripts with microheterogeneous 5' termini, but are at least 20-fold less efficient templates *in vitro* when compared to other viral transcription units (Fire et al., 1981).

The specificity of the *in vitro* reaction can be demonstrated by a comparison of the transcript obtained *in vivo* and *in vitro* from the adenovirus early region 4. The TATA consensus sequence of this transcription unit is represented by the sequence TATATATA. *In vivo* the transcripts produced downstream from this site have been found to exhibit a limited heterogeneity. Initiation preferentially occurs with an A residue, but to a lesser extent initiation upstream is found producing transcripts with a series of 5' U residues (Fig. 5). It is not known whether the repetitive nature of the extended TATA sequence or the presence of the six T residues at the start site is responsible for the heterogeneity of this initiation. However, it is clear from a

	?	-76 C - - G G T C A A T C T - -	-68	-31 A A - - T A T A T - -	-25	+1 mRNA START - - P_y A P_y
HISTONE H2A						
GROSSCHELL & BIRNSTIEL		-184 ↓		(HETEROGENEOUS)	+	
SV40 EARLY PROMOTER						
BENOIST & CHAMBON	-198 TO -125 ↓		?	(HETEROGENEOUS)	+	
HSV-TK					+	
MCKNIGHT	↓		−	(SLIGHTLY HETEROGENEOUS)	+	

FIG. 4. Effects of deletions in promoter region in *in vivo* transcription. Three studies of the *in vivo* biological activity of deletion mutants. The sequence deleted in each mutant is divided into 4 sets corresponding roughly to known regions of consensus sequences. Sequences around the cap s te (+1) are one set. Sequences around the TATA consensus sequences (−25 to −31) are another set. Sequences around the GGTCAATCT consensus sequence (−68 to −76) are a third set. Regions upstream from −80 are arbitrarily grouped as a fourth set. The effect of deletion of a region on transcription either after injection into the nuclei of oocytes (Grosschell and Birnstiel, 1980; S. McKnight, *personal communication*) or transfection into CV-1 cells (Benoist and Chambon, 1981) is indicated below each region. Symbols: +, wild-type levels of RNA; ↑, stimulation two to threefold higher than wild-type levels of RNA; ↓, synthesis of 1/20 the wild-type levels of RNA; −, not detectable RNA; and ?, no corresponding consensus sequence found in this region. The deletion of the TATA consensus sequence frequently yields macroheterogeneous capped termini distributed over several sites separated by as much as 40 nucleotides.

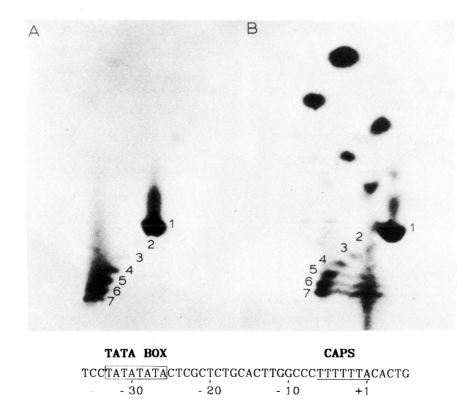

FIG. 5. Localization of 5' termini from the E4 region. RNAs were 5' labeled by decapping and kinasing (Fire et al., 1981). The RNA was selected by hybridization to promoter containing DNA sequences and T1 fingerprinted. (T1 nuclease cleaves only G residues.) Spot 1 and spots 2 to 7 derive from the major A and minor U termini, respectively. Identity of the spots was confirmed by redigestion with nuclease P1, RNases T2 and A, and chromatography on 540 or DEAE paper. **A:** Analysis of cytoplasmic RNA prepared from cycloheximide-treated adenovirus 5 infected HeLa cells at 5 hr postinfection. **B:** Analysis of in vitro transcribed RNA; 6.4 μg of Bgl II cleaved EcoRI B was transcribed in a 180 μl reaction mix (Fire et al., 1981). The prominent spots above the A terminus are not 2'-O-methylated and thus probably do not correspond to capped 5' termini. A number of the very minor spots were shown to have 2'-O-methylation and probably represent minor initiation events. **Bottom:** Sequence of EIV showing seven capped termini and TATA box. Note both in vivo and in vitro RNA have 5' termini distributed over the same set of sequences. At the moment of formation of the first phosphodiester bond, the recognition of these sequences by RNA polymerase II is similar in both reactions.

comparison of the fingerprint analysis of the 5' oligonucleotides of transcripts synthesized *in vivo* or *in vitro* that the extent of heterogeneity obtained *in vitro* is identical to that produced *in vivo* (Fig. 5) (Fire et al., 1981). Thus, as shown previously for the adenovirus major late promoter, which exhibits little or no detectable 5' heterogeneity (Weil et al., 1979; Manley et al., 1980), the *in vitro* transcription using a cell-free extract does provide an accurate reflection of the in vivo processes governing the position of initiation of RNA transcripts.

The lack of recognition of sequences further upstream from the TATA sequence in the soluble *in vitro* system will require further analysis. These results can perhaps be best interpreted, as suggested by Fire and coworkers (1981), by assuming that the TATA and flanking sequences have an affinity for RNA polymerase II. Initiation of transcripts could therefore be achieved independently from other upstream recognition signals in the presence of high concentrations of templates and RNA polymerase II found in the cell-free systems. This interpretation, however, implies that the whole cell extract lacks the factors required for the recognition of the upstream sequences necessary for initiation *in vivo*. Alternatively, the template itself may be required to acquire a chromatin-like structure, not achieved in the soluble extract, before these sequences are in the proper configuration for recognition.

The specificity exhibited by these cell-free systems is entirely template dependent. In the system of Roeder and his colleagues, the soluble extract, as well as the isolated polymerase of uninfected cells, serve equally well as those from infected cells to initiate transcripts on viral templates, for example, the major late transcription (Weil et al., 1979; Manley et al., 1980). Moreover, the mouse β-globin template is actively transcribed by extracts and polymerase isolated from human KB culture cells (Luse and Roeder, 1980), attesting to the lack of species and tissue specificity of these soluble systems. These observations emphasize again that additional factors, either missing or inactive in these systems, are likely to play a major role in regulation of synthesis of eukaryote genes. It is evident, however, that these initial efforts to reconstruct the components required for an accurate initiation of transcripts *in vitro* have defined sites of initiation of RNA synthesis and have provided the groundwork for future studies of the structure and function of promoter sequences in eukaryotes.

DEFINITION OF SPLICING SITES IN mRNA PRIMARY TRANSCRIPTS

As most eukaryote gene coding sequences are interrupted by intervening sequences or introns, removed at the RNA level, the molecular events governing the splicing of mRNA precursors provide the greatest impact on gene expression above transcription of the precursor itself. In viral systems, particularly adenovirus, SV40, and polyoma, RNA splicing provides different mRNAs, encoding different polypeptides from a single transcription unit (for a review, see Ziff, 1980). The expression of at least one adenovirus transcription unit, moreover, is regulated during the course of infection directly through the selection of alternative splicing patterns (Nevins and Wilson, 1981). On the cellular level, the regulation of the peptide coding capacity of immunoglobulin heavy chain is conducted in part by alterations in splicing patterns of the primary transcripts (see L. Hood, *this volume*). Although there is circumstantial evidence of the role of specific sequences within the RNA transcripts for splicing, there are no experimental data to indicate the molecular mechanisms that dictate the sites of splicing (see Lewin, 1980).

A common sequence used to define splicing sites within an RNA transcript can be easily recognized from a list of the sequences found adjacent to the splice junctions (Fig. 6) (Sharp, 1981). As initially recognized by Breathnach and coworkers (1978), both the 5' splicing site and the 3' splicing site can be assigned to reveal a limited consensus sequence centering on the first two nucleotides within the intervening sequence. At the 5' splicing site of all intervening sequences, the sequence GU is found. The 3' boundary of all but one intervening sequence is AG. Extending toward the center of the intervening sequence, the homology between the various splicing junctions begins to break down. Nevertheless, a consensus sequence of A·G|G·U·purine· A·G and a pyrimidine-rich tract followed by C·A·G| (where the vertical line denotes the site of cleavage) can be reached for the 5' and 3' splicing sites, respectively. Recently a model has been proposed in which a small nuclear RNA U_1 would provide the scaffolding structure required to join the ends of the intervening sequences (Lerner et al., 1980; Rogers and Wall, 1980). These workers posit that sufficient complementarity exists between the 5' end of the U_1 RNA and the 5' and 3' consensus sequences to hold these splicing sites together. Antibody preparations directed against this nuclear RNA have been isolated (Lerner and Steitz, 1979) and shown to inhibit the splicing of adenovirus mRNAs in isolated nuclei by about 45 to 80% (Yang et al., 1981). As it remains possible that the inhibition observed is due to an indirect consequence of the antibody preparation, further experimental support for this model awaits the development of a defined in vitro system that will provide a faithful splicing of mRNA precursors.

	INTERVENING SEQUENCES		CLASS OF INTRONS		
	5' SS	3' SS	0	I	II
OVALBUMIN	A U.A A A.U A A.G|G U G A G C C U A	C A A U U A C A G|G U.U G U.U C G.C*	+	Δ	+
	G A.A G C.U C A.G|G U A C A G A A A	U G U A U U C A G|U G.U G G.C A C.A*	+	Δ	+
	A U.C C U.G C C.A|G U A A G U U G C	G C U U U A C A G|G A.A U A.C U U.G*	–	Δ	+
	A.G A C.A A A.U G|G U A A G G U A G	U U C U U A A A G|G.A A U.U A U.C A*	+	+	Δ
	G U.G A C.U G A.G|G U A U A U G G G	G U U C U C C A G|C A.A G A.A A G.C*	+	Δ	+
	C U.U G A.G C A.G|G U A U G G C C C	U C C U U C C A G|C U.U G A.G A G.U*	+	Δ	+
OVOMUCOID	U C C U C C C A G|G U G A G U A A C	U U C C C C C A G|A U G C U G C C U*			
	G G G.G C U.G A G|G U G A G A A A G	U U U G U C G A G|G U G.G A C.U G C*	Δ	+	+
	C.U A C.A G C.A U|G U G U G U A C U	C C U C U U C A G|A.G A A.U U U.G G*	–	–	Δ
	A C U.G U U.C C U|G U A A G U A A U	C U U C C A C A G|A U G.A A C.U G C*	+	Δ	+
	C.C A C.A A A.G U|G U U A U U G U A	U C C U U U C A G|A.G A G.C A G.G G*	–	–	Δ
	G C U.G U G.A G U|G U A A G U A G C	C U U U U G C A G|G U U.G A C.U G C*	Δ	+	+
	U.G C A.G U C G U|G U A C G U A C A	C G C U U U C A G|G.G A A.A G C.A A*	+	–	Δ
HUMAN β GLOBINS	C.C U G.G G C.A G|G U U G G U A U C	C A C C C U U A G|G.C U G.C U G.G U*	+	+	Δ
RABBIT β "	C.C U G.G G C.A G|G U U G G U A U C	U U U U C U C A G|G.C U G.C U G.G U	+	+	Δ
MOUSE β "	C.C U G.G G C.A G|G U U G G U A U C	C U U U U U U A G|G.C U G.C U G.G U	+	+	Δ
HUMAN β GLOBINS	A A C.U U C.A G G|G U G A G U C U A	C C U C C A C A G|C U C.C U G.G G C*	Δ	–	+
RABBIT β "	A A C.U U C.A G G|G U G A G U U U G	U U C C U A C A G|C U C.C U G.G G C	Δ	–	+
MOUSE β "	A A C.U U C.A G G|G U G A G U C U G	#U U C C C A C A G|C U C.C U G.G G C	Δ	–	+
SV40 LARGE T	G C A.A C U.G A G|G U A U U U G C U#	G U A U U U U A G|A U U.C C A.A C C*	Δ	+	+
RAT INSULIN	A C.C C A.C A A.G|G U A A G C U C U	C C C U G G C A G|U G.G C A.C A A.C*	+	Δ	+
IMMUNOGLOBULINS					
λI L-VI	U C.A G C.U C A.G|G U C A G C A G C	U G U U U G C A G|G G.G C C.A U U.U*	+	Δ	+
λI J1-C1	C U.G U C.C U A.G|G U G A G U C A C	C A U C C U G C G|G C.C A G.C C C.A*	+	Δ	+
λII L-VII	U C.U G C.U C A.G|G U C A G C A G C	U G U U U G C A G|G A.G C C.A G U.U*	+	Δ	+
κ J1-1	A G.C U G.A A A.C|G U A A G U A C A	C U U C C U C A G|G G.G C U.G A U.G*	–	Δ	+
κ J2-2	A A.A U A.A A A.C|G U A A G U A G A				
κ J3-3	A A.A U A.A A A.C|C U A A G U A C A				
κ J4-4	A A.A U A.A A A.C|G U A A G U C U U				
κ J5-5	A A.A U C.A A A.C|G U A A G U A G A				
μ J H1-CH1	U C.U C C.U C A G|G U A A G C U G G	G U C C U C A G|A G.A G U.C A G*	+	Δ	+
μ CH1-CH2	C C.A U U.C C A G|G U A A G A A C C	U C A U U C C A G|C U.G U C C.C A*	+	Δ	+
μ CH2-CH3	G U.G C U.G C C.A|G U G A G U G G C	U G A U C G C A G|U G U.C C U.C C*	+	Δ	+
μ CH3-CH4	A A.C C C.A A U.G|G U A G G U A U C	C A U U U A C A G|A G.G U G.C A C*	+	Δ	–
μ CH4-CM1	A G.U C C.A C U.G|G U A A A C C C A	C C U U C A U A G|A G.G G G.A G.G*	+	Δ	–
μ CM1-CM2	C U G.U U C.A A G|G U A G U A U G G	C A C C U G C A G|G U G.A A A.U G A*	Δ	+	+
γ1 -CH1		U U C U U G U A G|C C.A A A.A C G.A			
γ1 CH1-HINGE	A G.A A A.U U.G|G U G A G A G G A	U C U C C A C A G|U G.C C C.A G G.G*	+	Δ	–
γ1 HINGE-CH2	U A.U G U.A C A.G|G U A A G U C A G	C A U C C U U A G|U C.C C A.G A A.G*	+	Δ	+
γ1 CH2-CH3	A A.A C C.A A A.G|G U G A G A G C U	C A C C C A C A G|G C.A G A.C C G.A*	+	Δ	+
γ2b -CH1		C U C U U C A G|C C.A A A.A C A			
γ2b CH1-HINGE	A A A.A C U U.G|G U G A G A G G A	U C U C C A C A G|A.C C C.A G C.G*	+	Δ	–
γ2b HINGE-CH2	A A.U G C.C C A.G|G U A A G U C A C	C C U C A U C A G|C U.C U U.A A C.C*	+	Δ	+
γ2b CH2-CH3	A A.A U U.A A A.G|G U G G G A C C U	A C C C C A C A G|G.G C U.A G U.C A*	+	Δ	+
BKV LARGE T	A G C.U C A.G A G|G U U U G U G C U	U U U U U A U A G|G U G.C C A.A C C*	Δ	+	+
PY LARGE T	G G C.U U C.C A G|G U A A G A A G G	U U C U U A C A G|G G C.U C U.C C C*	Δ	+	+
PY SMALL T	A U.A A U.C C A.A|G U A A G U A U C	U U C U U A C A G|G G.C U C.U C C.C*	–	Δ	+
AD5 E1A	A U.G A A.G A G.G|G U G A G G A G U	U U U U A A A A G|G U.C C U.G U G.U*	+	Δ	+
AD5 E1A	U U.G C U.A C A.G|G U A A G U G A A	U U U U A A A A G|G U.C C U.G U G.U*	+	Δ	+
	A 8 9 10 13 24 4|0 0 20 27 3 8 15 9	1 7 4 1 2 15 3 37|0 9 4 8 5 12 11			
	U 10 11 11 3 5 5|0 38 3 5 3 20 5 6	17 12 22 19 22 11 5 0|4 15 6 7 10 6			
	C 10 15 10 18 4 1|0 0 2 2 0 6 8 11	16 13 12 17 12 4 28 1|0 5 4 12 15 12 10			
	G 10 3 7 4 5 28|38 0 13 4 32 4 10 12	3 6 0 1 2 8 2 0 38|20 15 12 11 4 11			
CONSENSUS SEQUENCE	A G|G U R A G	Y – Y Y Y – C A G|			

FIG. 6. Consensus sequences around the 5' and 3' boundaries of intervening sequences. The 5' splice site (5' SS) and the 3' splice site (3' SS) have been aligned by placing the GU and AG dinucleotides, respectively, at the boundaries of the intervening sequences. The frequency of each base at a particular position is recorded in the matrix at the bottom. (Only sequences indicated by * were considered in forming the matrix; see Sharp, 1981, for references and further discussion.)

The necessity for other signals to direct splicing, perhaps within the intervening sequence, can be summarized by an examination of the frequency of the consensus sequences found at splicing junctions in a randomly ordered nucleotide sequence (Sharp, 1981). This analysis suggests a random occurrence of splicing sequences once every 2,000 nucleotides for the 5' terminus and 500 nucleotides for the 3' terminus of the intervening sequence. Moreover, the occurrence of transcription units containing as many as 51 intervening sequences (Vogeli et al., 1980), as well as those producing different mRNAs by alternative splicing patterns, present additional problems in the choice of splicing sites that cannot be simply met by the limited consensus sequence diagrammed in Fig. 5. Whether further homology between the 5' and 3' ends of the splicing junction facilitates the recognition required for specific splicing is not known. An experiment of Chu and Sharp (1980) would appear to rule out a rigorous requirement for such homology. Recombinant DNA molecules were constructed *in vitro* to contain the 5' portion of the SV40 T antigen gene connected by a hybrid intervening sequence to the 3' terminus of the mouse β-globin transcription unit. After transfection into cultured cells, this sequence gave rise to chimeric mRNAs that were the result of successful joining of viral and globin splicing sites.

The rules governing the splicing of eukaryote mRNAs appear to be complex. The problem is best represented as two sites of action. First, a mechanism to join the 5' and 3' ends of splicing junctions must exist, a role that might be fulfilled by the U_1 RNA or some other similar molecule. Second, the selection of splicing junctions in transcription units containing multiple intervening sequences and especially those that use alternative splicing patterns must be considered. A preferential order to the splicing of sequences along the mRNA precursor, as exhibited by the ovomucoid transcript (Tsai et al., 1980), and the tripartite leader sequence of the adenovirus major late transcript (Berget and Sharp, 1979), would appear to facilitate the joining of correct splice sites. This rather simplistic model, however, cannot account for the selection of different junctions within a transcript by alternative splicing patterns. It is critical, moreover, that any model for the processing of mRNA precursors in higher organisms accounts for the regulation of the selection of splicing junctions, as well as site of polyadenylation in transcripts produced by, as in the case of the immunoglobulin heavy chain, a single active transcription unit in a cell. As in the study of regulation of initiation of mRNA transcription, the future development of *in vitro* model systems exhibiting faithful splicing patterns of defined mRNA precursors promises a fruitful beginning to probe the mechanisms controlling the processing of eukaryote nuclear transcripts.

ACKNOWLEDGMENTS

This work was supported by grants from NSF (PCM78-23230), NIH (CA-26717 Program Project Grant) to the first author, and partially from an NIH Center for Cancer Biology at MIT grant (CA14051).

REFERENCES

Akusjärvi, G., and Persson, H. (1981): Controls of RNA splicing and termination in the major late adenovirus transcription unit at early times of infection. *Nature*, 292:420–425.

Benoist, C., and Chambon, P. (1981): In vivo sequence requirements of the SV40 early promoter region. *Nature*, 290:304–310.

Berget, S. M., and Sharp, P. A. (1979): Structure of late adenovirus 2 heterogeneous nuclear RNA. *J. Mol. Biol.*, 129:547–565.

Birkenmeyer, E. H., Brown, D. D., and Jordon, E. (1978): A nuclear extract of Xenopus laevis oocytes that accurately transcribes 5S RNA genes. *Cell*, 15:1077–1086.

Breathnach, R., Benoist, C., O'Hare, K., Gannon, F., and Chambon, P. (1978): Ovalbumin gene: Evidence for a leader sequence in mRNA and DNA sequence at the exon-intron boundaries. *Proc. Natl. Acad. Sci. USA*, 75:4853–4857.

Chow, L. T., Broker, T. R., and Lewis, J. B. (1979): Complex splicing patterns of RNAs from the early regions of adenovirus-2. *J. Mol. Biol.*, 134:265–303.

Chu, G., and Sharp, P. A. (1980): A gene chimera of SV40 and mouse β-globin is transcribed and properly spliced. *Nature*, 289:378–382.

Corden, J., Wasylyk, B., Buchwalder, A., Sassone-Corsi, P., Kedinger, C., and Chambon, P. (1980): Promoter sequence of eukaryotic protein-coding genes. *Science*, 209:1406–1414.

Fire, A., Baker, C. C., Manley, J. L., Ziff, E. B., and Sharp, P. A. (1981): In vitro transcription of adenovirus. *J. Virol.*, 40:703–719.

Gottesfeld, J. M., and Bloomer, L. S. (1980): Non random alignment of nucleosomes on 5S genes of X. laevis. *Cell*, 21:751–760.

Grosschell, R., and Birnstiel, M. L. (1980): Spacer DNA sequences upstream of the TATAAATA sequence are essential for promotion of H2A histone gene transcription in vivo. *Proc. Natl. Acad. Sci. USA*, 77:7102–7106.

Hu, S. -L., and Manley, J. L. (1981): DNA sequence required for transcription in vitro from the major late promoter of adenovirus 2. *Proc. Natl. Acad. Sci. USA*, 78:820–824.

Lerner, M. R., Boyle, J. A., Mount, S. M., Wolin, S. L., and Steitz, J. A. (1980): Are snRNPs involved in splicing? *Nature*, 283:220–224.

Lerner, M. R., and Steitz, J. A. (1979): Antibodies to small nuclear RNAs complexed with proteins produced by patients with systemic lupus erythematosus. *Proc. Natl. Acad. Sci. USA*, 76:5495–5499.

Lewin, B. (1980): Alternatives for splicing: Recognizing the ends of introns. *Cell*, 22:324–326.

Luse, D. S., and Roeder, R. G. (1980): Accurate transcription initiation on a purified mouse β-globin DNA fragment in a cell free system. *Cell*, 20:691–699.

Manley, J., Fire, A., Cano, A., Sharp, P. A., and Gefter, M. L. (1980): DNA-dependent transcription of adenovirus genes in a soluble whole cell extract. *Proc. Natl. Acad. Sci. USA*, 77:3855–3859.

Manley, J. L., Sharp, P. A., and Gefter, M. L. (1979): RNA synthesis in isolated nuclei: In vitro initiation of adenovirus 2 major late mRNA precursor. *Proc. Natl. Acad. Sci. USA*, 76:160–164.

Mathis, D., and Chambon, P. (1981): The SV40 early region TATA box is required for accurate in vitro initiation of transcription. *Nature*, 290:310–315.

Nevins, J. R., and Wilson, M. C. (1981): Regulation of adenovirus-2 gene expression at the level of transcriptional termination and RNA processing. *Nature*, 290:113–118.

Ng, S. Y., Parker, C. S., and Roeder, R. G. (1979): Transcription of cloned Xenopus 5S RNA gene by X. laevis RNA polymerase III in reconstituted systems. *Proc. Natl. Acad. Sci. USA*, 76:136–140.

Rogers, J., and Wall, R. (1980): A mechanism for RNA splicing. *Proc. Natl. Acad. Sci. USA*, 77:1877–1879.

Samal, B., Worcel, A., Louis, C., and Schedl, P. (1981): Chromatin structure of the histone genes of D. melanogaster. *Cell*, 23:401–409.

Sharp, P. A. (1981): Speculations on gene splicing. *Cell*, 23:643–646.

Shatkin, A. J. (1976): Capping of eukaryote mRNAs. *Cell*, 9:645–653.

Stadler, J., Larsen, A., Engel, J. D., Dolan, M., Groudine, M., and Weintraub, H. (1980): Tissue specific DNA cleavages in the globin chromatin domain introduced by DNase I. *Cell*, 20:451–460.

Tooze, J. (1980): *DNA Tumor Viruses*. Cold Spring Harbor Laboratory, New York.

Tsai, M., Ting, A. C., Nordstrom, J. L., Zimmer, W., and O'Malley, B. W. (1980): Processing of high molecular weight ovalbumin and ovomucoid precursor RNAs to messenger RNAs. *Cell*, 22:219–230.

Tsai, S., Tsai, M. J., and O'Malley, B. (1981): Specific 5' flanking sequences are required for faithful initiation of in vitro transcription of the ovalbumin gene. *Proc. Natl. Acad. Sci. USA*, 78:879–883.

Vogeli, G., Avvedimento, E. V., Sullivan, M., Maizel, J. V., Lozano, G., Adams, S. L., Pastan, I., and de Crombrugghe, B. (1980): Isolation and characterization of genomic DNA coding for a 2 type 1 collagen. *Nucleic Acids Res.*, 8:1823–1837.

Wasylyk, B., Derbyshire, R., Guy, A., Malko, D., Roget, A., Teoule, R., and Chambon, P. (1980): Specific in vitro transcription of conalbumin gene is drastically decreased by single-point mutation in t-a-t-a box homology sequence. *Proc. Natl. Acad. Sci. USA*, 77:7024–7028.

Weil, P. A., Luse, D. S., Segall, J., and Roeder, R. G. (1979): Selective and accurate initiation of transcription at the Ad-2 major late promoter in a soluble system dependent on purified RNA polymerase II and DNA. *Cell*, 18:469–484.

Weintraub, H., and Groudine, M. (1976): Chromosomal subunits in active genes have an altered conformation. *Science*, 193:848–858.

Wu, C. (1980): The 5' ends of Drosophila heat shock genes are exposed in chromatin. *Nature*, 286:854–860.

Yang, V. W., Lerner, M. R., Steitz, J. A., and Flint, S. J. (1981): A small nuclear ribonucleoprotein is required for splicing of adenoviral early RNA sequences. *Proc. Natl. Acad. Sci. USA*, 78:1371–1375.

Ziff, E. B. (1980): Transcription and RNA processing by the DNA tumor viruses. *Nature*, 287:491–499.

Ziff, E. B., and Evans, R. M. (1978): Coincidence of the promoter and capped 5' terminus of the RNA from the adenovirus major late transcription unit. *Cell*, 15:1463–1475.

Section IV

Intercellular Gene Transfer

A single human chromosome is estimated to contain upward of 100 million base pairs in its nucleotide sequence. Using standard methods for staining chromosome bands, an expert may be able to distinguish 20 to 30 differentiated regions along a chromosome, although the significance of the "stained" bands is not clear. However, with recombinant DNA methods, the nucleotide sequence of an isolated gene segment can be obtained relatively rapidly. If the appropriate gene can be isolated and sequenced, it can also be evaluated for point mutations and for insertions or deletions in the nucleotide sequence. Wholly synthetic gene fragments can also be readily obtained to study expression, to identify promoter zones, and to investigate the importance of specific flanking sequences for regulation of expression.

Thus far, there have been two major uses for these isolated gene sequences or their synthetic replicates: (1) to examine the molecular requirements for DNA expression under controlled conditions in cells or in cell-free media (see Section III); and (2) to map the location of the gene within a specific chromosome by hybridization methods, in which a DNA probe—perhaps a cDNA made from a purified mRNA for a gene product—is made radioactive and localized autoradiographically on a chromosome preparation (essentially *in situ* autoradiography). Hybridization depends on the complementary base pairings as already emphasized in the previous chapters.

Presently available methods also make possible a wide variety of genetic transfers between cells, which greatly enhance analytic capacity and improve mapping resolution. F. Ruddle describes these advances in gene mapping techniques and the use of specialized gene transfer methods. Before studying these exciting approaches, a more complete appreciation of them may be gained by reviewing briefly here some experimental tools and terms that are not yet widely used in neuroscience.

Major tools in chromosome mapping, as well as in DNA cloning, are microbacterial restriction enzymes, or more precisely "restriction endonucleases." These enzymes produce extremely specific cleavages of all types of DNA. For each enzyme, there is a characteristic set of 4 to 6 bases in sequence at the cleavage site; as one moves from the center of this cleavage zone, the nucleotides are arranged symmetrically (palindromically). The important feature for present concern is that when DNAs from different sources are exposed to the same restriction enzyme, the different DNA sequences will be cleaved wherever they exhibit the right series of base pairs. The fragments resulting from this cleavage are termed restriction fragments, and their size and charge can be compared by standard methods of electrophoretic separation. The comparisons can be greatly refined by "blotting," that is transferring, the DNA from the electrophoretic gels to paper filters on which hybridization autoradiography can again be used to probe for specific homologous sequences. Ruddle emphasizes the advances in analytic power obtained when these methods are applied to DNA preparations derived from interspecific somatic cell hybrids, and how this general approach can lead not only to more complete maps of the human genome, but eventually to the isolation of "disease genes" (see D. Housman and J. Gusella, *this volume*).

Applications of Somatic Cell Genetics and Gene Transfer Techniques for the Analysis of the Genetic Control of Development

Frank H. Ruddle

ABSTRACT

The genetic analysis of development in higher organisms cannot be easily accomplished using Mendelian methods alone. The potential contributions of somatic cell genetics and gene transfer systems are discussed. These include the genetic mapping of genes relevant to developmental processes and the creation of novel cells and organisms with altered genetic constitutions.

Mendelian genetics cannot be easily applied to the study of the genetic control of development. Mutations that affect the developmental process are often debilitating and frequently lethal, and as such they are difficult to propagate. Such difficulties are compounded in humans where naturally occurring mutants are infrequent and, of course, inaccessible for study. These problems have prompted the development of a somatic cell genetics system of analysis where the genetic regulation of development can be studied *in vitro*. Cultured cell lines may be used or particular differentiated cells may be studied in explants or cultures *in vitro* to obtain information about the genetics of complex organisms that would not be accessible by Mendelian genetics. Most of our knowledge of the human genome (see V. McKusick, *this volume*), for example, has come from the approaches used in somatic cell genetics, i.e., studies of genes transferred between somatic cells in culture, rather than from studies of genes transferred from one generation to the next. The methods of somatic cell genetics have now progressed to the point where there is real hope that we will be able to identify and study genes that determine the mechanisms of development, particularly the development and functioning of the brain.

A variety of methods have been developed to introduce genetic material into mammalian cells in culture (Fig. 1). In most approaches, genetic information is taken from a donor cell that can provide a selectable marker and is transferred to a recipient cell that is negative for that marker. Prototrophic donor cells (i.e., those lines able to synthesize a particular metabolite) and auxotrophic cells (lines unable to synthesize a particular metabolite) are essential to this methodology. Auxotrophic cells require supplementation of the growth medium; for example, thymidine-kinase-deficient cells require special growth media, (Ruddle and Creagan, 1975). Recipient cells that have taken up and expressed the transferred genetic information will have repaired any deficiency and can be positively selected in an appropriate growth medium. The amount of genetic information transferred depends on the technique and the purpose of the experiment.

In mapping experiments that seek to establish the position of genes in the genome, the whole chromosome set is transferred by cell hybridization. This technique may be simplified by

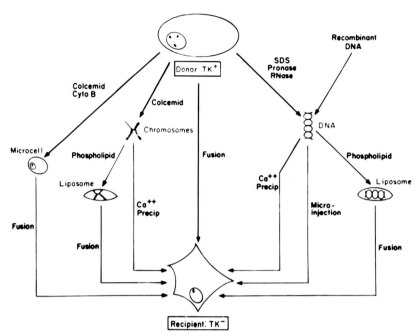

FIG. 1. Methods of gene transfer. TK, thymidine kinase. (K. M. Huttner and F. H. Ruddle, *unpublished data*.)

fragmenting the contents of the donor nucleus into micronuclei, isolated in microcells, allowing the introduction of one or a few intact chromosomes into the recipient cells. Alternatively, using techniques developed by McBride and Ozer (1973), individual chromosomes can be isolated. These are ingested by the recipient cells and fragments of the chromosomes are incorporated into the genome of the recipient cell (Klobutcher and Ruddle, 1979, 1981). In this case, regions of the chromosome that surround genes of interest are preserved intact so that the regulatory control of the gene may be maintained in the recipient. The transfer of genetic information has also been demonstrated with isolated segments of DNA (Wigler et al., 1977). The purified DNA, usually a cloned DNA sequence, is transferred as a calcium phosphate precipitate that can be taken up by the recipient cell by endocytosis. The DNA may also be inserted directly into the nucleus by microinjection techniques (Capecchi, 1980). Gene mapping is the assignment of a gene controlling a particular characteristic to a precise location on a defined chromosome. Gene mapping or linkage analysis in Mendelian genetics depends on the segregation of genes among the progeny of a genetic cross: genes that are linked on the same chromosome will tend to segregate together. The somatic cell genetic approach, on the other hand, depends on the segregation of genes in the cellular progeny of hybrid cells formed by the fusion of cells with two different genotypes (Fig. 2).

The technique of cell fusion, developed by Y. Okada and H. Harris and reviewed in Harris (1970) uses inactivated virus or, more usually now, high concentrations of polyethylene glycol (Pontecorvo, 1975) to induce the plasma membranes of two or more cells to coalesce (Ruddle and Creagan, 1975). Initially, a heterokaryon is formed, in which the several nuclei are surrounded by a single cytoplasm and plasma membrane. When the nuclei divide, all of the chromosomes are collected into a single nucleus to form a synkaryon, or hybrid cell. Appropriate genetic markers are used to select for hybrid cells and against parental cells (see E. Haber, *this volume*). The large multiple genomes of the hybrids are highly unstable and chromosomes are often lost with successive cell divisions. In human-mouse hybrids the human chromosomes

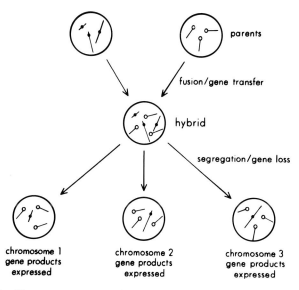

FIG. 2. Chromosome segregation in hybrid cells. (From Ruddle, 1980.)

are lost preferentially and at random. Eventually stable clones may be isolated, each containing only one or few different human chromosomes (Fig. 2).

By cytological procedures it is now possible to identify individually the 24 human chromosomes (22 autosomal pairs plus X and Y chromosomes) and to distinguish them from the chromosomes of other species (Ruddle and Creagan, 1975). The expression of a particular characteristic, such as an enzyme activity, in the cloned progeny of hybrid cells can therefore be correlated with the presence of a particular chromosome. By transferring fragments of a chromosome, the gene in question can be mapped to a particular location on the chromosome. The limits of resolution here are 1 to 10 centimorgans (1 centimorgan represents the distance on the chromosome between two genes that will recombine with a frequency of 1%) (Ruddle, 1981). These procedures have enabled human genes to be assigned to their chromosomes at the rate of three or four per month and are largely responsible for the expansion of our knowledge of the human genome (see V. McKusick, *this volume*).

MAPPING BY cDNA CLONING

The somatic cell genetic approach has been extended by combining the cell hybridization technique with gene cloning, enabling genes to be detected at the level of the genotype rather than by expression of the gene at the level of the phenotype. Radioactive nucleotides to be used as probes are derived from cloned genetic fragments. These probes can be used to detect complementary nucleotide sequences in the DNA of cell hybrids by the method of Southern (1975) (Fig. 3). In this approach, the DNA is fragmented by a restriction enzyme, the fragments are separated by electrophoresis and transferred ("blotted") onto nitrocellulose filters and exposed to the labeled nucleic acid probe. The complementary nucleotide sequences hybridize, and the DNA fragments containing the complementary sequences are detected as labeled bands by autoradiography. In different, but related, species, the nucleotide sequences of the genes for the same product are often sufficiently similar so that the labeled probe can detect both genes. The DNA segments on either side of these genes, however, may differ in sequence and in arrangement of restriction enzyme sites in the two species. Thus, the genes from two different

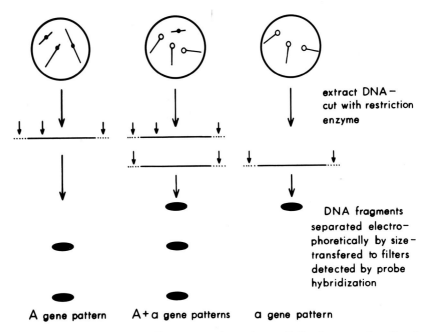

FIG. 3. Restriction fragment mapping. Chromosome 1 correlates with the A gene pattern therefore the A gene can be assigned to chromosome 1. (From Ruddle, 1980.)

species can be distinguished by the size of the fragment containing the complementary DNA. In rarer cases, the size of the restriction fragment(s) carrying the particular gene may vary among different members of the same species. This phenomenon, known as restriction fragment length polymorphism, can be used directly as a genetic marker (see D. Housman and J. Crusella, *this volume*).

Using the Southern blot technique, the presence of the gene is correlated with the presence of a particular chromosome. This approach has been used successfully to map the genes coding for immunoglobulins in the mouse; for example, the genes for mouse λ light chains were mapped to chromosome 16 by D'Eustachio and colleagues (1981). Previously the genes for κ light chains had been assigned to chromosome 6 and the heavy chain gene cluster to chromosome 12. With this approach, the multiple genes for interferon in humans have been mapped predominantly on chromosome 9, where there are 9 copies of the gene for leukocyte interferon and 1 copy of the gene for fibroblast interferon (Owerbach et al., 1981). Other copies of the fibroblast interferon genes are located on human chromosomes 2 and 5. Evidence from studies on the arginosuccinate synthetase genes indicates that the 12 copies of these genes are scattered throughout the genome rather than clustered on a single chromosome (Beaudet et al., 1981). A variety of different patterns of gene organization can be expected as more genes are mapped to other sites in the genome.

In combination with somatic cell genetics, the Southern blotting technique enables any cloned gene to be rapidly and accurately mapped. This approach will have obvious applications to neuroscience as clones become available for genes coding for products with specific functions in the nervous system. The particular advantage of this technique is that any kind of gene, whether structural or regulatory, can be mapped because the method does not require gene expression. This enables genes involved in neuronal function, for example, to be mapped in hybrid cells that would not normally express those genes.

GENE TRANSFER

When DNA is coprecipitated with calcium phosphate, particles that vary in size from less than 1 micron to 1 to 2 microns are formed. These particles bind to the surface of cells and are ingested into the cell by endocytosis (Loyter et al., 1982a). Most cells in a cultured population will take up DNA, but only a small fraction, approximately 1 to 100 per million, will subsequently express the transferred genetic information. By electron microscopy, the ingested DNA precipitates have first been found in cytoplasmic vesicles, or free in the cytoplasm, and later transferred to the nucleus of the recipient cell (Loyter et al., 1982b).

Usually a defined prototrophic gene marker is transferred to recipient cells that are auxotrophic for that marker. Transformed recipient cells that express the transferred gene can be detected by selection in an appropriate growth medium. Initially the transformants are unstable, that is, if the selective medium is removed, the transformed cells will lose the genetic marker at an appreciable rate, often 10% of cells per cell generation. Over longer periods of time, however, the same transformed cell population will give rise to stable transformants, that is, cells that retain the genetic marker even in the absence of selection. Although most cells initially take up the DNA only a minute fraction will become stably transformed.

The events of transformation by calcium-phosphate-precipitated DNA have been studied by Ruddle and coworkers using the thymidine kinase (TK) gene from herpes virus (Scangos and Ruddle, 1981; Scangos et al., 1981). The DNA, which was cloned in the plasmid pBR322, was mixed with a 1,000-fold excess of a randomly sheared, high-molecular-weight carrier DNA prepared from the recipient cells. The DNA was coprecipitated with calcium phosphate and introduced into recipient cells deficient in thymidine kinase (Fig. 4). Transformant cells expressing the enzyme were isolated by growth in selective medium and cloned. Each of the transformed clones was characterized as being stable or unstable by growing the cells for a period of time in a nonselective medium and assaying for the transformed phenotype by growth in selective medium.

The arrangement of the TK gene in the genome of each of the transformed clones was also analyzed by Southern blotting using the original cloned gene as a probe (Fig. 5). It was found that the patterns of DNA fragments containing TK genetic material could be correlated with

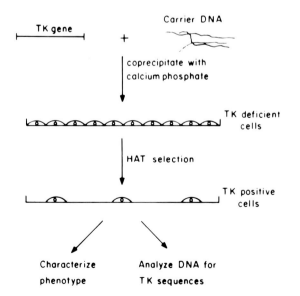

FIG. 4. Experimental design of the calcium-phosphate-mediated gene transfer system. HAT, hypoxanthine, aminopterin, thymidine.

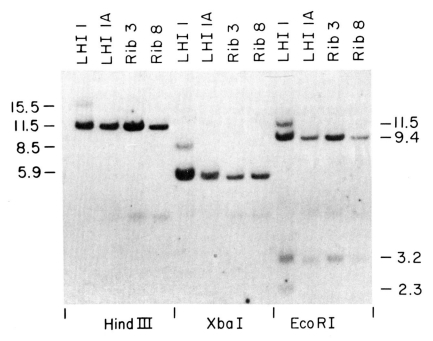

FIG. 5. Filter hybridization analysis of DNA from a single transformant and derivatives. LHI1 is an unstable transformant; LHI1A, stable derivatives; and Rib 3 and 8, two microcell hybrids formed between the stable transformant and Chinese hamster line RJK. (From Scangos et al., 1981.)

the stability of the transformed phenotype. In unstable transformations there were two or more DNA fragments carrying plasmid sequences. Stable transformants derived from the unstable clones showed a simpler pattern; usually there were fewer plasmid fragments. Significantly, the TK plasmid fragments detected in the stable clones were of different sizes from the TK plasmid that was introduced into the cells; in different clones the incorporated fragments were either smaller or larger in size than the original TK plasmid. This indicates that the transferred DNA was processed during the stabilization of the phenotype; parts of the plasmid containing the pBR322 sequences were probably lost and other, adventitious, DNA was added to the TK genetic material. There is good evidence that the DNA added to the TK gene was derived from the carrier DNA (Perucho et al., 1980). The Southern blotting patterns of TK fragments in different clones would therefore depend on the location of restriction enzyme sites in the carrier DNA that was recombined with the TK gene.

These experiments indicate that complex genetic structures are formed between the prototrophic marker genes and the carrier DNA (Fig. 6). These structures have been named *transgenomes* by Ruddle and *pekalasomes* by Perucho and colleagues (1980). The transgenomes, generated by ligation of the prototrophic gene with the carrier DNA, may be linear or circular and are often greater than 100,000 Kb in size. In the unstable transformed cells the transgenomes probably have the capacity to replicate independently of the host genome. In this form they can easily be lost during mitosis. On stabilization, the transgenomes become integrated into the chromosomal DNA of the host cell. The prototrophic gene is linked to the host DNA through the flanking carrier DNA sequences, retaining the restriction digestion pattern of the unincorporated transgenome.

There is considerable evidence that the transgenome is integrated into the host DNA in stable transformants. The position of the integrated prototrophic DNA can be mapped to a precise

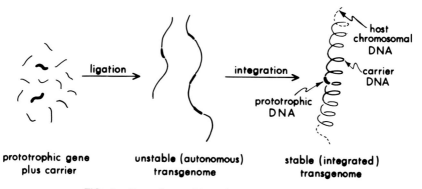

FIG. 6. Formation and fate of transforming DNA.

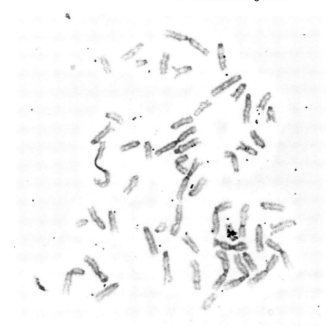

FIG. 7. *In situ* hybridization detection of integrated sequences in a stable transformant. (From Huttner et al., 1981.)

location in the host genome using the combination of somatic cell genetics and Southern blotting described previously. Furthermore, the location of the integrated gene on the host chromosome can be directly visualized by *in situ* hybridization using a labeled nucleic acid probe (Fig. 7) (Huttner et al., 1981). In addition, other cloned genes may be mixed into the calcium phosphate precipitate and, although these genes are not under selection, they will be cotransferred at a high rate to stable transformants, presumably as part of the transgenome.

It is also possible to transfer the marker genes in the absence of carrier DNA, producing somewhat different results. The transferred DNA is again ligated; several copies of the prototrophic gene are joined in tandem with occasional modifications of the arrangement of restriction sites. The unstable stage of the process is very short in duration. Without carrier, unstable transformants are rarely isolated, and most clones that are recovered are stable in phenotype. The transferred genes are integrated into one or a few sites in the host chromosomes and several copies of the gene may be integrated at one site. The fact that transformed cells produced in the absence of carrier DNA do not survive for long periods as unstable transformants suggests that the carrier DNA may transfer some function into the recipient cell, possibly enabling the transferred genetic material to propagate autonomously in the nonintegrated form.

An alternative technique of gene transfer is microinjection, a procedure that originated in some part from neurophysiology. Using fine microcapillary pipettes, DNA solutions can be injected directly into the nucleus of a recipient cell. Up to several hundred copies of a plasmid may be introduced into each nucleus, and an expert operator can inject 500 to 1,000 cells/hr. The Southern blotting patterns obtained from cells transformed in this way are similar to those obtained by the calcium phosphate technique in the absence of carrier DNA. Thus, the injected plasmids are ligated to form multimers, which are rapidly integrated into the host chromosomes. The transformation frequencies obtained with the microinjection technique, however, are appreciably higher than with the calcium phosphate procedure. These frequencies can be increased still further by the incorporation of other DNA sequences into the microinjected plasmid. For example, the plasmid pST6 (Gordon et al., 1980) has the origin of replication of the virus SV40 inserted into pBR322 together with the TK gene; microinjection of this plasmid produces transformation frequencies of 0.1 to 0.2%. Incorporation of the origin of replication and the "small t" region of polyoma virus gives frequencies of 2.5 to 5%. The small t region codes for a protein that has been implicated in the transformation of cells by polyoma (F. Ruddle, *unpublished results*). In similar experiments, Capecchi (1980) has reported transformation frequencies as high as 20%. Using the direct injection technique, transformation frequencies are sufficiently high that transformants can be identified by direct inspection of the injected cells for transfer of a particular gene marker.

Microinjection has also been employed to insert genetic material into one-cell mammalian embryos, which are then allowed to develop into viable offspring (Gordon et al., 1980). Given the high rates of transformation obtained by microinjection of cultured cells, the same degree of transformation might be expected in the whole organism. This enables the effect and expression of defined genes to be studied in the context of the normal development of a complex organism. The experimental approach is outlined in Fig. 8.

One-cell fertilized mouse embryos were isolated at the pronuclear stage of development. These cells are easy to isolate and can be maintained in culture for brief periods. They have the further advantage that foreign DNA injected at the one-cell stage will maximize the chance that this material will be distributed among all the progeny cells as development proceeds. The embryos were fixed in position with a holding pipette, and a plasmid suspension was injected into one of the two pronuclei. Several injected embryos were then reimplanted into the oviduct of a pseudopregnant foster mother and allowed to develop to term. With this technique, over 50% of the injected embryos produced viable offspring; of these, approximately 5% carried the injected DNA sequences. As many as 30,000 copies of a plasmid, the equivalent of two average-sized mouse chromosomes, were injected into a single embryo. The embryos appear to be rather tolerant to large amounts of foreign DNA, and the loss of viability is probably due to the trauma of handling rather than to the toxicity of the injected DNA.

SIGNIFICANCE FOR NEUROBIOLOGY

The approaches described here can all be applied to the study of neuronal systems. The combination of somatic cell genetics, gene cloning, and Southern blotting will provide a means to determine the map positions and organization of genes that are relevant for neurobiology. Gene transfer techniques, such as the calcium phosphate method, can be used to transform

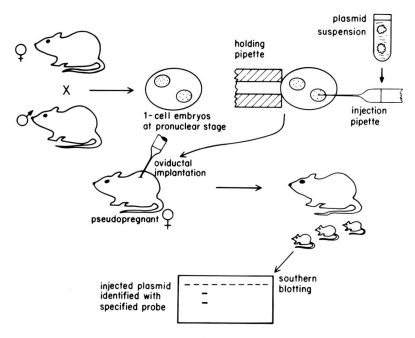

FIG. 8. Genetic transformation of whole organism.

established cell lines such as neuroblastoma of glioma cells to test the effects of specific genes in the environment of these cell types. Such genes could be readily introduced into cells by cotransformation with a selectable gene marker. The microinjection techniques offer even greater potential for the transfer of defined genes into cultured cell lines or even into primary cell cultures explanted directly from animals. Cloned genes could be introduced directly into the nucleus, or purified mRNA molecules (Liu et al., 1979) can be injected into the cytoplasm to study the expression of these molecules. Finally, the introduction of foreign genes into whole organisms by microinjection of embryos will become an important technique for studying the effects of specific genes on morphogenesis and the processes of development. For this approach to realize its full potential, it will be necessary to demonstrate site-specific integration of the transferred genes. This has been demonstrated quite clearly in yeast (Jackson and Fink, 1981). If it can be shown to occur in mammalian cells, this will open up a new system to examine the effect of specific genes on complex developing systems.

ACKNOWLEDGMENTS

This work was supported by NIH grant GM09966.

REFERENCES

Beaudet, A. L., Su, T. -S., Bock, H. -G., D'Eustachio, P., Ruddle, F. H., and O'Brien, W. E. (1981): Use of a cloned cDNA to study human argininosuccinate synthetase. In: *American Society of Human Genetics Meeting Abstracts*.

Capecchi, M. R. (1980): High efficiency transformation by direct microinjection of DNA into cultured mammalian cells. *Cell*, 22:479–488.

D'Eustachio, P., Bothwell, A. L. M., Takaro, T. K., Baltimore, D., and Ruddle, F. H. (1981): Chromosomal location of structural genes including murine immunoglobulin light chains: Genetics of murine light chains. *J. Exp. Med.*, 153:793–800.

Gordon, J. W., Scangos, G. A., Plotkin, D. J., Barbosa, J. A., and Ruddle, F. H. (1980): Genetic transformation of mouse embryos by microinjection of purified DNA. *Proc. Natl. Acad. Sci. USA*, 77:7380–7384.

Harris, H. (1970): *Cell Fusion*, Clarendon Press, Oxford.

Huttner, K. M., Barbosa, J. A., Scangos, G. A., Pravtcheva, D., and Ruddle, F. H (1981): DNA-mediated gene transfer without carrier DNA: Recipient cell regulation of donor genetic material. *J. Cell Biol.*, 91:153–156.

Jackson, J. A., and Fink, G. R. (1981): Gene conversion between duplicated genetic elements. *Nature*, 292:306–311.

Klobutcher, L. A., and Ruddle, F. H. (1979): Phenotype stabilisation and integration of transferred material in chromosome-mediated gene transfer. *Nature*, 280:657–660.

Klobutcher, L. A., and Ruddle, F. H. (1981): Chromosome mediated gene transfer. *Annu. Rev. Biochem.*, 50:533–554.

Liu, C. -P., Slate, D. L., Gravel, R., and Ruddle, F. H. (1979): Biological detection of specific mRNA molecules by microinjection. *Proc. Natl. Acad. Sci. USA*, 76:450–456.

Loyter, A., Scangos, G. A., Juricek, D., Keene, D., and Ruddle, F. H. (1982a): Mechanisms of DNA entry into mammalian cells. II. Phagocytosis and direct penetration of calcium phosphate DNA co-precipitate visualized by electron microscopy. *Exp. Cell Res., (in press)*.

Loyter, A., Scangos, G. A., and Ruddle, F. H. (1982b): Mechanisms of DNA uptake by mammalian cells: The fate of exogenously added DNA monitored by the use of fluorescent dyes. *Proc. Natl. Acad. Sci. USA*, 79:422–426.

McBride, O. W., and Ozer, H. L. (1973): Transfer of genetic information by purified metaphase chromosomes. *Proc. Natl. Acad. Sci. USA*, 70:1258–1262.

Owerbach, D., Rutter, W. J., Shows, T. B., Gray, P., Goeddel, D. V., and Lawn, R. M (1981): Leukocyte and fibroblast interferon genes are located on human chromosome 9. *Proc. Natl. Acad. Sci. USA*, 78:3123–3127.

Perucho, M., Hanahan, D., and Wigler, M. (1980): Genetic and physical linkage of exogenous sequences in transformed cells. *Cell*, 22:309–317.

Pontecorvo, G. (1975): Production of mammalian somatic cell hybrids by means of polyethylene glycol treatment. *Somatic Cell Genet.*, 1:397–400.

Ruddle, F. H. (1981): New era in mammalian gene mapping: Gene mapping by somatic cell genetics and recombinant DNA methodologies. *Nature*, 294:115–119.

Ruddle, F. H., and Creagan, R. P. (1975): Parasexual approaches to the genetics of man. In: *Annual Review of Genetics*, Vol. 9, edited by H. L. Roman, A. Campbell, and L. M. Sandler, pp. 407–486. Annual Reviews, Palo Alto.

Scangos, G. A., Huttner, K. M., Juricek, D. K., and Ruddle, F. H. (1981): Deoxyribonucleic acid-mediated gene transfer in mammalian cells: Molecular analysis of unstable transformants and their progression stability. *Mol. Cell. Biol.*, 1:111–120.

Scangos, G., and Ruddle, F. H. (1981): Mechanisms and application of DNA-mediated gene transfer in mammalian cells: A review. *Gene*, 14:1–10.

Southern, E. M. (1975): Detection of specific sequences among DNA fragments separated by gel electrophoresis. *J. Mol. Biol.*, 98:503–517.

Wigler, M., Silverstein, S., Lee, L. S., Pellicer, A., Cheng, Y. -C., and Axel, R. (1977) : Transfer of purified herpes virus thymidine kinase gene to cultured mouse cells. *Cell*, 11:223–232.

Section V

Structure and Diversification of Mammalian Genes

The availability of well-characterized restriction enzymes and repair enzymes, and rapid new methods of nucleotide sequencing, greatly accelerated the advance of molecular genetics in the mid-1970s. Discoveries in this extremely productive period served to reactivate interest in the field and to raise even higher the expectations of the discoveries to come. The chapters in this section cover the fields in which these magnificent achievements occurred, and in each case neuroscientists will readily find ways to apply such research to studies of the nervous system.

It is commonly held that the biggest surprise in the recent era of molecular genetics was the discovery that genes are not organized as continuous strings of nucleotide bases, but are instead fragmented into coding sequences and intervening sequences. After transcription into RNA, the intervening sequences are spliced out, as is detailed in the chapters of Section III. Furthermore, the coding sequences of DNA are capable of a surprising amount of movement, both within a given chromosome and between chromosomes.

These exciting revelations make it clear that nature has been using recombinant engineering far longer than have scientists. Much more significantly, these discoveries of a dynamically reorganizable genome open conceptual doors for probing deeper into the relationships between molecular structure and function which are of immense importance for neuroscientific phenomena such as developmental control, evolutionary adaptation, and functional plasticity. If coding sequences of DNA within genes can be physically rearranged and reassembled into new genes, could each of the coding sequences represent a particular functional portion of the final gene product, which the cell could employ in various combinations in large multifunctional macromolecules that might be needed under particular environmental conditions?

L. E. Hood describes the enormously fruitful application of these concepts to the expression of genes controlling antibody production. The well-detailed general molecular structure of antibody proteins, determined by the protein sequencing methods available in the 1960s, revealed the functional roles of the heavy and light chain polypeptides and their variable and constant domains. More recent evidence has demonstrated convincingly that each of these functional units is produced individually by distinct coding sequences of DNA. Each of the characteristic properties of the immunoglobulin specificity of antigen binding type of immunoglobulin, and disposition as membrane receptors or as secreted product, is regulated by selective splicing or by physical rearrangements of the gene segments controlling these features. Hood poses the possibility that the molecular mechanisms giving lymphocytes the opportunity for such enormous flexibility in a variety of secretory products could also represent an essential route for variation and amplification of information in the nervous system, especially during development.

T. Maniatis and associates review the opportunities to gain insight into the developmental regulation of structure and function with a description of the molecular genetics underlying the expression of the human globin gene family. These studies have provided a detailed view of the precise sequence of events that unfold as hemoglobin-producing cells mature, in concert with differing needs of the organism for oxygen transport during development from embryo to fetus to adult. The temporal expression of adjacent gene segments at each stage of functional development reveals yet another strategy for adaptive regulation that maintains a stable overall genetic composition. It would not be far-fetched to consider which series of DNA maneuvers might be most useful to neurons—the globin model in which coding sequences are arranged in a linear sequence, reflecting their order of temporal expression during development, or the immunoglobulin model in which the coding sequences are physically rearranged and combined to produce wholly new genes and gene products. Regardless of their eventual pertinence to the systems of genetic expression operating in neurons during developmental specialization or epigenetic adaptation, the globin "story" reveals clearly the profound functional consequences of relatively minor nucleotide alterations in the coding sequences responsible for the heritable hemoglobinopathies. By such methods, heritable disorders of nervous system structure and function may be effectively investigated for similar minor molecular alterations.

Although genes may have been continuously redesigned over evolutionary time, the ability of the scientist to do this has great importance—both in circumventing the limited biological production of vital hormones, such as the interferons, insulin, and growth hormone, and in accelerating the pace of man-made "evolution" by forcing production of hormones with properties specified to order. In the final chapter of this section, H. M. Goodman and colleagues describe efforts to employ recombinant methodology to produce human insulin, and the insights this work has provided with regard to the evolution of this universally required hormone, its genomic location in humans, and the regulation of its expression *in vivo* and *in vitro*. This record of innovative problem solving has very great potential benefit to medical biology and represents a pioneering effort that will undoubtedly be employed for hormones and other cell products needed in the future.

Antibody Genes: Arrangements and Rearrangements

Leroy Hood

ABSTRACT

Combinatorial and mutational strategies that antibody gene families use for the amplification of information are reviewed. The possibility is raised that other complex eukaryotic systems, such as those found in the nervous system, may employ similar strategies. A newly developed microchemical facility is described and its use in isolating interesting genes for developmental neurobiology is discussed.

Antibodies have a unique position in biology and biochemistry, not only as extremely precise and versatile tools for recognizing different surface topologies, but also as molecules ideally suited for investigations in genetic control and the expression of diverse, yet homologous protein structures. The enormous amount of research invested in molecular immunology during the past 10 years has provided detailed knowledge about the structure, function, and genetic origins of antibodies. The antibody system provides an interesting model for thinking about a variety of other complex eukaryotic systems, including many in neurobiology.

MODULAR STRUCTURE OF ANTIBODY MOLECULES

Antibodies, such as the common immunoglobulin (Ig) G, consist of two light (MW 25,000) and two heavy (MW 50,000) polypeptide chains. The tetramer folds into several structurally independent domains. Each domain corresponds to a portion of the polypeptide chain approximately 110 amino acid residues long. Pairs of domains from two different chains associate tightly together by multiple noncovalent contacts. Each domain dimer has one or more distinct functions. The N-terminal or variable (V) domain dimer (V_L-V_H) carries the binding site for antigen, whereas the other or constant domains (C_L of the light chain and C_H1, C_H2, and C_H3 of the heavy chain) form structures responsible for various effector functions that ultimately lead to elimination and destruction of the antigen-antibody complex. Different heavy chain classes (α, γ, μ, δ, or ϵ) are responsible for different biological manifestations of the antigen-antibody reaction, e.g., IgE molecules carrying the ϵ chain take part in hypersensitivity reactions.

Different antibody molecules of the same organism, or even the same species, share identical constant regions. However, variable regions, i.e., the N-terminal 110-amino-acid residues of both the light and heavy chains, differ from one antibody molecule to the next. The variability of primary structure, which involves both amino acid substitutions and deletions, is not distributed throughout the polypeptide chain at random; instead, it is concentrated into three short segments. These hypervariable regions (Fig. 1) are known to fold in three dimensions to

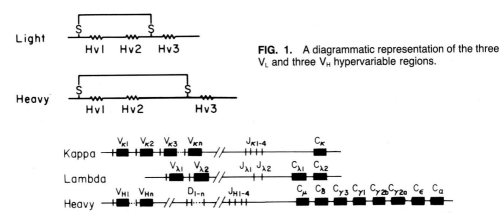

FIG. 1. A diagrammatic representation of the three V_L and three V_H hypervariable regions.

FIG. 2. A schematic representation of the three immunoglobulin gene families. V, J, D, and C represent the variable, joining, diversity, and constant gene segments, respectively. *Rectangles* represent exons or coding regions; *horizontal lines* indicate introns or intervening sequences; *vertical lines* represent leader sequences. (From Early and Hood, 1981a.)

constitute the walls of the antigen-binding site (Amzel et al., 1974). The rest of the variable regions, the so-called framework residues, are relatively invariant in their amino acid sequences.

ANTIBODY GENES ARE SPLIT

The origin and nature of immunoglobulin variability has been the subject of intensive study. Since the discovery that myelomas, i.e., plasma cell tumors, secrete homogeneous immunoglobulin molecules, numerous myeloma proteins have been isolated and their amino acid sequences determined. Although comparisons of these primary structures provided considerable insight into the nature of antibody variability, they could not answer the basic question about the genetic origin of immunoglobulin diversity. It became increasingly clear that research would have to turn to the study of antibody genes and mRNAs.

Several groups (Bernard et al., 1978; Seidman and Leder, 1978; Early et al., 1979) therefore embarked on isolating mRNA species coding for immunoglobulin light and heavy chains from myeloma cells. Once isolated, the mRNA was used as a template for the enzyme reverse transcriptase to produce complementary DNA (cDNA) copies. These copies, after being made double-stranded with *E. coli* DNA polymerase I, were inserted into plasmid or phage vectors, cloned, and amplified in bacteria. The immunoglobulin cDNAs were then sequenced. The cDNA copy was then used as a probe to isolate, from genomic libraries or banks, λ phage with inserted eukaryotic DNA, germline or somatic, which contained genes for immunoglobulins.

Three unlinked families of genes encode the antibody polypeptides—two for the light chains, λ and κ, and one for the heavy chains (Fig. 2). Light chains are coded for by three distinct types of gene segments: variable (V_L), joining (J_L), and constant (C_L); heavy chains are encoded by four different types of gene segments: V_H, J_H, C_H, and diversity (D_H) (Weigert et al., 1978; Early et al., 1980a). The antibody polypeptide chains are synthesized as precursors having an additional 18 to 22 (mostly hydrophobic) amino acids on their N-terminus, the signal peptide, which is cleaved off soon after biosynthesis (see Early and Hood, 1981a). Correspondingly, the V_L and V_H gene segments code for the signal peptide and most of the variable region, stopping at the beginning of the third hypervariable region (Tonegawa et al., 1978; Early et al., 1980a). In mouse λ chains, an intron separates part of the signal coding region from the

V gene segment (Tonegawa et al., 1978). The third hypervariable region and the rest of the variable region are encoded in a short J gene segment, which is located just to the 5' or upstream side of the C gene(s). The κ genes are organized in a manner similar to their λ counterparts. In contrast, the heavy chain genes have a fourth gene segment, D, which encodes the heart of the third hypervariable region (Fig. 2).

In the mouse genome there are some 100 to 300 V gene segments arranged into closely linked families of 4 to 20 (see Early and Hood, 1981a). The family consists of nucleotide sequences that are closely homologous to each other both in their coding and noncoding regions. Typically, a single mRNA would cross-hybridize (i.e., possess an extensive sequence homology) with all the members of the family. In contrast to a large V_κ family, the mouse genome contains only several, perhaps 2 or 3, V_λ gene segments. The J_κ cluster has 4 different gene segments that combine with the V_κ gene segments, and there are 4 gene segments in a separate J_H cluster to recombine with the V_H genes.

REARRANGEMENT OF V, D, AND J GENE SEGMENTS TO FORM V_L OR V_H GENES

Close examination of nucleotide sequences that flank the V_L and J_L gene segments revealed a possible mechanism for V-J gene joining (Fig. 3). At the 3' end of all the V_κ gene segments known so far, there is an inverted repeat (a palindrome) of about 7 nucleotides. Approximately 11 nucleotides further in the 3' direction there is an additional 10-nucleotide inverted repeat.

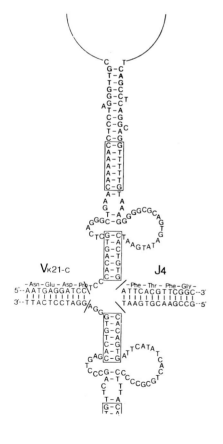

FIG. 3. A model for the joining of the V_k and J_k gene segments. *Diagonal lines* indicate recombinational sites at which the V_k and J_k gene segments would be joined precisely. (From Sakano et al., 1979).

These putative recognition sequences also are seen at the 5' end of the J_κ segment where there are the same 7mer and 10mer inverted repeats, only they are separated by approximately 22 nucleotides and are inverted in their 5'-3' orientations. Eleven nucleotides is approximately one turn of the DNA helix and 22 nucleotides represent two turns. This peculiar sequence arrangement allowed Early and co-workers (1980a) to predict that the joining of the V_L and J_L gene segments requires two different "joining" proteins, one recognizing the inverted repeat sequences separated by one turn of the double helix, the other binding to those sequences separated by two helix turns. When bound to target DNA sequences, these joining proteins may dimerize, thus bringing the 3' end of the V gene and the 5' end of the J gene to close proximity (Fig. 4). An enzyme could subsequently cleave and rejoin the DNA strands juxtaposing the V and J gene segments (Fig. 4). Such a joining mechanism would result in deletion of a DNA segment between the joined V and J pieces. Indeed, the rearranged V gene is known to lack the recognition and other intervening nucleotide sequences.

The application of the "one-turn to two-turn" joining mechanism for gene segments to heavy chain genes posed an interesting problem. Early and co-workers (1980a) observed that the arrangement of the recognition sequences at the 3' end of the V_H gene segments and the 5' end of the J_H gene segments differed from those of V_L and J_L gene segments in that the two inverted repeat sequences are separated by 22 or 23 nucleotides in both cases (two-turn recognition sequences). To preserve the same hypothetical joining mechanism—one-turn to two-turn joining—Early and his coworkers postulated that the D gene segment was flanked on both 5' and 3' sides by inverted repeat sequences separated by a spacer of 11 nucleotides (one-turn recognition sequences). The D gene segment has been recently isolated by Kurosawa and colleagues (1981) and Early and Hood (1981b). Sakano and co-workers (1981), Kurosawa and co-workers (1981), and Early and Hood (1981b) used a fragment of DNA containing the J_H sequences to probe genomic libraries of some myeloma, hybridoma, and normal B cells and found a family of D gene segments. The family contains some eight different D sequences, and its conspicuous feature is that the D gene segments have one-turn recognition sequences on their 5' and 3' sides. Thus V_H to D and D to J_H joining also employs one-turn to two-turn joining. The creation of a functional V_H gene thus proceeds in two steps, first by D-J_H and

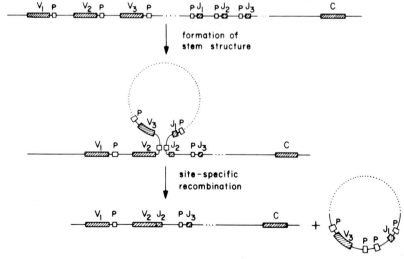

FIG. 4. A model for the formation of V genes through V and J gene segment rearrangements. P denotes the recognition sequences.

then by a (D-J_H)-V_H joining process. This process is probably mediated and catalyzed by the same putative joining proteins and enzymes that are responsible for the V_L-J_L joining.

The arrangement of the C gene segments is equally interesting. In mouse genome there is a single C_κ gene and probably two or three different C_λ genes. The C_H gene cluster is much more elaborate in that it consists of tandem repeats of groups of coding segments or exons interspersed with introns (Early et al., 1979; Sakano et al., 1979). A single exon corresponds to a single constant region domain; thus, the $C\alpha$ gene consists of three exons, and the C_μ gene consists of four exons (Fig. 5). Likewise, the C_γ genes each have three exons encoding the three C_H domains and a separate exon for the hinge region, ie., the flexible polypeptide segment that connects the Fab (antigen-binding) and Fc (crystallizable) regions of the antibody molecule.

THE CLASS SWITCH

A particular antibody-producing B cell first expresses IgM molecules and then, without changing antibody specificity (i.e., using the same V_H and V_L genes), switches to a production of an immunoglobulin of another class (e.g., IgA) (Fig. 6). This phenomenon, known as the class switch, has long puzzled immunologists. The detailed structure of the C_H gene cluster and the analysis of several rearranged heavy chain genes have provided an explanation of the class switch in molecular terms (Fig. 7) (Davis et al., 1980). During the ontogeny of a B cell, VDJ joining initially places the V_H gene to the 5' side of the C_μ gene and the B cell expressed

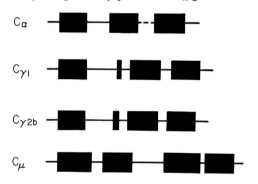

FIG. 5. A model of the gene organization for several mouse C_H genes. The *black rectangles* represent exons or coding regions that correspond precisely to the protein domain and hinge structures of the corresponding heavy chains.

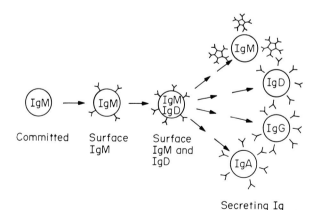

FIG. 6. A model of B-cell differentiation. The light chain VJ and the heavy chain VDJ that join initially lead to the expression of an IgM surface receptor. Subsequently, the B cell can simultaneously express IgM and IgD molecules using the same V domain. Then a B cell may undergo a class switch to express any one of the other classes or subclasses of immunoglobulins while still employing the same V domain.

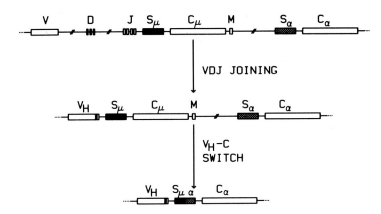

FIG. 7. A model of the two types of DNA rearrangements in B cells, VDJ joining and class switching. S_μ and S_α denote the switch regions of the C_μ and C_α genes.

$$mu_{secreted} \quad C\mu 4 \quad C^{secreted}_{terminus}$$
....DKST|GKPTLY[N]VSLIMSDTGGT[C]Y
              [CHO]

$$mu_{membrane} \quad C\mu 4 \quad C^{membrane}_{terminus}$$
....DKST|[E]G[E]VNA[EEE]GF[E]NLW<u>TT</u>AS<u>T</u>FIVLFLLSLFY<u>STTVTLF</u>[K]V[K]

FIG. 8. The tails of membrane and secreted IgM molecules. The *vertical lines* divide the tails from the C_μ 4 domains. In the tail sequence of the secreted molecule, the boxed asparagine (*N*) residue has a carbohydrate side chain. The boxed cysteine (*C*) forms a bridge with the cysteine residues of four other secreted μ proteins to form a pentamer. In the tail of the membrane-bound molecule, boxed glutamic acid *(E)* and lysine *(K)* residues are charged residues in the region outside the membrane. The 26-amino-acid boxed region consists of uncharged amino acids thought to form the section of the tail imbedded in the membrane. In this box, underlined amino acids are uncharged but somewhat more hydrophilic and may play a special role in trans-membrane signaling. (From Rogers et al., 1980.)

μ chains of IgM molecules. Subsequently, class switching replaces the C_μ gene with another C_H gene such as Cα. The class switch recombination event occurs in repetitive sequences, termed switch regions, which lie to the 5' side of most C_H genes. Although intimate details of this switch process are not understood, the DNA sequences of the switch regions possess an intricate pattern of internal homologies and repeats that must undoubtedly be involved in the class-switching recombinatorial events.

ALTERNATIVE RNA SPLICING PATTERNS

During B-cell development the B cell initially places monomeric IgM receptors on its surface. Subsequently, the B cell may secrete pentameric IgM molecules, effectors of humoral immunity, into the blood. Both the membrane and secreted IgM molecules have the same V domain. Analysis of the μ heavy chains from secreted and membrane IgM molecules demonstrated that, except for their C-terminal "tails," they were identical (Kehry et al., 1980). An analysis of the mRNAs for the membrane and secreted molecules revealed that the mRNA for the secreted IgM had a tail that coded for 20 hydrophilic amino acids whereas the mRNA for the membrane-bound IgM had a tail that coded for 41 hydrophobic amino acids (Fig. 8) (Alt et al., 1980; Rogers et al., 1980). Indeed, the secreted tail contained a half-cysteine residue that participated in pentamer formation of IgM molecules. This residue is not present in the membrane tail. Moreover, the tail of the membrane molecule had 26 consecutive hydrophobic and uncharged residues that almost certainly constitute the α helix of the transmembrane belt. A comparison of the mRNA structures with the C_μ gene demonstrated that the secreted tail of IgM is encoded just to the 3' side of the C_μ 4 domain (see Fig. 5), whereas the membrane tail is encoded by two membrane (M) exons, 1,850 nucleotides to the 3' side of the C_μ domain (Early et al., 1980b). Thus C_μ RNA transcripts must undergo two different patterns of RNA splicing—one to give the mRNA for the membrane IgM and the second to give the mRNA for the secreted IgM. Hence, alternative forms of virtually the same molecule can be generated by differential modes of RNA splicing. A similar type of mechanism explains how the B cell can simultaneously synthesize IgM and IgD molecules (Fig. 6).

ANTIBODY DIVERSITY BY COMBINATORIAL STRATEGIES

The mouse appears capable of synthesizing 10^8 or more different antibody molecules. This diversity appears to arise from several distinct mechanisms. First, within a given antibody gene family it appears that any V gene segment may join to any J (or D) gene segment. Our best guess is that the κ gene family has ~250 V_κ and four J_κ gene segments and the heavy gene family has ~250 V_H, >10 D, and four J_H gene segments. (The λ gene family is very small.) Thus, a combinatorial joining mechanism could generate 1,000 V_κ (250 × 4) and 10,000 V_H genes. Second, any light chain appears capable of associating with any heavy chain. Thus, combinatorial association of light and heavy chains appears capable of expanding 1,000 κ and 10,000 H chains to 10^7 antibody molecules (1,000 × 10,000). Combinatorial mechanisms are capable of expanding the information content of the antibody gene families enormously.

ANTIBODY DIVERSITY BY SOMATIC DIVERSIFICATION

Superimposed on the diversity generated by combinatorial strategies is additional variability generated by two distinct somatic mechanisms. The VJ or VDJ joining occurs not just at the precise boundaries of the V and J (and D) gene segments (see Fig. 3), but joining may occur at many different sites in the junctional region (Sakano et al., 1979; Weigert et al., 1980). This flexibility of VJ or VDJ joining generates sequence differences and insertions and deletions at the junctional boundaries that lie in the heart of the third hypervariable region—a critical region for antigen binding. Thus diversity in this region generates physiologically distinct antibody molecules. This mechanism is termed *junctional diversification*.

A fascinating somatic mutational mechanism also appears to operate on V genes. An analysis of the immune response in mice to the simple hapten phosphorylcholine has been carried out. The V_H regions from more than 35 hybridoma and myeloma proteins have been completely or partially sequenced (Hood et al., 1976; Gearhart et al., 1981). A family of four closely related V_H gene segments has been isolated from a sperm (germline) library and sequenced (Crews et al., 1981). Two somatic variant V_H genes have been isolated and sequenced together with their flanking regions (Kim et al., 1981). The following generalizations may be drawn from these studies:

1. The entire immune response in BALB/c mice to phosphorylcholine appears to be derived from a single V_H gene.
2. This gene is diversified by a somatic mutational mechanism that may generate variants differing by one to eight amino acid substitutions in their V_H regions.
3. The somatic mutational mechanism can create extensive variation that extends into the flanking sequences of the V_H gene.

4. The somatic mutational mechanism is correlated with class switching in that variants only arise in V_H regions associated with IgG or IgA molecules (see Fig. 6). The V_H regions from IgM molecules show no variation. This observation suggests that the somatic mutation mechanism is turned on at the time of the class switch, possibly by antigen. Since the somatic mutation mechanism appears to be turned off in terminally differentiated plasma cells (Fig. 6), it appears to operate only during a narrow window of B-cell development.
5. Many different selective forces could amplify the variant B cells. For example, it is attractive to postulate antigen might select variant B cells whose antibody receptors have a higher affinity for antigen. In this way, somatic mutation could fine tune the immune response. However, there also are other possible selective forces (see Crews et al., 1981).

In summary, antibody diversity is generated by a wide variety of different mechanisms. Germline information is employed in a combinatorial manner to generate a base-line repertoire onto which somatic mutation is superimposed.

AREA-CODE GENE FAMILIES

The combinatorial strategies for the amplification of information described in the immune system may be employed in other vertebrate systems such as those found in the nervous system. The evolution of split gene-multigene families with their combinatorial potential probably occurred much earlier than the appearance of the vertebrate immune system, perhaps about the time of emergence of metazoan organisms (Hood et al., 1977). Once having occurred, this system with all its potential and flexibility conferred on metazoan evolution an entirely new evolutionary possibility, namely, partial or complete duplication of the multigene families. In this way, new gene families could be generated that then could come to encode other complex aspects of eukaryotic phenotype. This evolutionary view would imply that the antibody gene

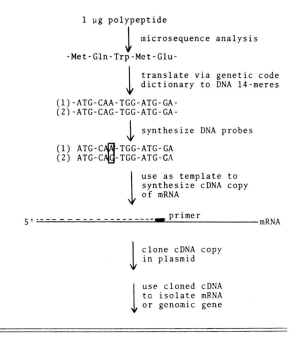

FIG. 9. A probe strategy for the isolation of rare-message genes. (From Hood et al., 1981.)

families are only one subset of a larger supragene family that encodes many different aspects of complex phenotypic traits in eukaryotes, such as cell receptors, tumor antigens, and retinotectal recognition molecules (see Hood et al., 1977). The other multigene families, or "area-code families," might be expected to employ, at least in part, some of the combinatorial strategies found in the antibody system. The hypothesis of area-code families focuses attention on protein molecules and gene segments of evolutionary interest, making testable predictions about their structure. A paramount experimental requirement for studying area-code gene families, however, is a micromethodology suitable for structurally characterizing minute amounts of gene products from area-code gene families and identifying the corresponding genes.

MICROCHEMICAL FACILITY

Hood and co-workers (1981) have developed a facility to analyze and synthesize proteins and genes using very small quantities of material. To date three new instruments have been developed—a gas-phase solid-state microprotein sequenator, a high-sensitivity mass spectrometer, and a DNA synthesizer. After a brief description of these instruments, a general strategy is given for isolating genes that make very small quantities of gene products (rare-message genes).

GAS-PHASE SOLID-STATE MICROPROTEIN SEQUENATOR

This instrument is now capable of extended sequence runs (e.g., 90 residues on 10 nmoles of myoglobin) with short cycle times (35 residues/day as compared to 12 to 14 residues/day for the commercial spinning-cup sequenator). Moreover, this instrument can sequence 25 residues on as little as 5 nmoles of myoglobin (~ 0.02 μg). We anticipate that this instrument will be capable of sequencing at the 1-nmole level in the near future. This sensitivity, if achieved, will permit direct sequencing of spots from two-dimensional gels, currently one of the most sensitive analytic techniques in biochemistry. However, to achieve this goal, a more sensitive instrument is needed for analyzing the PTH (phenylthiohydantoin) amino acid derivative than the currently employed high-pressure liquid chromatograph. For this purpose, a mass spectrometer is being developed in collaboration with the Jet Propulsion Laboratory in Pasadena.

HIGH-SENSITIVITY MASS SPECTROMETER

A new type of detector for the mass spectrometer allows analysis of samples at the femptogram level (Hood et al., 1981). Currently an automatic sample injection system is being created for this instrument. This automation, coupled with a computerized data analysis system, should provide a 5-min sample turn-around time and thus allow the multiplexing of several gas-phase solid-state sequenators to the mass spectrometer.

DNA SYNTHESIZER

Marvin Caruthers at the University of Colorado has developed a new technique to synthesize DNA employing solid-phase phosphite chemistry. This development has permitted the construction of an automated DNA synthesizer, now being tested, that has a half-hour cycle time and potentially appears capable of synthesizing fragments 20 to 40 nucleotides in length.

A STRATEGY FOR THE ISOLATION OF RARE-MESSAGE GENES

One of the major problems in developmental neurobiology is the identification, isolation, and characterization of genes that produce minute amounts of protein products. With the microchemical facility described here, a very general approach to this problem can be suggested (Fig. 9). The desired protein is identified, possibly with the aid of monoclonal antibodies, and isolated in microgram or submicrogram quantities from appropriate systems, e.g., acrylamide gel systems. This protein is then sequenced using the gas-phase microsequenator. The protein sequence is scanned to find a contiguous stretch of four to five amino acids that have minimal ambiguity when translated by the genetic code dictionary into DNA sequences. The DNA synthesizer is used to synthesize each of the alternative DNA sequences for this stretch of amino acids, and the mixture of oligonucleotides is added to a pool of mRNAs isolated from cells expressing the desired phenotype. The appropriate mRNA sequence hybridizes to the appropriate mRNA at its complementary position and serves as a primer to initiate the synthesis of a cDNA probe. This cDNA probe is then used to isolate the corresponding gene by conventional recombinant DNA techniques. This method appears to offer a general approach to gene isolation. These strategies offer an exciting approach to the analysis of a variety of systems in neurobiology including neural hormones, receptors, and interesting molecules delineated by monoclonal antibodies.

ACKNOWLEDGMENTS

Studies reported here were supported by grants from the NIH and NSF.

REFERENCES

Alt, F. W., Bothwell, A. L. M., Knapp, M., Siden, E., Mather, E., Koshland, M., and Baltimore, D. (1980): Synthesis of secreted and membrane-bound immunoglobulin μ heavy chain is directed by mRNAs that differ at their 3' ends. *Cell*, 20:293–301.
Amzel, L., Poljak, R., Saul, F., Jarga, J., and Richards, F. (1974): The three dimensional structure of combining-site ligand complex of immunoglobulin NEW at 3.5 Å resolution. *Proc. Natl. Acad. Sci. USA*, 71:1427–1430.
Bernard, O., Hozumi, N., and Tonegawa, S. (1978): Sequences of mouse immunoglobulin light chain genes before and after somatic changes. *Cell*, 15:1133–1144.
Crews, S., Griffin, J., Huang, H., Calame, K., and Hood, L. (1981): A single V_H gene sequence encodes the immune response to phosphorylcholine: Somatic mutation is correlated with class switching. *Cell*, 25:59–66.
Davis, M. M., Kim, S. K., and Hood, L. (1980): Immunoglobulin class switching: Developmentally regulated DNA rearrangements during differentiation. *Cell*, 22:1–2.
Early, P., Davis, M. M., Kaback, D. B., Davidson, N., and Hood, L. (1979): Immunoglobulin heavy chain gene organization in mice: Analysis of a myeloma genomic clone containing variable and α constant regions. *Proc. Natl. Acad. Sci. USA*, 76:857–861.
Early, P., and Hood, L. (1981a): Mouse immunoglobulin genes. In: *Genetic Engineering Principles and Methods, Vol. 3*, edited by Jane Setlow and Alexander Hollaender, pp. 157–188. Plenum Press, New York.
Early, P., and Hood, L. (1981b): Allelic exclusion and nonproductive immunoglobulin gene rearrangements. *Cell*, 24:1–3.
Early, P., Huang, H., Davis, M., Calame, K., and Hood, L. (1980a): An immunoglobulin heavy chain variable region gene is generated from three segments of DNA: V_H, D and J_H. *Cell*, 19:981–992.
Early, P., Rogers, J., Davis, M., Calame, K., Bond, M., Wall, R., and Hood, L. (1980b): Two mRNAs can be produced from a single immunoglobulin μ gene by alternative RNA processing pathways. *Cell*, 20:313–319.
Gearhart, P., Johnson, N., Douglas, R., and Hood, L. (1981): IgG antibodies to phosphorylcholine exhibit more diversity than their IgM counterparts. *Nature*, 291:29–34.
Hood, L., Huang, H., and Dreyer, W. J. (1977): The area-code hypothesis: The immune system provides clues to understanding the genetic and molecular basis of cell recognition during development. *J. Supramol. Struct.*, 7:531–559.
Hood, L., Hunkapiller, M., and Dreyer, W. J. (1981): Microchemical instrumentation. ICN-UCLA Symposium on Cellular Recognition, *J. Supramol. Struct. Cell Biochem.*, 17:27–36.
Hood, L., Loh, E., Herbert, J., Barstad, P., Eaton, B., Earky, P., Fuhrman, J., Johnson, N., Kronberg, M., and Schilling, J. (1976): Structure and genetics of mouse immunoglobins: An analysis of NZB myeloma protein and sets of BALB/c myeloma proteins binding particular haptens. *Cold Spring Harbor Symp. Quant. Biol.*, 41:817–836.
Kehry, M., Ewald, S., Douglas, R., Sibley, C., Raschke, W., Fambrough, D., and Hood, L. (1980): The immunoglobulin μ chains of membrane-bound and secreted IgM molecules differ in their C-terminal segments. *Cell*, 21:393–406.

Kim, S., Davis, M., Simm, E., Patten, P., and Hood, L. (1981): Antibody diversity: Somatic mutation may be extensive and is localized in and around the rearranged V_H gene. *Cell,* 27:573–581.

Kurosawa, Y., von Boemer, H., Haas, W., Sakano, H., Trauneker, A., and Tonegawa, S. (1981): Identification of D segment of immunoglobulin heavy-chain genes and their rearrangement in T lymphocytes. *Nature,* 290:565–570.

Rogers, J., Early, P., Carter, C., Calame, K., Bond, M., Hood, L., and Wall, R. (1980): Two mRNAs with different 3' ends encode membrane and secreted forms of immunoglobulin μ chain. *Cell,* 20:303–312.

Sakano, H., Kurosawa, Y., Weigert, M., and Tonegawa, S. (1981): Identification and nucleotide sequence of a diversity DNA segment (D) of immunoglobulin heavy chain genes. *Nature,* 290:565–565.

Sakano, H., Rogers, J. H., Huppi, K., Brack, C., Traunecker, A., Maki, R., Wall, R., and Tonegawa, S. (1979): Domains and the hinge region of an immunoglobulin heavy chain are encoded in separate DNA segments. *Nature,* 277:627–633.

Seidman, J. G., and Leder, P. (1978): The arrangement and rearrangement of antibody genes. *Nature,* 276:790–795.

Tonegawa, S., Maxam, A. M., Tizard, R., Bernard, O., and Gilbert, W. (1978): Sequence of a mouse germ-line gene for a variable region of an immunoglobulin light chain. *Proc. Natl. Acad. Sci. USA,* 75:1485–1489.

Weigert, M., Gatmaitan, L., Loh, E., Schilling, J., and Hood, L. (1978): Rearrangement of genetic information may produce immunoglobulin diversity. *Nature,* 276:2785–2790.

Weigert, M., Perry, R., Kelley, T., Hunkapiller, M., Schilling, J., and Hood, L. (1980): The joining of V and J gene segments creates antibody diversity. *Nature,* 283:497–499.

Molecular Genetics of Human Globin Gene Expression

Thomas Maniatis, Pamela Mellon, Vann Parker, Nicholas Proudfoot, and Brian Seed

ABSTRACT

The application of molecular cloning procedures to the study of human globin genes has provided a detailed understanding of their structure and chromosomal arrangement. Comparison of the structure of normal α- and β-globin genes with the corresponding genes isolated from individuals with α- or β-thalassemia has provided insights into the molecular basis of genetic disorders in globin gene expression. New methods for the rapid isolation of mutant globin genes and the analysis of their transcription in cells in culture is discussed.

The human globin gene family represents a unique system for studying differential gene activity during development. Detailed structural analyses of globin polypeptides and messenger RNA (mRNA) and extensive clinical investigations of inherited disorders in hemoglobin expression have provided information that is not available for any other eukaryotic gene system. One approach to the study of human globin gene expression is to isolate and characterize normal globin genes using molecular expression, and then to compare the structure and expression of normal and mutant globin genes in these assays (Maniatis et al., 1980). Mutant globin genes can be produced by *in vitro* mutagenesis of cloned genes or isolated from DNA of individuals with thalassemia.

HEMOGLOBIN: ITS STRUCTURE AND EVOLUTION

Hemoglobin is a tetramer consisting of two pairs of identical polypeptide chains, α-globins and β globins, and a prosthetic group, heme, which binds oxygen. Early in evolution, the primary biological function of a globin, namely to transfer oxygen from air into body tissues, was a single process based on a single kind of protein chain. In vertebrates, however, increasing demands placed on the oxygen-carrying systems have led to evolution of an interacting set of molecular catalysts specialized for various stages of the process. Hemoglobin must have a lower oxidation potential than the environment. Myoglobin, in turn, is adapted to transfer oxygen efficiently from hemoglobin to the tissues. In addition, specialized globins that function during specific stages of embryonic and fetal development have evolved. Comparisons of the primary structure of a large number of different globin polypeptides indicate that globin genes evolved by gene duplication followed by sequence divergence.

Globins were among the first proteins for which the three-dimensional structure was determined. Kendrew and his co-workers (1961) obtained the three-dimensional structure of sperm whale myoglobin in atomic detail by means of x-ray crystallography. X-ray studies of the horse hemoglobin tetramer by Perutz and colleagues (1968) have shown that each of the hemoglobin chains has the same overall shape as the myoblobin molecule, even though 80% of the amino acids have changed. In these proteins, an intricate molecular morphology exists, analogous to species morphology. Overall structural features have been preserved by natural selection and have persisted despite the extensive replacement of amino acid residues. Such features are those essential to the biochemical functioning of these proteins or to their steric interactions with other molecules.

ONTOGENY OF GLOBIN EXPRESSION

Ontological studies of globin gene expression have shown that the α-like and β-like globin genes are encoded by a group of genes and that they are expressed sequentially during development. The earliest embryonic hemoglobin tetramer was termed Gower 1 and consists of an ε (β-like) and a ζ (α-like) polypeptide chain. Beginning at approximately eight weeks of gestation the embryonic chains are gradually replaced by the adult α-globin chain and two different fetal β-like chains designated $^G\gamma$ and $^A\gamma$ (Figs. 1 and 2). The γ chains differ only in the presence of glycine and alanine at position 136. During the transition period between embryonic and fetal development, hemoglobin Gower 2 ($\alpha_2\epsilon_2$) and hemoglobin Portland ($\zeta_2\gamma_2$) are detected. Eventually, hemoglobin F ($\alpha_2\gamma_2$) becomes the predominant hemoglobin tetramer throughout the remainder of fetal life. Beginning just prior to birth, the γ-globin chains are gradually replaced by the adult β- and δ-globin polypeptides. At six months after birth, 97 to 98% of the hemoglobin is hemoglobin A ($\alpha_2\beta_2$), whereas hemoglobin A_2 ($\alpha_2\delta_2$) accounts for approximately 2%. Small amounts of hemoglobin F (1%) are also found in adult peripheral blood. The site of erythropoiesis changes from the yolk sac in the early embryo to the developing

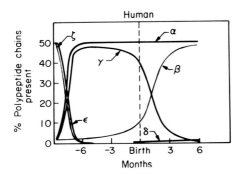

FIG. 1. Temporal expression of globin polypeptides during human development. The relative levels of embryonic, fetal, and adult globin polypeptides are plotted as a function of time during early development. See Bunn and co-workers (1977) for details.

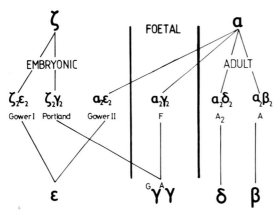

FIG. 2. Composition of various hemoglobin tetramers observed during human development.

liver, spleen, and bone marrow in the fetus, and finally to the bone marrow in adults. Because the ratio of hemoblobin A to hemoglobin F is the same in all fetal erythroid tissues, the switch from fetal to adult globin production is not correlated with the site of erythropoiesis.

In addition to the gene switching described above, globin gene expression is regulated within a particular developmental stage. For example, the patterns of expression of the δ and β genes during adult red cell maturation are quite different. Delta-globin in mRNA can be detected in nucleated erythroid precursor cells but not in mature reticulocytes. Thus, the small amount of δ-globin polypeptide found in circulating reticulocytes was synthesized in more immature cells in the bone marrow. Furthermore, bone marrow nuclei contain 10-fold less δ-globin mRNA precursor than β-globin precursor, suggesting that the difference in expression of the two genes is at the level of transcription or RNA processing (Wood et al., 1978; Kantor et al., 1980). The biological significance of the restriction of δ gene expression to immature cells is unknown. Obviously, the α-like and β-like globin gene families have coordinated programs for differential gene expression. The primary difference between the two gene families is that two switches in gene expression (embryonic to fetal to adult) are observed for the β-like genes, whereas a single switch results in activation of adult α-globin production early in fetal life.

GENE LIBRARIES

Initial advances in globin gene mapping and isolation were made possible by the development of procedures for synthesizing and cloning double-stranded DNA copies of poly(A)-containing mRNAs (for a review, see Maniatis, 1980). This major methodological advancement, together with the development of phage gene libraries (Maniatis et al., 1978) and the application of the Southern (1975) blotting procedure to mapping single-copy nucleotide sequences in the genome libraries (Jeffreys and Flavell, 1977a), made possible the construction of detailed physical maps of globin genes.

The creation of gene libraries relies on the use of DNA cloning vectors such as those developed by Blattner and co-workers (1977). These vehicles, so-called Charon vectors, were derived from the bacteriophage λ, a viral particle whose genetics have been thoroughly studied. Bacteriophage λ is particularly well suited for adaptations to make it useful as a lytic cloning vehicle. One-third of its DNA, which forms a continuous block in the middle of the genome, can be replaced without the phage losing its ability to grow in bacteria. Point mutations, substitutions, and deletions were introduced into wild-type λ to alter the distribution of restriction sites (i.e., sites recognized and cleaved by the restriction nucleases, endonucleases specific for double-stranded DNA molecules) and to eliminate sites in the essential regions of the phage genome. The phage was engineered in such a way as to permit molecular cloning with minimal manipulations of a variety of sizes of DNA fragments. There are multiple restriction sites that permit cloning with more than one restriction enzyme. Vectors and clones can be grown to a high yield, and cloned DNA can be readily recovered.

The strategy for construction of gene libraries in Charon phages can be briefly summarized as follows (also see Maniatis, 1980). The high molecular weight DNA representing a eukaryotic genome is fragmented either by shearing followed by S1 nuclease treatment, or by a nonlimit endonuclease digestion with restriction enzymes. Both procedures generate molecules with blunt ends, making insertion into phage DNA possible. Molecules approximately 20,000 base pairs long are selected by preparative sucrose-gradient centrifugation and rendered resistant toward the restriction enzyme Eco RI by treatment with Eco RI methylase. Then, synthetic DNA linkers bearing Eco RI recognition sites are covalently attached by blunt-end ligation using T4 ligase and cohesive ends are generated by digestion with Eco RI. These molecules are covalently joined to phage λ DNA, and the hybrid DNA molecules are then packaged into viable phage particles *in vitro*. Amplification of these phages produces a permanent library of eukaryotic DNA sequences. Once established, the library can be screened by *in situ* hybridization to radioactively labeled probes such as globin cDNA, and the gene segments in question can then be isolated, amplified, and studied in detail.

FINE STRUCTURE OF THE GLOBIN GENE

The rabbit (Jeffreys and Flavell, 1977b) and mouse (Tilghman et al., 1977) β-globin genes provided one of the first examples of intervening, nontranscribed sequences (introns) existing within an expressed gene segment (exon). Two introns have been identified in all of the functional globin genes thus far studied. In particular, the five expressed human β-like globin genes are interrupted by two introns at identical locations: the first, 120 to 133 base pairs in length, is located between codons 30 and 31; and the second, 850 to 900 base pairs in length, is between codons 104 and 105. Similarly, the location of introns in the human (Liebhaber et al., 1980) and mouse (Nishioka and Leder, 1979) α-globin genes are identical and analogous to the positions of introns in β-like globin genes (Fig. 3). It thus seems reasonable to assume that these interruptions antedate the emergence of separate α- and β-globin genes about 500 million years ago (Nishioka and Leder, 1979). In the case of mouse, rabbit, and human β-globin genes, both the exon and intron sequences are transcribed to produce a detectable nuclear mRNA precursor, which is processed or spliced to give a mature globin mRNA. Furthermore, the 5' ends of the mouse and rabbit β-globin nuclear precursors are coterminal with mature mRNA (Weaver and Weissman, 1979; Grosveld et al., 1981). These nuclear precursors are most probably the primary globin mRNA transcripts and the sequence encoding the 5' end of the mature mRNA probably corresponds to the transcription initiation site.

The complete nucleotide sequence of the five human β-like globin genes have been determined, and detailed comparison of these and other mammalian globin gene sequences has been presented (Efstratiadis et al., 1980). This sequence comparison revealed interesting homologies in regions that are potentially involved in globin gene transcription and splicing. Particularly, alignment of the sequences on the 5' sides of the human β-like globin genes revealed two blocks of sequence homology, which are present in analogous positions adjacent to most eukaryotic genes. The first homology block is an AT-rich sequence, originally identified in the *Drosophila* histone gene cluster, called the Goldberg-Hogness box (Goldberg, 1979) (also see P. A. Sharp and M. Wilson, *this volume*). A comparison of a number of different β-like globin genes has revealed that the AT-rich sequence CATAAA is found 31 ± 1 base pairs on the 5' side of the mRNA capping sites, but that the sequence shared by all of the β-like genes is

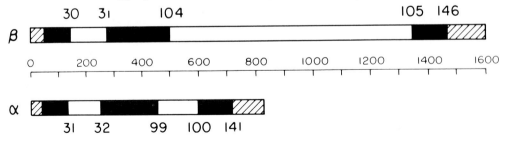

FIG. 3. Structure of human globin genes. *Solid and open boxes* represent coding (exon) and noncoding (intron) sequences, respectively. The α-like globin genes contain introns of 95 and 125 base pairs located between codons 31 and 32 and 99 and 100. The β-like genes contain introns of approximately 125 to 150 and 800 to 900 base pairs located between codons 30 and 31 and 104 and 105. The distance between genes is indicated by the scale above the maps in Kilobase pairs. *Hatched areas* designate 5' and 3' noncoding regions of mature mRNA.

ATA. This sequence was therefore designated the ATA box. The second homology block (designated the CCAAT box) is located 77 ± 10 base pairs on the 5' side of each gene. A possible role of these sequences in transcriptional initiation, RNA processing, or both is being studied with the use of *in vitro* and *in vivo* assays for globin gene expression (Dierks et al., 1981a,b; Grosveld et al., 1981a,b; Mellon et al., 1981).

Previous comparisons of noncoding sequence in human, mouse, and rabbit β-like globin genes indicated that these regions diverged by deletion and addition as well as by simple base substitution. Examination of the nucleotide sequences surrounding putative deletion sites suggests that short (two to eight nucleotides) sequences and direct repeats are involved in the generation of deletions (Efstratiadis et al., 1980). This pattern is remarkably similar to that observed for preferred deletion sites in the *lac i* gene of *E. coli* (Farabaugh and Miller, 1978).

ORGANIZATION OF GLOBIN GENE CLUSTERS

The entire β-globin and α-globin clusters have been isolated in sets of overlapping bacteriophage recombinants that were obtained from libraries of random, high molecular weight human DNA (Frisch et al., 1980; Lauer et al., 1980). The linkage arrangement of these gene clusters is shown in Fig. 4. Although the size of the two gene clusters is widely different, the arrangement of the genes in both clusters is analogous in that they appear in the order of their expression during development, and they are all transcribed from the same DNA strand in the 5'-3' direction. A similar pattern of gene organization has been found in the rabbit (Lacy et al., 1979) and mouse (Jahn et al., 1980) β-like globin gene clusters. All of the known human globin polypeptides can be accounted for by the gene segments found; however, both gene clusters contain additional sequences that are detected by globin gene hybridization probes but cannot be identified with any of the known globin polypeptides. Two genes, designated ψ-β_1 and ψ-β_2, fall into this category. Structural analysis of the ψ-α_1 gene indicates that it is a pseudogene, that is, a gene that displays significant homology to a functional gene but has mutations that prevent its expression (Proudfoot and Maniatis, 1980). Similar pseudogenes have been identified at corresponding positions in the rabbit (Lacy and Maniatis, 1980) and mouse (Jahn et al., 1980) β-like globin gene clusters.

THE NATURE OF GLOBIN PSEUDOGENES

Nucleotide sequence analysis of rabbit and mouse β and human and mouse α pseudogenes has demonstrated a variety of structural differences between each gene and its functional counterpart (see Fig. 5 for an example). Each pseudogene that has been analyzed exhibits 75 to 80% sequence homology with its corresponding normal gene. The only exception to this rule is the 5' side of the mouse β pseudogene, which is not homologous to the adult mouse β gene. None of these pseudogenes can encode a functional protein due to the presence of small deletions or insertions that result in alterations of the translational reading frame. In many cases, these frameshifts lead to the presence of in-phase termination codons. In addition, one or more of the intron-exon junctions of rabbit ψ-β_2, mouse βH_3, and ψ-α_1 are different from the sequence common to splicing junctions in globin genes and all expressed genes studied to date (for the common junction sequence, see P. Sharp and M. C. Wilson, *this volume*, Fig. 6). Thus, even if these pseudogenes are transcribed, it is unlikely that they would produce a normal mRNA.

In all of the mammalian globin gene clusters thus far characterized, pseudogenes are found between the embryonic (or fetal) genes and the adult genes. This could mean that pseudogenes have some as yet unidentified function in globin gene clusters or that they are the products of

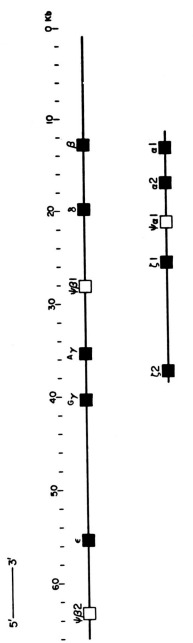

FIG. 4. Linkage arrangement of the human α- and β-globin gene clusters. The *black boxes* correspond to genes encoding known globin polypeptides and the *white boxes* indicate the location of globin pseudogenes. The distance between genes is indicated by the scale above the maps in kilobase pairs. The β-cluster (*top line*) is located on chromosome 11; the α-cluster (*bottom line*) is located on chromosome 16.

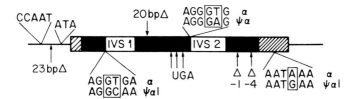

FIG. 5. Schematic diagram of the ψ-α₁ pseudogene. Differences between the ψ-α₁ and the normal α-globin gene are indicated as follows: (1) the number of base pairs separating the highly conserved CCAAT and ATA sequences is 23 less in ψ-α₁ as a result of a deletion. (2) The highly conserved CT sequences at exon 1/IVS 1 and exon 2/IVS (intervening sequence) 2 junction have been mutated to a GC and GA, respectively, making it unlikely that α-globin primary transcript would be properly spliced. (3) A 20-base pair deletion was found in exon 2, resulting in a shift of the translational reading frame that brings three UGA termination signals in phase. (4) There are two small deletions in exon 3. (5) The highly conserved AATAAA sequence thought to be involved in poly(A) addition has been mutated to AATGAA.

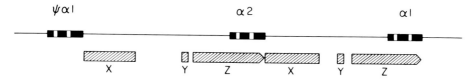

FIG. 6. The distribution of sequence homologies within a region of the human α-like globin cluster. Regions of sequence homology, designated X, Y, and Z, are indicated by *cross-hatched boxes*. These homologies were detected by heteroduplex analysis of the cloned α-globin gene clusters (Lauer et al., 1980).

gene duplication and subsequent sequence divergence. The variation in human α-globin gene number observed in present-day populations and the location of ψ-α₁ within the α-like globin gene cluster are consistent with the latter possibility; ψ-α₁, α₂, and α₁ are separated from each other by approximately 4,000 base pairs, which is the size of the α₁-α₂ duplication unit, and the nucleotide sequence of ψ-α₁ indicates that it is α-like rather than δ-like. It therefore seems possible that ψ-α₁ was once part of a set of three functional α-globin genes.

As in the case of other pseudogenes, the mouse α-globin pseudogene has frameshift mutations that would result in premature translational termination. However, unlike the other pseudogenes, the mouse gene is missing both introns. The mechanism by which this pseudogene arose and its location with respect to the normal α-globin gene cluster are unknown.

GENE DUPLICATIONS AND HOMOLOGY REGIONS IN GLOBIN CLUSTERS

A common feature of globin gene clusters is the occurrence of two immediately adjacent genes, which are coordinately expressed during a given developmental stage (Efstratiadis et al., 1980). Examples of this are the human δ-β, $^G\gamma$-$^A\gamma$, α₁-α₂, and ζ₁-ζ₂ globin gene pairs. The δ and β genes are highly homologous in the coding regions, but the noncoding sequences within and surrounding the two genes have diverged considerably. In contrast, the two members of the $^G\gamma$-$^A\gamma$ gene pair are virtually identical to one another throughout their coding, intervening, and flanking sequences (Slightom et al., 1980a). Restriction mapping and heteroduplex analysis of the α₁ and α₂ genes indicate that the sequences within and flanking these two genes are virtually identical. Each α-globin gene is located within an approximately 4,000-base-pair region of homology interrupted by two small regions of nonhomology (Fig. 6).

The extensive sequence homology within and flanking the $^G\gamma$-$^A\gamma$ (Slightom et al., 1980) and α₁-α₂ gene pairs appears to be the product of a mechanism for gene matching during evolution. Based on the nearly identical distribution of restriction sites surrounding the α-globin genes in a number of primate species, it has been suggested that the α-globin gene duplication occurred before the time of primate divergence (Zimmer et al., 1980). Differences between the α-globin amino acid sequences of various primate species are consistent with sequence drift following primate divergence. However, intraspecies comparisons show much less divergence, indicating that the α-globin genes within a species have been corrected against one another. Maintenance of homology among a family of evolving genes within a species has been termed "concerted evolution." Gene conversion and expansion-contraction of gene number by homologous but unequal crossing-over have been proposed as mechanisms for concerted evolution.

The precise end points of the α-gene duplication unit have been located by nucleotide sequence analysis. The left end point of the duplication is located immediately adjacent to the putative poly(A) addition site of ψ-α_1, whereas the right end point is found 12 base pairs on the 3' side of the poly(A) addition site of α_1. The 15-base-pair sequence on the 3' side of the poly(A) addition sites of $\psi\alpha_{-1}$, α_2, and α_1 consists of a repeated pentanucleotide (GCCTG), separated by TGTGT. The occurrence of this sequence in all three genes and its location with respect to the end points of the α-globin gene duplication suggest that this sequence might be associated with the mechanism by which the genes were duplicated or corrected.

Evidence that α-globin gene sequence matching could occur by expansion and contraction of gene number by unequal crossing-over is provided by the frequent occurrence of chromosomes containing one or three adult α-globin genes in some human populations. Comparison of the end points of the deletion associated with α-thalassemia-2 with the location of blocks of homologous sequence within the α_1-α_2 gene duplication suggests that the deletion results from unequal crossover between homologous sequences. Interestingly, deletions that are indistinguishable from those found in α-thalassemia-2 occur in the cloned gene cluster during propagation in *E. coli* (see Lauer et al., 1980, for discussion).

The $^G\gamma$ and $^A\gamma$ genes on one chromosome were shown to be identical in the region on the 5' side of the center of the large intron, yet show greater divergence on the 3' side of that position (Slightom et al., 1980). Examination of the boundary between the conserved and divergent regions revealed a block of a poly(TG) sequence. It may well be that this simple sequence is a preferred site for initiation or termination of recombination events that lead to unidirectional gene conversion.

Cross hybridization experiments between the intragenic sequences of the human β-gene clusters and α-gene clusters revealed a nonglobin repeat sequence that is interspersed within the globin gene clusters and also repeated many times in the human genome. Nucleotide sequence analysis of the repetitive sequences within the β-globin gene cluster indicates that they are members of a particular repeat sequence family, the Alu family, which is reiterated approximately 300,000 times in the human genome (see Frisch et al., 1980, for review). The repeats show sequence homology with an abundant class of small nuclear RNAs with double-stranded heterogeneous nuclear RNA and are homologous to a sequence found near the replication origin of SV40, polyoma, and BK DNA tumor viruses. At present, there is little information regarding the expression or function of these interesting repetitive elements *in vivo*.

GENETIC DISORDERS IN GLOBIN GENE EXPRESSION

The analysis of globin genes from individuals with genetic defects in α- or β-thalassemia, respectively, has made it possible to establish the molecular basis of these genetic abnormalities (see Maniatis et al., 1980, for review). Most α-thalassemias result from the deletion of one or more of the adult α-globin genes, and the deletions have been mapped by DNA blotting procedures (see Fig. 7); β° and β+ are the most frequently occurring mutations in β-globin gene expression. Characterized by reduced levels of β-globin production is β+-thalassemia,

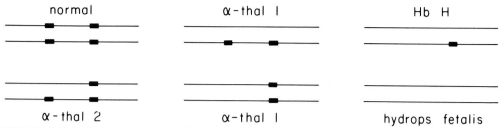

FIG. 7. Schematic representations of genotypes associated with α-thalassemia syndromes. The pair of *horizontal lines* for each syndrome represents the chromosome 16 homologues. A *black rectangle* indicates the presence of a functional α-globin gene. Absence of a black rectangle signifies either a gene deletion or nondeletion defect leading to a nonfunctional α-globin gene.

whereas in β° there is no detectable β-chain synthesis. Beta-globin gene deletions that affect the expression of other globin genes in addition to β constitute a third class.

The primary defect in β+-thalassemias appears to be a reduced level of transcription or inefficient processing of mRNA precursors. An analysis of pulse-labeled α- and β-globin mRNA precursors indicates that discrete intermediates accumulate in B+-thalassemic individuals, and only a fraction of labeled RNA is chased into mature β-globin mRNA. When an RNA blotting procedure was used, an abnormal 650-nucleotide intermediate was observed in one patient whereas a normal, 1,300-nucleotide intermediate was found to accumulate in another. These studies strongly suggest that some β+-thalassemias result from mutations that alter RNA processing (Kantor et al., 1980; Maquat et al., 1980).

The molecular defects in β°-thalassemia appear to be quite heterogeneous. In one case, the defect appears to be a single base change that gives rise to an in-phase termination codon (Chang and Kan, 1979). In another case, the β° phenotype is associated with a 600-base-pair deletion that removes part of the large gene (Flavell et al., 1979; Orkin et al., 1980). Finally, one β° gene appears to be defective by virtue of a single base change at an intron/exon splice junction.

Most cases of β-thalassemia result in moderate to severe anemia and are not associated with detectable deletions in the β-like globin cluster. In these cases, the level of fetal globin expression in adults is normal or only slightly increased, suggesting that the switch from fetal to adult globin synthesis is operative. In other cases of genetic disorder known as δ-β-thalassemia and hereditary persistence of fetal hemoglobin (HPFH), the absence of β-globin chains is compensated for by continued expression of γ-globin chains in the adult. Genomic blotting mapping studies have demonstrated the variety of molecular rearrangements that can alter γ-globin gene expression and have led to formation of new models for the mechanism of hemoglobin switching. These studies revealed that deletions may act in *cis* configuration (i.e., on the same chromosomal strand) over considerable distance to influence differential gene expression within the β-like globin gene cluster (see Maniatis et al., 1980, for review).

ASSAYS FOR GLOBIN GENE EXPRESSION

The availability of cloned globin genes makes it possible to investigate the conditions under which the globin genes are transcribed into RNA and spliced to produce a functional messenger molecule. There are basically two types of assays for the expression of cloned genes, namely, *in vitro* assays using [^{32}P]-labeled RNA precursors and cellular transcription assays, which take advantage of virus vectors such as SV40, and their expression in monkey cell cultures.

Two *in vitro* transcription systems have been described. One system consists of a cytoplasmic extract that requires the addition of purified RNA polymerase II for activity, the other system consists of a concentrated whole cell extract with endogenous RNA polymerase II activity (Weil et al., 1979; Manley et al., 1980). Typically, the RNA transcripts derived from a reaction mix of the extract, polymerase, and a cloned globin gene are analyzed by gel electrophoresis and autoradiography to determine their length. The transcripts can alternatively be digested with T1 and T2 nucleases and resulting oligonucleotides analyzed by two-dimensional mapping (homochromatography) and nearest-neighbor analysis (Luse and Roeder, 1980; Proudfoot et al., 1980). The cloned globin genes are accurately transcribed in the *in vitro* assays, but splicing and polyadenylation (i.e., addition of homopolymeric poly(adenosyl) tracts to the 3' end of the RNA molecule) have not been observed. By deleting the nucleotide sequences 5' to the mRNA capping sites of a number of eukaryotic genes, including globin genes, and by analyzing the products of *in vitro* transcription, it was possible to show that the loss of the so-called Goldberg-Hogness, the TATA sequence, abolishes normal *in vitro* transcription (see P. A. Sharp and M. C. Wilson, *this volume*; Mathis and Chambon, 1980; Dierks et al., 1981a,b; Grosveld et al., 1981b).

Transcription experiments with β- and α-like globin genes suggested that the 5' ends of *in vitro* gene transcripts are near their respective capping sites. To define the 5' ends precisely, fragments of different length were prepared from the first exon of the β-, ε-, and α-globin genes. These fragments were end-labeled and the strands were separated and used as primers in extension experiments with reverse transcriptase. When the length of the resulting RNA transcripts was examined, it was found that the 5' ends of all three genes are coterminal with their mRNA capping sites (Proudfoot et al., 1980).

On the basis of the *in vitro* transcription experiments, two basic conclusions can be made. (1) All the human globin genes assayed appear to comprise individual transcription units. (2) The embryonic, fetal, and adult genes are transcribed with roughly equal efficiencies in an extract prepared from HeLa cells, which do not ordinarily express globin genes. Thus, the interaction of different globin promoters with RNA polymerase II *in vitro* is approximately the same, and the mechanisms that mediate tissue-specific transcription do not operate *in vitro*.

There are two different experimental strategies employed in cellular transcription assays. In one, globin genes are introduced into mouse Ltk$^-$ cells grown in culture by cotransformation with the herpes virus thymidine kinase gene (Mantei et al., 1979; Wigler et al., 1979; Wold et al., 1979). The second assay involves the insertion of globin genes into SV40 virus vectors and propagation of the recombinant molecule in monkey cells in the presence of an SV40 helper virus (Hamer and Leder, 1979; Mulligan et al., 1979). The recent development of an SV40-vector/host system takes advantage of a monkey cell line transformed with a replication-defective SV40 DNA molecule (Gluzman, 1980). This stable cell line, Cos-7, is permissive for SV40 growth and supports the replication of bacterial plasmids containing the SV40 origin (Fig. 8) (Myers and Tjian, 1980). The human α- and β- globin genes have been inserted into SV40 origin vectors, and their replication and transcription studied in Cos-7 cells. Forty-eight hours after introducing α-globin gene recombinants into the Cos-7 cell line, approximately 40,000 copies of the globin gene can be detected per transfected cell. Analysis of the mRNA from these cells indicates that the cloned globin genes are accurately transcribed, and their mRNAs are spliced and polyadenylated.

To identify further sequences that are required for a globin gene expression in cells in culture, sequences on the 5' side of the α-globin gene were deleted and the genes then assayed for α-globin gene transcription in Cos-7 cells. A region between 87 and 55 base pairs upstream from the 5' end of the gene was found to be required for accurate transcription (Mellon et al., 1981).

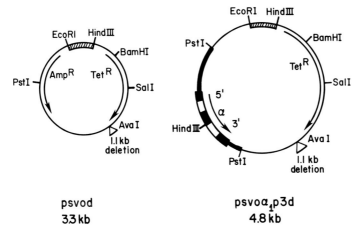

FIG. 8. Restriction maps of an SV40-ORI vector and an SV40/α-globin gene recombinant. The SV-ORI vector, pSVOd, was derived from plasmids described by Myers and Tjian (1980) and Lusky and Botchan (1981). The human α_1-globin gene was inserted at the Pst 1 site as a 1.5-Kb DNA fragment.

This region includes the sequence CCAAT, which is found near all globin genes that have been sequenced thus far.

Procedures now exist that make it possible to introduce single base changes in specific regions of cloned DNA. Thus, with this procedure and the assays described above, it should be possible to obtain a fine structure genetic map of globin gene promoters.

ISOLATION AND CHARACTERIZATION OF MUTANT GLOBIN GENES

To complement studies of *in vitro* mutagenized genes, procedures have been developed for isolating and studying globin genes from individuals with naturally occurring genetic defects in globin gene expression. The gene isolation procedure currently in use involves constructing bacteriophage λ libraries of human DNA and screening the libraries by hybridization with a [^{32}P]-labeled globin gene. Although this screening procedure is used routinely to isolate individual eukaryotic genes, it is laborious to isolate the same gene from many different individuals. A novel procedure has been developed to circumvent this difficulty. It involves the use of a positive selection for recombinant bacteriophage bearing cloned globin gene sequences (B. Seed, *unpublished data*). The selection relies on homologous reciprocal recombination between a very small probe plasmid and recombinant phage that carry globin genes. The probe plasmid contains a suppressor tRNA gene and a short segment of nonrepeated DNA proximal to the β-globin gene. Recombination between phage and plasmid yields phage bearing an integrated copy of the probe plasmid. The suppressor tRNA gene of the integrated plasmid then confers on amber-mutated phage the ability to grow in suppressor-free hosts; amber mutations create the premature termination codon UAG. This defect in the gene-protein translation process can be suppressed by a special class of tRNA molecules called suppressor tRNAs, which insert a particular amino acid in response to the termination codon. The recombination occurs at high enough frequency to allow direct recovery of globin genes from genomic libraries of recombinant phage, and should be applicable to any gene for which a cloned segment of homologous sequence is available (Fig. 9).

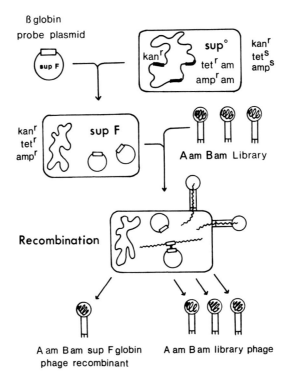

FIG. 9A. Schematic diagram of a recombination selection for cloned genes. The probe plasmid contains a globin DNA segment and a suppressor tRNA gene. The suppressor gene confers ampicillin and tetracycline resistance on cells containing a second plasmid (p3), which has amber-mutated ampicillin and tetracycline genes. Probe plasmid-containing cells are infected with a genomic phage library constructed from a doubly amber-mutated (am) vector. Recombination takes place only between the probe plasmid and phage bearing sequences homologous to the probe segment.

One gene, isolated from an individual with β-thalassemia using conventional procedures, has been studied in detail (Fig. 10). Comparison of the complete nucleotide sequence of this gene with that of a normal β-globin gene revealed only two differences out of 2,000 base pairs of DNA (Lawn et al., 1980). One difference is a G to T transversion located within the large intervening sequence of the β-globin gene, 74 base pairs away from the 5′ junction, with coding DNA. The second difference is a G to A transition with the GT sequence located at the 5′ end of the large intervening sequence. This GT is an invariant feature of all the intervening sequences and is therefore thought to be necessary for splicing (Sharp, 1981). To investigate the possibility that transcripts from the cloned β-globin gene are abnormally spliced, the gene was inserted into an SV40-origin vector, and the expression of the mutated globin gene analyzed in Cos cells as described above.

In summary, the application of molecular cloning techniques to the study of human globin genes has provided a detailed understanding of their structure and chromosomal arrangement. In addition, the development of assays for cloned-globin gene expression has provided the means for establishing the functional significance of the structural data. Thus, human globin genes can be subjected to fine structure analysis similar to that used in the study of prokaryotic genes. The methods that have been developed for the study of globin genes should be generally applicable to other eukaryotic genes, including those of the nervous system.

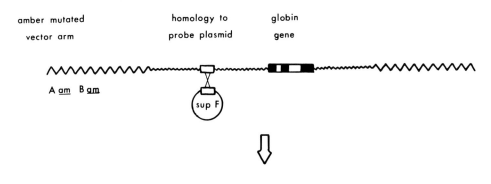

FIG. 9B. Structure of a phage recombinant. Insertion of the probe plasmids results in a duplication of the probe region. The suppressor tRNA gene in the plasmid then allows the amber-mutated phage to grow in nonsuppressing cells.

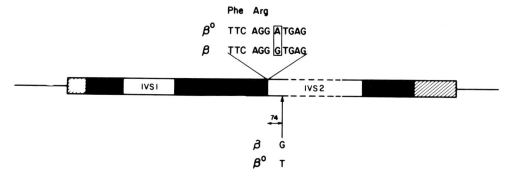

FIG. 10. Schematic diagram of one type of β°-thalassemia gene. The nucleotide sequence of the entire β-globin gene was determined and only two differences were observed. One was a G to A change at the junction between exon 2 and IVS 2; the other was a G to T change 74 base pairs into the large IVS. The latter base change is unlikely to be associated with the genetic disease.

ACKNOWLEDGMENTS

Studies reported here were supported by grants from the NIH and NSF.

REFERENCES

Blattner, F. R., Williams, B. G., Blechl, A. E., Denniston-Thompson, K., Faber, H. E., Furlong, L. A., Grunwald, D. J., Kiefer, D. O., Moore, D. D., Schumm, J. W., Sheldon, E. L., and Smithies, O. (1977): Charon phages: Safer derivatives of bacteriophage λ for DNA cloning. *Science*, 196:161–169.
Bunn, F. H., Forget, B. G., and Ranney, H. M. (1977): *Human Hemoglobins*. W. B. Saunders, Philadelphia.

Chang, J. C., and Kan, Y. W. (1979): β°-Thalassemia, a nonsense mutation in man. *Proc. Natl. Acad. Sci. USA*, 76:2886–1889.
Dierks, P., van Ooyen, A., Mantei, N., and Weissmann, C. (1981a): DNA sequence preceding the rabbit β-globin gene are required for formation in mouse L cells of β-globin RNA with the correct 5' terminus. *Proc. Natl. Acad. Sci. USA*, 78:1411–1415.
Dierks, P., Wieringa, B., Marti, D., Reiser, J., van Ooyen, A., Meyer, F., Weber, H., and Weissmann, C. (1981b): Expression of β-globin genes modified by restructuring and site-directed mutagenesis. *ICN-UCLA Symposium. J. Supramol. Struct. Cell Biochem.*, (in press).
Efstratiadis, A., Posakony, J. W., Maniatis, T., Lawn, R. M., O'Connell, C., Spritz, R. A., DeRiel, J. K., Forget, B. G., Weissman, S. M., Slightom, J. L., Blechl, A. E., Smithies, O., Baralle, F. E., Shoulders, C. C., and Proudfoot, N. J. (1980): The structure and evolution of the human β-globin gene family. *Cell*, 21:653–668.
Farabaugh, P. J., and Miller, J. H. (1978): Genetic studies of the lac repressor VII on the molecular nature of spontaneous hot spots in the loci gene of E. Coli. *J. Mol. Biol.*, 126:847–863.
Flavell, R. A., Bernards, R., Kootes, J. M., and De Boer, E. (1979): The structure of the human β-globin gene in β-thalassemia. *Nucleic Acids Res.*, 6:2749–2760.
Frisch, E. F., Lawn, R. M., and Maniatis, T. (1980): Molecular cloning and characterization of the human β-like globin genes. *Cell*, 19:959–972.
Gluzman, Y. (1980): SV40 transformed simian cells support the replication of early SV40 mutants. *Cell*, 23:175–182.
Goldberg, M. L. (1979): Sequence analysis of Drosophila histone genes. Ph.D. thesis, Stanford University, Stanford, California.
Grosveld, G. C., Koster, A., and Flavell, R. (1981a): A transcription map for the rabbit β-globin gene. *Cell*, 23:573–584.
Grosveld, G. C., Shewmaker, C. K., Jat, P., and Flavell, R. A. (1981b): Localization of DNA sequences necessary for transcription of the rabbit β-globin gene *in vitro*. *Cell*, 25:215–226.
Hamer, D. H., and Leder, P. (1979): Expression of the chromosomal mouse β-maj globin gene cloned in SV40. *Nature*, 281:35–40.
Jahn, C. L., Hutchison, C. A., III, Phillips, S. J., Weavers, S., Haigwood, N. L., Voliva, C. F., and Edgell, M. H. (1980): DNA sequence organization of the β-globin complex in the BALB/c mouse. *Cell*, 21:159–168.
Jeffreys, A. J., and Flavell, R. A. (1977a): A physical map of the DNA regions flanking the rabbit β-globin gene. *Cell*, 12:429–39.
Jeffreys, A. J., and Flavell, R. A. (1977b): The rabbit β-globin gene contains a large insert in the coding sequence. *Cell*, 12:1097–1108.
Kantor, J. A., Turner, P. H., and Nienhaus, A. W. (1980): β$^+$ Thalassemia: Mutants which offset processing of the β-globin mRNA precursor. *Cell*, 21:149.
Kendrew, J. C., Watson, H. C., Strandberg, B. E., Dickerson, R. E., Phillips, D. C., and Shore, V. C. (1961): The amino acid sequence of sperm whale myoglobin. A partial determination by X-ray methods and its correlation with chemical data. *Nature*, 190:666–670.
Lacy, E., Hardison, R. C., Quan, D., and Maniatis, T. (1979): The linkage arrangement of four rabbit β-like globin genes. *Cell*, 18:1273–83.
Lacy, E., and Maniatis, T. (1980): The nucleotide sequence of a rabbit β-globin pseudogene. *Cell*, 21:545–553.
Lauer, J., Shen, C. K. J., and Maniatis, T. (1980): The chromosomal arrangement of human α-like globin genes: Sequence homology and α-globin gene deletions. *Cell*, 20:119–130.
Lawn, R. M., Efstratiadis, A., O'Connell, C., and Maniatis, T. (1980): The nucleotide sequence of the human β-globin gene. *Cell*, 21:647–651.
Liebhaber, S. A., Goossens, M. J., and Kan, Y. W. (1980): Cloning and complete nucleotide sequence of human 5'-α-globin gene. *Proc. Natl. Acad. Sci. USA*, 77:7054–7058.
Luse, D. S., and Roeder, R. G. (1980): Accurate transcription inciation on purified mouse β-globin DNA fragment in a cell-free system. *Cell*, 20:691–699.
Lusky, M., and Botchan, M. (1981): Inhibition of SV40 replication in similar cells by specific pBR322 DNA sequences. *Nature*, 293:79–81.
Maniatis, T. (1980): Recombinant DNA procedures in the study of eukaryotic genes. In: *Cell Biology, A Comprehensive Treatise, Vol. 3*, edited by L. Goldstein and D. Prescott, pp. 564–582. Academic Press, New York.
Maniatis, T., Fritch, E. F., Lauer, J., and Lawn, R. M. (1980): The molecular genetics of human hemoglobins. *Annu. Rev. Genet.*, 14:145–178.
Maniatis, T., Hardison, R. C., Lacy, E., Lauer, J., O'Connell, C., Quon, D., Sim, G. K., and Efstratiadis, A. (1978): The isolation of structural genes from libraries of eukaryotic DNA. *Cell*, 15:687–701.
Manley, J. L., Fire, A., Lang, A., Sharp, P. A., and Gefter, M. L. (1980): DNA-dependent transcription of adenovirus genes in a soluble whole-cell extract. *Proc. Natl. Acad. Sci. USA*, 77:3855–3859.
Mantei, N., Boll, W., and Weissman, C. (1979): Rabbit β-globin mRNA production in mouse L-cells transformed with cloned rabbit β-globin chromosomal DNA. *Nature*, 281:40–46.
Maquat, L. E., Kinniburgh, A. J., Beach, L. R., Honig, G. R., Lazerson, J., Ershler, W. B., and Ross, J. (1980): Processing of human β-globin mRNA precursor to mRNA is defective in three patients with β$^+$-thalassemia. *Proc. Natl. Acad. Sci. USA*, 77:4287–4291.

Mathis, D. J., and Chambon, P. (1980): The SV 40 early region TATA box is required for accurate *in vitro* initiation of transcription. *Nature*, 290:310–315.
Mellon, P., Parker, V., Gluzman, Y., and Maniatis, T. (1981): Identification of sequences required for transcription of the human α_1-globin gene using a new SV40 host-vector system. *Cell*, 27:279–288.
Mulligan, R. C., Howard, B. H., and Berg, P. (1979): Synthesis of rabbit β-globin in cultured monkey kidney cells following infection with a SV40 β-globin recombinant clone. *Nature*, 277:108–114.
Myers, R. M., and Tjian, R. (1980): Construction and analysis of SV40 origins defective in tumor antigen binding and DNA replication. *Proc. Natl. Acad. Sci. USA*, 77:6491–6495.
Nishioka, Y., and Leder, P. (1979): The complete sequence of the chromosomal mouse α-globin gene reveals elements conserved throughout vertebrate evolution. *Cell*, 18:875–882.
Orkin, S. H., Kolodner, R., Michelson, A., and Husson, R. (1980): Cloning and direct examination of a structurally abnormal human β° thalassemia gene. *Proc. Natl. Acad. Sci. USA*, 77:3558–3562.
Perutz, M., Muirhead, H., Cox, J. M., and Goaman, L. C. G. (1968): Three-dimensional Fourier synthesis of horse oxyhaemoglobin at 2.8 A° resolution: The atomic model. *Nature*, 219:131–139.
Proudfoot, N. J., and Maniatis, T. (1980): The structure of a human α-globin pseudogene and its relationship to a α-globin gene duplication. *Cell*, 21:537–544.
Proudfoot, N. J., Shander, M. H. M., Manley, J., Gefter, M., and Maniatis, T. (1980): The structure and *in vitro* transcription of human globin genes. *Science*, 209:1329–1336.
Sharp, P. A. (1981): Speculations on RNA splicing. *Cell*, 23:646–646.
Slightom, J. L., Blechl, A. E., and Smithies, O. (1980): Human fetal Gγ-and Aγ genes: Complete nucleotide sequences suggest that DNA can be exchanged between these duplicated genes. *Cell*, 21:627–638.
Southern, E. M. (1975): Detection of specific sequences among DNA fragments separated by gel electrophoresis. *J. Mol. Biol.*, 98:503–517.
Tilghman, S. M., Tiemeier, D. C., Seidman, J. G., Peterlin, B. M., Sullivan, M., Maizel, J. V., and Leder, P. (1977): Intervening sequence of DNA identified in the structural portion of a mouse β-globin gene. *Proc. Natl. Acad. Sci. USA*, 75:725–729.
Weaver, R. F., and Weissman, C. (1979): Mapping of RNA by a modification of the Brok Sharp procedure. The 5' termini of 15S β-globin mRNA precursor and mature 10S β-globin mRNA have identical map coordinates. *Nucleic Acids Res.*, 6:1175–1193.
Weil, P. A., Luse, D. S., Segall, J., and Roeder, R. G. (1979): Selective and accurate initiation of transcription at the Ad2 major late promotor in a soluble system dependent on purified RNA polymerase II and DNA. *Cell*, 18:469–484.
Wigler, M., Sweet, R., Sim, G. K., Wold, B., Pellicer, A., Lacy, E., Maniatis, T., Silverstein, S., and Axel, R. (1979): Transformation of mammalian cells with genes from procaryotes and eukaryotes. *Cell*, 16:77.
Wold, B., Wigler, M., Lacy, E., Maniatis, T., Silverstein, S., and Axel, R. (1979): Expression of an adult rabbit β-globin gene stably inserted into the genome of mouse L-cells. *Proc. Natl. Acad. Sci. USA*, 76:5684–5688.
Wood, W. G., Old, J. M., Roberts, A. V. S., Clegg, J. B., and Weatherall, D. J. (1978): Human globin gene expression: Control of β, and chain production. *Cell*, 15:437–46.
Zimmer, E. A., Martin, S. L., Beverly, S. M., Kan, Y. W., and Wilson, A. C. (1980): Rapid duplication and loss of gene coding for the α-chains of hemoglobin. *Proc. Natl. Acad. Sci. USA*, 77:2158.

Molecular Genetic Neuroscience, edited by
F. O. Schmitt, S. J. Bird, and F. E. Bloom.
Raven Press, New York © 1982.

Structure, Evolution, and Expression of Mammalian Insulin Genes

*Howard M. Goodman, Paul Berg, Steve Clark,
*Barbara Cordell, *Don Diamond, Chi Nguyen-Huu,
Yuet W. Kan, and Roger V. Lebo

ABSTRACT

Using recombinant DNA technology, the insulin encoding genes from several vertebrate species have been isolated and characterized. The complete nucleotide sequences of the genes were determined and by virtue of this information (1) the structural organization of the ancestral form of the insulin gene has been deduced; (2) the human insulin gene has been located to the short arm of chromosome 11; (3) the rat insulin genes have been shown to be undermethylated only in tissues programmed for insulin expression, whereas in nonexpressing tissue the genes are methylated; and (4) the molecular details of rat insulin expression have been dissected by reintroducing the gene into mammalian cells in culture.

The mature insulin molecule consists of two polypeptide chains, an A chain of 21 amino acids and a B chain of 30 amino acids, which are linked by two disulfide bonds. The immediate product of the gene, however, is a larger precursor called preproinsulin. This polypeptide includes an amino terminal signal peptide and a C peptide sequence connecting the B and A chain polypeptides. The signal peptide (see D. Anderson et al., *this volume*) is involved in directing the newly synthesized preproinsulin into the secretory system of the cell and is removed rapidly after synthesis. In both rat and human preproinsulins, the signal peptide is 24 amino acids long, including the amino terminal methionine. The C peptide is 35 amino acids in length and is removed by specific proteolytic cleavage at the dibasic amino acids that link it to both the A and B chain sequences in the precursor. The insulin gene, therefore, codes for the preproinsulin precursor rather than for the mature insulin molecule.

The insulin genes were one of the first mammalian genetic systems to be investigated using recombinant DNA techniques. Initially, cDNA clones were constructed corresponding to the mRNA of rat insulin from islets of Langerhans (Ullrich et al., 1977), and these clones were used to isolate cDNA clones for human insulin (Bell et al., 1979). Subsequently, the cDNA clones were used as radiolabeled hybridization probes to detect genomic clones in rat and human gene libraries (Cordell et al., 1979; Lomedico et al., 1979; Bell et al., 1980). Isolation of the genes for rat and human insulin has provided an important system for studying genetic organization in mammals. Since insulin is a secreted product of the endocrine system, it may provide a model for the organization and expression of genes coding for similar peptides involved in neuronal function. Most recently, the rat insulin I gene has been inserted into the virus SV40 to study the expression of the cloned gene in a eukaryotic cell system. This chapter outlines present knowledge of the structure and evolution of the insulin genes, which has been derived

---

**Present address:* Department of Molecular Biology, Massachusetts General Hospital, Boston, Massachusetts 02114

from a comparison of the genes in different species. In addition, information concerned with the expression of the insulin gene in mammalian cells is described.

STRUCTURE AND EVOLUTION OF INSULIN GENES

Most vertebrate species have a single gene encoding insulin. In a few species, however, including rat, mouse, and two species of fish, amino acid sequence studies indicate the presence of two different insulin molecules, and it was suggested that these molecules were the products of two nonallelic insulin genes. Hybridization of rat DNA with the cloned cDNA for rat insulin mRNA demonstrates that there are indeed two insulin genes in the rat. The two genes, rat insulin I and rat insulin II from two species, have been cloned independently in two different laboratories (Cordell et al., 1979; Lomedico et al., 1979). Determination of the DNA sequence of the two genes revealed that they have different structures (Fig. 1). Both genes have an intron of 119 base pairs in the region corresponding to the 5'-untranslated region of the mRNA, which precedes the initiation codon for preproinsulin. The position of this intron is identical in the two genes, and there is also extensive homology between the nucleic acid sequences of the

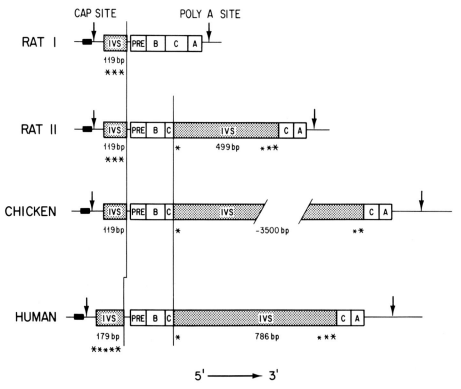

FIG. 1. Schematic comparison of human, chicken, and rat insulin genes. The topology of the two rat insulin genes (I and II), the single chicken insulin gene, and the single human insulin gene is displayed. The coding sequences for the peptide chains (pre-, B, C, and A) of preproinsulin are represented by the *clear boxes*. Intervening sequences (IVS) are distinguished by the stippled areas, with the length of each intervening sequence indicated below. The extent and position of nucleotide homology between intervening sequences of each species is represented by the size of the *asterisks*—the larger the asterisk the greater the homology. *Vertical lines* indicate the positions at which intervening sequences occur. Also indicated are the sites for polyadenylation and capping *(arrows)* as well as a potential site for the initiation of transcription *(small black box)*. *Horizontal lines* indicate noncoding regions.

introns. The rat insulin I gene has no other intron. The insulin II gene, however, has an additional intron of 499 base pairs in the region encoding the C peptide.

The sequences of the rat genes were compared to the sequence of the insulin gene from humans (Bell et al., 1980) and that from chickens (Perler et al., 1980). Both these genes each have two introns (Fig. 1) and therefore have the same general structure as rat insulin II gene. The first intron in both genes is in an almost identical position to the first intron in the rat genes. The second intron in the region coding for the C peptide is in an identical position in the human, chicken, and rat insulin II genes. There is a considerable difference, however, in the size of this intron: in the human it is 786 base pairs in length and in the chicken it is over 3,500 base pairs in length. The rat insulin II gene therefore probably represents the general structure of vertebrate insulin genes. Presumably the rat insulin I gene was generated by duplication of the insulin gene, followed by a very precise deletion of the second intron, which retained the coding sequence of the C peptide. A similar exact deletion of an intron has been observed in an α-globin pseudogene (Nishioka et al., 1980). Both rats and mice possess two genes for insulin, whereas hamsters only have a single insulin gene. This suggests that the gene duplication event occurred some 25 to 30 million years ago, after the ancestors of rats and hamsters had diverged but before rats and mice diverged from each other.

The amino acid sequences of human and rat insulin are highly conserved, so not surprisingly, the nucleic acid sequences of the coding regions of the rat and human insulin genes show considerable homology. The DNA sequences of noncoding regions of the two genes, particularly in the introns, are much less homologous. There are, however, some sequences that appear to be evolutionarily conserved. These are found in the region of the TATAAA, or Hogness, sequence, at the cap site, and adjacent to the splice sites in the introns (Fig. 1). Presumably, the conservation of these sequences during evolution indicates some common function or regulatory role for these sequences (see P. Sharp and M. Wilson, and T. Maniatis, et al., *this volume*).

METHYLATION OF RAT INSULIN GENES

In the rat, insulin I and II genes are expressed in the normal pancreas to approximately equal extents. In an insulinoma, a pancreatic β-cell tumor that produces insulin, there is evidence for differential expression of the two genes (H. H. Schöne, *personal communication*). It was hypothesized that the genes might be differentially methylated in the tumor, causing them to be differentially expressed since the demethylation of genes appears to reflect their transcriptionally active state. To investigate this question, the DNA methylation patterns of different rat tissues were examined by digestion of the DNA with the restriction enzymes Msp I or Hpa II. These enzymes are isochizomers, that is, they both recognize and cleave at the same nucleotide sequence, CCGG. Hpa II will only cut if the cytosine residues are not methylated at the 5' position. Msp I will cut both methylated and unmethylated DNA with equal efficiency. The location and methylated state of CCGG sites in the two genes can be deduced from Southern blotting experiments (Fig. 2) and are summarized in Fig. 3. Sites within the immediate gene region are unmethylated (there is a cluster of three CCGG sites at the 3' end of the coding region in both genes; the resolution of the blots is not sufficient to determine which of these sites is unmethylated), whereas sites several Kb outside the genes are methylated. In the normal pancreas and in the insulinoma, some of the CCGG sites in the insulin genes are not methylated as shown by the presence of fragments after digestion of the DNA with the enzyme Hpa II, which hybridize with an insulin gene probe. The patterns of methylation in the tumor and in the normal tissue are, in fact, quite similar, suggesting that DNA methylation is not responsible

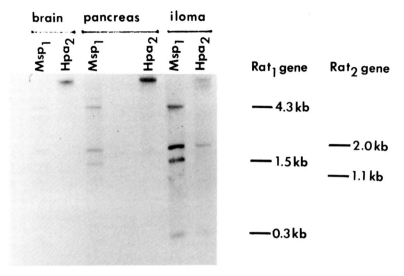

FIG. 2. Southern blot analysis of the rat insulin genes in expressing and nonexpressing tissue. High molecular weight genomic DNAs were isolated from normal rat brain and pancreas as well as from a rat insulinoma (iloma), cleaved with Msp I or Hpa II and analyzed by the Southern blotting procedure (Southern, 1975) by hybridization with a probe made from the cloned rat insulin I gene. The predicted Msp I fragments for the rat insulin I and II genes are schematically illustrated.

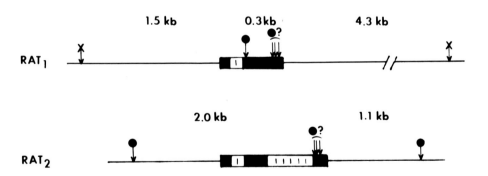

FIG. 3. Restriction map of the rat insulin I and II genes and schematic representation of the unmethylated sites in expressing tissue. The Msp I restriction map and fragment sizes for rat insulin I and II genes are illustrated with each *arrow* indicating a Msp I site. Unmethylated sites are indicated by the *closed circles* and methylated sites by the *X*. The *question mark* indicates that the methylation pattern of individual sites within the cluster cannot be determined. *Open boxes* represent introns occurring within each gene; *black boxes* indicate exons.

for differential expression of the genes in the two tissues. Since insulin has been reported to be synthesized in the rat brain (Havrankova et al., 1978), it was of interest to determine the methylation pattern of the insulin genes in this tissue. The same sites in the insulin genes in rat brain DNA, however, do appear to be methylated (B. Cordell, *unpublished observation*). A similar result has been found for β-globin genes in human brain DNA (van der Ploeg and Flavell, 1980). (For discussion of DNA methylation, see A. Rich, *this volume*.)

CHROMOSOMAL MAPPING OF INSULIN GENES

The chromosomal location of the human insulin gene has been mapped by a novel technique that forms an alternative to somatic cell genetic approaches (see F. Ruddle, *this volume*). The fluorescence-activated cell sorter was used to separate fluorescently labeled human metaphase chromosomes into some 12 to 16 different fractions and the chromosomes in each fraction were identified by karyotype analysis. The DNA in each fraction was then extracted, digested with the restriction enzyme, EcoRI, and subjected to Southern blot analysis using the cloned insulin gene as a probe. In this case, the fraction containing chromosomes 9, 10, 11, and 12 was positive for the insulin gene, whereas the adjacent fraction containing chromosomes 7 and 8 and the X chromosome was negative. While this work was in progress, the gene for insulin was mapped to chromosome 11 using human mouse cell hybrids in a somatic cell genetic approach (Owerbach et al., 1980).

The location of the insulin gene on chromosome 11 was mapped more precisely using chromosome translocations from the Genetic Cell Depository. In the case studied, the short arm of one chromosome 11 had been translocated to the X chromosome and the long arm of the X chromosome to chromosome 11 (Fig. 4). When these chromosomes were fractionated by the fluorescence-activated cell sorter, the two parts of the translocated chromosome 11 were separated into quite different fractions: the short arm going to the fraction marked A in Fig. 4 and the long arm to fraction B. The normal homologue of chromosome 11 of course remained in its original fraction. Southern blot hybridization analysis of the DNA from these fractions showed that fraction A contained the insulin gene sequences. The human insulin gene is therefore located on the short arm of chromosome 11. The position of the break in chromosome 11 is known, thus confining the location of the insulin gene to region 1 in the terminus of the short arm of chromosome 11 between bands 2 and 3.

In the translocation experiment the total DNA from the cells contained two fragments, 16 and 14 Kb in length, which hybridized to the insulin gene probe. Since there is good evidence for only one insulin gene in humans, the two fragments must represent different alleles of the insulin gene. The translocated arm of chromosome 11 was found to carry the 16-Kb insulin fragment, whereas the normal chromosome 11 possessed the 14-Kb fragment, indicating that each homologue of chromosome 11 carried one of the two alleles of the insulin gene. This variation in fragment length for the insulin gene is widespread in human populations and in all cases is due to an insertion of DNA at the same position, 1.3 Kb before the initiation site of the insulin gene. The insertion can vary in size from several hundred base pairs to 3.4 Kb and was found in 6 out of 27 normal individuals. The variation in organization of the DNA adjacent to the insulin gene is therefore a genetic polymorphism and is an example of restriction fragment length polymorphism (see D. Housman and J. Crusella, *this volume*), which has been observed in a number of regions of the genome, particularly the α-globin cluster (Jeffreys, 1979). The insulin gene polymorphism has been tested for correlation with diabetes (Rotwein et al., 1981): the insertion was found in none of 8 individuals with insulin-dependent, juvenile-onset, type 1 diabetes, but it was found in 6 of 13 individuals with insulin-independent, late-onset, type 11 diabetes, suggesting a possible link with the latter disease, although these data are not yet statistically significant.

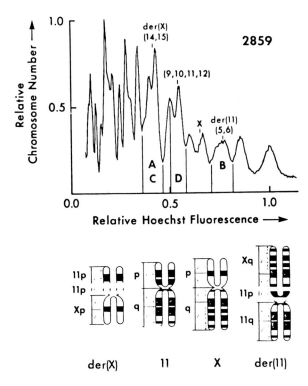

FIG. 4. Fractionation of human chromosomes for localization of the insulin gene. Fluorescence distribution generated by the fluorescence-activated cell sorter of chromosomes from a cell line (GM2859) carrying a translocation of chromosomes X and 11 (depicted above graph). Fractionated chromosomes of interest are labeled above the total distribution. The sorted fractions, labeled A and B, were employed for isolation of DNA and hybridization with [^{32}P]-labeled insulin-specific probe after electrophoretic separation of DNA fragments obtained by restriction digestion of the DNA (From Southern, 1975).

EXPRESSION OF INSULIN GENES

To study the expression of the insulin gene, a cloned fragment of the rat insulin 1 gene was inserted into the viral vector SV40 and introduced into monkey kidney cell cultures by transfection. Prokaryotes possess neither spliced genes nor the enzymatic machinery for splicing RNA transcripts. The mechanisms of gene transcription must therefore be studied in eukaryotic cell system. There are a number of available techniques for transferring cloned genes into cultured cells (see F. Ruddle, *this volume*) but transfection using SV40 offers the advantage that the immediate environment of the gene, that is, the flanking DNA sequences, can be manipulated with great precision. It is therefore possible to test the effects of controlling elements in the virus DNA or alterations in the inserted gene sequence on the transcriptional pattern of the gene.

SV40 is a small circular DNA virus, approximately 5,000 base pairs in length, which is infectious for monkey cells. The total DNA sequence of the virus is known and its transcriptional units have been well characterized. There is a single origin of replication and two transcriptional units: the "late" gene region, which is transcribed in a clockwise direction, and the "early" gene region, which is transcribed counterclockwise (the orientation in which the genome is usually represented). Each transcriptional unit produces a number of different mRNA molecules by differential splicing. Part of the late gene region of the virus can be removed and replaced by a cloned gene fragment. Although this procedure destroys the late genes, the virus can still be propagated in the presence of another SV40 virus that is defective for the early gene functions.

A 1.5-Kb fragment of rat insulin I gene was inserted into the late region of SV40. This gene fragment consisted of approximately 440 base pairs of rat DNA upstream from the cap site, 42 base pairs from the cap site to the 119-base-pair intron, 400 base pairs of sequence coding for rat preproinsulin I and the 3' untranslated region of the mRNA and about 500 base pairs of rat DNA after the poly(A) addition site (for a discussion of transcriptional unit organization, see J. Darnell and M. Wilson, *this volume*). To study the control of initiation and splicing, the fragment was inserted in two orientations with the direction of transcription going clockwise, that is, in the same direction as the transcription of the SV40 late gene region, or with the direction of transcription going counterclockwise, the same direction as the early gene region (Fig. 5).

The possible products of transcription in these recombinants are shown in Fig. 6. In the late orientation (SVGT1-Rt-ins-l), there are several possible products: transcription may begin either

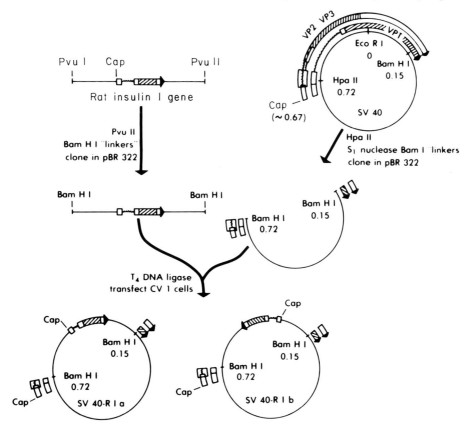

FIG. 5. Strategy for cloning the rat insulin I gene into the SV40 genome. Bam HI linkers were ligated to a previously subcloned 1.5-Kb Pvu II rat insulin I gene fragment (Cordell et al., 1979). SV40 genomic DNA was cleaved with Hpa II (at 0.72 map units) and a Bam HI linker was added at this site using standard procedures (Ullrich et al., 1977). Cleavage of the resulting SV40 DNA with Bam HI (at 0.15 and now 0.72 map units) removed the entire late viral coding region that was then replaced with the rat insulin I gene fragment (in either orientation). The 5' and 3' termini of both the SV40 and insulin transcripts are indicated by *cap* and the *solid arrowheads*, respectively. For both SV40 and insulin genes, coding and untranslated sequences are represented by the *cross-hatched boxes* and *open boxes*, respectively.

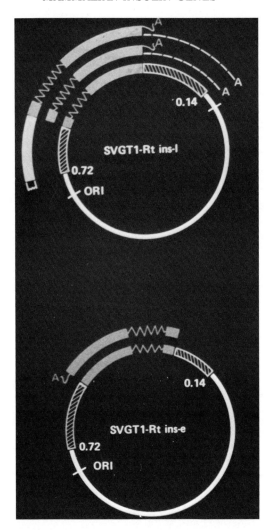

FIG. 6. Schematic representation of predicted transcripts from SV40 rat insulin hybrid viral genomes. Both orientations (l-late and e-early) of the rat insulin I gene cloned into the late region of SV40 DNA are shown (also see Fig. 5). *Solid boxes* represent coding sequences: *cross-hatched boxes*, noncoding regions; and *wavy lines*, intervening sequences. *Open boxes* and *thin lines ending in A* indicate the 5' and 3' termini, respectively. ORI, origin.

at the initiation site of the insulin gene or at the promoter site of the SV40 late gene with transcription continuing into the insulin gene. In either case, transcription may stop at the termination site of the insulin gene or continue to the termination site of the late gene. All of these products will contain the 119-base-pair intron of the insulin gene. In the early orientation (SVGT1-Rt ins-e), only the normal transcript of the insulin gene is possible.

A third recombinant virus (SVGT3-Rt ins-*l*, not shown) was constructed with the insulin gene in the late gene orientation but incorporating more of the SV40 late gene sequence. The SV40 late gene sequence in this construct includes two different splice sites for the late gene transcripts. As before, transcription may be initiated at either the SV40 or insulin gene initiation sites but the transcript starting at the SV40 promoter may also be spliced at the SV40 late gene splice sites.

Monkey kidney cell cultures were infected with these three recombinant viruses. The fidelity of initiation and termination of transcription and the splicing of the transcripts was investigated by examining the products of transcription. The method of analysis has been described by Favalord and coworkers (1980). Briefly, the mRNA molecules were extracted from the cells after infection and hybridized to radiolabeled, single-stranded DNA fragments of the insulin gene, in high concentrations of formamide, a condition that favors the formation of DNA-RNA hybrids. Introns in the DNA that have been spliced out of the mRNA molecule will form single-stranded loops. Treatment of the DNA-RNA hybrid with the enzyme S1 nuclease will digest all single-stranded regions, including loops. Electrophoresis on gels at neutral pH will show only the double-stranded DNA-RNA hybrids, but when the same hybrids are run on gels under alkaline conditions the DNA and RNA chains are separated: the presence of two or more labeled DNA fragments indicates that introns have been spliced out of the mRNA molecule. Alternatively, the enzyme exonuclease VII can be used to digest only single-stranded terminal regions, leaving single-stranded loops intact.

The S1 nuclease digestion technique was used to test the fidelity of splicing of insulin gene transcripts in cells infected with the recombinant SV40 viruses. A 361-base-pair fragment of the insulin gene, which included the 119-base-pair intron and a 215-base-pair segment of the coding region adjacent to the intron, was radiolabeled and hybridized to mRNA extracted from an insulinoma or from monkey cells infected with each of the recombinant SV40 viruses. The hybrids were digested with S1 nuclease and analyzed by gel electrophoresis under alkaline conditions. In all cases, a 215-base-pair fragment of the DNA probe was generated, indicating the RNA transcript was spliced exactly at the correct position. This indicates that the sequences defining the splice sites in the insulin gene can be recognized correctly even when the gene is inserted into SV40 and transcribed in monkey cells.

Initiation of the insulin gene transcript was examined with the exonuclease VII technique. A long fragment of the insulin gene was radiolabeled as indicated in Fig. 7 and hybridized with mRNA from infected cells. The DNA-RNA hybrids were digested with exonuclease VII and run on alkaline gels. Messenger RNA molecules that were initiated correctly should produce a labeled DNA fragment 560 bases in length, the distance from the cap site to the end of the DNA probe. With mRNA from insulinoma, a 560-base labeled fragment was obtained (Fig. 7) exactly as expected, indicating correct initiation of transcription. (The fragment of approximately 350 bases in this gel is an artifact, due to melting of the DNA-RNA hybrid and further digestion by the nuclease.)

With mRNA from cells infected with the recombinant SV40 viruses, the situation was quite different: only 10% of the mRNA molecules appeared to have been initiated correctly. The majority of the mRNA produced longer DNA fragments in the exonuclease VII analysis (Fig. 7), indicating that these transcripts were initiated at points in the rat DNA sequence that preceded the correct initiation site. This occurred with the insulin gene in either the early or late gene orientation. Although discrete fragments are visible in this experiment (Fig. 7), an undefined smear is the more usual result, suggesting that the initiation of transcription of the insulin gene in the recombinant viruses does not generally occur at specific DNA sequences.

Similar results were obtained by S1 nuclease experiments: correct initiation should produce a fragment of 43 bases, which is equivalent to the leader sequence, that is, from the cap site to the 5' splice site of the intron. The mRNA from the insulinoma produced a leader fragment of 43 bases as expected, but mRNA from the recombinant SV40 viruses gave fragments ranging in size from 43 to 200 bases in length, again indicating the lack of any specific initiation site.

More detailed analysis of the transcripts produced by the insulin-SV40 recombinant viruses showed that a number of RNA molecules hybridized with both insulin and SV40 DNA probes.

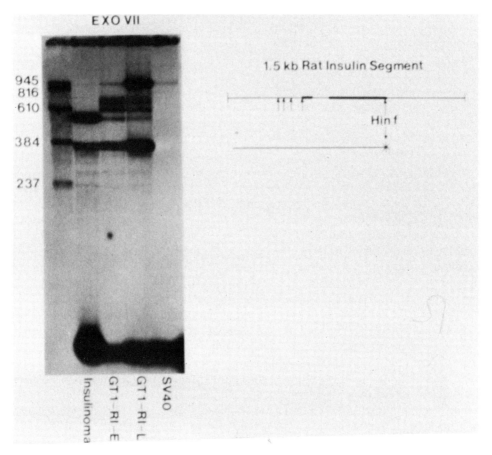

FIG. 7. Analyses of the initiation sites of SV40 rat insulin gene recombinant transcripts. Poly(A) RNA was isolated from cells infected with wild-type SV40 virus or SV40 rat insulin I gene recombinant viruses using both orientations (GT1-RI-E or GT1-RI-L) of the cloned inserts, as well as from an insulinoma producing normal insulin. An end-labeled probe using the single Hin fI site located in the 3' coding area of the rat insulin gene was prepared. This probe was used to form hybrids with the isolated poly(A) RNA, after which the hybrids were digested with exonuclease VII under conditions that digest single-stranded DNA in the 5' to 3' direction (Favalord et al., 1980). The resulting fragments were sized by polyacrylamide gel electrophoresis along side markers (945, 810, 610, 384, and 237 nucleotides). The rat insulin I gene is schematically represented with a set of *arrows* indicating the heterogeneous initiation site for the SV40 rat insulin recombinant transcript.

These could not be explained by any simple mechanism, and to examine them further, cDNA copies of the mRNA molecules were cloned; cDNA clones were obtained from the rat insulinoma, which were essentially full length copies of rat insulin gene I and II mRNA.

Cloning was then used to obtain cDNA clones of the aberrant mRNA molecules produced by the insulin SV40 recombinant viruses. There were several examples of cDNA clones containing both SV40 and insulin sequences. In all of these, a splice had occurred from a 5' (or donor site) in an SV40 sequence to the 3' (or acceptor site) of the intron in the insulin sequence. These unusual splices occurred even though the 5' donor splice site in the insulin gene was much closer to its normal 3' acceptor site. For example, clone 33-20 (Fig. 8), derived from mRNA transcribed from the recombinant SV40 virus containing the insulin gene in the early gene orientation, was found to have been initiated at the early gene promoter region of SV40. The normal "large T" splice of SV40 early transcripts was spliced out and the RNA continued

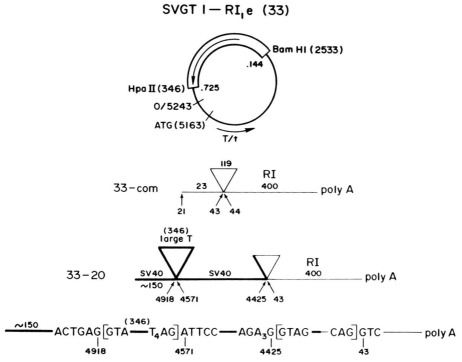

FIG. 8. Analysis of RNA splicing of SV40 rat insulin recombinant transcript. One of the two orientations of SV40 rat insulin gene recombinant (SVGT1-RIe) was analyzed with respect to the accuracy of removal of intervening sequences. Poly(A) RNA was isolated from CV-1 monkey cells after infection with the recombinant virus and a cDNA copy prepared and cloned using standard procedures. Insulin-containing cDNA clones were sequenced. Several of the clones (33-com) showed correct transcript splicing (removing a 119bp intervening sequence) but initiated at a position downstream from the normal insulin start site (position 21). One clone (33-20) initiated in the early gene region of SV40 contained the usual splices shown. The *bottom line* shows the nucleotide sequence at the junction of splice sites for clone 33-20. *Thick line*, SV40 genome; *thin line*, rat insulin I gene segment. ATG, initiation site of T/t antigen gene of SV40; O, origin.

to position 4,425 of SV40 where a splice occurred to the normal acceptor site at position 43 of the insulin gene, a distance of 2,593 bases. Splicing at position 4,425 in the SV40 early gene has not been previously observed. Other clones (e.g., 33-com; Fig. 8) from this infection show the expected pattern of splicing within the insulin gene, although this example had initiated transcription at a position downstream from the normal insulin start site. Similar examples of aberrant splicing were found in cDNA clones from the recombinant viruses that had the insulin gene inserted in the late gene orientation. Generally, initiation occurred at the start sites for the late genes of SV40 and the RNA was spliced from both normal and abnormal sites in the SV40 sequence to the acceptor site in the insulin gene.

Overall, approximately half of the mRNA molecules transcribed from the recombinant viruses showed correct splicing within the insulin gene; the other half showed aberrant SV40-insulin splicing. A small fraction, about 5 to 10% of the mRNA molecules, were initiated at the correct site in the insulin gene, but the majority started at other sites in the insulin gene or even in the SV40 sequences. From other experiments it was found that 80 to 90% of the mRNA molecules terminated at the stop site in the insulin gene; the remaining transcripts continued to stop sites in the SV40 sequences. The transcription of the insulin gene in this situation does not appear to be as rigorously controlled as in the pancreas or insulinoma. This may be due to the effects of the adjacent sequences in the virus or to the different genetic environment of the gene in monkey kidney cells. Alternatively, the bizarre splicing events may be the result of the large number of copies of the gene in each cell, of the order of 100,000 per cell.

Despite the variations in RNA transcription, large quantities of proinsulin are produced by the monkey cells after infection by the SV40 insulin recombinant viruses (Fig. 9). Material with the size of proinsulin was immunoprecipitated from the culture medium using antibodies against insulin. In fact, proinsulin was the most predominant component in the culture medium; the other major products were SV40 proteins. Very little or no insulin, however, was detectable in the medium or in the cells. It appears that preproinsulin is synthesized from the mRNA transcripts of the insulin gene, the signal peptide is removed by proteolysis, and the proinsulin is secreted from the cell. The monkey kidney cell is therefore capable of recognizing and processing the signal peptide but lacks the necessary enzymes for converting proinsulin to the mature insulin molecule, a process that may then be limited to the pancreas. Radiochemical sequence analysis of the secreted proinsulin molecule showed that removal of the signal peptide did not occur at the correct position, suggesting that the machinery for processing signal peptide sequences is more generally distributed.

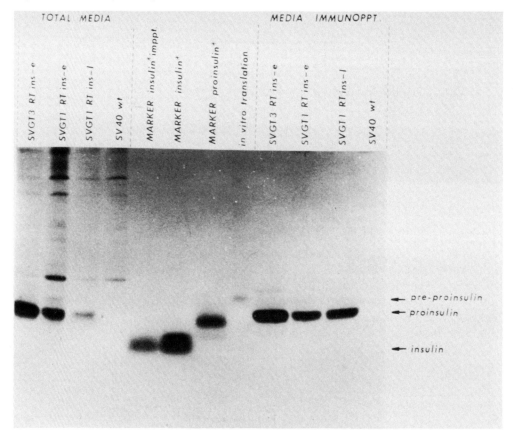

FIG. 9. Characterization of the insulin protein synthesized from infection with SV40 rat insulin recombinant viruses. Cells were infected with either of 3 recombinant SV40 rat insulin I gene viruses or wild-type SV40. About 48 hr after infection, CV-1 monkey cells were labeled with [³H]-leucine. Labeled media were collected and the proteins were analyzed either directly (total media) or after immunoprecipitation with anti-insulin guinea pig sera (media immunoPPT). Prepro-, pro-, and insulin specific markers appear in the center panel. Proteins were analyzed by polyacrylamide gel electrophoresis.

The results reported here illustrate the power of the recombinant DNA technology and offer a paradigm for the analysis of any genetic system and its products. In a relatively short space of time, beginning with the isolation of a cDNA clone for rat insulin (Ullrich et al., 1977), more has been discovered about the genetics, evolution, and expression of this clinically important protein than had been learned in the nearly 60 years since Banting and Best discovered insulin in 1923. The application of these approaches to the study of neuroscience can only increase our knowledge of neuronal mechanisms and brain function to unimaginable levels.

ACKNOWLEDGMENTS

H. M. Goodman and Y. W. Kan were investigators and B. Cordell was an associate investigator of the Howard Hughes Medical Institute Laboratory during the period this work was carried out. A portion of this research was supported by grants from the National Institutes of Health (AM 16666 and CA 14026) and the National Science Foundation (PCM 7808950). P. Berg is a recipient of a grant from NIH GM 13235-15 and the National Cancer Institute CA 15513-06. C. Nguyen-Huu is a recipient of a grant from the American Cancer Society.

REFERENCES

Bell, G. I., Pictet, R. L., Rutter, W. J., Cordell, B., Tischer, E., and Goodman, H. M. (1980): Sequence of the human insulin gene. *Nature*, 284:26–32.
Bell, G. I., Swain, W. F., Pictet, R., Cordell, B., Goodman, H. M., and Rutter, W. J. (1979): Nucleotide sequence of a cDNA clone encoding human preproinsulin. *Nature*, 282:525–527.
Cordell, B., Bell, G., Tischer, E., DeNoto, F. M., Ullrich, V., Pictet, P., Rutter, W. J., and Goodman, H. M. (1979): Isolation and characterization of a cloned rat insulin gene. *Cell*, 18:533–543.
Favalord, J., Treisman, R., and Kamen, R. (1980): Transcriptional maps of polyoma virus-specific RNA: Analysis by two-dimensional nuclease S1 and mapping methods in imaging. *Methods Enzymol.*, 65:718–749.
Havrankova, J., Schmechel, D., Roth, J., and Brownstein, M. (1978): Identification of insulin in rat brain. *Proc. Natl. Acad. Sci. USA*, 75:5737–5741.
Jeffreys, A. J. (1979): DNA sequence variants in the gamma, alpha, alpha and beta-globin genes of man. *Cell*, 18:1–10.
Lomedico, P., Rosenthal, N., Efstratiadis, A., Gilbert, W., Kalodner, R., and Tizard, R. (1979): The structure and evolution of the two nonallelic rat preproinsulin genes. *Cell*, 18:545–558.
Nishioka, Y., Leder, A., and Leder, P. (1980): Unusual α globin-like gene that has clearly lost both globin intervening sequences. *Proc. Natl. Acad. Sci. USA*, 77:2806–2809.
Owerbach, D., Bell, G. I., Rutter, W. J., and Shews, T. B. (1980): The insulin gene is located on chromosome 11 in humans. *Nature*, 286:82–84.
Perler, F., Efstradiadis, A., Lomedico, P., Gilbert, W., Kolodner, R., and Dodgson, J. (1980): The evolution of genes: The chicken preproinsulin gene. *Cell*, 29:555–566.
Rotwein, P., Chyn, P., Chirgwin, J., Cordell, B., Goodman, H., and Permutt, A. (1981): Polymorphism in the 5' flanking region of the human insulin gene and its possible relation to type 2 diabetes. *Science*, 213:1117–1120.
Southern, E. M. (1975): Detection of specific sequences among DNA fragments separated by gel electrophoresis. *J. Biol. Chem.*, 98:503–517.
Ullrich, A., Shine, J., Chirgwin, J., Pictet, P., Tischer, E., Rutter, W. J., and Goodman, H. M. (1977): Rat insulin genes: Construction of plasmids containing the coding sequences. *Science*, 196:1313–1319.
van der Ploeg, L. H. T., and Flavell, R. A. (1980): DNA methylation in the human gamma, alpha, beta-globin locus in erythroid and nonerythroid tissues. *Cell*, 19:947–958.

Section VI

Antibodies: Synthesis and Use as Tools

A fruitful collaborative arrangement has long existed between immunologists, cell biologists, and biochemists in their common quests for immune reagents with which to purify, assay, or localize specific cellular components. A major moving force in the explosion of studies on neuropeptides, for example, has been the heavy reliance on specific antisera for radio immunoassays and for immunocytochemical localizations.

These collaborative efforts have been importantly influenced by the recent methodological advances in molecular and cellular genetics, and have helped to solve critical problems. To obtain antisera to specific cellular components, such as cell surface markers, enzymes, or structural macromolecules, or to secretory products, such as hormones and transmitters, the classic requirement has been a purified substance to use as the immunogen. This requirement is obviously justified since an impure immunogen will raise heterogeneous antisera containing several subpopulations of antibodies to each of the contaminants as well as the desired immunogen. Sera with multiple antibodies can cause considerable confusion since the contaminating antisera can confound both radioimmunoassays and immunocytochemistry, sometimes in totally different ways depending on the titers and affinity of the antibodies present. Although methods such as immunoadsorption purification of sera have been developed to circumvent the complications of impure immunogen preparations, the best and most reliable solution to the problem has been purification before immunization.

The monoclonal antibody production method eliminates almost all such problems. When successfully cloned and sub-cloned, antibody-producing hybridoma cells each yield a single, immunologically pure antibody, which reads one and only one immunogenic epitope. These multiple advantages of monoclonal antibodies, detailed by E. Haber, provide the opportunity to raise specific antisera despite the absence of pure immunogens. C. J. Barnstable describes the use of monoclonal antibodies to distinguish specific cellular membrane antigens in the retina, starting not only with rather crude membrane preparations, but with antigens whose chemical structure and cellular function remain unknown.

R. A. Lerner describes an alternative approach to the production of selective antisera: the use of synthetic fragments of a larger peptide chain, selected on the basis of their predicted immunogenic potency, to raise, in whole animals, antisera that can read the natural larger peptide. What makes this approach even more pertinent to molecular genetic research is the ability to determine peptide sequences to be used as candidates for synthetic immunogen regions directly from the nucleotide sequences of a DNA coding region without ever having to isolate, purify, and sequence the gene product.

These several new approaches confirm and solidify the important continuing work relationship between immunology and neuroscience, which will unquestionably remain important for decades to come.

Antibodies to Chemically Synthesized Peptides From DNA Sequences as Probes of Gene Expression

Richard A. Lerner

ABSTRACT

Because of the ease with which genes can be cloned and sequenced, the classic direction of genetics is often reversed: instead of searching for a gene for a known protein, we are now in the position of having a gene, but not knowing the protein that it synthesizes. However, if a peptide predicted from the nucleic acid sequence is chemically synthesized, an antibody can be raised against the peptide. Such antibodies often react with the native molecule, thus offering a unique probe for locating and characterizing the gene product.

For many decades biologists and geneticists have been interested in tracing phenotype back to genotype. Because of the elegant advances in nucleotide sequencing and recombinant DNA technology, it is fair to say that the flow of the genetic question is changing. More often than not we find ourselves in the position of having a specific genotype sequence and asking the question, What is the phenotype? In other words, it is not unusual to have a complete nucleotide sequence but little information on the correct protein that goes with the structure. As people begin to clone and understand messenger RNAs in the brain, this problem will become ever more frequent.

There are several methods to bridge this gap. A classic approach would be to take the cDNA for the mRNA and clone it in a bacterial plasmid expressor system, using an active bacterial promotor to initiate transcription. The prokaryote might synthesize the protein product of the mRNA, which could then be purified. Perhaps the protein could be fused to a bacterial protein and the bacterial information would allow a simple purification. Once purified, the protein could be used as an immunogen with which to make antibody specific to the product of the mRNA. The specific antibody would then be used to fish the protein out of the tissue.

Sutcliffe and associates (1980) have offered an alternative solution to this problem. The genetic code is used to predict the protein sequence from the completed nucleotide sequence of an mRNA. Then, using rules that are currently evolving to select a specific region of the predicted protein, the peptide is chemically synthesized and used as an immunogen in rabbits to produce an antibody. The antibody is then used to find the native molecule in cells.

Two examples are used here to illustrate this technology. Work with retroviruses has shown that antibodies made in this way are sequence specific. Obviously, all antibodies are sequence specific, but in this instance they are specific for a region of the protein that the experimenter selected in advance, and so in a real sense they are biochemical reagents. A more thorough study has been done on the second example of this approach, the influenza virus hemagglutinin.

The crystal structure of the protein is known, and in a structure-function sense, the synthetic-peptide antibody can be correlated with the three-dimensional structure.

For this discussion, the Moloney leukemia virus is treated as an mRNA molecule. Figure 1 presents a genetic map of this RNA genome correlated to the physical map derived from its nucleotide sequence. Moving from the left, or the 5' end, the first gene that one encounters is the *gag* gene, which stands for group antigen gene. This gene encodes a polyprotein of 65,000 daltons, which is cleaved to proteins of molecular sizes of 15,12,30, and 10 kD. These proteins make up the core of the virus and may have enzymological functions. The next gene is *pol*, encoding the reverse transcriptase, which in this system is thought to be a single chain of 70,000 daltons. The last gene is the *env* gene. It encodes a polyprotein, which in the murine system is cleaved into a glycoprotein having an apoprotein size of approximately 50,000 daltons. That protein is glycosylated to make a hydrophobic molecule with an apparent molecular weight of 70,000 daltons. It is linked, presumably by disulfide bonds, to a very hydrophobic molecule called p15E, and that linkage is thought to anchor the molecule in the plasma membrane of infected or certain normal cells.

The genes were then examined in terms of the complete nucleotide sequence for this virus (Shinnick et al., 1981). The mRNA in this case is 8,332 nucleotodies long, with a cap at the 5' end and a distance of 621 nucleotides before the first coding triplet is found. Next encountered is the coding region of p15, p12, p30, and p10, the *gag* proteins, followed by a single termination triplet, the TAG amber codon. The reading frame continues for 3,600 nucleotides. There must be another protein encoded here because 3,600 nucleotides is obviously much larger than the 2,100 nucleotides necessary to encode the 70,000-dalton reverse transcriptase. Then, overlapping for some 57 nucleotides, comes the *env* gene. It overlaps the *gag-pol* gene, albeit in a different reading frame, and beginning with a 33-amino-acid signal peptide, continues through gp70 and p15E.

The sequence information provides two sorts of findings. The first confirmed what had been expected from the biology of the virus and the protein chemistry that had been done in the preceding years. The second was unexpected and could only have been found by nucleic acid studies. For example, the *gag-pol* junction is separated by a single stop triplet, and the *pol-env* coding regions overlap. Overlapping genes have been the rule in viruses whose sequences have been determined, but it is not yet known if the generality can be extended to cellular genes. Viruses may need this feature because of their limited coding capacity.

The structure of the *env* gene provided another surprise. The hydrophobic anchor protein, p15E, had a larger coding region (shown in Fig. 2) than was thought necessary for a protein

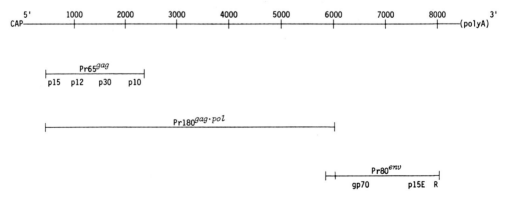

FIG. 1. The genetic map of a replication competent retrovirus.

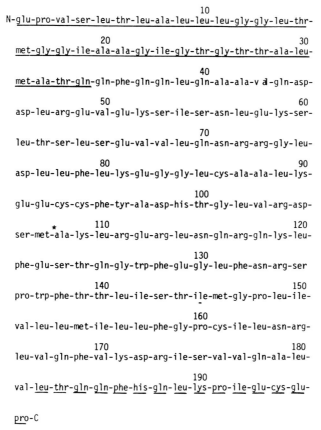

```
                            10
N-glu-pro-val-ser-leu-thr-leu-ala-leu-leu-gly-gly-leu-thr-
             20                              30
met-gly-gly-ile-ala-ala-gly-ile-gly-thr-gly-thr-thr-ala-leu-
                              40
met-ala-thr-gln-gln-phe-gln-gln-leu-gln-ala-ala-val-gln-asp-
             50                              60
asp-leu-arg-glu-val-glu-lys-ser-ile-ser-asn-leu-glu-lys-ser-
                              70
leu-thr-ser-leu-ser-glu-val-val-leu-gln-asn-arg-arg-gly-leu-
             80                              90
asp-leu-leu-phe-leu-lys-glu-gly-gly-leu-cys-ala-ala-leu-lys-
                             100
glu-glu-cys-cys-phe-tyr-ala-asp-his-thr-gly-leu-val-arg-asp-
          *  110                             120
ser-met-ala-lys-leu-arg-glu-arg-leu-asn-gln-arg-gln-lys-leu-
                             130
phe-glu-ser-thr-gln-gly-trp-phe-glu-gly-leu-phe-asn-arg-ser
            140                             150
pro-trp-phe-thr-thr-leu-ile-ser-thr-ile-met-gly-pro-leu-ile-
                             160
val-leu-leu-met-ile-leu-leu-phe-gly-pro-cys-ile-leu-asn-arg-
            170                             180
leu-val-gln-phe-val-lys-asp-arg-ile-ser-val-val-gln-ala-leu-
                             190
val-leu-thr-gln-gln-phe-his-gln-leu-lys-pro-ile-glu-cys-glu-
pro-C
```

FIG. 2. The predicted protein sequence of the retroviral gene product, p15E. The first 34 amino acids are underlined and correspond to those predicted by the DNA sequence. The 15 C-terminal amino acids are individually underlined and indicate the small polypeptide synthesized and used to produce antiserum.

with its known characteristics. The first 34 amino acids predicted by the DNA sequence agreed with the amino acid sequence determined by protein sequence analysis. The amino acid composition and size of the p15E molecule were certainly compatible with the N-terminal portion of the sequence predicted by the DNA. The question was what was going on at the C-terminus. To answer this, a peptide was chemically synthesized corresponding to the C-terminus of the predicted molecule, and an antibody was then made to the synthetic polypeptide. The antiserum was tested against a second, larger synthetic polypeptide for activity. The reasoning was that the antibody to the small peptide was unlikely to see the native structure, but since the larger molecule had a better chance of being in the *in vivo* conformation, an antiserum that recognized the larger molecule could be isolated and used in *in vivo* experiments.

However, it was soon realized that such an elaborate technique was unnecessary. Nonetheless, an antibody was made to the synthetic peptide, along with monoclonal antibodies to gp70 and p15E. These three immunological reagents were used to follow three separate domains of the *env* polyprotein during its cleavage and processing. Figure 3 shows what targets these antisera find in [^{35}S]methionine-labeled infected cells. Antibody to gp70 precipitates a precursor molecule of 80,000 daltons and a small amount of a 17,000-dalton molecule, which is precipitated because it is linked to gp70 by a disulfide bond. The antibody to P15E precipitates the precursor and three more molecules, two forms of the 17,000-dalton molecule and a small amount of a protein that is about 2,000 daltons smaller. The antibody to the synthetic peptide brings down

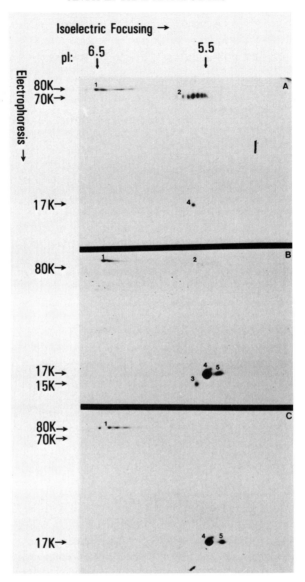

FIG. 3. Two-dimensional gel electrophoresis of Pr80^{env} and its products. Extracts of [^{3}H]leucine-labeled SCRF 60A cells were reacted with **(A)** anti-gp70, **(B)** anti-p15E, or **(C)** anti-R serum (antipeptide antiserum). 1, 80,000-dalton precursor of gp70; 2, gp70; 3, 4, and 5, smaller molecules also precipitated by antibodies to gp70 and p15E.

the precursor and the two 17,000-dalton species, but not the smaller species. The background is very low in the latter immune precipitates, and this finding should lay to rest one frequently asked question: If you make an antibody to a small synthetic peptide, will the conservation of domains or of certain proteins cause the antibody to read many of the proteins in a cell? Obviously, for this antipeptide serum the answer is no. Despite the large number of proteins in the cell, the antibody brought down only three simple proteins; further analysis showed that each of these three proteins shared meaningful immunoreactivity. Thus, the N-terminal amino acid sequence of one of the 17,000-dalton species was found to coincide with that of p15E.

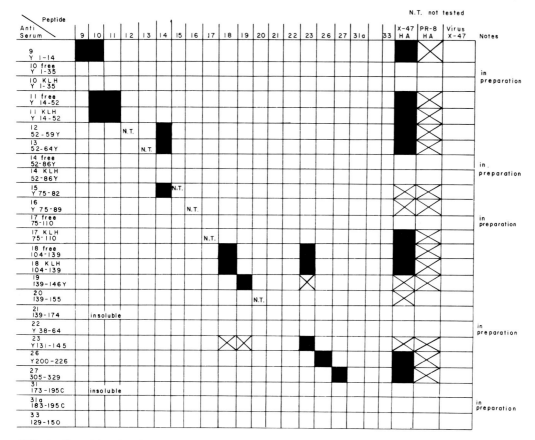

FIG. 4. Reactivity of antipeptide sera with HA-1 peptides and intact hemagglutinin. Larger peptides consisting of 34 or more amino acids were injected in two forms, uncoupled ("free") and coupled to keyhole limpet hemocyanin (KLH) through cysteine residues. Peptides 16 and 17 were insoluble in aqueous buffers and were injected uncoupled. All other peptides were coupled to KLH for immunization. The numbers under each antiserum designation in the left-hand column indicate the sequence position of the peptide used for immunnization. Titers in ELISA are expressed as the serum dilution that binds 50% of 5 pmoles of antigen (*black boxes*). A non-cross-reacting rabbit antiserum (prepared against a hepatitis B envelope peptide) was used in each assay to assess background activity (not shown). *Crosses* indicate no reactivity. *Open boxes* indicate that the particular antiserum/antigen combination has not been tested. Antibodies 21 and 31 were insoluble and could not be tested. ×-47 HA, purified hemagglutinin isolated by bromelain cleavage; ×-47 virus, intact virus; PR-8 HA, hemagglutinin of a related virus.

The species in the virus that reacts with anti-p15E serum does not cross-react with antipeptide. Therefore, the C-terminus of the molecule is removed during virus maturation. The only way to determine this, short of sequencing the entire short-lived precursor, is to use an antibody that is specific for the C-terminus of the molecule, which is present in the cell but not in the mature virus.

These data provide direct information on events in processing the *env* precursor: (1) removal of the single peptide, (2) cleavage to yield gp70 and the precursor to p15E (these two proteins are linked by disulfide bonds to each other), and (3) removal of a short piece from the right end of the p15E precursor molecule. Signal peptides are ordinarily removed from the N-terminus of the molecule. How general is the removal of a C-terminal piece of protein during maturation?

To make a retrovirus, eight or nine cleavages must be performed. Each cleavage sequence is distinct from the next. Since any cell, or at least most cells can be used as hosts in which these cleavages occur, one expects that the eukaryotic cells have a diversity of enzymes capable of cleaving polyproteins in quite specific ways. Therefore, it might not be proper to take a polyprotein sequence, scan for lysine-arginine doublets, and automatically decide that these are the sites for cleavage (see D. Steiner, *this volume*). For example, if one examines the sites working from the left to the right on the leukemia virus message, one finds the cleavage between p15 and p12 to be tyrosine-proline, that between p12 and p30 to be phenylalanine-proline, and that between p30 and p10 to be leucine-alanine. The *env* signal is cleaved at threonine-alanine, the site between gp70 and p15E is lysine-arginine-glutamic acid, and the site between p15E and the C-terminal piece is leucine-valine.

In relating the genetic structure of leukemia virus to its protein structure, antibodies made against synthetic peptides can be used as biochemical reagents. The next question is how widely useful such reagents are. The influenza hemagglutinin was synthesized in pieces to investigate this. Wilson and colleagues (1981) have solved the structure at 3-Å resolution, and eight different published nucleotide sequences for the hemagglutinin gene were used to determine the primary protein sequence of that molecule. The test molecule is the major hemagglutinin of the influenza virus. It is made up of two peptide chains, HA-1 and HA-2, that are synthesized together as a polyprotein that is cleaved at an arginine (Wilson et al., 1981).

Small peptide pieces of varying lengths were synthesized, covering some 70% of the hemagglutinin HA-1 protein sequence (Fig. 4). On the top are listed the peptides. A black square means that an antibody to the peptide indicated precipitated the test molecule. In every case, with the exception of the insoluble peptides, it was easy to make an antibody to the peptide with relatively standard techniques. The key point is that about two-thirds of the antibodies that were tested were able to precipitate the native molecule. The antipeptide sera precipitated X-47 HA-1 (the target virus) but not the hemagglutinin of a related virus of a different sequence (PR-8).

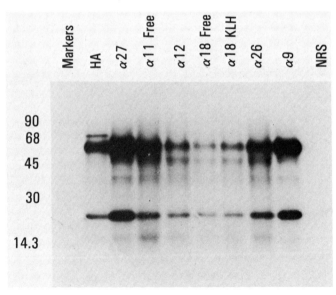

FIG. 5. Immune precipitation of hemagglutinin from X-47 virus by antipeptide antibodies. [^{125}I]X-47 virus (1 × 10^6 cpm) was reacted with 10 μl of normal rabbit sera (NRS) or antipeptide antibodies as indicated. Precipitants were collected with *Staphylococcus aureus* and electrophoresed on a 5 to 17% SDS-polyacrylamide gel prior to autoradiography. Molecular weight markers are in kilodaltons; HA-1, 60 kD; HA-2, 25 kD. (From Lerner et al., 1981.)

What do the precipitates look like? Figure 5 displays the immune precipitates on an SDS gel. The target, called HA, was a trimeric form of the hemagglutinin containing both HA-1 and HA-2 chains. HA-1, the larger major species, is glycosylated. A small amount of contaminating neuraminidase is seen above HA-1. The antibodies bring down HA-1 and HA-2, but not neuraminidase. Two points are important: The first is another example that an antibody to a synthetic peptide can precipitate the native molecule, in this case even in its trimeric conformation. (HA-2 is precipitated because it is bonded tightly to HA-1.) The second is that with a molecule of known structure, the immunogenicity of specific parts of the structure can be assessed. For example, antipeptide-27 recognizes a relatively structureless portion of the C-terminus of HA-1; antipeptide 9 recognizes the 14 N-terminal amino acids of HA-1.

With the average protein molecule, success is possible with about two-thirds of the synthetic peptides used as immunogens. This assessment includes failures from poor peptide selection, such as hydrophobic domains that are not soluble. Further investigations should permit selection of peptides with higher probabilities of success, for example, a number of computer programs predict hydrophilic and hydrophobic domains of protein molecules (J. Kyte and R. Doolittle, *personal communication*).

It is clear that the diversity of the immunological repertoire is infinite, suggesting that if one can figure out what to present to the immune system, an antibody can be elicited. However, if a very globular protein, such as myoglobin, is injected into an experimental animal, only four protein domains, or subregions, are found to be immunogenic (Atassi, 1975). That is, one can account for all the antibodies in the rabbit by these four regions of the molecule. The same sort of results are found for lysosyme and albumin. These molecules probably have a large number of antigenic determinants that are not seen as immunogenic when the whole molecule is injected into an animal. One way of looking at this is to say that the immune repertoire is infinite, but the result is constrained by a less than perfect presentation system. To tap this relatively infinite repertoire, it is necessary to inject the pieces of that molecule, rather than an entire molecule. This will allow cells to be probed for the products of their genes.

ACKNOWLEDGMENTS

I wish to thank my colleagues, Thomas M. Shinnick, J. Gregor Sutcliffe, Nicola Green, and Hannah Alexander with whom much of the work was carried out.

REFERENCES

Atassi, M. Z. (1975): Antigenic structure of myoglobin: The complete immunochemical anatomy of a protein and conclusions relating to antigenic structures of proteins. *Immunochemistry*, 12:423–438.

Lerner, R. A., Green, N., Alexander, H., Liu, F-T., Sutcliffe, J. G., and Shinnick, T. M. (1981): Chemically synthesized peptides predicted from the nucleotide sequence of the hepatitis B virus genome elicit antibodies reactive with the native envelope protein of Dane particles. *Proc. Natl. Acad. Sci. USA*, 78:3403–3407.

Shinnick, T. M., Lerner, R. A., and Sutcliffe, J. G. (1981): Nucleotide sequence of Moloney leukemia virus. *Nature*, 293:543–548.

Sutcliffe, J. G., Shinnick, T. M., Green, N., Liu, F-T., Niman, H. L., and Lerner, R. A. (1980): Chemical synthesis of a polypeptide predicted from nucleotide sequence allows detection of a new retroviral gene product. *Nature*, 287:801–805.

Wilson, I. A., Skehel, J. J., and Wiley, D. C. (1981): Structure of the haemagglutinin membrane glycoprotein of influenza virus at 3Å resolution. *Nature*, 289:366–373.

Applications of Monoclonal Antibodies in Biological Research

Edgar Haber

ABSTRACT

Milstein and his colleagues caused a revolution in immunology by applying somatic cell fusion to antibody-secreting B cells and myeloma tumor cells, with the result that clones of cells secreting large quantities of monoclonal antibodies may be propagated indefinitely. The facility of clonal selection *in vitro* and the unique recognition properties of monoclonal antibodies provide powerful biological tools not available with conventional immunization and polyclonal antibodies. The following examples of applications of monoclonal antibodies are described: (1) resolution of similar molecules in immunoassay, (2) selection of a common antigenic site on dissimilar molecules, (3) differentiation between the active and inactive sites on the surface of a biologically potent protein, and (4) isolation of a biologically active protein from a mixture.

Antibodies represent one of the most precise molecular-recognition tools ever applied to biochemical research. Classic studies of Landsteiner and Heidelberger, performed some 40 years ago, document the fine specificity with which an antiserum recognizes minute details of molecular structures. For example, antiserums readily distinguish positional isomers such as ortho- or meta-derivatives of the benzene ring. This is perhaps the ultimate specificity of recognition any biochemical probe can achieve.

Despite their remarkable attractiveness, conventional antisera have many limitations. An antiserum is a mixture of several to many hundreds of single antibody molecules all capable of binding to immunizing antigen. These antibodies vary considerably in their affinity and specificity for antigen as well as in the portion of the surface of that molecule (epitope) that they will bind. The properties of antisera vary from immunization to immunization and from animal to animal, and their limited supply leads to a lack of reproducible properties.

To better define the antibody combining site, a restricted immune response was sought so that structural studies could be performed on conventional antibodies elicited by immunization. Such a response was found in rabbits immunized with pneumococcal vaccine (Haber, 1970). Restriction implies that 1 to perhaps 12 antibodies are produced rather than the many hundreds seen after immunization with most antigens. The relatively small number of antibodies present allows for fractionation of single antibody species either by conventional means, such as ion-exchange chromatography, or by affinity chromatography. The restricted response seems to be a consequence of a relatively simple, repeating structure of pneumococcal polysaccharides: the two most widely used types, type III and type VIII, consist of repeating units of a disaccharide cellobiuronic acid (glucuronic acid-1,4β-glucose-1,4β-galactose).

In the examination of one such antiserum specific for type VIII pneumococcal polysaccharide, four antibodies were obtained in sufficient yield and homogeneity to permit detailed study

(Chen et al., 1976). The relative affinity of these antibodies for cellobiose, a component of the sequence of the type VIII polysaccharide, was assayed in an inhibition of binding experiment (Fig. 1). Each of the four homogeneous antibodies, 3322 A to 3322 D, may be fully inhibited in its binding to the antigen by different concentrations of the competing cellobiose disaccharide. The apparent role of the cellobiose component of the polysaccharide in antibody binding varies considerably among these four antibodies. Cellobiose is most effective in inhibiting the binding of antibody A to the polysaccharide; it is less effective for B and C, and least effective for D.

When their primary structure is examined, it is apparent that each of the antibodies has a different amino acid sequence. This is particularly evident when the first complementarity region of the light chain is examined (Fig. 2, residues 29 to 33). Not only do variations in sequence occur (i.e., different amino acid residues occur in the same position in different polypeptide chains), but also variations in length of the complementarity region. Antibody A has the longest complementarity region, B an intermediate-sized region, and C the shortest. It is of singular interest that antibodies B and C, which appear to have binding characteristics

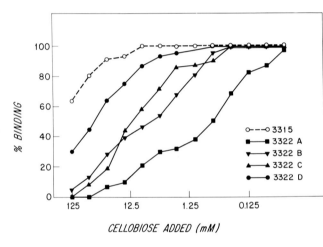

FIG. 1. Inhibition of binding of anti-S8 antibodies to [^{125}I]BGG-S8 by cellobiose. Percentage of binding is corrected by nonspecific binding and normalized to control binding. (From Chen et al., 1976, with permission.)

```
       0   1                                       10                                          20
3315   ALA GLN ILE VAL MET THR GLN THR PRO SER SER VAL SER ALA ALA VAL GLY GLY THR VAL THR ILE
3322A          ALA ——— LEU ————————————————————————————————————————————————————————————————————
3322B      ALA TYR ASP ————————————————————————————————————————————————————————————————————————
3322C      ASP VAL ———————————————————— ALA ——— PRO ———————————————————————————————————————————
3322D  BLOCKED AMINO-TERMINUS

                                               30  31  32  32A 32B 33                      40
3315   ASN CYS GLN SER SER GLN SER VAL TYR GLU ASN GLY ARG LEU SER TRP PHE GLN GLN LYS PRO
3322A  SER ——————— ALA ———————————————————— ASN ——— ASN ——————— ALA ——— TYR ———————————
3322B  LYS ——————— ALA ——————————— LEU THR ASN GLN TYR --- ——— ALA ——— TYR ———————————
3322C  ——————————— ALA ——————————— ILE GLY SER ASP --- --- ——— VAL ——— TYR ———————————————
```

FIG. 2. Amino-terminal sequence (positions 0 to 40) of rabbit anti-S8 antibody light chains. A *solid line* indicates that the sequence is identical to the top sequence (3315). A *broken line* indicates a deletion of that position. (From Chen et al., 1976, with permission.)

similar to cellobiose, manifest very different amino acid sequences within the complementarity-determining region.

Although limited amounts of monoclonal antibodies could be obtained by selective immunization with special antigens such as the pneumococcal or streptococcal polysaccharides, it was not possible to generalize this method to all antigens, nor could a single monoclonal antibody be reproduced once identified. The demonstration of Köhler and Milstein (1975) and Köhler et al. (1976) that somatic cell fusion may be carried out between normal antibody-secreting B cells from the spleen of an immunized donor and a selected myeloma tumor line has created a revolution in biology (Kennett et al., 1980; Haber et al., 1981).

Antibodies derived from a single clone isolated by somatic cell fusion have a unique primary structure and consequently a unique specificity of epitope recognition. The apparent immortality of the cell lines that produce these antibodies also implies that their properties are likely to be reproducible. The library of antibodies that may be examined as the result of somatic cell fusion is likely to be as large as that from conventional immunization, but the antibodies obtained are single molecular species rather than heterogeneous mixtures.

A typical experimental protocol (Köhler and Milstein, 1975) for a somatic cell fusion starts with immunizing mice with antigen or hapten-carrier conjugate. Balb/c mice are used almost exclusively since most of the myeloma cells used in fusions are also of Balb/c origin. Approximately 35 days after the first injection, the mice are boosted with a second injection of antigen; 5 days later the spleens are removed and prepared for the fusion. Spleen cells and myeloma cells (about 10^7 of each) are centrifuged together and resuspended in Dulbecco's Modified Eagle's Medium (DMEM), which had been made 50% in polyethyleneglycol 1500. The fusion occurs immediately and the cells are then incubated for 24 hr in the so-called RF medium containing 10% fetal-calf serum. After the incubation, the cells are transferred into a selective medium that contains hypoxanthine, aminopterine, and thymidine (HAT), which ensures that only the fused cells of a hybrid spleen-myeloma origin survive. Since the unfused spleen cells are unable to divide in the culture and die out spontaneously, there are only two kinds of perpetually growing cells, parental myelomas and spleen-myeloma hybrid cells. The myeloma clone used in fusion is defective in the metabolic pathway responsible for nucleotide and DNA synthesis. Whereas the wild-type cells possess two alternative synthetic pathways, the de novo pathway from sugars and amino acids and the so-called scavenger pathway that builds up nucleotides from existing nucleosides and externally supplied thymidine and hypoxanthine, the myeloma cells used for fusions are lacking the scavenger pathway. In the experiment, the de novo pathway is inhibited by aminopterine; the HAT medium selects against parent myelomas by inhibiting their de novo pathway while at the same time supplying thymidine and hypoxanthine for the scavenger pathway. Thus, only the fused hybridoma cells, which acquired the scavenger pathway genes from the spleen cells, eventually survive. Once the hybridoma cells are selected by their growth in the HAT medium, they can be screened for antibody production, and those that secrete antibody molecules can be propagated in cell cultures or grown as ascites in mice.

The molecular complexity of a mixture of hybridoma-produced monoclonal antibodies resembles that of a conventional antiserum. Structural diversity of a set of monoclonal antibody light chains specific for the simple hapten p-azophenylarsonate is shown in Fig. 3 (Margolies et al., 1981). It is apparent that the light chain variability is similar to that observed when a complex antiserum is fractionated (Fig. 2). The antibody polypeptide chains differ not only in amino acid sequence, but also in the size of the light chain complementarity-determining region examined.

The ability to select a single antibody that possesses the specificity and affinity of interest provides a powerful tool. Recent applications include (1) selection of antibodies that resolve

		10		20
Prototype	Id +	D I Q M T Q T T S S L S A S L G D R V	T	I S C
36 - 60	Id -	— V V —————— P L T ——— V T I — Q	P	A S ———
31 - 64	Id -	— V V —————— P L T ——— V I I — Q	P	A S ———
31 - 41	Id -	E N V L ——— S P A I M ——— P — E K		— M T —
45 - 112	Id -	————— S P ——————— E		— S L T —
45 - 165	Id -	—— V — S — S P ——— A V — A — E K		— M —
45 - 49	Id -			
44 - 1-3	Id -	Blocked		

		CDR1		
		30 34 a b c d e f	35	40
Prototype	Id +	R A S Q D I S N Y L N - - - - -	W Y Q Q K P	D G
36 - 60	Id -	K S —— R L L D S D G K T Y L N -	— L L — R —	G Q
31 - 64	Id -	K S —— S L L D S D G K T Y L S -	— L L — R —	G Q
31 - 41	Id -	——— S S V S S Y F - - - - -		
45 - 112	Id -	(-)—(-)— E —(-)(G)——(S) - - - - -	— L ——	
45 - 165	Id -	K S —— S L L — S R T R K N Y L T	—————	G Q
45 - 49	Id -	- - - - -		

FIG. 3. Amino acid sequences of murine A/J anti-*p*-azophenylarsonate idiotype negative monoclonal antibody light chains. Insertions in the first complementarity determining region (CDR1) are arbitrarily designated by letters carboxyl-terminal to position 34 and ending at the invariant tryptophan at position 35. (From Margolies et al., 1981, with permission.)

two similar molecules by virtue of specific recognition of limited structural differences; (2) a selection of antibodies that recognize common structures among different molecules; (3) selection of antibodies that bind a functionally important part of a molecule and inhibit biological activity as well as antibodies that bind but do not alter activity; and (4) the selection of an antibody to one component of a heterogeneous mixture of antigens to be used as an aid in fractionation (Cuello, et al., 1979; Berzofsky, 1980; Bundesen, et al., 1980; Ivanyi and Davies, 1980).

RESOLVING SIMILAR MOLECULES

The hallmark of cell death is the loss of integrity of the cell membrane, which defines the unique intracellular environment. Leakage of intracellular proteins into the extracellular fluid and then into the plasma has provided a marker for tissue necrosis. Because these enzymes are shared by many tissues, specific localization of the organ affected is not possible. It would be highly desirable, for example, to identify an intracellular marker for cardiac damage to aid in the diagnosis of myocardial infarction. The cardiac myocyte shares most intracellular proteins with the cells of other tissues, but light chains of cardiac myosin have a unique structure (Sarkar et al., 1971). Cardiac myosin differs in amino acid sequence from myosin of skeletal or smooth muscle. Measurement of cardiac myosin light chain concentration should provide a useful diagnostic marker in myocardial infarction (Khaw et al., 1978). Plasma radioimmunoassays, however, lack specificity in that antisera invariably cross-react with skeletal or smooth muscle light chains, probably because of common determinants among these molecules. For specific diagnosis, an antibody that recognizes only unique determinants of cardiac myosin light chains is essential. Monoclonal antibodies resulting from somatic cell fusion provide the requisite specificity (Katus et al., 1980).

Balb/c mice were immunized with a mixture of human cardiac myosin light chains 1 and 2. Three days after booster immunization, the spleen cells were fused with cells from the NS-1 myeloma line, and the hybrids were screened for antibody production. As depicted in Fig. 4, three types of antibodies were found that differed in their ability to distinguish the three myosin light chains. (Ehrlich et al., 1979; Zurawski et al., 1979). Antibody 3C7 typifies the most common type. There is complete identity of reactivity between cardiac and skeletal myosin light chains, indicating recognition of common antigenic determinants. Antibody 2B2 is representative of a less frequent type. These monoclonal antibodies demonstrate a partial cross-reactivity between skeletal and smooth muscle light chains. Perhaps only a part of a common antigenic determinant is recognized. The least common monoclonal response is typified by the antibody 1E6. This antibody shows no evidence of cross-reactivity between skeletal and smooth muscle light chains, and thus recognizes the unique determinants of the cardiac light chain. It can provide the basis for the development of an entirely specific immunoassay for cardiac cell damage.

Specificity of recognition of an antigen may be enhanced further by using two monoclonal antibodies each specific for a different epitope on the antigen's surface (Hurrell et al., 1981; Katus et al., 1982). The principle of such a "sandwich" type of radioimmunoassay is illustrated in Fig. 5 (Katus et al., 1979, 1980; Haber et al., 1982). Monoclonal antibody A is immobilized on a solid support such as a Sepharose bead. A standard procedure for antibody immobilization involves activation of Sepharose with cyanogen bromide and reaction of activated Sepharose

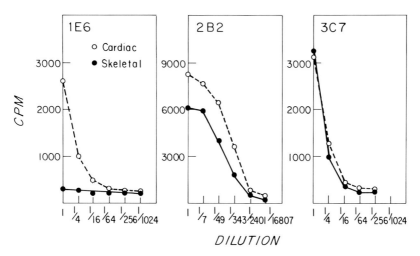

FIG. 4. Anti-myosin light chain hybridoma culture supernatants. Binding of hybridoma antibodies to cardiac or skeletal muscle light chains immobilized on plastic microtiter plates. Dilution of the cell culture supernatant is plotted against counts per minute of [^{125}I]goat antimouse Fab antibody. (From Haber et al., 1981, with permission.)

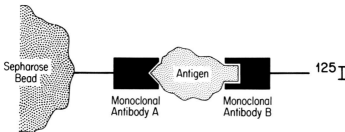

FIG. 5. Schematic representation of sandwich assay employing two different monoclonal antibodies specific for different epitopes on the same antigen. (From Haber et al., 1981, with permission.)

with antibody under mildly alkaline conditions in the cold. When an antigen solution is mixed with Sepharose-bound antibodies, all antigen molecules that possess the epitope binding to antibody A are bound, whereas antigen molecules not possessing the appropriate determinant do not bind. Monoclonal antibody B, specific for a different epitope on the molecule, may be labeled with radioisotope. It should be apparent that the labeled antibody will bind only to those immobilized antigen molecules that bear the appropriate determinant. Thus, after washing procedures, the radioactivity bound to the beads represents only the presence of those antigen molecules that are recognized both by monoclonal antibodies A and B.

If this hypothetical construct is correct, then the exposure of antigen bound by monoclonal antibody A to radiolabeled monoclonal antibody A should result in no specific binding because of steric hindrance. Figure 6 provides an example. Two monoclonal antibodies specific for different epitopes on the myoglobin molecule were selected (Katus et al., 1979, 1980; Hurrell et al., 1981). Antibody 1B8 was immobilized on Sepharose, and antibody 1F10 was labeled with radioactive isotope ^{125}I. Antibody 1B8 bound myoglobin, which in turn bound radioiodinated antibody 1F10. Increasing concentrations of myoglobin effectively displaced antibody 1F10 from the Sepharose particles; however, if the myoglobin bound to antibody 1B8 on the Sepharose beads is exposed to labeled antibody 1B8, no binding of the label can be demonstrated, presumably because the epitope recognized by 1B8 is fully masked by the immobilized 1B8.

Table 1 shows how this principle may be applied to enhancing the specificity of cross-reactive antibodies (Katus et al., 1982). Three cardiac-myosin light chain antibodies are examined.

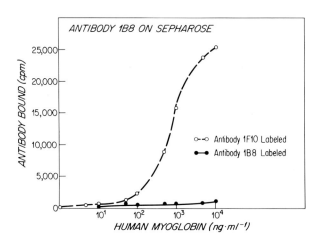

FIG. 6. Sandwich assay employing two monoclonal antibodies specific for different epitopes on myoglobin. Concentration of myoglobin is plotted against counts per minute of the respective [^{125}I]antibody bound to Sepharose. (From Haber et al., 1981, with permission.)

TABLE 1. *Assay of myosin light chains*

Antibody	Cross reactivities between skeletal and cardiac myosin light chains (%)
Immunoassay (single antibodies)	
1C5	100
4F10	17
2B9	25
Sandwich assay (two antibodies)	
1C5 + 2B9	25
1C5 + 4F10	17
2B9 + 4F10	4

Antibody 1C5 is completely cross-reactive with skeletal-muscle light chains and is similar in properties to 3C7 in Fig. 4. The antibodies 4F10 and 2B9 display only partial cross-reactivity between the two different light chains, similar to 2B2 in Fig. 4. Theoretical considerations imply that by varying combinations among these antibodies one may obtain a greater selectivity than through the use of a single antibody. The experiment shows that the combination of a fully cross-reactive antibody with one that is partially cross-reactive does not enhance specificity. When two partially cross-reactive antibodies specific for different epitopes are combined, however, a decrease in cross-reactivity results. Thus, the combination between 2B9 and 4F10 (antibodies showing 19 and 30% cross-reactivity, respectively) in the type of sandwich assay illustrated in Fig. 5 results in a final effective cross-reactivity of only 3%.

RECOGNIZING COMMON EPITOPES

At times it is important to select monoclonal antibodies that recognize common epitopes on structurally different molecules. An illustration is the desirability of obtaining an antibody for the treatment of intoxication with the digitalis glycosides. The pharmacologic activity of digoxin and digitoxin, the two most commonly used drugs of this type, resides in the aglycone steroid portion of the molecule; the glycosidic residues modify the degree of water solubility, distribution *in vivo*, and potency. Both drugs exert positive ionotropic and electrophysiologic effects on the heart; digitalis toxicity in hospitalized patients is one of the most prevalent adverse drug reactions seen in clinical practice, and this high frequency of life-threatening digoxin toxicity relates to the narrow margin between the therapeutic and toxic doses of the digitalis glycosides. Digoxin and digitoxin differ very little in their chemical structure (digoxin possesses a single hydroxyl group at position 12 of the C ring in the steroid nucleus that is not present in digitoxin), but most of the elicited antibodies tend to differentiate between these molecules, and binding affinities may vary by as much as a factor of 50 (Smith et al., 1970). Specific antibody Fab (antigen-binding) fragments have been used effectively in the reversal of overwhelming digoxin intoxication (Smith et al., 1976). It would be convenient to have a reagent that was equally effective in the reversal of digitoxin intoxication. A hybridoma protein has been selected that has both the appropriate affinity ($K_a = 5 \times 10^9$) as well as specificity (it does not differentiate between digoxin and digitoxin) to be used interchangeably in treating either digoxin or digitoxin toxicity (Mudgett-Hunter et al., 1980). The affinity is also sufficient for the antibody to be used either in the digoxin or digitoxin plasma radioimmunoassay.

DIFFERENTIATION BETWEEN ACTIVE AND INACTIVE SITES ON A PROTEIN

An antitoxin (i.e., an antibody molecule that specifically inactivates the biological effects of a toxin) should be most effective when it is directed at the site on the toxin molecule that is essential for its biological effect. Two monoclonal antibodies that have similar affinities for tetanus toxin are 2B8B3 with a K_a of 3.5×10^8 and 1E3D8 with a K_a of 5.4×10^8 (Zurawski et al., 1980). Although both molecules bind the toxin, only one behaves as an antitoxin in the *in vivo* mouse toxicity test. As can be seen from Table 2, antibody 1E3D8 failed to protect mice against the tetanus toxin over a wide range of toxin doses. All the animals died within 2 days. On the other hand, antibody 2B8B3 resulted in complete survival at the two lower doses of toxin and survival, but paralysis, at an intermediate dose. At higher toxin doses, the antitoxin effect was overcome. It is conceivable that an antibody directed against the same epitope as 2B8B3, but having a higher binding activity for the toxin, would be protective even at high toxin doses, especially if the antibody activity were considerably higher than that of the toxin to a putative toxin receptor. It would appear that antibody 2B8B3 is specific for that part of

TABLE 2. *Toxin neutralization in vivo by ascites fluid containing monoclonal antibodies*[a]

FDA toxin units per ml ascites	Mouse survival (days)	
	Clone 2B8B3	Clone 1E3D8
2.6	2	2
0.64	4/T	2
0.16	T	2
0.04	S	2
0.01	S	2

[a]S, survival; T, survival with partial paralysis.

the toxin molecule required for binding to target cells, whereas antibody 1E3D8, although capable of binding to the toxin molecule, cannot impair its biological activity. Thus, although these antibodies have a nearly identical association constant for tetanus toxin, only one has specificity for the epitope associated with toxin action.

BINDING A SINGLE COMPONENT IN A MIXTURE

The antibody molecule should provide an exceedingly selective vehicle for affinity chromatography in the course of protein fraction. However, prior to the advent of monoclonal antibodies, a purified protein was required before a monospecific antibody could be elicited. The somatic cell fusion technique now permits immunization with mixed antigens and the subsequent selection of clones that will bind only one of them. The principle is illustrated in Fig. 7. To purify antigen 2, a mouse is immunized with a mixture of antigen 1, 2, 3, ... n. After somatic cell fusion has occurred and the resultant hybridomas cloned, cell culture supernatants are screened for the ability to inhibit the biological activity of antigen 2. When an appropriate clone is found, it is amplified by reintroduction as an ascites-producing tumor *in vivo*. The antibody specific for antigen 2 is recovered in large amounts from the ascites fluid of hybridoma-injected mice. (Typically, antibody concentration in the ascites fluid ranges from 2 to 4 mg/ml, and virtually pure antibody preparation can be obtained by passing the ascites fluid through an ion-exchange column of DEAE-cellulose.) Antibody is then immobilized on an affinity chromatography support. The mixture of antigen 1, 2, 3, ... n is passed over this affinity column. Only antigen 2 will bind; the others are recovered in the void volume.

This general approach has been used to purify Mullerian-inhibiting substance (Mudgett-Hunter et al., 1981), an embryonic regulator essential in sexual differentiation that has been exceedingly difficult to purify. Following secretion by the embryonic testis, this factor causes atrophy of the Mullerian duct, defining the embryo as a male. After five stages of conventional fractionation, at least five proteins remain. This mixture was used to immunize Balb/c mice. After somatic cell fusion, the supernatants of hybrid-containing wells were examined for their ability to inhibit the biological activity of a partially purified preparation of Mullerian-inhibiting substance. Of 384 hybrids obtained in a fusion experiment with the NS-1 myeloma line, 26 were capable of binding [^{125}I] mixed-protein antigen (all the protein components of the partially purified mixture). One of these wells was selected (1D1) and subjected to diluting cloning. Of 96 resultant wells, 7 had supernatants that were capable of binding the labeled mixed antigen, but only one of these wells (1B2) both bound labeled antigen and inhibited the activity of the embryonic factor. This clone was amplified by the ascites method to produce sufficient quantities of antibody for affinity chromatography. A Sepharose-antibody 1B2 column fully bound Mul-

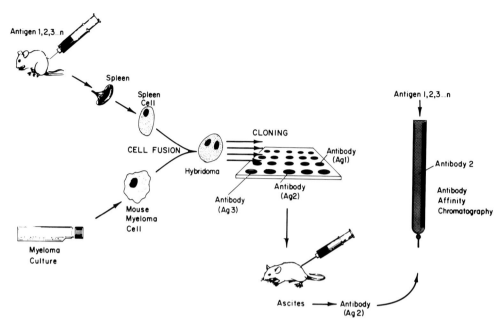

FIG. 7. Schematic representation of use of the hybridoma technique to isolate a single component from a mixture. The criterion for clonal selection is the ability of culture fluid to inhibit the biological activity of the antigen of interest. *In vivo* amplification of the clone in syngeneic mice is needed to obtain sufficient antibody for affinity chromatography. (From Haber et al., 1981, with permission.)

lerian-inhibiting substance from the partially purified mixture. Elution with a chaotropic agent, potassium thiocyanate, resulted in recovery of activity in the column eluate. Alternatively, to reduce the number of subclones assayed, the capacity of a mixed hybridoma culture to block biological activity could be assessed prior to subcloning.

The successful application of this approach requires that the protein of interest have a measurable biological property that is inhibitable by an antibody. The protein to be purified should also represent a significant fraction of the mixture, either with respect to mass or to immunodominance. It is unlikely, for example, to expect to find specific antibody-producing clones if the antigen in question represents less than 1% of the mixture.

Somatic cell fusion of antibody-producing cells and the use of the monoclonal antibodies produced permit innovative approaches to difficult problems in biological research. Antibodies need no longer be considered mixtures of varying properties but may now be obtained as defined molecular entities. The enormous diversity of the immune response may still be used, however, in selecting precisely the antibody that has the properties of interest. It has been demonstrated here how that specificity may be employed to resolve very similar molecules, to recognize different epitopes on the surface of a single molecule, and to select molecules from a mixture. These are but a few of the many applications that the hybridoma technique is likely to have in biological research.

ACKNOWLEDGMENTS

This work was supported in part by an NIH Digoxin Program Project grant (HL-19259 and USPHS grant HL-26215 Ischemia SCOR. This review includes work by the following researchers: Michael N. Margolies, Meredith Mudgett-Hunter, Ban-An Khaw, Hugo A. Katus, John G. Hurrell, Vincent R. Zurawski, Jr., and Patricia K. Donahoe.

REFERENCES

Berzofsky, J. A., Hicks, G., Fedorko, J., and Minna, J. (1980): Properties of monoclonal antibodies specific for determinants of a protein antigen, myoglobin. *J. Biol. Chem.*, 255:11188–11191.

Bundesen, G., Drake, R. G., Kelly, K., Worsley, I. G., Friesen, H. G., and Sehon, A. H. (1980): Radioimmunoassay for human growth hormone using monoclonal antibodies. *J. Clin. Endocrinol, Metab.*, 51:1472–1474.

Chen, F. W., Cannon, L. E., Margolies, M. N., Strosberg, A. D., and Haber, E. (1976): Purification specificity and hypervariable region sequence of antipneumococcal polysaccharide antibody elicited in a single rabbit. *J. Immunol.*, 117:807–813.

Cuello, A. C., Galfre, G., and Milstein, C. (1979): Detection of substance P in the central nervous system by a monoclonal antibody. *Proc. Natl. Acad. Sci. USA*, 76:3532–3536.

Ehrlich, P. H., Zurawski, V. R., Jr., Khaw, B. A., and Haber, E. (1979): Hybridoma antibodies against human cardiac myosin. (Abstract) *Circulation*, (Suppl.II), 59–60:139.

Haber, E. (1970): Antibodies of restricted heterogeneity for structural study. *Fed. Proc.*, 29:66–71.

Haber, E., Donahoe, P., Ehrlich, P., Hurrell, J., Katus, H., Khaw, B. A., Margolies, M. N., Mudgett-Hunter, M., and Zurawski, V. R. (1981): Resolving antigenic sites and purifying proteins with monoclonal antibodies. In: *Monoclonal Antibodies in Endocrine Research*, edited by R. E. Fellows and G. Eisenbarth. pp. 1–11. Raven Press, New York.

Haber, E., Katus, H. A., Hurrell, J., Matsueda, G. R., Ehrlich, P. H., Zurawski, V., Jr., and Khaw, B. A. (1982): Monoclonal antibodies specific for cardiac myosin: In vivo and in vitro diagnostic tools in myocardial infarction. In: *Monoclonal Hybridoma Antibodies: Techniques and Applications.* CRC Uniscience, *(in press)*.

Hurrell, J. G., Katus, H. A., Haber, E., and Zurawski, V. R. (1981): Monoclonal antibodies directed against human myoglobin: Characterization and applications in a bideterminant radioimmunoassay. *J. Immunol. Methods*, 45:249–254.

Ivanyi, J., and Davies, P. (1980): Monoclonal antibodies against human growth hormone. *Mol. Immunol.*, 17:287–290.

Katus, H. A., Hurrell, J., Ehrlich, P., Zurawski, V., Khaw, B. A., Bahar, I., and Haber, E. (1980): Radioimmunoassay for human cardiac light chains using monoclonal light chain antibodies. *Circulation*, (Suppl.III), 62:216.

Katus, H. A., Hurrell, J., Matsueda, G., Ehrlich, P., Zurawski, V., Khaw, B. A., and Haber, E. (1982): Increased specificity in human cardiac myosin radioimmunoassay utilizing two monoclonal antibodies in a double sandwich assay. *Mol. Immunol.*, 19:451–455.

Katus, H. A., Khaw, B. A., Misuzawa, E., Gold, H., and Haber, E. (1979): Circulating cardiac myosin light chains in myocardial infarction: Detection by radioimmunoassay. *Circulation*, 139(Suppl.II):59–60.

Kennett, R. H., McKearn, T. J., and Bechtol, K. B., eds. (1980): *Monoclonal Antibodies.* Plenum Press, New York.

Khaw, B. A., Gold, H. K., Fallon, J. T., and Haber, E. (1978): Detection of serum cardiac myosin light chains in acute experimental myocardial infarction: Radioimmunoassay of cardiac myosin light chains. *Circulation*, 58:1130–1136.

Köhler, G., Howe, C. S., and Milstein, C. (1976): Fusion between immunoglobulin secreting and nonsecreting myeloma cell lines. *Eur. J. Immunol.*, 6:292–295.

Köhler, G., and Milstein, C. (1975): Continuous cultures of fused cells secreting antibody of predefined specificity. *Nature*, 256:495–497.

Margolies, M. N., Marshak-Rothstein, A., and Gefter, M. L. (1981): Structural diversity among anti-p-azophenylarsonate monoclonal antibodies from A/J mice: Comparison of Id^- and Id^+ sequences. *Mol. Immunol.*, 18:1065–1077.

Mudgett-Hunter, M., Budzik, G. P., Sullivan, M. D., and Donahoe, P. K. (1981): Monoclonal antibody to Mullerian-inhibiting substance. *Fed. Proc.*, 40:995.

Mudgett-Hunter, M., Margolies, M. N., Rosen, E. M., and Haber, E. (1980): Hybridoma antibodies to the cardiac glycoside digoxin. *Fed. Proc.*, 39:928.

Sarkar, S., Sreter, F. A., and Gergely, J. (1971): Light chains of myosins from white, red and cardiac muscles. *Proc. Natl. Acad. Sci. USA*, 68:946–950.

Smith, T. W., Butler, V. P., and Haber, E. (1970): Characterization of antibodies of high affinity and specificity for the digitalis glycoside digoxin. *Biochemistry*, 9:331–336.

Smith, T. W., Haber, E., Yaetman, L., and Butler, V. P. (1976): Reversal of advanced digoxin intoxication with Fab fragments of digoxin-specific antibodies. *N. Engl. J. Med.*, 294:797–800.

Zurawski, V. R., Jr. Ehrlich, P. H., Khaw, B. A., Katus, H., and Haber, E. (1979): Monoclonal antibodies to human cardiac myosin light chains: Potential for improvement of a radioimmunometric assay for myocardial infarction. *Circulation*, (Suppl.II), 59–60:12.

Zurawski, V. R., Hurrell, J. G. R., Latham, W. C., Black, P. H., and Haber, E. (1980): Monoclonal antibodies to tetanus toxin produced by clones of murine hybridomas. *Fed. Proc.*, 39:1204.

Molecular Genetic Neuroscience, edited by
F. O. Schmitt, S. J. Bird, and F. E. Bloom.
Raven Press, New York © 1982.

Analysis of Retina and Other Neural Tissues Using Cell-Specific Antibodies

Colin J. Barnstable

ABSTRACT

Monoclonal antibodies have been produced that interact specifically with photoreceptors, Müller cells, or ganglion cells in rat retina. These antibodies were defined by their patterns of reactivity on frozen sections of adult tissue. In addition, the patterns of antigen development were studied by examining tissues from animals at various postnatal ages. Some of the antibodies have been used to label cells dissociated from adult or juvenile retinae, thus opening up the possibility of isolating and working with purified cell populations *in vitro*.

The identification of specific cell populations within the brain is one of the primary goals of neurobiology. To unravel the synaptic organization of the brain, the structure and function of the interconnected neurons must be determined. Until recently, neurobiologists were dependent on the morphological approaches developed by Ramón y Cajal and the histologists of the nineteenth century. In a few cases, such as the histofluorescence technique used to identify catecholamine-containing cells, neurons and their processes were identified by their neurotransmitter content. However, the discovery that many neuropeptides are widely distributed in the central and peripheral nervous systems has led to a dramatic increase in our knowledge of neuronal organization. The use of specific antibodies against neuropeptides and other neuronal products and the development of sensitive immunocytochemical procedures has enabled us to define a large number of different neuronal cell types by structure and function. The application of the monoclonal antibody technique to the study of nerve cell organization and development provides a powerful tool for the identification of cell types since it is not dependent on any known property or compound.

Monoclonal antibodies have been used with great success to define functional populations of cells in other systems, particularly the immune system (Melchers et al., 1978), transplantation antigens (Barnstable et al., 1978; Brodsky et al., 1979), and tumor immunology (Hellstrom et al., 1980). They provide homogeneous reagents that specifically and reproducibly detect particular antigenic determinants. Furthermore, monoclonal antibodies lack some of the disadvantages of conventional antibodies, particularly their heterogeneity and cross-reactivity, and they can be raised against virtually any antigen, although the production of cell-specific antibodies depends ultimately on the existence of antigenic differences in different cell types rather than on the method of producing the antibodies. The assumption that different cell types do possess antigenic differences, however, is reasonable and is borne out by the experiments reported here. The antigens that might define different neuronal cell populations would not necessarily be related to the transmitter function of the cells and might, in fact, define a functional organization of the nervous system that would underlie the more obvious organization based on neurotransmitter content. The experimental uses of antibodies are varied; they can be used to define the significance of their antigens in terms of distribution, function, and developmental history prior to further characterization of the molecules by the molecular genetic procedures described throughout this volume.

The studies reported here were designed to answer several questions:

1. Is it possible to make antibodies that recognize discrete cell populations in any part of the central nervous system? This question is obviously crucial to the whole approach.
2. Can such antibodies be used to label dissociated or isolated cells? It is likely that, at first, few antibodies will be directed against molecules of functional interest, such as those involved in cell-cell recognition, which are probably in low concentration or on only a few cells. Antibodies recognizing a specific population of cells could be used to isolate those cells, which would then be used to produce further antibodies in a second round of immunization.
3. Can the antibodies be used to study the development of the nervous system *in vivo* and *in vitro*? Each antibody recognizes a single determinant that can be followed during development, avoiding complex physiological or morphological measurements. The comparison of development *in vivo* and *in vitro* may define cell interactions or factors that are involved in specific developmental events.

The system chosen for study was the mammalian visual system and, in particular, the adult rat retina. The retina has the advantage of having only seven basic neuronal cell types, several of which have well characterized synaptic connections and physiological responses. Furthermore, the retina has a layered structure (Fig. 1), enabling cell types to be identified from their position, and it can be readily dissociated into single cells for physiological and biochemical studies.

For the production of antibodies, mice were immunized with a crude membrane preparation from adult rat retina and their spleen cells fused with a mouse myeloma cell line (for details

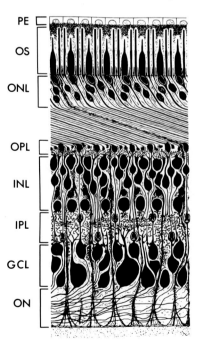

FIG. 1. Diagram of human retina. PE, pigment epithelium; OS, outer and inner segments of photoreceptors; ONL, outer nuclear layer; OPL, outer plexiform layer; INL, inner nuclear layer; IPL, inner plexiform layer; GCL, ganglion cell layer; ON, optic nerve fibers. (Adapted from Polyak, 1941.)

of the procedure, see Barnstable, 1980, 1982). A membrane preparation rather than a whole cell homogenate was deliberately chosen to bias the production of antibodies toward those directed against cell-surface components. The clones were tested for antibody activity against a homogenized preparation of rat retina, using a radioactive binding assay. The procedure used to produce the target homogenate probably enriched this preparation with photoreceptor and glial cells, with a loss of some of the neuronal material. The selection of positive hybrids was therefore probably biased toward those directed against glia or photoreceptor cells. The antibodies chosen for further study appeared to be relatively specific for neural tissue, that is, they did not react with either rat thymocytes or lung fibroblasts. The specificity of these antibodies was tested further by indirect fluorescence immunocytochemistry using sections of lightly fixed (1% paraformaldehyde, 0.1% glutaraldehyde) adult rat retina.

Three monoclonal antibodies specifically stained Müller cells, the glial cell peculiar to the retina. These cells are the only cell type to extend radially across the retina (Fig. 2), from the outer limiting membrane, where they end in microvillus-like processes at the level of the inner segment of the photoreceptors, to the inner limiting membrane where the cells extend bulbous

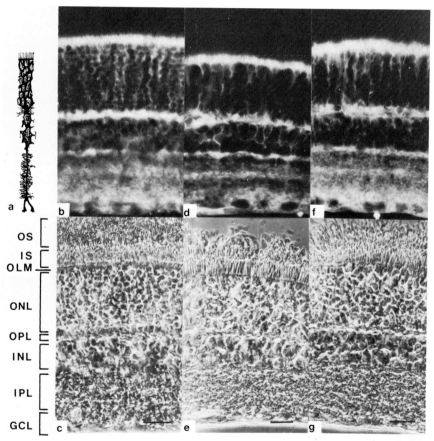

FIG. 2. Immunofluorescent labeling of sections of adult rat retina by antibodies RET-G1, RET-G2, and RET-G3. **a:** a Müller cell taken from a Golgi-stained retina as drawn by Cajal (1955); **b,c:** RET-G1 antibody; **d,e:** RET-G2 antibody; **f,g:** RET-G3 antibody; **b,d,f:** fluorescence; **c,e,g:** phase contrast. OS, outer segments; IS, inner segments; OLM, outer limiting membrane; ONL, outer nuclear layer; OPL, outer plexiform layer; INL, inner nuclear layer; IPL, inner plexiform layer; GCL, ganglion cell layer. Scale bar represents 20 μm. (From Barnstable, 1980.)

feet around the ganglion cell bodies. In both the inner and outer plexiform layers Müller cells send out very fine tangential processes. The immunofluorescent staining patterns of the three antibodies—RET-G1, RET-G2, and RET-G3—are shown in Fig. 2 and reflect these morphological features. The triple band of staining in the inner plexiform layer was similar to that seen by autoradiographic staining of Müller cells following [^{3}H]γ-aminobutyric acid uptake by isolated retinae in vitro (Barnstable, 1982). Antibody RET-G1 also stained isolated Müller cells after enzymatic dissociation of the retina (Fig. 3): only cells with the morphology of Müller cells were stained; other cell types were unstained.

Although the staining patterns of the three anti-glial antibodies appeared to be roughly similar, there were marked differences in the specificities of the antibodies. Antibody RET-G1 reacted with glial cells in all regions of the brain, and in fact, in a binding assay, cortex and cerebellum were found to express more of the RET-G1 antigen per milligram of membrane protein than retina. This indicates, as might be expected, that Müller cells are related to other glial cell types. In contrast, antibodies RET-G2 and RET-G3 reacted only with Müller cells, although there were fine differences in the patterns of staining. Antibody RET-G3 gave a punctate pattern in the inner plexiform layer, whereas antibody RET-G2 gave even labeling. None of the anti-glial antibodies reacted with Müller cells in animals other than rat.

Developmental differences in the staining patterns of Müller cells were also observed. In a retina of an 8-day-old rat, antibody RET-G2 stained the outer limiting membrane and Müller fibers in the outer nuclear layer and ganglion cell layer. In addition, stained Müller cell bodies, which are very difficult to detect in the adult, were visible in the inner nuclear layer. There was very little staining, however, in the inner plexiform layer. In contrast, antibody RET-G1

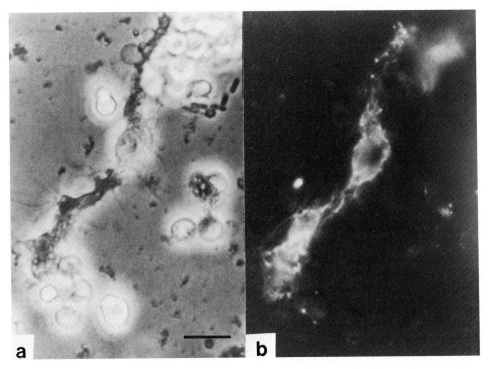

FIG. 3. Immunofluorescent labeling of enzymatically dissociated Müller cells by RET-G1 antibody. **a:** phase contrast; **b:** fluorescence. (From Barnstable, 1980.)

stained retinae of the same age much more strongly in the inner plexiform layer than in the other layers. At this stage of development, the inner plexiform layer is a region of active synaptogenesis. In monolayer cultures of neonatal retina, antibody RET-G1 gave bright spots of specific labeling; these spots coincide with branch points in putative neuronal processes that occur over underlying glial cells. Thus, the data from both *in vivo* and *in vitro* experiments suggest that antigen RET-G1 may be involved in some way with neuronal process formation.

Three of the monoclonal antibodies—RET-P1, RET-P2, and RET-P3—labeled photoreceptors in the adult rat retina (Fig. 4). The patterns of staining, however, were strikingly different for each antibody. For example, antibody RET-P1 stained photoreceptor outer segments, inner segments, and cell bodies. There was clear ring fluorescence around most of the cell bodies in the outer nuclear layer, a staining pattern that was quite different from that of the anti-Müller cell antibodies in this layer (Fig. 2). Antibody RET-P2, however, labels only photoreceptor outer segments and antibody RET-P3 labels only photoreceptor cell bodies. In both cases, no other parts of the photoreceptor were labeled. These results clearly indicate that particular cell

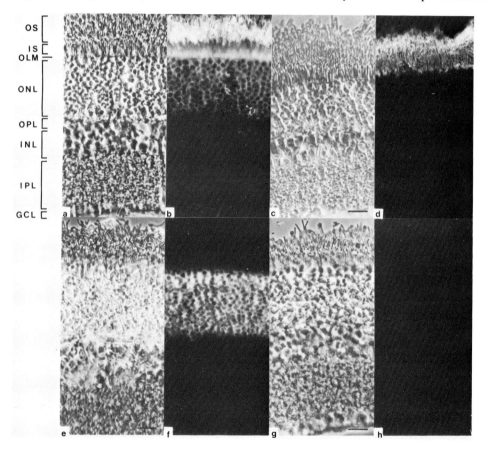

FIG. 4. Immunofluorescent labeling of sections of adult rat retina by antibodies RET-P1, RET-P2, and RET-P3. **a,b:** RET-P1 antibody; **c,d:** RET-P2; **e,f:** RET-P3; **g,h:** control using a monoclonal antibody against a human cell-surface glycoprotein. **a,c,e,g:** phase contrast; **b,d,f,h:** fluorescence. OS, outer segments; IS, inner segments; OLM, outer limiting membrane; ONL, outer nuclear layer; OPL, outer plexiform layer; INL, inner nuclear layer; IPL, inner plexiform layer; GCL, ganglion cell layer. Scale bar represents 20 μm. (From Barnstable, 1980.)

membrane components are confined to discrete regions of the cell membrane. There must, therefore, be mechanisms for directing newly synthesized components to particular regions of the cell surface.

All three antibodies against photoreceptors were specific for the retina, as tested by inhibition of a binding assay using appropriate tissue homogenates (Barnstable, 1980). Both antibodies RET-P2 and RET-P3 were specific for the retina of the rat and related species such as rabbit, whereas reactivity with antibody RET-P1 was widely distributed in vertebrates. This provided the opportunity to test whether antibody RET-P1 was specific for rods or cones, an experiment not possible in the rat retina because it consists almost entirely of rods. When sections of retina from the tiger salamander, which has both rods and cones, were stained with this antibody (Fig. 5), only the rods were stained, indicating that antibody RET-P1 is specific for rods. In the salamander, the entire cell body was labeled, including the axons, whereas in the rat the outer plexiform layer was unlabeled. The reason for this difference is unclear; it may be an artifact of antibody penetration or represent a real difference in antigen compartmentalization.

In order for monoclonal antibodies to be valuable reagents for defining the detailed structures of cells, it is essential that they can be used at the electron microscopic level. Unfortunately, considerable development in the preservation of antigenicity and the integrity of tissue sections is still required for immunocytochemistry to be a general technique for electron microscopy independent of the manner in which the antibodies were prepared. Some progress has been made using the antibody RET-P1. Electron micrographs (Fig. 6) using ferritin labeling clearly showed labeling on the outer surface of the plasma membrane of the outer segment. Where the outer segment was broken during processing, staining of the inner face of the disks of the outer segment was also visible with antibody RET-P1. The free floating disks of the photo-

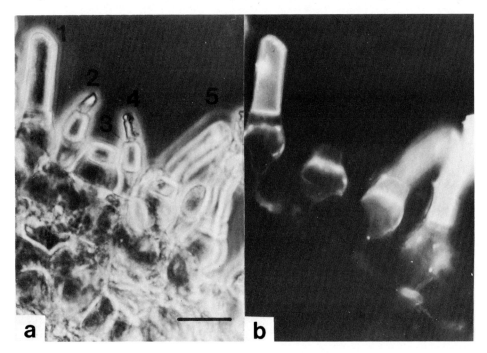

FIG. 5. Immunofluorescent labeling of tiger salamander photoreceptors by antibody RET-P1. Sections were prepared from retinas isolated from dark-adapted animals. **a:** phase contrast; **b:** fluorescence. Cells labeled 1, 3, and 5 are rods; 2 and 4 are cones. Scale bar represents 20 μm. (From Barnstable, 1980.)

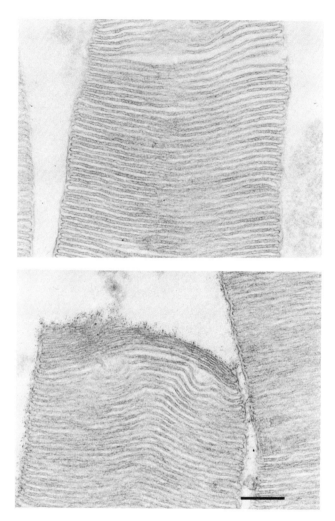

FIG. 6. Labeling of adult rat photoreceptors by antibody RET-P1 using ferritin conjugated second antibody. **Top:** control outer segment with buffer substituted for RET-P1. **Bottom:** outer segments showing ferritin grains outside the plasma membranes. The broken edge at bottom shows ferritin grains adjacent to an inside face of a disk. Scale bar represents 250 μm. (From Fekete and Barnstable, 1982.)

receptor outer segment are formed from invaginations of the plasma membrane. The inner surface of the disk is therefore the topological equivalent of the outer surface of the plasma membrane.

Developmental studies were also carried out with the antibodies against photoreceptors. In the retinae of 1-day-old rats, the cellular organization is quite different from the adult. Antibody RET-P1 stained a population of cells on the ventricular surface of the retina (Barnstable, 1981). The cells were bipolar and sent one process to the middle of the retina and the other to the ventricular surface ending in a brightly labeled structure. The number of cells staining with antibody RET-P1 in this layer increased with time and by 8 days after birth the photoreceptor cell layer had filled in and had begun to assume its adult appearance. There were, however, some cell bodies staining with antibody RET-P1, which still remained in the other layers of the retina. Staining with antibody RET-P2, which labels outer segments, first appeared at day 5 to 6 after birth, at the stage when the outer segments were starting to form and just before the first detectable synthesis of the visual pigment, opsin. Staining with antibody RET-P3, which stains cell bodies, did not appear until retinal development was almost complete, at day 13 to 14 after birth, when the eyes were about to open. Initial studies of neonatal retinal cells in culture indicated that there were cells present that stained with antibody RET-P1. Some of these showed the expected morphology of immature photoreceptors, but these did not mature further during a brief culture period.

A third type of monoclonal antibody was made against material from cerebral cortex. This antibody reacted with many cells of unknown type in cortex and in cerebellum. In sections of adult retina, only the inner plexiform layer and most, but not all, cell bodies in the ganglion cell layer were labeled. There was no staining of cell bodies in the inner nuclear cell layer. This antibody can be used to label dissociated ganglion cells as shown in Fig. 7. To determine the nature of the cell bodies stained by this antibody, a fluorescent dye, granular blue, was injected into the superior colliculus of a rat. Only the ganglion cells in the retina make direct connections, via the optic nerve, with the superior colliculus. After a sufficient period of time for the dye to reach the retina by retrograde transport, the retina was removed. Dissociated cells from the retina were labeled with the antibody and examined in the fluorescence micro-

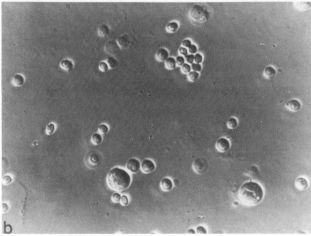

FIG. 7. Labeling of dissociated ganglion cells from a retina of a 10-day-old rat. Only a small percentage of cells, all of which were in the larger size class, were labeled. **a:** Fluorescence; **b:** phase contrast.

scope. Cells containing the transported dye also stained with the antibody, indicating that this antibody is indeed specific for ganglion cells (C. J. Barnstable and U. Drager, *unpublished results*). The unlabeled cell bodies in the ganglion cell layer seen in tissue section are probably displaced amacrine cells.

The data described here indicate that it is possible to raise monoclonal antibodies that recognize discrete cell populations in a nervous system. This has also been clearly shown in other systems, particularly in the leech where a variety of monoclonal antibodies were produced recognizing different individual cells or groups of cells (Zipser and McKay, 1981). In addition, monoclonal antibodies have been produced against retinal cells of the chicken (Eisenbarth et al., 1979). One such antibody bound to an antigen was distributed in a topographic gradient in the retina (Trisler et al., 1981). In mammals, monoclonal antibodies have been generated against a number of neuronal antigens, including components of the synapse. (For a review of this field, see McKay et al., 1981.) In addition, there are monoclonal antibodies that react specifically with central neurons (Cohen and Selvendran, 1981) or peripheral neurons (Vulliamy et al., 1981).

The monoclonal antibodies described here can also be used in a variety of experimental approaches to define the possible function and developmental history of the antigens that they detect. For example, the antibodies have been shown to stain dissociated cells. This will enable these cells to be isolated either for physiological studies or for further rounds of immunization and antibody production. These antibodies can also be used to monitor the development of specific cell types *in vivo* or *in vitro*. It is clear from the experiments described here that different antigens recognized by monoclonal antibodies, even on the same cell type, have quite different developmental histories. In future studies, however, it may be necessary to produce monoclonal antibodies against immature tissue rather than adult tissue to detect antigens present at earlier stages of development. Furthermore, the antibodies can be used in the biochemical characterization of the antigens, either directly by immunoprecipitation and affinity chromatography or indirectly by providing a radioimmunoassay to monitor the purification of the antigens. It is likely that the monoclonal antibodies described here are detecting antigens that are present in the cell membrane or on the cell surface, although definite proof of this will require biochemical and ultrastructural studies on cells in intact nervous tissue and on living cells in culture.

The elegance and attraction of both monoclonal antibody and recombinant DNA techniques lie in their ability to simplify complex systems by means of cloning, an element inherent in both approaches. In recombinant DNA technology, the components of a system, transformed into genetic information in bacteria, are cloned directly. This approach has the advantage of rapidly providing extensive and accurate structural information about the cloned molecules. In monoclonal antibody technology, on the other hand, it is the detecting reagents, the antibodies that distinguish individual components, that are cloned. Although this approach does not lead easily to structural characterization, monoclonal antibodies have great potential as probes for the function of particular antigens. Furthermore, it may be possible to select for monoclonal antibodies directed against molecules that mediate a particular function. The two approaches are therefore complementary, and real progress in our understanding of the nervous system will come through a combination of both technologies. In this way, we may be able to approach the molecular genetic basis of neuronal function and development.

ACKNOWLEDGMENTS

I thank Lynn Mangini and Elizabeth Silvestro for technical assistance and Kathy Cross for photographic assistance. This work was supported by NEI grants EY 01995, EY 3735, and NINCDS grant NS 17309.

REFERENCES

Barnstable, C. J. (1980): Monoclonal antibodies which recognize different cell types in the rat retina. *Nature*, 286:231–235.
Barnstable, C. J. (1982): Immunological studies of the retina. In: *Neuroimmunology*, edited by J. Brockes, pp. 183–214. Raven Press, New York.
Barnstable, C. J. (1981): Developmental studies of rat retina cells using cell-type specific monoclonal antibodies. In: *Monoclonal Antibodies against Neural Antigens*, edited by R. McKay, M. Raff, and L. Reichardt, pp. 219–230. Cold Spring Harbor Laboratory, New York.
Barnstable, C. J., Bodmer, W. F., Brown, G., Galfré, G., Milstein, C., Williams, A. F., and Ziegler, A. (1978): Production of monoclonal antibodies to group A erythrocytes, HLA and other human cell surface antigens. *Cell*, 14:9–12.
Brodsky, F. M., Parham, P., Barnstable, C. J., Crumpton, M. J., and Bodmer, W. F. (1979): Monoclonal antibodies for analysis of the HLA system. *Immunol. Rev.*, 47:3–62.
Cohen, J., and Selvendran, S. Y. (1981): A neuronal cell-surface antigen is found in the CNS but not in peripheral neurons. *Nature*, 291:421–423.
Eisenbarth, G. S., Walsh, F. S., and Nirenberg, M. (1979): Monoclonal antibody to a plasma membrane antigen of neurons. *Proc. Natl. Acad. Sci. USA*, 76:4913–4917.
Fekete, D. M., and Barnstable, C. J. (1982): Ultrastructural localization of rat photoreceptor membrane antigens. *J. Cell Biol. (in press)*.
Hellstrom, K. E., Brown, J. P., and Hellstrom, I. (1980): Monoclonal antibodies to tumor antigens. *Contemp. Top. Immunobiol.*, 11:117–137.
McKay, R., Raff, M., and Reichardt, L., eds. (1981): *Monoclonal Antibodies against Neural Antigens*. Cold Spring Harbor Laboratory, New York.
Melchers, F., Potter, M., and Warner, N. L. (1978): Lymphocyte hybridomas. *Curr. Top. Microbiol. Immunol.*, 81:9–23.
Polyak, S. (1941): *The Retina*. University of Chicago Press, Chicago.
Ramón y Cajal, S. (1955): *Histologie du systeme nerveux de l'homme et des vertebres, Vol. 2*. CSIC, Madrid.
Trisler, G. D., Schneider, M. D., and Nirenberg, M. (1981): A topographic gradient of molecules in retina can be used to identify neuron position. *Proc. Natl. Acad. Sci. USA*, 78:2145–2149.
Vulliamy, T., Rattray, S., and Mirsky, R. (1981): Cell-surface antigen distinguishes sensory and autonomic peripheral neurons from central neurons. *Nature*, 291:418–420.
Zipser, B., and McKay, R. (1981): Monoclonal antibodies distinguish identifiable neurons in the leech. *Nature*, 289:549–554.

Section VII

Posttranslational Processing

Up to this point, our considerations have been the complex dynamic process of genetic expression and the DNA-RNA molecular processing steps of transcription. The final mRNA is then delivered to the ribosome–endoplasmic reticulum–Golgi complex in secretory cells, and the translation of the nucleotide sequences into the corresponding amino acid sequences of the secretory product can begin. The chapters in this section deal with the cytoplasmic events that operate on secretory products after translation.

D. F. Steiner examines the well-studied posttranslational processing of the proinsulin precursor to insulin and offers generalizations to other multi-chain hormones. H. Gainer examines the processing of oxytocin and vasopressin from their prohormone forms in analogous operations and describes how these processing steps take place within the secretory storage vesicle as it is carried by intraneuronal transport from the cell body down the axon for release at the terminal. P. W. Robbins describes the steps by which glycoproteins, those ubiquitous molecules marking the surfaces of every eukaryotic cell, are processed and appropriately placed. Finally, J. F. McKelvy and coworkers turn our attention to the regulatory metabolic processing of secretory peptides that occurs after the final transmitter or hormonal product has been synthesized and released. They consider the peptide-specific cleavage enzymes, which may either terminate biological potency or redirect the remaining fragments to other actions on different target cells.

Having completed our detailed excursion through the evolving principles of molecular and cellular genetics, we are prepared to consider how awareness of these principles and the strategies for studying them in other cell types may facilitate our pursuit of problems in neuroscience.

Proteolytic Processing of Secretory Proteins

Donald F. Steiner

ABSTRACT

Limited proteolysis is a widely occurring mechanism in protein biosynthesis. Protein precursors can be classified according to their functions, localization within cell compartments, and enzymatic cleavage mechanisms. The presecretory proteins represent an important class of very rapidly turning over precursors that play an early role in the sequestration of secretory products. Proinsulin is representative of the general class of proproteins that are processed posttranslationally within their secretory cells during the formation and maturation of secretory granules. Defects can occur at several stages in the secretory mechanism and may provide further information on processing mechanisms as well as new insights into metabolic and neural disorders.

The islets of Langerhans, small organs distributed throughout the vertebrate pancreas, are fascinating because they produce a variety of peptide hormones in rather interesting ways. In recent years, several of the hormonal products of cells in the islets of Langerhans have been found in the brain also (somatostatin, glucagon, and pancreatic polypeptide) (Burger et al., 1980; Snyder, 1980), and thyrotropin-releasing hormone (TRH), commonly associated with the hypothalamus, has been found in the islets (Martino et al., 1978). Thus, in terms of peptide production, one could view the islets of Langerhans as a highly condensed version of the brain. The major islet cell type is the insulin-producing beta cell. Glucagon-producing alpha cells are present at the periphery of the islets. Pancreatic polypeptide, a 36-amino-acid peptide whose function is unknown (Schwartz et al., 1980), is produced by cells that are also peripheral to the islets, especially in the duodenal part of the pancreas. The cells that produce somatostatin (D cells) are present at low levels.

The normal biosynthesis of insulin has been studied for many years, and much is known about its precursor forms (see H. Goodman et al., *this volume*) and the mechanisms of their production and cleavage. Figure 1 summarizes the structure of the precursor form of insulin, which is called preproinsulin. It consists of a single polypeptide chain containing within it the B and A chain regions of the hormone, connected by a large polypeptide segment (C peptide) to form proinsulin. At the earliest stage of biosynthesis, the chain contains an additional 24 amino acids at the N-terminus. The formation and maturation of preproinsulin has turned out to be a reasonable model for many of the small regulatory peptides of the brain and gut.

Figure 2 summarizes the biosynthetic machinery of the typical secretory cell, which can also be generalized to some extent to nerve cells, even though the anatomical organization of the nerve cell is much more specialized. The biosynthesis of the precursor occurs in the rough endoplasmic reticulum (RER) by membrane-bound ribosomes. These bound ribosomes transfer the protein product into the cisternal spaces to segregate it from the cytosol of the cell, permitting it to be secreted. This process was elucidated by the discovery of Milstein and his coworkers

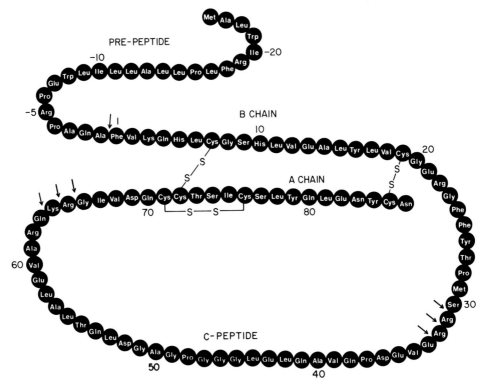

FIG. 1. Covalent structure of rat preproinsulin II (from Chan et al., 1979).

(1972) that translation of immunoglobulin mRNA produced a chain longer than the mature species and that the extension served as a signal for the production of the ribosome–membrane complex. This assembly of ribosome to membrane then leads to the segregation of the nascent peptide chain across the microsomal membrane into the cisternal space. (For a recent review, see D. Anderson et al., *this volume*, and Steiner et al., 1980.) The newly synthesized peptide is cleaved to proinsulin by removal of the 24-amino-acid N-terminal extension by a microsomal protease and is then carried in small vesicles to the Golgi apparatus where it is packaged, presumably along with the enzymes that will complete the processing of proinsulin into insulin.

Early in their development, the packets, or secretion granules, have a characteristic uniform light granulation, and later assume a different structural organization with a condensed, often crystalline-appearing inclusion, surrounded by a clear space. Electron microscopic studies have indicated that the small central inclusion is in fact a small crystal of insulin (Blundell et al., 1972), and the space around it presumably contains the C-peptide that has been cleaved from proinsulin. In response to various stimuli, these two products, derived by proteolysis of proinsulin, are cosecreted from the cell into the circulation by exocytosis (Rubenstein et al., 1977). Thus, the gene products produced by the pancreatic beta cell include not just insulin but equimolar amounts of another peptide product, the C-peptide, as well as small amounts of proinsulin and intermediates. Similar phenomena are seen in many other peptide-producing cells: a variety of peptides are produced that are derived from the partial or complete processing of precursor forms. The physiology of any given secretory cell cannot be understood until all of its products are known.

The turnover of preproinsulin is very rapid; the subsequent maturation of proinsulin is much slower, about 1 hr in the normal mammalian islet, a lag that is reasonably typical of the

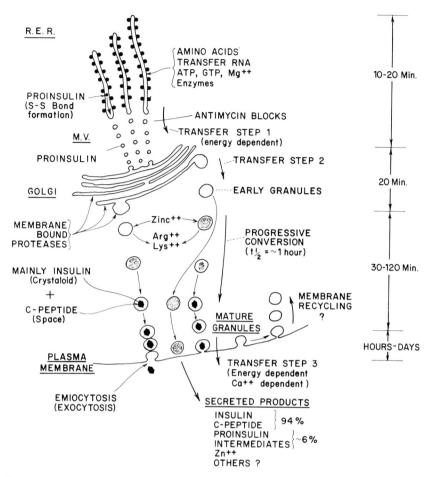

FIG. 2. Intracellular formation and maturation of preproinsulin to insulin and C-peptide in the beta cells of the islets of Langerhans. R.E.R., rough endoplasmic reticulum; M.V., microvesicles.

processing of many other hormonal precursor forms. Figure 3 illustrates the processing of proinsulin to insulin. In this continuous labeling study of isolated islets (Steiner, 1967), the formation of proinsulin begins to plateau after about 2 hr, and no detectable insulin is produced before 20 to 30 min. This lag represents the time required for the passage from the RER to the Golgi area, where the packaging into secretion granules occurs. Because that transfer step is energy dependent, it can be inhibited with a variety of inhibitors of energy metabolism; for example, if antimycin is added to islets shortly after pulse labeling of proinsulin, the subsequent conversion to insulin is blocked (Steiner et al., 1970).

By the use of pulse-chase experiments, the production and processing of preproinsulin can be followed after mRNA translation is activated by glucose in islets of Langerhans or insulin-producing tumors of rats. Pulse labeling in the presence of glucose reveals that even at 2 min, there is a rapid accumulation of proinsulin with only a small amount of preproinsulin; after 2 min of chase, the proinsulin is still present, but the preproinsulin has disappeared (Patzelt et al., 1978). In normal islets, the conversion of proinsulin to insulin-related peptides begins to occur at 30 min and continues up to 2 hr. The tumor cells follow the same pattern except that the processing of proinsulin to insulin is even slower.

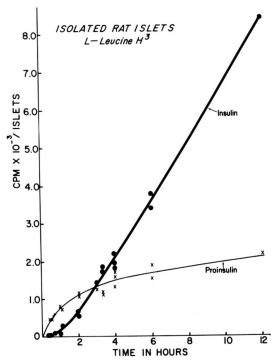

FIG. 3. Precursor product relationship of proinsulin to insulin during continuous labeling of isolated rat islets (from Steiner, 1967).

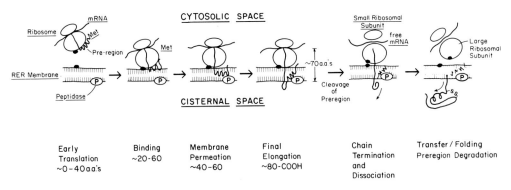

FIG. 4. Proposed model for the intracellular segregation of presecretory proteins. See Steiner et al. (1980) for further detail on the intramembranous structure of the prepeptide loop.

Even in very short pulses, preproinsulin is only present as a very minor component. One explanation for this finding is illustrated in the model shown in Fig. 4, which is a variation of the signal hypothesis as originally proposed by Blobel and Dobberstein (1975). According to their hypothesis, the nascent protein, particularly the prepeptide region, is thought to enhance the binding of the large ribosomal subunit to a site on the membrane, allowing the hydrophobic amino acid sequence in the central region of these prepeptides to partition into the bilayer (for details, see Steiner et al., 1980). This leads to the formation of a transmembrane loop, which retains its initial amino terminus on the outside of the membrane. This loop structure is then

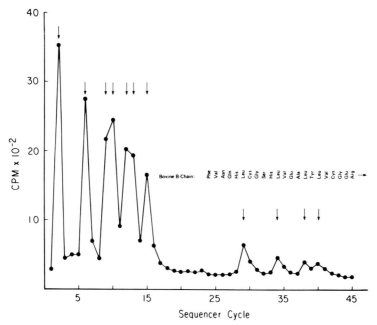

FIG. 5. Automated sequential Edman degradation of bovine preproinsulin labeled during cell-free translation of pancreatic mRNA in the presence of [³H]leucine. *Arrows* indicate location of leucine residues in bovine preproinsulin (from Lomedico et al., 1977).

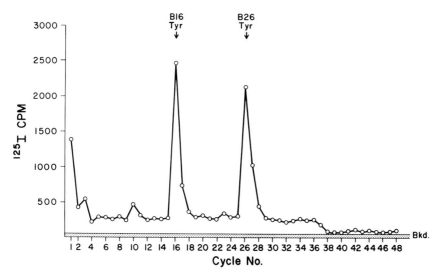

FIG. 6. Microsequencing of ¹²⁵I-labeled human proinsulin produced in *E. coli*. Less than 0.01 pmoles of protein were required for this analysis (from Chan et al., 1981a).

cleaved by a peptidase within the cisternal space, producing a new N-terminus that allows the growing peptide chain to enter the luminal space. Since the overall length of preproinsulin is just about sufficient to form this complete structure before chain termination occurs, it is apparent that intact preproinsulin will be found only in a small number of cases where chain termination occurs before cleavage. Thus, preproinsulin appears to defy the usual direct precursor-product

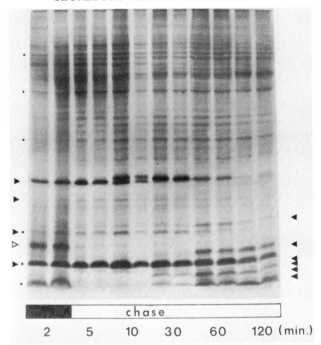

FIG. 7. Autoradiographs of a slab gel SDS electrophoretogram of labeled rat islet proteins showing the locations of bands corresponding to precursors of the major islet cell secretory products. On the left margin (from top to bottom) *arrows* indicate the positions of (1) 18-kD proglucagon, (2) 16-kD unknown precursor, (3) 12.5 kD prosomatostatin, (4) preproinsulin, and (5) proinsulin. *Arrows on right margin* indicate positions of intermediate end product peptides derived from the primary precursors. *Asterisks* indicate positions of molecular weight markers. (For details, see Patzelt et al., 1979, 1980b.)

relationship, and yet, clearly, there is every reason to believe that all translation begins at the initiator methionine at the beginning of the prepeptide region. The excised prepeptide regions are degraded so rapidly that they are difficult to demonstrate (Patzelt et al., 1978).

If one translates islet mRNA preparations and, after immunoprecipitation, displays the products of translation on a tube gel, a sharp peak of protein product is seen which is larger than proinsulin (Chan et al., 1976). To study the amino acid sequence of this peptide, preproinsulin, specific radioactive amino acids were incorporated into it, using a cell-free expression system, and then using the Edman sequenator to find the locations of known amino acids along the peptide chains (Tager et al., 1973; Chan et al., 1976). Figure 5 shows the location of leucines within bovine preproinsulin (Lomedico et al., 1977). There are many leucine residues in the prepeptide, but only four in the B chain. The B chain leucines can be identified, although they are displaced by 24 cycles due to the presence of the prepeptide extension. This method not only reveals the length of the primary translation product (i.e., the presecretory form) but also the location of the leucines (or any other labeled amino acid). Reiteration of the method with other amino acids can establish the full sequence and has become a widely used method for such studies.

Another application of microsequencing that has been quite useful is illustrated in Fig. 6. In this experiment, human proinsulin synthesized in transfected, cloned bacteria was sequenced (Chan et al., 1981b). In this case, an insulin cDNA was inserted into an expression vector, and a human preproinsulin hybrid protein was made. (See H. Goodman et al., *this volume*). Proinsulin-like material exported into the periplasmic space by the bacteria was then partially

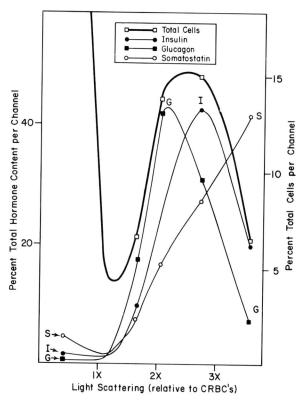

FIG. 8. Analysis of viable rat islet cells by means of fluorescence-activated cell sorting. Light scattering (abscissa) is measured relative to the scatter intensity of fixed chicken red blood cells (cRBC). Cells from various regions of the cell peak were collected in acetic acid and analyzed for their hormone content by immunoassay to indicate the relative distributions of A, B, and D cells.

purified, iodinated to high specific activity with ^{125}I, and immunoprecipitated. The amino acid sequence data show that labeled tyrosines are present at B16 and B26 and are quite sharply defined, indicating not only that the periplasmic product is human proinsulin, but also that the cleavage of the hybrid prepeptide occurring in the bacterial system has been as faithful as in the eukaryotic system. Furthermore, the amino terminus is in the correct location relative to the tyrosines. Since many neuropeptides have tyrosine residues, this same method might be usefully applied, since it can work with less than 0.01 picomole of peptide material.

Some of the other islet peptides about which there is less information are glucagon (29 residues) and somatostatin (14 residues) (see E. Habener, this volume). Tager and Steiner (1973) isolated a peptide that consisted of glucagon with an additional sequence of eight amino acids at the carboxyl end. A pair of basic residues (which often mark the cleavage sites in these precursor forms) connected the C-terminal extension to glucagon's carboxyl terminus. Gut glucagon, which is a higher molecular weight form of glucagon-like material, also has this carboxyl-terminal extension, as well as some additional N-terminal amino acids, making a sequence of 100 amino acids. (Patzelt et al., 1979).

Patzelt and colleagues (1979, 1980b) employed slab gel electrophoresis to the analysis of islet proteins labeled with various amino acids (Fig. 7). In addition to proinsulin (whose production was minimized by using low glucose media and labeling with methionine, which is in low abundance in proinsulin) other processed proteins were found to be turned over with

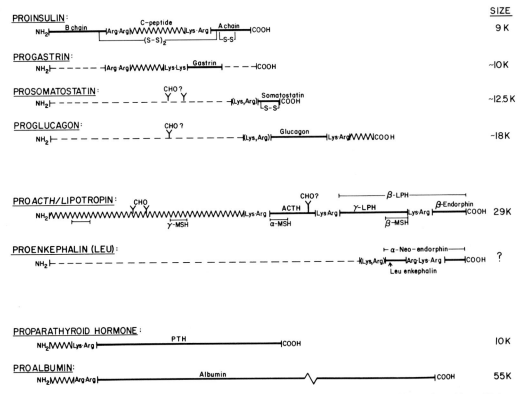

FIG. 9. Diagrammatic structure of several peptide precursor forms (proproteins). (Reproduced from Steiner et al., 1980.)

essentially the same kinetics as proinsulin. One such protein of 18,000 daltons was found to undergo some posttranslational modification, disappearing at about the same time that proinsulin was converted to insulin. A 13,000-dalton peptide that appeared and later disappeared turned out to be an intermediate cleavage product of proglucagon (Patzelt et al., 1979). To complete the identification of proglucagon, it was demonstrated that the band reacted with antibodies to glucagon, and peptide material excised from slab gels had a tryptic fingerprint similar to glucagon, containing peptides that were identical to those derived from the tryptic digestion of glucagon. A similar approach was used for the identification of a 12,500-dalton peptide as prosomatostatin (Patzelt et al., 1980a). Thus, this approach, pulse labeling of somewhat purified tissue, can identify precursor forms and help to sort out the patterns and interrelationships of these peptides, prior to the application of cloning techniques for structural and genetic analysis.

Another technique useful in islet peptide research, the fluorescence-activated cell sorter, has proved to be an invaluable tool in separating the various cellular elements in the immune system and might also be useful in studying the central nervous system. Preliminary results with suspensions of dispersed islet cells are also yielding promising results (Fig. 8). When separated by forward angle light scatter, which is largely a measure of cell size, islet cells appear as a single, rather broad peak. However, when hormone assays are carried out across this peak, the glucagon producers (alpha cells) are found on the low side, the insulin producers (beta cells) are in the center, and the somatostatin producers (D cells) are more abundant toward the high side of the scatter profile. When the sorted cell fractions are labeled with [^{35}S]methionine, they can be shown to synthesize the expected hormones and their precursor forms, in some cases strengthening the cellular assignments of those precursors and also aiding in the appropriate assignment of intermediate forms (Patzelt et al., 1980a).

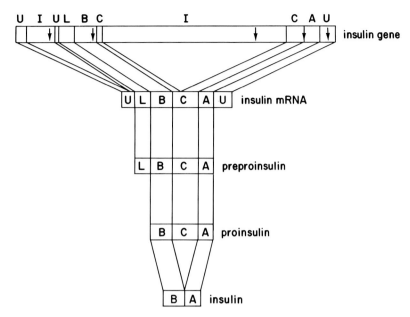

FIG. 10. Summary of the processing (both mRNA and protein) involved in the expression of the insulin gene. A, B, and C, peptide regions of proinsulin; L, leader sequence; U, untranslated regions of the mRNA; and I, intervening sequences in the gene. *Arrows* indicate sites of allelic (Ullrich et al., 1980) or mutational (Kwok et al., 1981; Robbins et al., 1981) differences that have been found in the human insulin gene structure. Those occurring within the coding sequence have been found to lead to altered processing and products. (This figure kindly provided by Dr. Howard S. Tager.)

Several generalizations about these precursors are now possible (see Fig. 9). Proinsulin has turned out to be the smallest precursor. Presumably the topography of the ribosome membrane junction demands a minimum peptide chain length in order for successful transfer across the RER membrane and segregation. Such a simple explanation might account for the production of many other small hormonal and regulatory peptides from larger precursors. Thus, in the course of evolution, when a fragment of an available protein proved useful in regulation, the secretory cells tended to continue producing it via the original gene product from which it arose. Another generalization from these several identified precursors is that the regions that undergo processing are bracketed by pairs of basic amino acid residues. This clearly points to the existence of a common mechanism of conversion of these propeptides involving a trypsin-like enzyme and a carboxypeptidase that can remove the basic residues after they have been exposed by the endopeptidase.

In special cases, other proteases may also participate in processing. For example, the pre-peptides are cleaved by an entirely different mechanism localized to the microsomes (Steiner et al., 1980). Relaxin, whose gene has been cloned recently (Hudson et al., 1981), is a homologue of proinsulin. Its connecting peptide is 109 residues long, which makes it more similar in length to the serine proteases that some believe may be ancestral precursor forms of the insulin-like peptides (Chain et al., 1981a). Prorelaxin also has a very interesting chymotryptic-like cleavage in its maturation that involves a leucine-serine bond, rather than the usual basic dipeptide cleavage. A similar minor cleavage at a leucine residue has also been found to occur in several C-peptides, including those of the pigs, rats, and humans during their maturation. Thus, in the rat, about 20% of the proinsulin produced in the islets is normally cleaved at that leucine during its conversion to insulin (Tager et al., 1973). A major pathway of conversion in relaxin biosynthesis appears to be only a relatively minor cleavage pathway in insulin formation. These observations indicate that the secretory granules that process these proteases may contain not only trypsin-like and carboxypeptidase B-like enzymes, but also low levels of chymotrypsin-like activity, and perhaps other proteases as well (Steiner et al., 1980).

To summarize the processing involved in expressing the insulin gene (Fig. 10), a rather large structural unit in the chromosomal DNA codes for a rather small but very potent and critically important peptide hormone. A number of defects and errors can occur in the formation and maturation of insulin (Kwok et al., 1981; Robbins et al., 1981), and in some cases these are associated with diabetic syndromes. It seems reasonable to predict that some brain disorders may turn out to be due to defective processing of the precursors of one or more of the myriad peptides normally produced in the brain, none of whose functions are yet well understood.

ACKNOWLEDGMENTS

Studies reported here were supported by grants from the NIH (AM 13914 and AM 20595), the Novo Research Institute, the Kroc Foundation, the Lolly Coustan Memorial Fund, and Cetus Corporation. Ms. Myrella Smith provided able assistance in preparing this manuscript.

REFERENCES

Blobel, G., and Dobberstein, B. (1975): Transfer of proteins across membranes. I. Presence of proteolytically processed and unprocessed nascent immunoglobulin light chains on membrane bound ribosomes of murine myeloma. *J. Cell Biol.*, 67:835–851.
Blundell, T., Dodson, G., Hodgkin, D., and Mercola, D. (1972): Insulin: The structure in the crystals and its reflection in chemistry and biology. *Adv. Protein Chem.*, 26:279–402.
Burger, A., Kosterlitz, H. W., and Iversen, L. L., eds. (1980): *Neuroactive Peptides*, pp. 80, 152. The Royal Society, London.
Chan, S. J., Keim, P., and Steiner, D. F. (1976): Cell-free synthesis of rat preproinsulins: Characterization and partial amino acid sequence determination. *Proc. Natl. Acad. Sci. USA*, 73:1964–1968.
Chan, S. J., Kwok, S. C. M., and Steiner, D. F. (1981a): The biosynthesis of insulin: Some genetic and evolutionary aspects. *Diabetes Care*, 4:4–10.
Chan, S. J., Noyes, B. E., Agarwal, K. L., Ackerman, E., Quinn, P. S., Keim, P. S., Sigler, P. B., Heinrikson, R. L., and Steiner, D. F. (1979): Elucidation of the primary structures of the initial translation products of the rat insulin genes. In: *Proceedings of the 10th IDF Congress*, edited by W. K. Waldhäusl, pp. 113–118. Excerpta Medica Foundation, Vienna.
Chan, S. J., Weiss, J., Konrad, M., White, T., Bahl, C., Yu, S.-D., Marks, D., and Steiner, D. F. (1981b): Biosynthesis and periplasmic segregation of human proinsulin in Escherichia coli. *Proc. Natl. Acad. Sci. USA*, 78:5401–5405.
Hudson, P., Haley, J., Cronk, M., Shine, J., and Niall, H. (1981): Molecular cloning and characterization of cDNA sequences coding for rat relaxin. *Nature*, 291:127–131.
Kwok, S. C. M., Chan, S. J., Rubenstein, A. H., Poucher, R., and Steiner, D. F. (1981): Loss of a restriction endonuclease cleavage site in the gene of a structurally abnormal human insulin. *Biochem. Biophys. Res. Commun.*, 98:844–849.
Lomedico, P. T., Chan, S. J., Steiner, D. F., and Saunders, G. F. (1977): Immunological and chemical characterization of bovine preproinsulin. *J. Biol. Chem.*, 252:7971–7978.
Martino, E., Lernmark, Å., Seo, H., Steiner, D. F., and Refetoff, S. (1978): High concentration of thyrotropin-releasing hormone in pancreatic islets. *Proc. Natl. Acad. Sci. USA*, 75:4265–4267.
Milstein, C., Brownlee, G. G., Harrison, T. M., and Mathews, M. B. (1972): A possible precursor of immunoglobulin light chains. *Nature New Biol.*, 239:117–120.
Patzelt, C., Labrecque, A. D., Duguid, J. R., Carroll, R. J., Keim, P., Heinrikson, R. L., and Steiner, D. F. (1978): Detection and kinetic behavior of preproinsulin in pancreatic islets. *Proc. Natl. Acad. Sci. USA*, 75:1260–1264.
Patzelt, C., Nielsen, D., Carroll, R., Quinn, P., Lernmark, Å., Tager, H. S., and Steiner, D. F. (1980a): Studies on the biosynthesis of the other peptide hormones of the rat islets of Langerhans. *Biochem. Soc. Trans.*, 8:411–413.
Patzelt, C., Tager, H. S., Carroll, R. J., and Steiner, D. F. (1979): Identification and processing of proglucagon in pancreatic islets. *Nature*, 282:260–266.
Patzelt, C., Tager, H. S., Carroll, R. J., and Steiner, D. F. (1980b): Identification of prosomatostatin in pancreatic islets. *Proc. Natl. Acad. Sci. USA*, 77:2410–2414.

Robbins, D. C., Blix, P. M., Rubenstein, A. H., Kanazawa, Y., Kosaka, K., and Tager, H. S. (1981): A human proinsulin variant at arginine 65. *Nature*, 291:679–681.

Rubenstein, A. H., Steiner, D. F., Horowitz, D. L., Mako, M. E., Block, M. B., Starr, J. I., Kuzyya, H., and Melani, F. (1977): Clinical significance of circulating proinsulin and C-peptide. *Recent Prog. Horm. Res.*, 33:435–468.

Schwartz, T. W., Gingerich, R. L., and Tager, H. S. (1980): Biosynthesis of pancreatic polypeptide. Identification of a precursor and a co-synthesized product. *J. Biol. Chem.*, 255:11494–11498.

Snyder, S. H. (1980): Brain peptides as neurotransmitters. *Science*, 209:976–983.

Steiner, D. F. (1967): Evidence for a precursor in the biosynthesis of insulin. *Trans. NY Acad. Sci., ser. 2, no. 1*, 30:60–68.

Steiner, D. F., Clark, J. L., Nolan, C., Rubenstein, A. H., Margoliash, E., Melani, F., and Oyer, P. E. (1970): The biosynthesis of insulin and some speculations regarding the pathogenesis of human diabetes. In: *Nobel Symposium 13, on the Pathogenesis of Diabetes Mellitus*, edited by E. Cerasi and R. Luft, pp. 57–80. Almqvist and Wiksell, Stockholm.

Steiner, D. F., Quinn, P. S., Patzelt, C., Chan, S. J., Marsh, J., and Tager, H. S. (1980): Proteolytic Cleavage in the Posttranslational Processing of Proteins. In: *Cell Biology: A Comprehensive Treatise*, edited by L. Goldstein and D. M. Prescott, pp. 175–201. Academic Press, New York.

Tager, H. S., Emdin, S. O., Clark, J. L., and Steiner, D. F. (1973): Studies on the conversion of proinsulin to insulin. II. Evidence for a chymotrypsin-like cleavage in the connecting peptide region of insulin precursors in the rat. *J. Biol. Chem.*, 248:3476–3482.

Tager, H. S., and Steiner, D. F. (1973): Isolation of a glucagon-containing peptide: Primary structure of a possible fragment of proglucagon. *Proc. Natl. Acad. Sci. USA*, 70:2321–2335.

Ullrich, A., Dull, T. J., Gray, A., Brosius, J., and Surls, I. (1980): Genetic variation in the human insulin gene. *Science*, 209:612–615.

Molecular Genetic Neuroscience, edited by
F. O. Schmitt, S. J. Bird, and F. E. Bloom.
Raven Press, New York © 1982.

The Processing of Cell-Surface Glycoproteins

Phillips W. Robbins

ABSTRACT

A specific mannose oligosaccharide is transferred to asparagine residues cotranslationally during the course of glycoprotein synthesis. This oligosaccharide is then processed along various pathways depending on cell type. The same lipid-linked oligosaccharide donor is present in all eukaryotic cells. A primary focus for current research is on variations in the processing pathway and possible roles that oligosaccharides may serve in glycoprotein functioning.

With the 1970s came the realization that glycoproteins constitute an integral component of the cell surface. Not only do they serve as receptors for intercellular messengers, but they have also been implicated in cellular adhesion. Since their carbohydrate moieties display marked heterogeneity, they may also be involved in cellular recognition during development. Whether the final structures of these moieties provide some element of functional regulation in determination of the phenotype of the cell remains unclear. However, before possible regulatory controls can be assigned to glycosylation, it is necessary to understand fully how these structures are constructed.

A summary of the maturation of glycoproteins is illustrated in Fig. 1. During translation of the primary peptide sequence at the rough endoplasmic reticulum, the initial linkage of a carbohydrate structure is added to specific asparagine residues. After transport of the polypeptide

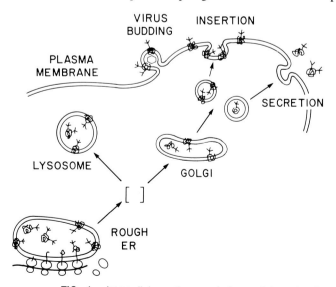

FIG. 1. Intracellular pathways of glycoprotein maturation.

to the Golgi apparatus, the asparagine-linked oligosaccharide is further modified; moreover, some proteins may undergo fatty acylation and glycosylation of serine and threonine residues within the Golgi. Following proteolytic cleavage, the mature form of glycoprotein is subsequently delivered to the cell surface where it is either incorporated into the plasma membrane or secreted. This chapter focuses on the glycosylation events that occur on the rough endoplasmic reticulum and the modification of that structure in the Golgi apparatus.

GLYCOSYLATION AT THE ROUGH ENDOPLASMIC RETICULUM

The glycosylation of nascent polypeptides at the rough endoplasmic reticulum consists of the attachment of a large lipid-linked oligosaccharide to specific tripeptide sequences with asparagine residues at the N-terminus. Glycosylation occurs on any asparagine residue followed by an amino acid other than aspartic acid; these two residues are followed by serine or threonine (Marshall, 1972). This site, Asn-X-$^{Ser}_{Thr}$, would appear at first to be too limited in informational content to provide a specific recognition signal for glycosylation. Although more complex mechanisms directing the site of glycosylation may be used, there are two clear ways by which inappropriate glycosylation may be deterred. First, a search of the protein atlas demonstrates that proteins secreted into the lumen of the rough endoplasmic reticulum, either to be secreted or to appear on the cell surface, have relatively few Asn-X-$^{Ser}_{Thr}$ sequences as compared to cytoplasmic proteins. There appears, therefore, to be discrimination against this simple sequence in proteins whose biogenesis includes glycosylation. The second method, although less well understood, concerns the sequestering of these signal sequences within the tertiary structure of the polypeptide. If the formation of folds within the nascent polypeptide is more rapid than the rate of glycosylation, then sites not accessible to the enzymatic machinery are either incompletely glycosylated or not glycosylated at all according to the kinetics of the reaction. Although largely hypothetical, the removal of potential glycosylation sites by protein folding seems to be an attractive and likely mechanism for the limitation of polypeptide modification.

The first step in glycosylation is the transfer of a previously completed triantennery structure composed of three glucose (Glc), nine mannose (Man), and two N-acetylglucosamine (GlcNAc) residues in a pyrophosphate linkage to the polyisoprenoid, dolichol. The structures of this lipid-linked oligosaccharide and two oliogosaccharides linked to mature glycoproteins at an asparagine residue are depicted in Fig. 2. Recent evidence suggests that the $Glc_3Man_9GlcNAc_2$ structure, first reported by Li and coworkers (1978), is the universal donor for the glycosylation at the rough endoplasmic reticulum. First, this structure is the major lipid-linked oligosaccharide in the eukaryotic cell. Moreover, in all eukaryotes that have been examined including yeast, dictyostelium, insects, and mammals, this structure is precisely conserved and therefore falls into the category of common coenzymes such as coenzyme A and nicotinamide adenine dinucleotide (NAD). The biosynthesis of the lipid-linked oligosaccharide donor is extremely rapid and requires pulse labeling to detect the short-lived intermediate products in the pathway. In Chinese hamster ovary cells, three oligosaccharide species can be detected after 2½ min of labeling with [^{3}H]mannose: the complete oligosaccharide, $Glc_3Man_9GlcNAc_2$, and two precursors, $Man_8GlcNAc_2$ and $Man_5GlcNAc_2$ (Hubbard and Robbins, 1980). Chase of the labeled mannose for periods as short as 2½ min with excess unlabeled mannose results in the accumulation of the label entirely within the complete branched structure, attesting to the rapid biosynthesis of the oligosaccharide donor. The synthesis of the donor oligosaccharide is summarized in Fig. 3.

What becomes apparent from the biosynthesis of the lipid-linked donor oligosaccharide is the many steps at which regulation of glycosylation of nascent peptides could occur. The

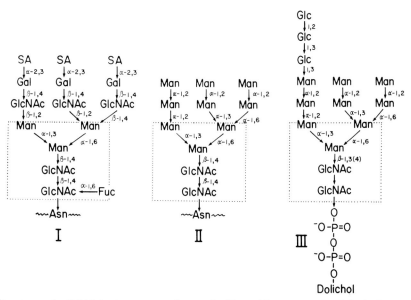

FIG. 2. Structures of a lipid-linked precursor oligosaccharide and two asparagine-linked oligosaccharides from mature glycoproteins. I: A complex oligosaccharide from the vesicular stomatitis virus glycoprotein (Reading et al., 1978). II: A high-mannose oligosaccharide from bovine thyroglobulin (Ito et al., 1977). III: The major lipid-linked oligosaccharide from Chinese hamster ovary cells (Li et al., 1978). *Dotted box* indicates common core structure. Bonds between residues are labeled. Asn, asparagine; Fuc, fucose; Gal, galactose; GlcNAc, N-acetylglucosamine; Man, mannose; SA, sialic acid.

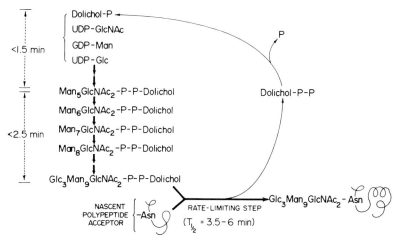

FIG. 3. Proposed pathway for assembly of the lipid-linked precursor oligosaccharide in Chinese hamster ovary cells. The regeneration of dolichol phosphate after transfer of the completed oligosaccharide to protein has not yet been demonstrated. GDP, guanosine diphosphate; UDP, uridine diphosphate; GlcNAc, N-acetylglucosamine; Man, mannose. (From Hubbard and Robbins, 1980.)

kinetics of formation of this structure imply, however, that synthesis is controlled by the rate-limiting step, linkage to the growing peptide chain. The kinetics of formation and transfer of the lipid-linked oligosaccharide, however, can only suggest that the specificity of glycosylation at the rough endoplasmic reticulum is indeed regulated by availability of nascent peptide substrate.

This type of control has been demonstrated more directly by Hubbard and Robbins (1980). If labeling of the oligosaccharide is followed with a chase of unlabeled sugar precursors, the radioactivity as free lipid-linked oligosaccharide is rapidly turned over, presumably due to transfer to the nascent polypeptides within the lumen of the rough endoplasmic reticulum. However, when protein synthesis is inhibited with cycloheximide, no label is drawn from the oligosaccharide pool and the turnover of free oligosaccharide is effectively halted.

The reciprocal experiment, inhibition of protein synthesis in the presence of radioactively labeled sugar precursors, effectively inhibits the accumulation of labeled oligosaccharide sugar precursors, demonstrating further that synthesis of the lipid-linked oligosaccharide is dependent on ongoing protein translation. The interpretation of these experiments is of course open to explanation in terms of pleiotropic effects of inhibition of total protein synthesis. Cessation of protein synthesis might limit, through the rapid turnover of oligosaccharide biosynthetic enzyme pathways, the formation of the oligosaccharide itself. Taken together, however, these data, in the absence of specific inhibition of transfer of the oligosaccharide from the lipid-linked donor structure, support the conclusion that the extent and hence specificity of this primary step in glycosylation is in fact dependent on the primary and perhaps tertiary structure of the substrate, the nascent glycopeptide.

Although the accessibility of specific sequences within the growing peptide chain appears to play the decisive role in the extent of glycosylation, the structure of the lipid-linked oligosaccharide itself must be correct for effective transfer of the oligosaccharide. Employing an *in vitro* microsomal preparation, Turco and coworkers (1977) have assayed the extent of transfer of oligosaccharide structures deficient in sugar residues. Their results demonstrate that the presence of three glucose residues on the third arm of the triantennary oligosaccharide structure is the key element to the recognition of the complete lipid-linked oligosaccharide and successful glycosylation in the endoplasmic reticulum.

These observations clearly define the mechanisms required during the initial glycosylation of nascent polypeptide chains; however, less is known of the precise location of these events in the endoplasmic reticulum. The ribosomes actively engaged in the translation of glycoproteins are bound to the outside of the membrane of the endoplasmic reticulum. The question remains whether the synthesis of the lipid-linked oligosaccharide and subsequent linkage to the growing polypeptide chain occurs on the cytoplasmic side or within the lumen of the endoplasmic reticulum. Snider and associates (1980) attempted to identify the location of the enzymes required for the addition of the N-acetylglucosamine, mannose, and glucose in the construction of the lipid-linked oligosaccharide. Their data show that protease treatment of isolated rat liver microsomal vesicles inactivates these enzymes of oligosaccharide synthesis. This finding, together with the fact that no latent enzymatic activity was found protected within the lumen of the endoplasmic reticulum vesicles, indicates that the oligosaccharide-lipid is assembled on the cytoplasmic side of the microsomal membrane. The question remains open as to what is transported across the membrane. Further experiments are required to ascertain whether the oligosaccharide is transported into the lumen of the endoplasmic reticulum independently as a lipid-linked structure or after covalent linkage with the growing polypeptide chain.

After transfer to the polypeptide chain and during the migration of the glycoprotein to the Golgi apparatus, the oligosaccharide undergoes a series of processing steps, including removal of glucose and mannose residues (Li and Kornfeld, 1978; Hubbard and Robbins, 1979). The purpose of these cleavage steps is to reduce the high-mannose oligosaccharide (see Fig. 2) to a core structure of $Man_5GlcNAc_2$ that serves as a base for further glycosylation to form the complex-type chains observed for glycoproteins in the cell surface. The first step in the processing pathway is the removal of the first glucose residue by glucosidase I. This cleavage is extremely rapid and efficient, leaving virtually no detectable molecules containing the terminal glucose at steady state. An independent enzyme, glucosidase II, is responsible for the cleavage of the second and third glucose residues. This step proceeds more slowly, and molecules with single glucose residues are occasionally detected at the cell surface. After complete cleavage of the glucose residues, the sequential removal of the four α-1-2 mannose residues begins. Cleavage of these residues probably starts at the endoplasmic reticulum but is completed at the Golgi apparatus.

The mechanism by which the oligosaccharide is enzymatically trimmed down illustrates an important aspect in the regulation of glycosylation. The initial modification of the asparagine-linked oligosaccharide is performed by enzymes, glucosidases, and possibly α-mannosidase, which are specific processing enzymes that do not simply hydrolyze carbohydrate linkages. These enzymes recognize specific substrates, requiring the correct oligosaccharide structure and linkage to either protein or lipid. If one mannose residue is randomly removed from the oligosaccharide structure, the enzymatic cleavage of glucose residues is reduced up to threefold. When two mannose residues are absent there is essentially no recognition of the substrate by the glucosidase. At least in the case of the vesicular stomatitis virus (VSV) glycoprotein, the mannose residues also appear to be cleaved in a specific sequence, beginning with the terminal residue in the middle branch and proceeding until the last α-1-2 mannose residue on the chain originally linked to the three glucose residues is removed (Kornfeld et al., 1978). Thus, the substrate specificity provides a rigorous pathway for the initial processing of the asparagine-linked oligosaccharide. If interrupted, further modification of the oligosaccharide structure is halted, resulting in the inhibition of the formation of complex-type oligosaccharides found on mature glycoproteins.

Some experiments suggest that passage to the Golgi apparatus may be conducted through clathrin-coated vesicles. VSV, a budding RNA virus encoding a glycoprotein that is transported to the cell surface and ultimately implanted within the cell membrane to form the envelope of the viral genome, has provided a model system to study the mechanism for transport of glycoproteins. After pulse-chase labeling and subsequent isolation of clathrin-coated vesicles, Rothman and coworkers (1980) observed two waves of labeled oligosaccharide structures within these vesicles. Initially unprocessed asparagine-linked oligosaccharides are present in the coated vesicles. The amount of labeled unprocessed high-mannose oligosaccharide declines within 10 min of the chase with cold precursor, and a consequent rise in labeled, processed carbohydrate structures is observed. Although *in vitro* experiments are required to confirm this observation, the initial findings implicate clathrin-coated vesicles as the vehicle of transport to the Golgi apparatus.

In the Golgi apparatus, extensive processing of the carbohydrate moiety is conducted, resulting in the complex chains found on mature glycoproteins at the cell surface. Somatic mutants of Chinese hamster cell lines have provided much of the detail of this process (Li and Kornfeld, 1978). For example, the block resulting in the failure to remove mannose residues leads to the accumulation of a substrate that cannot be acted on further in the processing pathway. To summarize that pathway briefly (Fig. 4), N-acetylglucosamine is added to the α-1-3 branch of the oligosaccharide and the remaining two terminal mannose residues are cleaved and replaced by an N-acetylglucosamine residue. This reaction is catalyzed by an enzyme distinct from that which links N-acetylglucosamine to the initial α-1-3 branch. To these three termini, galactose and the sialic acid residues are added. In addition, fucose residues are added (see Fig. 2, structure I). There appear to be two important distinctions to be made between these steps in the glycosylation pathway from those taken previously. First, in the Golgi apparatus, these sugars are added one at a time by specific glycosyl transferase rather than linked as a complete

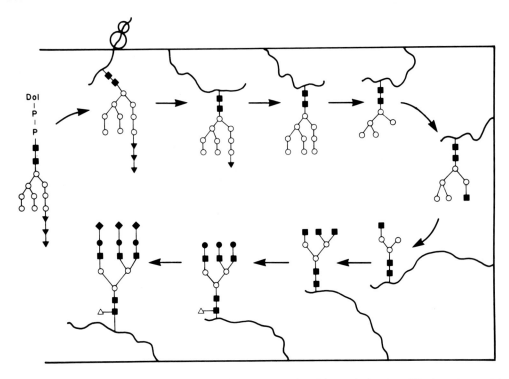

FIG. 4. Schematic representation of oligosaccharide processing of the vesicular stomatitis virus glycoprotein after translation. Symbols: ■, N-acetylglucosamine; ○, mannose; ▼, glucose; ●, galactose; ◆, sialic acid; △ fucose. (From Kornfeld et al., 1978.)

structure as it occurs in the rough endoplasmic reticulum (Waechter and Lennarz, 1976). Second, the completion of the mature complex oligosaccharide structure is dependent on the availability of the individual glycosyl transferase in the cell and the accessibility of the protein-linked oligosaccharide to these enzymes.

FURTHER GLYCOSYLATION ON THE GOLGI APPARATUS

In addition to the modification of oligosaccharide structures covalently linked to asparagine residues, a second type of glycosylation occurs on specific serine and threonine residues of the glycoprotein. The polypeptide signal for this site of glycosylation is not at this time well defined. The initial step in this pathway is always the addition of N-acetylgalactosamine. This is followed by a series of single-step transfer reactions, similar to those involved in the addition of the "outer" sugars (N-acetylglucosamine, galactose, sialic acid, and fucose) to complex asparagine-linked oligosaccharides. This glycosylation results in a wide variety of structures, some at least as complex as those of the carbohydrate structures attached to the asparagine linkages. However, Ivatt (1981) suggests that the addition of specific carbohydrate residues may not be as random as would be expected for a system of competing glycosyl transferases. Glycosyl transferases may in fact form specific complexes with one another, effectively shuttling a substrate from one enzyme to another. Further experimentation is required to ascertain whether this type of control mechanism is important in regulating the ultimate structure of the glycoprotein at the cell surface.

SUMMARY OF THE TIME COURSE OF GLYCOSYLATION

Sindbis virus provides a useful model for summarizing the various pathways of glycosylation in the eukaryotic cell, including fatty acylation. The Sindbis virus is made up of three proteins, translated from a single 26S mRNA. The core protein becomes associated with the viral genome and is cleaved before the appearance of the signal sequence for the precursor of one of the two envelope proteins (E_1 and E_2). After cleavage of the core protein, the ribosomes become attached to the endoplasmic reticulum and translate the two glycoproteins incorporated into the viral particle. Since the cell is essentially taken over by the viral infection in a synchronous manner, the formation of oligosaccharide structures may be viewed in a temporal fashion. Figure 5 diagrams the temporal pattern of carbohydrate modification, fatty acylation, and cleavage of the glycoproteins of this virus. It is interesting to note that fatty acylation precedes proteolytic cleavage of the precursor polypeptide (pE_2) (Schmidt and Schlesinger, 1980). This suggests that the fatty acylation could be responsible for a change in conformation of the precursor that leads to the cleavage of the precursor pE_2 to E_2. On the other hand, it is clear that during the maturation of Sindbis proteins the stepwise addition of N-acetylglucosamine, galactose, fructose, and sialic acid to form complex-type oligosaccharides spans the fatty acid addition and proteolytic cleavage steps. In contrast, the recent investigation of the processing of the pro-adrenocorticotropic-hormone-endorphin (Philips et al., 1981) demonstrates that the conversion of high-mannose to complex oligosaccharide structures is complete before the cleavage of the precursor polypeptide.

CONCLUSION

The phenotypic relevance of glycosylation remains to be determined. With the mechanisms governing glycosylation becoming understood and somatic mutants blocking the glycosylation pathway at various steps now available, it may soon be apparent what role, if any, the pattern of glycosylation may play in the morphological development of complex cell-cell interactions such as those found in the nervous system. It should be noted, however, that conflicting observations as to the importance of precise glycosylation may be drawn from several systems. At one specific site of Sindbis virus E_1 protein, the extent of glycosylation is variable; in chicken cells, the site is glycosylated 75 to 80% of the time; and in baby hamster kidney cells, it is not glycosylated at all. Significantly, growth of retrovirus (RNA tumor virus) is not affected by the block of complex oligosaccharide structures in the somatic mutants of Chinese hamster

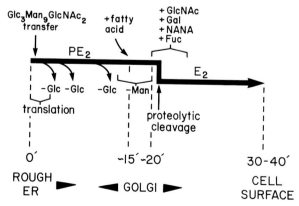

FIG. 5. Intracellular migration and processing of the E_2 glycoprotein of Sindbis virus.

cells. In contrast, one of the most consistent correlations of cellular transformation is the occurrence of highly sialylated glycoproteins on the cell surface. Ivatt and coworkers (1980) have compared the glycosylation pathways of avian sarcoma virus in transformed rodent cells with those of rat cells of normal phenotypes and concluded that the altered glycosylation pattern occurs not at the initial selection of glycosylation sites but at the expense of specific high-mannose structures, which are further modified to complex oligosaccharide structures. Stanley and Sudo (1981), employing somatic mutants of Chinese hamster cells, have suggested that dramatic changes in cellular morphology of these cells may result from a block in the linkage of a specific galactose residue prior to the generation of the complex asparagine-linked structures. Whether this mutation reflects a significant aspect of cellular recognition or inadvertent loss of function of particular cellular glycoproteins through insolubility remains to be clarified. It should be noted that the production of infectious VSV is unaffected in these cells, although the viral glycoprotein of the envelope of the virus particle shares the same glycosylation deficiency as the mutant cell.

In summary, it is clear that the glycosylation of proteins has the potential for regulation of cellular recognition. It is not apparent, however, that this potential is used in the epigenetic determination of cellular interactions.

ACKNOWLEDGMENTS

This work was supported by grants CA14142 and CA14051 (to P. W. Robbins and S. E. Luria) from the National Cancer Institute, Department of Health, Education and Welfare.

REFERENCES

Hubbard, S. C., and Robbins, P. W. (1979): Synthesis and processing of protein-linked oligosaccharides in vivo. *J. Biol. Chem.*, 254:4568–4576.
Hubbard, S. C., and Robbins, P. W. (1980): Synthesis of the N-linked oligosaccharides of glycoproteins: Assembly of the lipid-linked precursor oligosaccharide and its relation to protein synthesis in vivo. *J. Biol. Chem.*, 255:11782–11793.
Ito, S., Yamashita, K., Spiro, R. G., and Kobata, A. (1977): Structure of a carbohydrate moiety of a Unit A glycopeptide of calf thyroglobulin. *J. Biol. Chem. Tokyo*, 81:1621–1631.
Ivatt, R. J. (1981): Regulation of glycoprotein biosynthesis by the formation of specific glycosyltransferase complexes. *Proc. Natl. Acad. Sci. USA*, 78:4021–4025.
Ivatt, R. J., Hubbard, S. C., and Robbins, P. W. (1980): Processing of cell surface glycoproteins in normal and transformed fibroblasts. In: *Tumor Cell Surfaces and Malignancy*, edited by R. Hynes and C. F. Fox, p. 857. A.R. Liss, New York.
Kornfeld, S., Li, E., and Tabas, I. (1978): The synthesis of complex-type oligosaccharides: II. Characterization of the processing intermediates in the synthesis of the complex oligosaccharide units of the vesicular stomatitis virus G protein. *J. Biol. Chem.*, 253:7771–7778.
Li, E., and Kornfeld, S. (1978): Structure of the altered oligosaccharide present in glycoproteins from a clone of Chinese hamster ovary. *J. Biol. Chem.*, 253:6426–6431.
Li, E., Tabas, I., and Kornfeld, S. (1978): The synthesis of complex-type oligosaccharides: I. Structure of the lipid-linked oligosaccharide precursor of the complex-type oligosaccharides of the vesicular stomatitis virus G protein. *J. Biol. Chem.*, 253:7776–7770.
Marshall, R. D. (1972): Glycoproteins. *Annu. Rev. Biochem.*, 41:673–702.
Philips, M. A., Budarf, M. L., and Herbert, E. (1981): Glycosylation events in the processing and secretion of pro-ACTH-endorphin in mouse pituitary tumor cells. *Biochemistry*, 20:1666–1675.
Reading, C. L., Penhoet, E. E., and Ballou, C. E. (1978): Carbohydrate structure of vesicular stomatitis virus glycoprotein. *J. Biol. Chem.*, 253:5600–5612.
Rothman, J. E., Bursztyn-Pettegrew, H., and Fine, R. E. (1980): Transport of the membrane glycoprotein of vesicular stomatitis virus to the cell surface in two stages by clathrin-coated vesicles. *J. Cell Biol.*, 86:162–171.
Schmidt, M. F. G., and Schlesinger, M. J. (1980): Relation of fatty acid attachment to the translation and maturation of vesicular stomatitis and Sindbis virus membrane glycoproteins. *J. Biol. Chem.*, 255:3334–3339.
Snider, M. D., Sultzman, L. A., and Robbins, P. W. (1980): Transmembrane location of oligosaccharide-lipid synthesis in microsomal vesicles. *Cell*, 21:385–392.

Stanley, P., and Sudo, T. (1981): Microheterogeneity among carbohydrate structures may be important in recognition phenomena. *Cell*, 23:763–769.

Turco, S. I., Stetson, B., and Robbins, P. W. (1977): Comparative rates of transfer of lipid-linked oligosaccharides to endogenous glycoprotein acceptors in vitro. *Proc. Natl. Acad. Sci. USA*, 74:4411–4414.

Waechter, C. I., and Lennarz, W. J. (1976): Lipid-linked sugars in glycoprotein synthesis. *Annu. Rev. Biochem.*, 45:95–112.

Precursor Processing and the Neurosecretory Vesicle

Harold Gainer

ABSTRACT

The organization of the neuron is considered as it relates to the processing of peptide hormones. Discussed are posttranslational events involved in processing of the hypothalamic neurohypophysial prohormones, propressophysin and pro-oxyphysin, to vasopressin and oxytocin, respectively, and the acidic carrier proteins known as neurophysins. This processing occurs primarily in the axon during transport of the secretory vesicles to the nerve terminal.

Peptidergic neurosecretory cells produce neuropeptides by first synthesizing precursor proteins that are subsequently transformed to the smaller peptides by posttranslational cleavage. Evidence for this mechanism in neurons comes principally from experiments on various molluscan neurosecretory cells and on the magnocellular nuclei of the mammalian hypothalamus (Gainer et al., 1982). In contrast to other cells that produce and secrete peptides, neurons have very long axons that can comprise as much as 99% of the cytoplasm of the cell. These axons are completely devoid of ribosomes, so that translation of peptide precursors must take place in the cell body, and precursors and peptide products must be packaged, processed, and transported down the axon for secretion by the nerve terminals.

Axonal transport was first described by Weiss and Hiscoe (1948). Characteristics of different rates of transport and the organelles, functions, and proteins associated with them are discussed in Lasek and Shelanski (1981). Transport has been divided into five groups based on velocity. The highest velocity, 240 to 400 mm/day, is associated with membrane components that are destined principally for the secretory terminal and for the axon itself. These components travel as membrane vesicles or as tubular elements of the endoplasmic reticulum, and have a high glycoprotein content. The glycoproteins appear to be involved in membrane processes and cell-cell communication. The next fastest rate of transport, 34 to 68 mm/day, is associated with larger membrane-bound organelles (e.g., mitochondria). The neurosecretory vesicles carrying the peptide hormones fall in between these groups and are moved at a rate of about 140 mm/day. Two slow transport components are not membrane-bound and include various cytoskeletal and other proteins. Extensive discussion of the slow components of axoplasmic transport can be found in Lasek and Shelanski (1981).

The supraoptic nucleus in the hypothalamus is amenable to studies of the axonal transport and processing of peptide hormones because the cell bodies are tightly localized and most of the axons project to the posterior lobe of the pituitary (Fig. 1). The large terminal areas in the pituitary are specialized for storage and secretion of neuropeptides and contain a large number of vesicles. Ultrastructural examination of the median eminence and neural lobe shows the

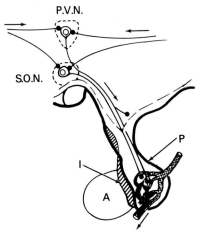

FIG. 1. The hypothalamic-neurohypophysial system. Separate neuronal perikarya in the supraoptic (S.O.N.) and paraventricular nuclei (P.V.N.) project their axons toward the posterior pituitary (P) where they release their peptide hormones and proteins into the blood from nerve endings. A, anterior lobe; I, intermediate lobe.

large number of neurosecretory vesicles (160 to 190 nm) being transported from the hypothalamus to the pituitary (Morris et al., 1978).

Axonal transport studies have been done in the rat even though the axons between the hypothalamus and pituitary are fairly short, 6 to 7 mm. By stereotaxically injecting radioactive amino acids, e.g., [^{35}S]methionine, directly over the supraoptic nucleus, it is possible to label proteins being synthesized in the cell bodies. After 4 hr, a substantial amount of labeled protein is transported to the posterior pituitary and essentially none to adjacent tissues. This transport can be completely blocked by treatment with colchicine, which disrupts microtubules. The distribution and type of labeled proteins can be examined over time following the injection. A micropunch technique (Palkovits, 1973) is used to obtain proteins from the median eminence and posterior pituitary, and labeled proteins are resolved by SDS-polyacrylamide gel electrophoresis (Fig. 2). Labeled neurophysin is first seen in the median eminence at 1 hr; it is maximal by 2 hr, and passes through so that by 24 hr there is very little left in the median eminence. Other labeled proteins pass through the median eminence at different rates. The terminals in the posterior pituitary are loaded with labeled neurophysin by 6 hr. After 8 days the stored labeled neurophysin is still retained, and, in addition, other more slowly transported proteins have begun to accumulate, e.g., actin and tubulin. The transport rate of proteins of a given molecular weight class can be established over time by comparing the amount of labeled protein in the median eminence to that in both the median eminence and posterior pituitary (Fig. 3). Note that peaks seen in Fig. 3 correspond to the major transport rates found in other neurons. Further resolution of the labeled proteins and their rate components is possible by employing two-dimensional electrophoresis and fluorography (Fig. 4). By this approach, it is possible to show that neurophysin is in a fast transport component (140 mm/day), and actin and tubulin are exclusively in the slow components of transport.

The neurosecretory vesicle represents one of the few organelles that can be clearly associated with the fast axonal transport component. This organelle is more than simply a vehicle in transport and secretion process; it also appears to play an active role in the biosynthesis of hormones and neurophysins.

In 1964, Sachs and Takabatake hypothesized that vasopressin was biosynthesized via a prohormone mechanism, and later proposed that the precursor contained both vasopressin and its associated neurophysin (Sachs et al., 1969). Later pulse-chase studies (for a review, see

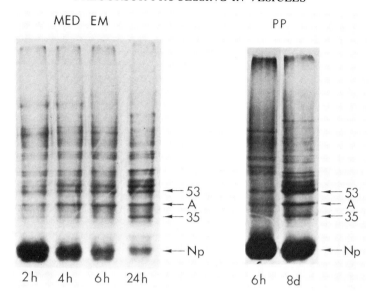

FIG. 2. Autoradiograph of gradient slab SDS gel (4 to 12% polyacrylamide) of protein transported to the median eminence (MED EM) and posterior pituitary (PP) at various times after pulse label of the supraoptic nucleus with [^{35}S]methionine. Median eminence samples at 2, 4, 6, and 24 hr and posterior pituitary samples at 6 hr and 8 days are shown. Arrows point to migration positions of actin (A), neurophysin (Np), and to prominent 53-kD and 35-kD-labeled protein bands. Note that most of the labeled neurophysin band passes through the median eminence in the earliest times, while the higher molecular weight protein bands (A, 53, and 35) become more labeled with time. The labeled protein bands also increase in the posterior pituitary with time (from Fink et al., 1981).

Brownstein et al., 1980) have led to the identification of prohormones in the rat, one for oxytocin and its neurophysin, and the other for arginine vasopressin and its neurophysin.

Understanding of the processing of the vasopressin precursor has been aided by using [^{35}S]cysteine to label proteins in the supraoptic nuclei, as this precursor is relatively rich in this amino acid as compared to other proteins, and by studying the Brattleboro rat that has diabetes insipidus and does not make this precursor. In particular these experiments have helped to establish that different precursor molecules and neurophysins are associated with vasopressin and oxytocin, which are synthesized in different cells in the supraoptic nucleus. The oxytocin- and vasopressin-related neurophysin precursors were identified by comparing the labeled proteins in the supraoptic nucleus and posterior pituitary of normal and Brattleboro rats (Fig. 5). Twenty minutes after injecting [^{35}S]cysteine near the supraoptic nuclei of Brattleboro rats, only one major labeled protein peak, pI (isoelectric point) 5.4, could be identified by isoelectric focusing. After 1 hr another protein, pI 5.1, appears prominent in this nucleus, and finally a protein of pI 4.6 can be demonstrated in the posterior pituitary after 24 hr. The sequence of processing of the oxytocin precursor is then from a polypeptide of pI 5.4 to 5.1 to 4.6 (neurophysin). The normal animal data (Fig. 5) indicate that the vasopressin precursor goes through a series of polypeptides of pI 6.1 to 5.6 to 4.8 (neurophysin). The precursors of oxytocin and vasopressin can be isolated, immunoprecipitated with antibodies to neurophysin, and characterized by two-dimensional gel electrophoresis (Fig. 6). (As noted by other workers studying other peptide precursor molecules, antibodies to oxytocin and vasopressin do not easily immunoprecipitate their precursors.)

The current status of information about these prohormones is given in Table 1. The vasopressin-neurophysin precursor (propressophysin) is a protein of about 19,500 MW with a pI equal to 6.1, and contains vasopressin, the vasopressin-associated neurophysin (pI = 4.8), and

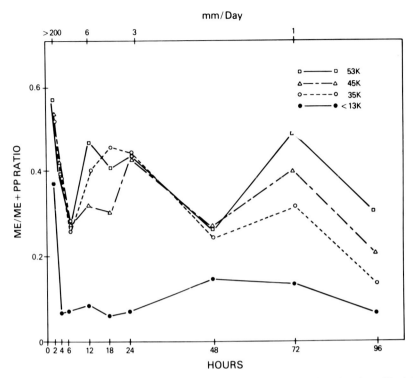

FIG. 3. Quantitative evaluation of the axonal transport of the 53-kD, 45-kD (comigrating with actin A), 35-kD and < 13-kD (Np) protein bands through the median eminence (see Fig. 2). The ratio of disintegrations per minute (DPM) in the median eminence (ME) to the total transported DPM (median eminence + posterior pituitary, PP) for each protein band, reflects the relative amount of the labeled protein band that has arrived halfway down the axon (ME) at various times after injection of [^{35}S]-methionine in the supraoptic nucleus, versus the total amount of the transported labeled protein. Therefore, a ratio equal to 1.0 indicates that the labeled protein band was in the ME but had not yet arrived at the nerve terminal (PP). As the labeled protein band enters the PP the rates will fall, and will rise again only when a new "wave" of transport of the labeled protein band passes through the ME. The fast component is characterized by the < 13-kD curve, which contains mostly Np at time points less than 24 hr. Note that in addition to being found in the fast component, the other three labeled molecular weight bands are also found in waves at 3 to 6 mm/day and 1 mm/day. (From Fink et al., 1981.)

a glycopeptide of about 8,000 MW. In contrast, the oxytocin-neurophysin precursor (pro-oxyphysin) does not contain carbohydrate, is smaller (MW about 15,000), has a pI = 5.4, and contains oxytocin, the oxytocin-associated neurophysin (pI = 4.6), and possibly another cysteine-containing peptide. It is interesting that the "intermediates" shown in Table 1 and Fig. 6 appear to contain all the peptide components found in their respective precursors. They differ from the precursors primarily in their isoelectric points, and their rates of appearance in pulse-chase studies (Fig. 5). Only further studies employing amino acid sequencing techniques will allow us to determine whether these "intermediates" are true products of processing or are independent gene products.

Recent studies on cell-free translation of bovine hypothalamic mRNA in wheat germ extracts and rabbit reticulocyte lysates (Koch and Richter, 1980) are in good agreement with the data shown in Table 1 for rat precursors. In conventional *in vitro* translation systems, the precursors are synthesized as preprohormones. The molecular weights of bovine prepropressophysin and prepro-oxyphysin are 21,000 and 16,500, respectively. When the translation is performed in the presence of microsomal membranes, the signal sequences are removed and the bovine

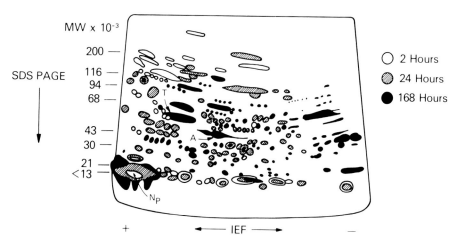

FIG. 4. Distribution of major labeled proteins (revealed by fluorography) on two-dimensional gels, 2 *(white area)*, 24 *(striped area)*, and 168 hr *(black area)*, following injection of [^{35}S]methionine into the supraoptic nucleus. Arrows indicate migration positions of tubulin (T), actin (A), and neurophysin (Np); + and − indicate poles for isoelectric focusing (IEP). (From Fink et al., 1981.)

propressophysin and pro-oxyphysin molecular weights are 19,000 and 15,500, respectively. As in the case of the rat prohormones, only the bovine propressophysin can be glycosylated. No pI values for the *in vitro* translated prohormones have been reported.

Another question important to understanding the processing of peptide hormones is the order of the peptide components within the prohormones. Given the fact that pro-oxyphysin contains only two known peptides (i.e., oxytocin and oxytocin neurophysin), and propressophysin contains, in addition to arginine vasopressin and arginine vasopressin neurophysin, a glycopeptide moiety, then the order of the constituent parts of these molecules allows for two possibilities for pro-oxyphysin, and six possibilities for propressophysin. These possibilities are illustrated in Fig. 7. Biosynthetic studies indicate that the only methionine residues in the rat precursor are in the neurophysin molecules (H. Gainer, J. T. Russell, and M. J. Brownstein *unpublished data*); there is known to be only one in each neurophysin close to the N-terminus (i.e., at residue 5 in rat vasopressin neurophysin and at residue 7 in rat oxytocin neurophysin). The position of this methionine in the alternative propressophysin and pro-oxyphysin structures is illustrated in Fig. 7. Since cyanogen bromide cleavage of proteins is specifically at methionine residues, then such treatment of the precursors illustrated in Fig. 7 should produce different peptide fragments depending on the order of the constituent peptides. When [^{35}S] cysteine pro-oxyphysin is treated with cyanogen bromide, a [^{35}S]-labeled neurophysin-sized molecule and a [^{35}S]-labeled oxytocin-sized molecule were generated. These results are consistent only with the top model in Fig. 7 where oxytocin is on the N-terminal side of neurophysin. Cyanogen bromide treatment of [^{35}S]-labeled propressophysin generated a large [^{35}S]-labeled fragment (close to the size of the propressophysin itself) and a [^{35}S]-labeled vasopressin-sized fragment. The only structure in Fig. 7 that would predict such fragments would be the one in which vasopressin is at or near the N-terminal of the propressophysin, followed by neurophysin, and then the glycopeptide nearest to the C-terminus. Such experiments are obviously indirect approaches to a question that would be best solved by knowledge of the amino acid sequences of the precursors. Until the sequences are known, these conclusions should be regarded only as tentative suggestions of the order of the prohormone constituents.

The question as to where conversion of prohormones takes place can be answered in part by axonal transport studies (Fig. 8). One hour after injection of [^{35}S]-cysteine over the supraoptic

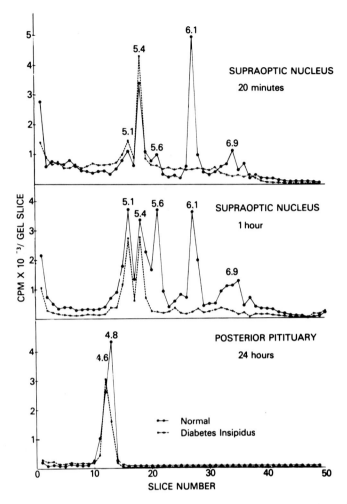

FIG. 5. Isoelectric focusing of [^{35}S]-labeled proteins extracted from the supraoptic nuclei and posterior pituitaries of normal and Brattleboro (homozygous for diabetes insipidus) rats after injecting [^{35}S]cysteine into the supraoptic nuclei. The pI values (isoelectric points) of the labeled protein peaks are indicated. The Brattleboro rats lack the pI 5.6, 6.1, and 6.9 peaks, and the pI 4.8 neurophysin. (From Brownstein and Gainer, 1977.)

nucleus, it is possible to immunoprecipitate from that region both precursor and intermediate molecules containing neurophysin, indicating that some processing does take place in the neuronal cell bodies. After 24 hr, all the precursors have been converted or transported down the axon and only some residual neurophysin remains in the supraoptic nuclei. When immunoprecipitable proteins containing neurophysin are examined in the median eminence, halfway down the axon, after 1 hr, prohormones, not products, are primarily found. This implies that a substantial amount of conversion takes place within the neurosecretory vesicles, as these are the only transport vehicles that have newly synthesized, neurophysin-containing material. Even in the posterior pituitary at 2 hr following injection, precursors as well as the products are present (Fig. 9). Thus, processing takes place in the cell body, axons, and terminals; in the last two cases it must occur within the secretory vesicles.

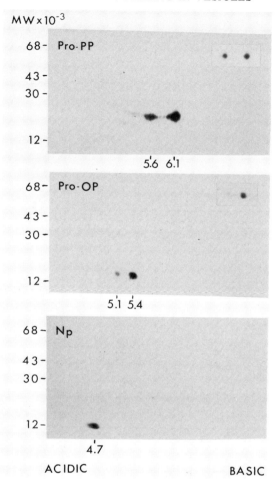

FIG. 6. Two-dimensional gels illustrating the labeled prohormones for oxytocin (pro-oxyphysin, Pro-OP) and vasopressin (propressophysin, Pro-PP) isolated from [^{35}S]-cysteine injected supraoptic nuclei (Table 1). The migration of labeled neurophysin (Np) is also shown. **Insets:** Immunoprecipitates using antisera directed against rat neurophysin.

Given the hypothesis that the neurosecretory vesicle is the major site for prohormone conversion (Fig. 10), then one would predict that appropriate converting enzymes should be present in the vesicles. Inspection of the amino acid sequences of all those prohormones for which such information is available (see D. Steiner, *this volume*) reveals that they all contain pairs of basic amino acid residues (i.e., lysine and/or arginine) at their expected sites of enzymatic cleavage. Hence it has been proposed (see D. Steiner, *this volume*) that the converting enzymes for prohormones should involve trypsin-like endopeptidases and carboxypeptidase-B-like enzymes. Fletcher and associates (1980) have described a tryptic-like enzymatic activity in secretory granules from anglerfish pancreatic islet tissue, which successfully converts prosomatostatin. The enzyme appears to be a vesicle-associated thiol proteinase, with a specificity for arginine residues. This enzymatic activity has a pH optimum of around 5.2, but is not affected by known inhibitors of cathepsin-B, a well-known lysosomal acid protease.

Indirect observations suggest an acidic (pH 5 to 6) milieu within the neurosecretory vesicle. First, vasopressin and oxytocin, which are believed to be bound to their respective neurophysin

TABLE 1. *Properties of rat neuropeptide/neurophysin precursors*

Molecule	pI	Molecular weight		Molecule contains[c]		
		15% SDS gel[a]	G75-GuHCl[b]	Np	AVP or OT	CHO
Propressophysin (Pro-PP)						
Precursor	6.1	19,500	20,500	+	+	+
Intermediate	5.6	18,000	—	+	+	+
Prooxyphysin (Pro-OP)						
Precursor	5.4	15,000	18,700	+	+	—
Intermediate	5.1	14,000	—	+	+	—
Neurophysin (Np)						
Neurophysin (AVP)	4.8	10–12,000	10–12,000	+	—	—
Neurophysin (OT)	4.6	10–12,000	10–12,000	+	—	—

[a]Molecular weight determined by electrophoresis of purified precursors, intermediates, and neurophysins (using preparative isoelectric focusing) on one-dimensional 15% SDS gels.

[b]Molecular weight determined by chromatography on G75 Sephadex columns in the presence of 6M Gu-HCl.

[c]Presence of neurophysin (Np) in isolated proteins was determined by direct immunoprecipitation with antirat Np, or by demonstrating binding of 10-KD tryptic product to lysine vasopressin affinity columns. Presence of arginine vasopressin (AVP) or oxytocin (OT) was determined by binding of tryptic peptide product to Np affinity columns, and by immunoprecipitation of tryptic peptides by antibodies to AVP (for Pro-PP) or OT (for Pro-OP). Presence of carbohydrate (CHO) on protein was determined by binding to concanavalin A affinity columns.

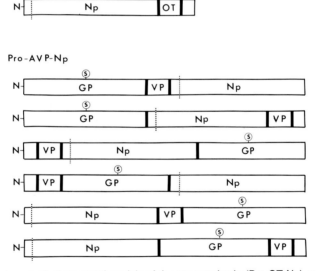

FIG. 7. Alternative hypothetical structural models of the pro-oxyphysin (Pro-OT-Np) and propressophysin (Pro-AVP-Np) precursors. The N-termini of the molecules are on the left side. *Vertical dotted lines* drawn through the molecules show the positions of the only known methionine residues. OT, oxytocin; Np, neurophysin; VP, vasopressin; GP, glycopeptide; S, sugar moiety.

carrier proteins in the vesicle, bind at an optimum pH of about 5.5. Second, morphological studies (Morris et al., 1978) indicate that the dense cores of neurosecretory vesicles in the posterior pituitary are most stable if fixation is performed at an acidic pH (i.e., around pH 5).

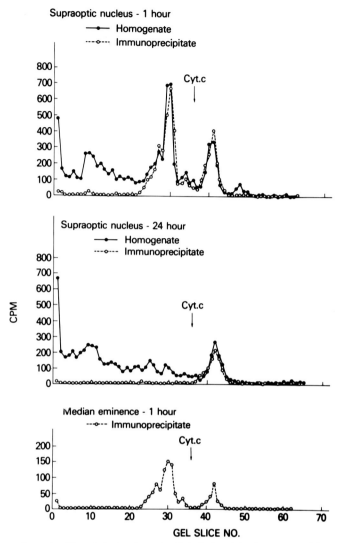

FIG. 8. Acid-area polyacrylamide gel electrophoresis of labeled proteins in extracts of the supraoptic nucleus 1 and 24 hr after [^{35}S]cysteine injection adjacent to the supraoptic nucleus *(solid lines)*. The labeled proteins in these extracts that were bound by antirat neurophysin antisera are shown by the *dotted lines*. Note that the immunoreactive labeled proteins arriving in the median eminence after 1 hr are largely in the higher molecular weight forms.

This issue of intravesicular pH is central since it defines at least one aspect of the microenvironment within the vesicles in which the converting enzymes are presumed to work.

For this purpose, relatively pure populations of stable vesicles need to be isolated, and various characteristics of these vesicles need to be determined (e.g., the internal pH). Russell (1981) has isolated intact vesicles using differential centrifugation and iso-osmotic density gradient centrifugation, using Percoll and metrizamide (Fig. 11). The distribution of marker proteins on Percoll gradients shows that a vesicle population can be obtained relatively free of mitochondria and lysozymes (Fig. 12). The specific activity of these isolated vesicles, i.e., milligrams vasopressin per millograms protein, is higher than for vesicles isolated by other procedures (Russell, 1981).

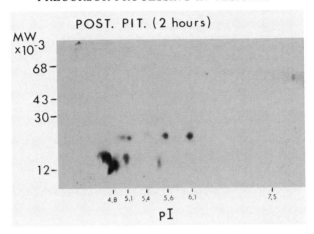

FIG. 9. Autoradiograph of a two-dimensional gel containing [^{35}S]cysteine proteins that were obtained from rat posterior pituitary 2 hr after injection of the labeled amino acid ([^{35}S]cysteine) into the supraoptic nucleus. Note, in addition to labeled neurophysin (around pI 4.8), prominent labeled spots corresponding to the pI 6.1 propressophysin and pI 5.6 intermediate, and the pI 5.1 pro-oxyphysin intermediate.

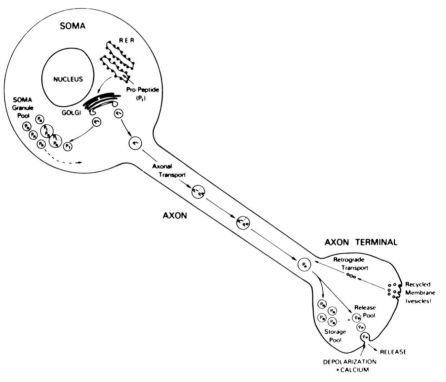

FIG. 10. Hypothetical model of biosynthesis, translocation, processing, and release of peptides in the hypothalamus-neurohypophysial system. The secretory vesicle is viewed as a principal site of the conversion of the prohormone (P_1) to smaller peptide and protein products (P_n). R.E.R., rough endoplasmic reticulum. (From Gainer et al., 1977.)

The internal pH of the vesicles can be determined by the intravesicular to extravesicular labeled ratio of methylamine (Russell and Holz, 1981). The volume of intra- and extracellular space is established using [^{3}H]H$_2$O and [^{14}C]dextran. The pH in the extracellular fluid is varied

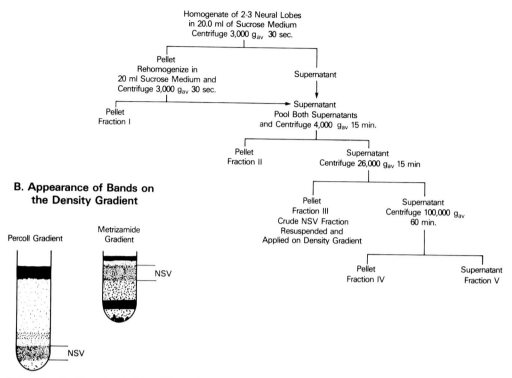

FIG. 11. **A:** Flow sheet of the differential centrifugation of the homogenates of bovine neurohypophyses. **B:** Appearance of particulate matter as bands on density gradient centrifugation of fraction III. See Russell (1981), for details of gradient formation.

and the ΔpH across the membrane is measured. A plot of the ΔpH versus the extracellular pH is linear, and extrapolates to the internal vesicular pH, about 5.5 (Fig. 13). This pH is the same as that which is optimal for binding of vasopressin and oxytocin to their respective neurophysins. Although this binding is quite weak—affinity constant 10^{-5}—it is compatible with the maintenance of high concentrations of these peptides in secretory vesicles. This pH in the secretory vesicles is also very similar to that reported for chromaffin granules (Johnson and Scarpa, 1979), which also process peptide hormones. Maintenance of the internal pH may be critical to both intravesicular organization and regulation of enzyme activity.

As with the chromaffin granule (Johnson and Scarpa, 1979), the pH of the neurosecretory vesicle appears to be maintained to a large extent by a Mg^{++}-dependent adenosine triphosphatase (ATPase). The membrane potential across the vesicular membrane can be determined by measuring the extravesicular and intravesicular distribution of thiocyanate. In the absence of adenosine triphosphate (ATP), the change in membrane potential is consistent with leakage of hydrogen ions out of the vesicle; in the presence of ATP the potential is reversed (Fig. 14). This is probably a result of an electrogenic hydrogen pump since the hydrogen ionophore FCCP prevents this shift in potential.

To study prohormone converting activity in isolated secretory vesicles, Y. Peng Loh, in our laboratory, has used pro-opiocortin as a model substrate. This prohormone's sequence is known (see D. Krieger, and E. Herbert et al., *this volume*), and it contains a variety of dibasic residues (e.g., Lys-Lys, Lys-Arg) that are not cleaved during processing *in situ*, as well as certain

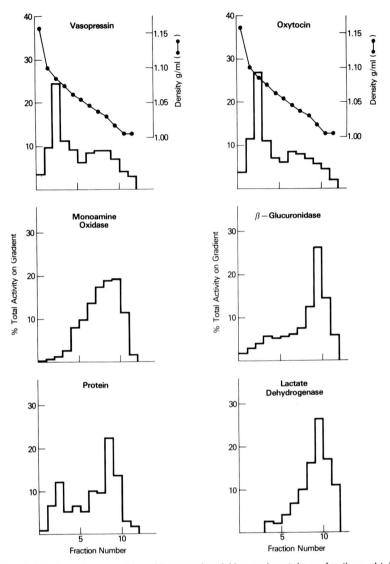

FIG. 12. The distribution of enzymatic and hormonal activities and protein on fractions obtained from a Percoll-sucrose gradient. Fraction III was suspended in 1.0 ml of sucrose medium and mixed with 11.5 ml of 30% Percoll solution in 0.25 M sucrose, 10 mM HEPES (pH 7.0). The mixture was centrifuged at 50,000 × g_{av} for 45 min, and fractions were collected by piercing the bottom of the tube. *Ordinate*, percentage of the total activity applied on the gradient. Note that nearly 40% of the hormonal activities are present in fractions 2 to 4, which contain negligible amounts of monoamine oxidase and β-glucoronidase. (From Russell, 1981.)

dibasic residues that are cleaved. This feature of pro-opiocortin is particularly useful in a test assay, since inappropriate proteases, such as cationic trypsin, will inappropriately degrade this precursor (Loh and Gainer, 1980). In this regard, it is important to note that there was no tryptic-like activity detected in secretory granule (vesicle) lysates using conventional peptides as substrates for the analysis of pancreatic trypsin or cathepsin-B.

However, when [³H]arginine-labeled toad pro-opiocortin was incubated in the presence of lysed rat neurointermediate lobe secretory granules (between pH 5 to 6), the prohormone was converted to appropriate peptide products (Fig. 15; the immunologically identified products are

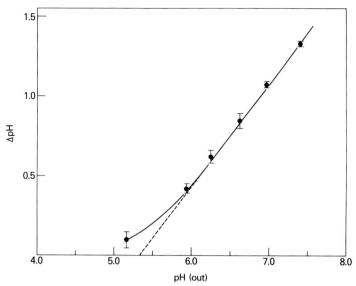

FIG. 13. Measurement of ΔpH in neurosecretory vesicles as a function of medium pH. The medium contained 0.25 M sucrose and 10 mM PIPES adjusted to the appropriate pH and neurosecretory vesicles (11.8 mg protein/ml). The measurement and calculation were performed as described in Russell and Holz (1981).

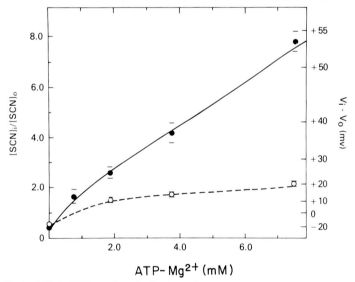

FIG. 14. The effect of Mg^{2+}-ATP on the distribution of thiocyanate (SCN^-) in isolated neurosecretory vesicles. The reaction mixture contained 0.25 M sucrose, 10 mM HEPES (pH 7.0), plus [^{14}C]thiocyanate and neurosecretory vesicles (10.6 mg protein/ml). Varying amounts of Mg^{2+}-ATP concentrations were also present. *Open circle*, incubation solution-contained FCCP; *closed circle*, no FCCP present. Incubations and calculations of the thiocyanate concentration ratio [SCN_i^-]/[SCN_o] as described for methylamine in Russell and Holz (1981).

shown in Fig. 15B). These products were stable in the lysate for up to 18 hr (see inset in Fig. 15A), and the labeled prohormone was not degraded at all when incubated in the absence of secretory granules. Similar results were obtained when bovine neurosecretory vesicles (isolated as described earlier, Figs. 11 and 12) were used as the source of converting enzyme.

The pH dependency of this converting activity is illustrated in Fig. 16. The optimum pH for converting activity is around 5, and is greater than 50% of maximum activity at the internal pH found in the vesicle (i.e., at pH 5.5). At pH values greater than 7, the converting activity

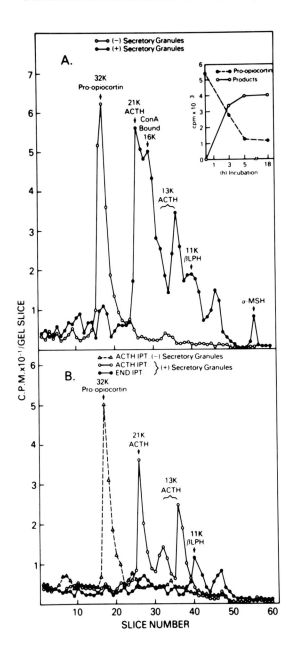

FIG. 15. Conversion of [³H]arginine pro-opiocortin, at pH 5.0, by lysed rat neurointermediate lobe secretory granules. **A:** Acid-urea gel profiles of labeled proteins and peptides following a 5-hr incubation (pH 5.0, 37° C) of the [³H]pro-opiocortin substrate in the absence and presence of lysed secretory granules. Note that conversion of the pro-opiocortin to peptide products occurs only in the presence of secretory granule lysate, and that the migration positions on the gel of the major labeled peptide peaks correspond to migration positions of peptides derived from pro-opiocortin (indicated by *arrows*). *Inset:* Time course of conversion of [³H]arginine pro-opiocortin to peptide products by lysed secretory granules at pH 5, 37° C. Under these conditions, conversion was complete by 5 hr. **B:** Immunological identification of peptide products seen in A by immunoprecipitation with adrenocorticotropic hormone (ACTH) and β-endorphin antisera and electrophoresis on acid-urea gels. (From Loh and Gainer, 1982.)

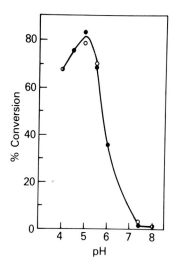

FIG. 16. Influence of incubation pH on the conversion of [³H]arginine pro-opiocortin to the appropriate processed products by granule lysates, as determined by acid-urea gel electrophoresis. Maximum conversion occurs at pH 5 (pH optimum). (From Loh and Gainer, 1982.)

is inhibited. Therefore, the pH of the microenvironment (i.e., the vesicle) appears to be critical for the activity of the converting enzyme. Since the pH of the last cisternae of the Golgi, in which most of the secretory materials are being packaged, is about 7, conversion may normally occur only after packaging. Indeed, it is conceivable that the rate of processing of the prohormone could be regulated *in situ* by the rate of acidification of the secretory vesicle (i.e., by the rate of activity of the Mg-ATPase in the vesicle membrane responsible for generating the proton gradient).

Converting activity appears to be present in both membrane and soluble fractions prepared from the granule lysates (Fig. 17). Preliminary evidence suggests that the granule membrane contains slightly more activity than the soluble fraction (note more unconverted 32-kD pro-opiocortin in Fig. 17B than in Fig. 17A). Figure 17 shows that the immunoprecipitable products produced by the two fractions are identical. Incubations using [³H]phenylalanine-labeled pro-opiocortin followed by immunoprecipitation with β-endorphin antiserum revealed the presence of β-lipotropin (β-LPH), but no peak that co-ran with β-endorphin. Instead an endorphin-related peptide (slice 46) was observed. This peptide is also present in the toad intermediate lobe after pulse labeling with [³H]phenylalanine followed by chase incubation. Its amino acid sequence is yet to be elucidated.

Characterization of this pro-opiocortin-converting activity thus far shows that it (1) is present in membrane and soluble fractions of the granule lysates, (2) has a pH optimum of 5, (3) appears to cleave at dibasic amino acid residues in the precursor, and (4) is inhibited by PCMB, a thiol blocker, and leupeptin and pepstatin A, two protease inhibitors, but not by the protease inhibitors, DFP, TLCK, and chloroquine, or EDTA, a divalent cation chelator. These inhibitor studies indicate that the converting enzyme activity in secretory vesicles is due to an acid, thiol-protease, distinct from any known cathepsin-B-like activity (Loh and Gainer, 1982).

ACKNOWLEDGMENTS

I would like to acknowledge the active collaboration and helpful discussion of Drs. Michael J. Brownstein, James T. Russell, and Y. Peng Loh.

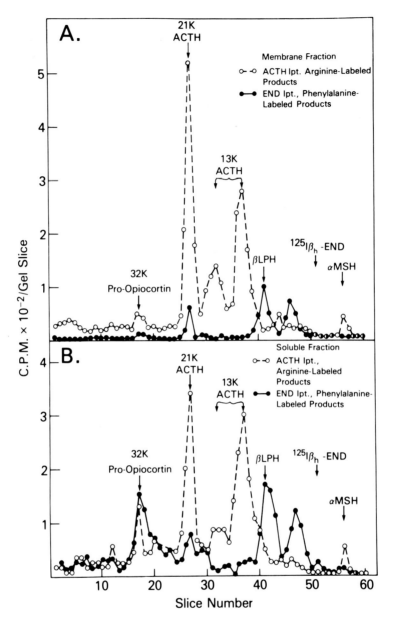

FIG. 17. A: Acid-urea gel profiles of [³H]arginine peptides immunoprecipitated by adrenocorticotropic hormone (ACTH) antiserum and [³H]phenylalanine peptides immunoprecipitated by endorphin antiserum formed after incubation of pro-opiocortin with the membrane fraction of lysed neuro-intermediate lobe secretory granules for 5 hr at pH 5.0. **B:** Acid-urea gel profiles of [³H]arginine peptides immunoprecipitated by ACTH antiserum and [³H]phenylalanine peptides immunoprecipitated by endorphin antiserum formed after incubation of pro-opiocortin with the soluble fraction of lysed neuro-intermediate lobe secretory granules for 5 hr at pH 5.0. The ordinate shows the counts per minute per gel slice, and the abscissa, the gel slice number. The identity of each peak and the marker for ¹²⁵I β-endorphin are indicated by *arrows*. (From Loh and Gainer, 1982.)

REFERENCES

Brownstein, M. J., and Gainer, H. (1977): Neurophysin biosynthesis in normal rats and in rats with hereditary diabetes insipidus. *Proc. Natl. Acad. Sci. USA*, 74:4046–4049.

Brownstein, M. J., Russell, J. T., and Gainer, H. (1980): Biosynthesis, axonal transport, and release of posterior pituitary hormones. *Science*, 207:373–378.

Fink, D. J., Russell, J. T., Brownstein, M. J., Baumgold, J., and Gainer, H. (1981): Multiple rate components of axonally transported proteins in the hypothalamoneurohypophysial system of the rat. *J. Neurobiol.*, 12:487–503.

Fletcher, D. J., Noe, B. D., Bauer, G. E., and Quigley, J. P. (1980): Characterization of the conversion of a somatostatin precursor to somatostatin by islet secretory granules. *Diabetes*, 29:593–599.

Gainer, H., Sarne, Y., and Brownstein, M. J. (1977): Biosynthesis and axonal transport of rat neurohypophysial proteins and peptides. *J. Cell Biol.*, 73:366–381.

Gainer, H., Loh, Y. P., and Neale, E. A. (1982): The organization of post-translational precursor processing in peptidergic neurosecretory cells. In: *Proteins of the Nervous System: Structure and Function.*, edited by B. Haber, J. R. Perez-Polo, and J. D. Coulter, pp. 131–145. Alan Liss Inc., New York.

Johnson, R. G., and Scarpa, A. (1979): Protonmotive force and catecholamine transport in isolated chromaffin granules. *J. Biol. Chem.*, 254:3750–3760.

Koch, G., and Richter, D. (1980): *Biosynthesis, Modification, and Processing of Cellular and Viral Polyproteins*. Academic Press, New York.

Lasek, R. J., and Shelanski, M. L. (1981): Cytoskeletons and the architecture of nervous systems. *Neurosci. Res. Program Bull.*, 19:1–153.

Loh, Y. P., and Chang, T. L. (1982): Pro-opiocortin converting activity in rat intermediate and neural lobe secretory granules, *FEBS Lett.*, 137:57–62.

Loh, Y. P., and Gainer, H. (1980): In vitro evidence that glycosylation of pro-opiocortin and corticotropins influences their proteolysis by trypsin and blood. *Mol. Cell. Endocrinol.* 20:35–44.

Loh, Y. P., and Gainer, H. (1982): Characterization of pro-opiocortin converting activity in purified secretory granules from rat pituitary neurointermediate lobe. *Proc. Natl. Acad. Sci. USA*, 79:108–112.

Morris, J. F., Nordmann, J. J., and Dyball, R. E. J. (1978): Structure-function correlation in mammalian neurosecretion. *Int. Rev. Exp. Pathol.*, 18:1–95.

Palkovits, M. (1973): Isolated removal of hypothalamic or other brain nuclei of the rat. *Brain Res.*, 59:449–450.

Russell, J. T. (1981): The isolation of purified neurosecretory vesicles from bovine neurohypophysis using iso-osmolar density gradients. *Anal. Biochem.*, 113:229–238.

Russell, J. T., and Holz, R. W. (1981): Measurement of ΔpH and membrane potential in isolated neurosecretory vesicles from bovine neurohypophyses. *J. Biol. Chem.*, 256:5950–5953.

Russell, J. T., Brownstein, M. J., and Gainer, H. (1981): Biosynthesis of Neurohypophysial Peptides: The order of peptide components in pro-pressophysin and pro-oxyphysin. *Neuropeptides*, 2:59–65.

Sachs, H., Fawcett, P., Takabatake, Y., and Portanova, R. (1969): Biosynthesis and release of vasopressin and neurophysin. *Recent Prog. Horm. Res.*, 25:447–491.

Sachs, H., and Takabatake, Y. (1964): Evidence for a precursor in vasopressin biosynthesis. *Endocrinology*, 75:943–948.

Weiss, P., and Hiscoe, H. B. (1948): Experiments on the mechanism of nerve growth. *J. Exp. Zool.*, 107:315–395.

Since submission of this chapter, the entire sequence of a cDNA encoding the nonapeptide arginine vasopressin and its neurophysin from bovine hypothalamus has been elucidated [Land, H., Schütz, G., Schmale, H., and Richter, D. (1982): Nucleotide sequence of cloned cDNA encoding bovine arginine vaso-pressin-neurophysin II precursor, *Nature*, 295:299–303.]. The sequence indicates a 19 amino acid pre-sequence followed by AVP connected to NpII by a Gly-Lys-Arg sequence. The carboxyterminal is a 39 amino acid sequence glycopeptide connected to NpII by a single Arg.

Neuropeptide Degradation

Jeffrey F. McKelvy, James E. Krause, and J. P. Advis

ABSTRACT

The mechanisms by which the actions of neuropeptides are terminated have not yet been elucidated, but evidence exists that peptidases may catalyze the breakdown of biologically active neuropeptides to inactive products. Biological significance also may lie in peptidases' ability to cleave certain "mature" neuropeptides to liberate new biological activities. These points are considered by reviewing the actions of three peptidases: enkephalinase, luteinizing hormone-releasing hormone endopeptidase, and post-proline-cleaving enzyme.

The action of peptidases on neuropeptides is of great interest in neurobiology, but the state of information is quite primitive. Despite this, there are some examples that can be used to study the regulation of peptidase activities. Such studies may in turn lead to an understanding of how such regulation is important to the nervous system, particularly by the application of molecular biological techniques, such as measuring levels of messenger RNA that code for the peptidases.

In general, the actions of most neuropeptides, let alone the mechanisms by which they exert their actions, are poorly understood. It is clear, however, that there must be a means for the cessation of these actions. One of the major means of regulating the concentration of low molecular weight amines, such as norepinephrine, dopamine, and serotonin, at sites of action is by high-affinity reuptake mechanisms. There are only two reports on the existence of carrier-mediated transport systems in neural tissue involving peptides. The first was from Udenfriend's laboratory and described the transport of carnosine (β-alanylhistidine) into brain slices (Abraham et al., 1963). The second involves the apparently carrier-mediated transport of the tripeptide pyroglutamyl-histidyl-proline amide (thyrotropin-releasing hormone, TRH) into cerebellar slices (Pacheco et al., 1981). The ability of slices of cerebellum to accumulate TRH with tissue-to-medium ratios as high as 10:1 has been observed and is dependent on temperature and sodium ion concentration. However, the observed affinity of approximately 0.1 µM is relatively low. The generality of this observation is not clear, since the uptake of TRH requires slice preparations and cannot be demonstrated in synaptosomal preparations. Also, it has so far only been shown to occur in the cerebellum.

In contrast to the difficulties encountered in observing peptide uptake, a frequent observation has been that whenever peptides are applied to a variety of tissue preparations, they are rapidly broken down. In addition, over the past several years there has been much descriptive information concerning the wide variety of peptidases found in neural tissue. An important question that must be raised is whether these peptidase activities are relevant to the regulation of the concentration of peptides at their sites of action. There are two possible approaches to answering this question. The first is to make an extract of brain, add a peptide, and observe the fragmentation of the peptide by the tissue. An attempt could then be made to purify the enzymes catalyzing the defined cleavages. The purified enzymes could then be studied to characterize the enzymatic activity, to elicit antibodies, and ultimately to determine whether the enzyme is important in some neural subsystem. The second approach is to study a neural subsystem involving peptides in which there is a probable functional role for a given peptide. The system could be subjected to physiological stimulation, and correlations made between peptidase activity exerted on the peptide and the function of the system. Examples of both approaches are described here.

ENKEPHALIN ENDOPEPTIDASE AND AMINOPEPTIDASE

Malfroy and colleagues (1978) have examined the corpus striatum, a region rich in enkephalin and opiate receptors, and have found an enzyme activity in a particulate fraction derived from this region, which cleaves methionine enkephalin (Tyr-Gly-Gly-Phe-Met) at the Gly^3-Phe^4 bond. Based on the likely similarity of the mechanism of action of the enzyme catalyzing this cleavage and carboxypeptidase A, a molecule was designed that they believed would inhibit the action of this activity. They observed the activity of a dipeptidyl carboxypeptidase that acts on enkephalin and favors aromatic cleavage. In addition, they observed that chronic morphine treatment gave rise to an "adaptive increase" in the level of activity of this enzyme, which they called enkephalinase. This is presumably in response to increased levels of natural opiate peptides caused by a lack of feedback from opiate receptors blocked by the exogenous morphine.

Based on the supposed similarity to carboxypeptidase A and the metal requirements for enzyme activity, the inhibitor thiorphan (Fig. 1) was synthesized (Roques et al., 1980). Thiorphan consists of a derivative of phenylalanine and a thiol group capable of chelating metals. Such an inhibitor should fit into the active site of enkephalinase, if enkephalinase is mechanistically similar to carboxypeptidase A. *In vitro* data showed that it did inhibit the enkephalinase; an *in vivo* test measured delay in the withdrawal of mice tails from a beaker of hot water after the mice had been administered either thiorphan plus the exogenous analgesic D-Ala^2-

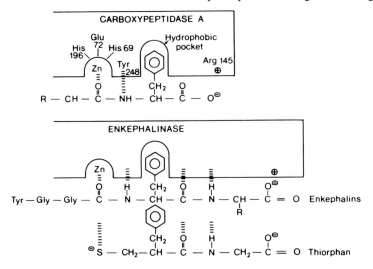

FIG. 1. Structure of thiorphan, an active-site directed inhibitor of enkephalinase (enkephalin dipeptidyl carboxypeptidase), and a model of the enkephalinase activity site and its similarity to that of carboxypeptidase A. It is postulated that inhibition occurs by complexing of the ionized sulfhydryl group of thiorphan with a zinc atom in the metalloendopeptidase. (From Roques et al., 1980.)

Met[5]-enkephalin or the analgesic alone. The animals that had been injected with the exogenous analgesic exhibited a small increase in withdrawal latency over controls, indicating some analgesic activity. After the administration of thiorphan and D-Ala[2]-Met[5]-enkephalin, there was a prolongation of this latency, presumably due to the effect of thiorphan in maintaining concentrations of the exogenous opiate.

Although this approach provides important information, it is also necessary to learn something about the properties of the enzymes catalyzing these reactions. Some interesting data have been obtained from studies of properties of purified peptidases that act on enkephalin. It is well known that a free N-terminal tyrosine residue is essential for opiate activity and it is also known that there is an enzyme in neural tissue that catalyzes the removal of N-terminal tyrosine, the so-called enkephalin aminopeptidase (Jacquet et al., 1976). In studies examining corpus striatum and the degradation of enkephalin, Malfroy and colleagues (1978) found no indication that enkephalin aminopeptidase activity responded to the physiology and neuropharmacology of that system.

An alternative approach involved purifying the tyrosine-removing enkephalinase from whole bovine brain and characterizing the catalytic properties of the enzyme (Hersh and McKelvy, 1981). Figure 2 shows the basis on which it was isolated, namely, the removal of N-terminal tyrosine by the enzyme, liberating free tyrosine, which in a coupled assay will generate the colored product quinone diimine and allow a rapid, continuous assay. This assay was used in purification of the enzyme to homogeneity. It was shown that the enzyme acts on both Met- and Leu-enkephalins with a K_m in the micromolar range, and also that it is a soluble activity since it was purified from the supernatant of lysed brain tissue. However, significant activity was also extractable with detergent from the pelleted cell debris. Purification of the enzyme from the pellet gave a final homogenous enzyme with a virtually identical specific activity and K_m for enkephalin.

This enzyme was isolated on the basis of its ability to hydrolyze the tyrosine residue from the enkephalin pentapeptides. D. Krieger *(this volume)* points out that some des-tyrosine derivatives of the endorphins are biologically active. Des-Tyr-γ-endorphin and des-Tyr-α-endorphine can be shown to be behaviorally active in several behavioral assays. However, without the N-terminal tyrosine, they do not bind to opiate receptors and therefore do not exert their behavioral activities on the basis of opiate receptor action.

The ability of the enkephalin aminopeptidase to act on endorphins as well as enkephalins was tested and the results are shown in Fig. 3 (Hersh et al., 1980), which shows the time course of release of tyrosine from Met-enkephalin and from α-, β-, and γ-endorphine. It was

Aminopeptidase:

Tyr-Gly-Gly-Phe-Met → Tyr
+ Gly-Gly-Phe-Met

L-**Amino acid oxidase:**

Tyr + O_2 → H_2O_2
+ *p*-hydroxyphenylpyruvate

Horseradish peroxidase:

H_2O_2 + *o*-dianisidine
→ quinone diimine + O_2

FIG. 2. Assay scheme used in the purification of bovine brain enkephalin aminopeptidase. (From Hersh and McKelvy, 1981.)

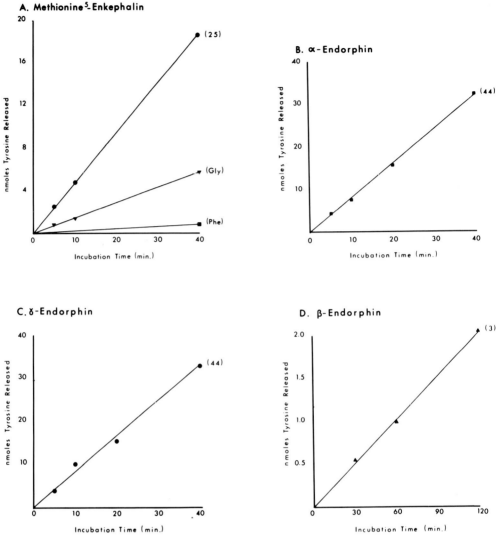

FIG. 3. N-terminal tyrosin cleavage from Met^5-enkephalin, and α-, γ-, and β-endorphins catalyzed by homogeneous bovine brain aminopeptidase. *Numbers in parentheses* are the number of experiments performed. (From Hersh et al., 1980.)

found that α- and γ-endorphin also served as good substrates for the enzyme. In addition, the following differences were noted: when enkephalin was used as substrate, internal residues were removed as well as tyrosine. When endorphins were used there was no evidence for the removal of anything but N-terminal tyrosine; and in the case of β-endorphin, the rate of liberation of tyrosine was very much lower than for the other endorphins.

These data are summarized in Table 1, in which the turnover numbers for the substrates are presented. It can be seen that both α- and γ-endorphin exhibit substantial turnover numbers, about a third that of Met-enkephalin, and that β-endorphin is not a good substrate in this assay. These are solution measurements of enzyme activity and were not made in a physiological context. Since this activity exists in both a soluble and a membrane-bound form and is able to act on a cleavage product of β-lipotropin-generating fragments that have been shown to be

TABLE 1. *Rate of tyrosine release from Met^5-enkephalin and α-, β-, and γ-endorphin by bovine brain aminopeptidase*[a]

Substrate	Rate of tyrosine release (nmoles per min per mg enzyme)	Turnover number (min^{-1})[b]
Met^5-enkephalin	20.4 (6.1)	2,040
α-Endorphin	6.5	634
γ-Endorphin	6.9	689
β-Endorphin	0.10	10

[a]The data from Fig. 3 were used to calculate the rate of tyrosine release. The value in parentheses is the rate of glycine release.
[b]For hydrolysis of the Tyr^1-Gly^2 bond; molecular weight of the enzyme, 100,000 (Hersh and McKelvy, 1981).
From Hersh et al., 1980.

behaviorally active, it may be possible to associate known physiological characteristics with this enzyme. It is clear that regulation of opiate peptide activity could focus on this N-terminal tyrosine based on the structure-activity requirements for opiates and for the des-Tyr derivatives. The results that have been obtained thus far suggest that further exploration of this activity is warranted.

LUTEINIZING HORMONE-RELEASING HORMONE

The question of soluble versus particulate peptidase activities is a good model to pursue in considering a peptide as an informational molecule communicating between nerve cells. Both types of activities may have physiological relevance. For example, a membrane-bound peptidase might act to terminate the action of a peptide signal whereas a soluble form of the enzyme could not. However, the observation of peptidase activity in crude membrane fractions or even fractions prepared by classic subcellular fractionation should be interpreted with caution. Thermodynamic parameters based on binding data must be calculated confidently to conclude which fractions in a subcellular fractionation actually have the higher affinity binding and the higher binding capacity. In addition, there are other considerations; for example, several peptidergic neurons have been shown by morphological techniques to have long dendrites with immunoreactive peptide products and a paucity of granules definable by electron microscopic immunocytochemistry. In the case of luteinizing hormone-releasing hormone (LHRH) neurons, these dendrites can extend for considerable distances (A. Silverman, *personal communication*). The question of release and metabolism of peptides in soluble pools within the cell is one that must be entertained.

An example of how a neural subsystem has been used in asking whether peptidase activity is relevant in regulation involves peptide-secreting neurons in the hypothalamus, which are under the influence of a variety of regulatory signals. Figure 4 illustrates a neuron system in the hypothalamus that projects axons to a neurohemal region in the median eminence where it secretes a peptide product that is then delivered through a short vascular route to the anterior pituitary, including hormone release. The particular model system that was chosen deals with hypothalamic regulation of pituitary gonadotropin secretion by neurosecretion of the decapeptide amide, LHRH. The cell bodies of the LHRH-secreting neurons lie in the medial preoptic area and project their axons down to the median eminence of the hypothalamus, terminating in the neurohemal zone, which leads to the connection between hypothalamus and pituitary. During hypothalamic regulation of pituitary gonadotropin secretion, there are profound changes in the

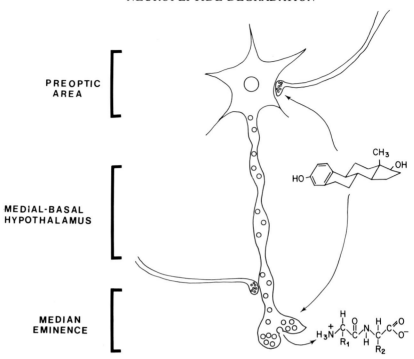

FIG. 4. Schematic view of an LHRH-secreting neuron projecting from the preoptic area to the median eminence in the rat, and under the influence of steroid hormones and synaptic input. For brevity, the output of the neuron is shown to be peptidergic, and estradiol is used as an example of steroid action.

activity of these neurons as they integrate steroid signals from the periphery (estradiol in Fig. 4), phasic timing signals, such as serotonin from the raphe and norepinephrine from the locus coeruleus, and probably peptide signals as well. This orchestration of several kinds of signals onto these final common pathway neurons results in the regulation of pituitary activity necessary for reproductive cyclicity in the organism. This is a system in which molecular biological tools are needed for obtaining necessary regulatory information. The structure of the precursor for the signal (LHRH) that is delivered to the target cells is not known, although attempts to clone the gene coding for the precursor are already under way. In general, biochemical markers must be defined for these systems, and probes must be obtained to look at the level of expression of these markers in physiologically relevant situations.

Experiments were carried out to determine if peptidases contributed to setting the levels of LHRH in these neurons. A micropunch technique was used to sample the areas of cell bodies, axons, and nerve terminals in the system. Broken cell preparations were made and their ability to degrade LHRH was determined by high-performance liquid chromatographic (HPLC) and amino acid analysis of the fragments. The first case that was studied was the entry of an animal into puberty (Advis et al., 1982a,b). It was known from other investigations that the content of LHRH in the hypothalamus drops during the time when the pituitary is stimulated to release luteinizing hormone, which activates the gonads. Figure 5 reflects the dynamics of LHRH content in the nerve terminals under the influence of these regulatory signals. An assay of the activity of enzymes cleaving LHRH during this time course showed that the ability to degrade LHRH paralleled LHRN content except during the period immediately preceding the LH surge (late proestrus), when the activity decreased. This could contribute to the generation of LHRH levels appropriate for the LH surge. Thus, it is probable that degradation contributes to the set

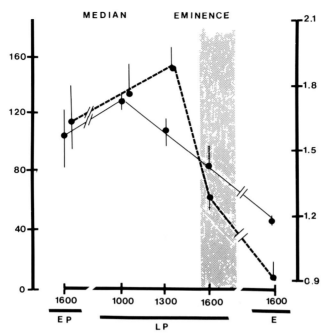

FIG. 5. Relationship between LHRH degradation (●——●, pmoles/μg protein/30 min), LHRH content (●----●, ng/ ME) and the first estrous cycle at puberty in the female Sprague-Dawley rat median eminence. EP, early proestrus; LP, late proestrus; E, estrus (clock time in hours).

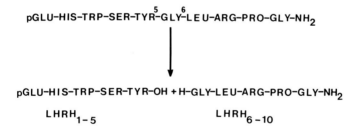

FIG. 6. Site of cleavage of the LHRH molecule by a hypothalamic LHRH metalloendopeptidase.

point for LHRH in the terminals of the actively discharging neurons, since there is a high correlation between the amount of LHRN in the terminal zone and activity of the enzyme. The activity of another enzyme, which cleaves the Pro^9-Gly^{10} bond in LHRH (post-proline-cleaving enzyme) was invariant over the same time course (data not shown).

It was determined by characterization of the products of degradation that the peptidase activity was catalyzing a single cleavage at the Tyr^5-Gly^6 bond, shown in Fig. 6 (Krause et al., 1981). The observed physiological changes could be accounted for by production of the pentapeptide $LHRH_{1-5}$.

Characterization of the peptidase activity indicates that it is a metalloendopeptidase, since it is inhibited by chelators that preferentially bind zinc ion. It also requires a thiol group for activity and does not appear to be a chymotryptic-type activity because it is not inhibited by either the specific inhibitor for chymotrypsin TPCK or the serine protease inhibitor DFP. It is a novel activity compared with other known peptidases described in brain.

The gonadotropin release system was manipulated in an artificial paradigm of steroid feedback to obtain further correlation between the activation of LHRH-secreting neurons and the level of expression of the peptidase (see Fig. 7) (Advis et al., 1982c). In this experimental system at the critical time for integration of signals leading to gonadotropin release, the action of the unique peptidase that was inhibited following the administration of progesterone (top panel) allows the content of LHRH to rise transiently, giving rise to the pituitary surge (bottom panel). The specificity of this response to progesterone is attested to by the lack of change in activity of post-proline-cleaving enzyme (middle panel).

An additional experiment was performed in which the noradrenergic transmission into the system was uncoupled by addition of the inhibitor DDC (diethyl dithiocarbamate). Under these conditions, neither the expected change in LHRH content nor the regulation of the peptidase were observed.

A rapid, continuous assay for the Tyr^5-Gly^6 cleavage activity is presently being developed. This will allow partial purification of the enzyme and preparation of monoclonal antibodies

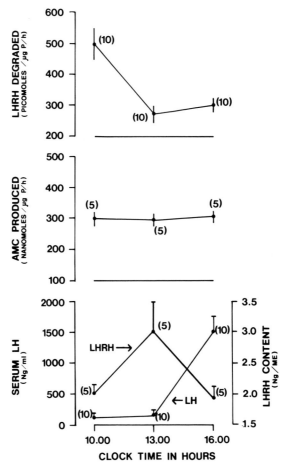

FIG. 7. Relationship between **(top)** LHRH degradation, **(middle)** post-proline-cleaving enzyme activity (as production of aminomethyl coumarine, AMC, by the hydrolysis of Z-Gly-Pro-coumarine-amide), and **(bottom)** serum LH and median eminence LHRH content during positive feedback of progesterone (administered at 10 a.m.) in ovariectomized estrogen-primed adult female rats. Numbers in parentheses are the number of experiments performed.

against it, using the assay to screen. These antibodies can then be used to investigate the biosynthesis of the enzyme and ultimately to obtain a nucleic acid probe for it. It will then be possible to ask whether changing the level of expression of the enzyme is a consequence of any or all the regulatory signals exerted on this system in situ.

POST-PROLINE-CLEAVING ENZYME

Post-proline-cleaving enzyme is a proline endopeptidase that cleaves on the carboxyl side of proline in a variety of neuropeptides. It will cleave TRH as a deamidase, as shown in Fig. 8. It also cleaves the Pro^9-Gly^{10} bond in LHRH and the Pro^4-Gln^5 bond in substance P (Hersh and McKelvy, 1979; Blumberg et al., 1980). These cleavage sites were determined by purifying the enzyme to homogeneity and observing its action on known neuropeptides. In addition, many other neuropeptides, such as α-melanotropin (α-MSH) and neurotensin, also serve as substrates. This enzyme is capable of cleavage at specific prolin residues that have a positively charged amino acid residue penultimate to the proline. The N-terminal tetrapeptide of the substance P molecule may have a separate biological activity from the substance P undecapeptide. The tetrapeptide seems to promote neurite outgrowth, both in neuroblastoma cells and cultured dorsal root ganglion (Narumi and Fujita, 1978; Narumi and Maki, 1978). It also activates macrophages and mast cells in some cases (Bar-Shavit et al., 1980). Furthermore, the biological activity on nerve cells of substance P can be accounted for by the C-terminal heptapeptide amide (Bury and Mashford, 1976; Blumberg and Teichberg, 1979).

Iversen's group has been pursuing the identification of peptidases acting on substance P. Although they find no evidence for the strong expression of this activity, they have found a different endopeptidase that is active in cleaving substance P at three internal sites (Lee et al., 1981). In kinetic characterization of post-proline-cleaving enzyme, it was discovered that the inhibitory constant of substance P in this reaction was the order of 1 μM and that the value of K_{cat} over K_m was about 10^8, which is enormously high (Blumberg et al., 1980; S. Blumberg and J. F. McKelvy, *unpublished data*). Therefore, the notion that the action of an enzyme like

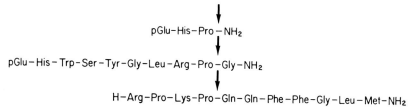

FIG. 8. *Arrows* indicate sites of cleavage of **(top)** TRH, **(middle)** LHRH, and **(bottom)** substance P by post-proline-cleaving enzyme.

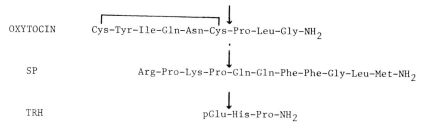

FIG. 9. Sites of cleavage of "mature" neuropeptides believed to give rise to new biological information: oxytocin cleavage to yield MSH release inhibiting factor (MIF); substance P cleavage to yield basic N-terminal tetrapeptide; TRH cleavage to yield His-Pro diketopiperazine.

this could generate both a synaptically active fragment and another fragment with different biological activities seems feasible. Model systems are now being sought in which to test this hypothesis. In addition to substance P, cleavage sites in other peptides have also been shown to generate fragments with new biological activities (Fig. 9). Thus, cleavage of oxytocin yields the tripeptide prolyl-leucyl-glycinamide, which has activity as an MSH release-inhibiting factor (MIF) (Celis et al., 1971), and TRH is cleaved by pyroglutamyl peptidase to yield histidyl-proline diketopiperazine (Prasad et al., 1977), which is active in decreasing motor activity under conditions where TRH has no effect (Bhargava and Matwyshyn, 1980). Thus, it is probable that peptidase action is followed by the assumption of new conformations by the cleavage products and the generation of new information. This point is also made by S. Reichlin *(this volume)* using somatostatin and its prohormones as examples.

The immunocytochemical localization of the post-proline-cleaving enzyme is now being pursued. This enzyme appears to be localized to neuron cell bodies; there is also evidence for its localization in glial cells in certain regions of the central nervous system. Although the enzyme acts on a large number of neuropeptides, it is found in highest concentration in the supraoptic nucleus and the intermediate lobe of the pituitary. Two of the substrates for the enzyme are oxytocin and vasopressin, whose site of origin is in the supraoptic nucleus, suggesting that studies on the relationship of this enzyme activity to activation of the system by osmotic and other stimuli may prove fruitful.

It is hoped that a molecular biological approach, currently in progress, will lead to the availability of nucleic acid probes complementary to the messenger RNA coding for post-proline-cleaving enzyme. Then a subsystem of neurons that contains both the substrate and the enzyme could be studied by changing the activity of the system and asking whether the expression of the enzyme is altered. Thus, the tools appear to be at hand to enable a critical study of the relevance of peptidase action to neuropeptide-mediated events in the nervous system.

ACKNOWLEDGMENTS

This work was supported by NSF grant GB/4043 and NIH Research Career Development Award AM/00331.

REFERENCES

Abraham, D., Pisano, J. J., and Udenfriend, S. (1963): Uptake of carnosine and homocarnosine by rat brain slices. *Arch. Biochem. Biophys.*, 104:160–165.
Advis, J. P., Krause, J. E., and McKelvy, J. F. (1982a): LHRH peptidase activities in discreet hypothalamic regions and anterior pituitary of the rat: Apparent regulation in the prepubertal period and first estrous cycle at puberty. *Endocrinol.*, 110:1238–1247.
Advis, J. P., Krause, J. E., and McKelvy, J. F. (1982b): LHRH peptidase activities in the female rat: Characterization of an assay based on high-performance liquid chromatography. *Analytical Biochem., (in press)*.
Advis, J. P., Krause, J. E., and McKelvy, J. F. (1982c): Evidence for the physiological regulation of an LHRH endopeptidase. *Nature, (in press)*.
Bar-Shavit, Z., Goldman, R., Stabinski, Y., Gottlieb, P., Fridkin, M., Teichberg, V., and Blumberg, S. (1980): Enhancement of phagocytosis—a newly found activity of Substance P residing in its N-terminal tetrapeptide sequence. *Biochem. Biophys. Res. Commun.*, 94:1445–1457.
Bhargava, H. N., and Matwyshyn, G. A. (1980): Influence of thyrotropin releasing hormone and histidyl-proline diketopiperazine on spontaneous locomotor activity and analgesia induced by Δ^9-tetrahydrocannabinol in the mouse. *Eur. J. Pharmacol.*, 68:147–154.
Blumberg, S., and Teichberg, V. (1979): Biological activity and enzymic degradation of substance P receptor. *Biochem. Biophys. Res. Commun.*, 90:347–354.
Blumberg, S., Teichberg, V. I., Charli, J. L., Hersh, L. B., and McKelvy, J. F. (1980): Cleavage of substance P to an N-terminal tetrapeptide and a C-terminal hepapeptide by a post-prolin cleaving enzyme from bovine brain. *Brain Res.*, 192:477–486.
Bury, R. W., and Mashford, M. L. (1976): Biological activity of C-terminal partial sequences of substance P. *J. Med. Chem.*, 19:854–856.

Celis, M. E., Taleisnik, S., and Walter, R. (1971): Regulation of formation and proposed structure of the factor inhibiting the release of melanocyte-stimulating hormone. *Proc. Natl. Acad. Sci. USA*, 68:1428–1433.

Hersh, L. B., and McKelvy, J. F. (1979): Enzymes involved in the degradation of thyrotropin releasing hormone (TRH) and luteinizing hormone releasing hormone (LH-RH) in bovine brain. *Brain Res.*, 168:553–564.

Hersh, L. B., and McKelvy, J. F (1981): An aminopeptidase from bovine brain which catalyzes the hydrolysis of enkephalin. *J. Neurochem.*, 36:171–178.

Hersh, L. B., Smith, T. E., and McKelvy, J. F. (1980): Cleavage of endorphins to des-Tyr endorphins by homogeneous bovine brain aminopeptidase. *Nature*, 286:160–162.

Jacquet, V. F., Marks, N., and Li, C. H. (1976): Behavioral and biochemical properties of opioid peptides. In: *Opiates and Endogenous Opioid Peptides*, edited by H. W. Kosterlitz, pp. 411–414. North-Holland, Amsterdam.

Krause, J. E., Advis, J. P., and McKelvy, J. F. (1982): Identification of a physiologically regulated LHRH endopeptidase activity in female rat hypothalamus as a Tyr^5-Gly^6 endopeptidase. *J. Neuroscience, (in press)*.

Lee, C-M., Sandberg, B. E. B., Hanley, M. R., and Iversen, L. L. (1981): Purification and characterization of a membrane-bound Substance P-degrading enzyme from human brain. *Eur. J. Biochem.*, 114:315–327.

Malfroy, B., Swerts, J. P., and Schwartz, J. C. (1978): High-affinity enkephalin-degrading peptidase in mouse brain and its enhanced activity following morphine. *Nature*, 276:523–526.

Narumi, S., and Fujita, T. (1978): Stimulatory effects of Substance P and Nerve Growth Factor (NGF) on neurite outgrowth in embryonic chick dorsal root ganglia. *Neuropharmacology*, 17:73–76.

Narumi, S., and Maki, Y. (1978): Stimulatory effects of Substance P on neurite extension and cyclic AMP levels in cultured neuroblastoma cells. *J. Neurochem.*, 30:1321–1326.

Pacheco, M. F., Woodward, D. J., McKelvy, J. F., and Griffin, W. S. T. (1981): TRH in the rat cerebellum: II. Uptake by cerebellar slices. *Peptides*, 2:283–288.

Prasad, C., Matsui, T., and Peterkofsky, A. (1977): Antagonism of ethanol narcosis by histidyl-prolin diketopiperazine. *Nature*, 268:142–144.

Roques, B. P., Fournie-Zaluski, M. C., Soroca, E., Lecomte, J. M., Malfroy, B., Llorens, C., and Schwartz, J. C. (1980): The enkephalinase inhibitor thiorphan shows antiociceptive activity in mice. *Nature*, 288:286–288.

Section VIII

Molecular Genetics of Opioid Peptides

Many of the points discussed in the previous section are exemplified in the posttranslational processing of opioid peptides and other peptide hormones. These peptides and their processing also raise provocative questions regarding issues for both neuroscience and molecular genetics.

Numerous peptides with opioid activity are found in various regions of the central nervous system. A large peptide precursor, common to a number of compounds with opioid activity, is pro-opiomelanocortin (POMC). In addition, the multivalent proenkephalin, discussed by S. Udenfriend, apparently encodes leucine enkephalin and several methionine enkephalins. However, although the peptide sequence suggests that each enkephalin is cleaved from and acts independently of, the precursor, some larger precursor fragments have greater activity in certain physiological systems and may well have a separate and distinctly different function from that of enkephalin. The structure and function of POMC and proenkephalin suggest that other small peptides may also share a common precursor.

The tissue-specific processing of POMC, described by E. Herbert and colleagues, results in widely differing end-products with separable biological activities. Yet differential processing may not be unique to this precursor. Rather, other large precursors like proenkephalin or a dynorphin precursor may also undergo tissue-related processing to yield end-products tailored for their localization and/or their function. That more than one gene for POMC may exist (indicated by Herbert and coworkers), that chromosomal localization of genes may be related to regulation of their expression, and that different genes for a given molecule may be expressed in different tissues suggests a multi-tiered mechanism of control over the final product of a given gene. Such diverse mechanisms of regulation of the expression of peptide genes and their differential processing in different cells is likely to be reflected in the distribution of these peptides and their physiological role. D. T. Krieger reviews the distribution of opioid peptides and their receptors and the physiological functions that they mediate; A. Goldstein discusses the possible significance of multiple opioid ligands and multiple opioid receptors.

Certain basic principles of molecular genetics have been elucidated by the examination of large peptides and their processing. For example, the observation by J. D. Baxter and associates of differential splicing of pre-mRNAs to yield different mature mRNAs for both growth hormone and prolactin indicates a mechanism for the regulation of gene expression distinct from initiation of transcription. In addition, although there is reason for thinking that introns in the genes of large polypeptides may delimit functional domains of final proteins, the absence of introns between apparent functional domains of POMC and within growth hormone and related compounds (see Krieger, and Baxter et al.) tends to suggest that this is not an overriding feature of intron location. Furthermore, as emphasized by Goldstein in discussing the amino acid sequence of dynorphin (and as is also suggested by the amino acid sequence of proenkephalin described by Udenfriend), pairs of basic residues, which often signal a cleavage site in a large precursor, need not necessarily do so. Rather, the double basic residues may indicate some form of processing—or none at all.

Endorphins and Enkephalins: Historical Aspects of Opioid Peptides

Dorothy T. Krieger

ABSTRACT

Multiple peptides with opioid activity have been described in the central nervous system. This chapter provides historical perspectives on their discovery, their possible functions, and their synthesis and processing. The advent of recombinant DNA techniques has provided important information with regard to the cDNA and mRNA structure of the precursors of these peptides, evolutionary modifications, and the possibility of their coding by multiple genes.

In the last decade, advances in several seemingly disparate fields have resulted in (1) the discovery of endogenous opioids in brain, pituitary, and extracranial tissues; (2) the demonstration that opioids are formed from larger precursor molecules; and (3) the discovery that these precursor molecules are synthesized independently and are differentially processed in the various tissues in which they occur (Table 1). A beginning has also been made toward the characterization of the gene(s) coding for the precursor molecule(s). Some of the highlights of these discoveries are surveyed here.

TABLE 1. *Historical overview of description and characterization of opioid-like peptides and their receptors*

Discovery	Year
β-Lipotropin (β-LPH) characterized in pituitary; physiological characterization of multiple opiate receptors	1964–67
Specific characterization of opiate receptor	1971–73
Isolation of first endogenous ligand of opiate receptor (enkephalin) in brain (recognition as part of β-LPH)	1975
Opiate-like material described in pituitary (not characterized)	1975
Endorphins isolated from pituitary (demonstration of biological potency)	1976
ACTH described in brain of hypophysectomized animals	1977
Precursor molecule for ACTH/β-LPH demonstrated (pituitary)	1977
Multiple opiate receptors demonstrated in binding studies	1977
Brain synthesis of ACTH/LPH precursor molecule reported	1978
Immunoassay and immunocytochemical demonstration of brain ACTH, LPH, endorphin	1978
Enkephalin described in pituitary (in magnocellular nuclei)	1978
Enkephalin described in adrenal medulla	1978
Enkephalin precursor described	1979
Pituitary dynorphin characterized (in magnocellular nuclei)	1979
Multiple forms of endorphin characterized	1979
Multiple forms of enkephalin characterized	1980

BIOLOGICAL FUNCTIONS OF ENDOGENOUS OPIATES

Before discussing the historical aspects it might be well to outline briefly the postulated functions of the various endogenous opiates. Opiates are defined as substances that bind to opiate types of receptors: endorphins are molecules with a particular chemical configuration and are derived from the last 31 amino acids or shorter segments of β-lipotropin (β-LPH); enkephalins are substances that contain the Tyr-Gly-Gly-Phe-Met or -Leu sequence, but are structurally unrelated to endorphin (Figs. 1 and 2).

OPIATES AND PAIN

The first major putative function suggested for opiates was their possible role in pain. It is now established that there are descending inhibitory pathways, one of the sources of which is the periaqueductal gray, which synapses on dorsal horn cells; the cells in the periaqueductal gray are richly endowed with opiate receptors and also have a high concentration of endorphin- and enkephalin-like material. Stimulation of this area increases the concentration of endorphin- and enkephalin-like material and other thus far uncharacterized material in cerebrospinal fluid. The role of the pituitary endorphin-enkephalin system in pain is still uncertain. It has been reported that various types of stress can cause analgesia, and that the effects of some of these can be blocked by hypophysectomy, dexamethasone, or naloxone. However, hypophysectomy decreases pain sensitivity, and adrenalectomy, which should increase the concentrations of some of these putative endorphins, increases pain sensitivity. Thus, whereas the role of brain pathways that deal with pain seems to be well understood, the role of the pituitary system is less clear.

OPIATES AND SCHIZOPHRENIA

The area of opioid research that has received the greatest amount of attention has been that of mental disease. It has been hypothesized that either increased or decreased amounts of endogenous opiates can be observed in schizophrenia. Thus, for instance, Terenius and co-workers (1979) have reported increased levels of substances they call type I and type II opiates that decrease with successful therapeutic treatment. The exact chemical structure of these opiates is not known.

OPIATES AND OBESITY

Interesting new data have been reported relating endogenous opiates to feeding behavior and obesity. One line of evidence links this behavior to increased pituitary β-endorphin concentrations; the other, to increased brain β-endorphin levels. Initially it was reported that genetically obese animals, the so-called Zucker rats, have high levels of β-endorphin-like material in the intermediate lobe (Margules et al., 1978). In addition, these animals also had elevated central nervous system levels (Gibson et al., 1981), particularly in the paraventricular nucleus of the hypothalamus, one of the areas implicated in feeding behavior. On the other hand, Rossier and colleagues (1979) have shown that obesity appears before there is any change of pituitary β-endorphin levels. In the brain, it has been demonstrated that the intraventricular infusion of β-endorphin increases feeding in normal animals and that, during fasting, levels of β-endorphin in the brain decrease.

OPIATES AND BLOOD PRESSURE

The facts that opiates decrease blood pressure and naloxone has been successfully used in the treatment of hypovolemic shock (Faden and Holaday, 1979) have been clinically interpreted

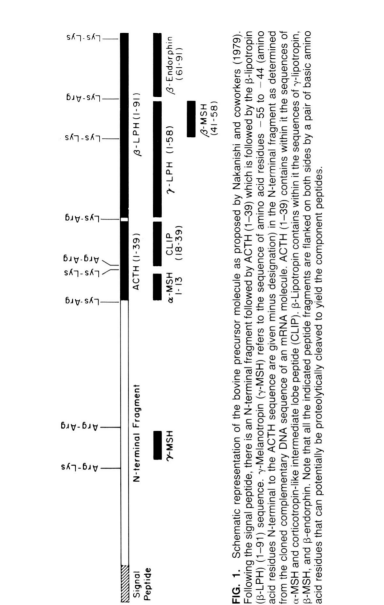

FIG. 1. Schematic representation of the bovine precursor molecule as proposed by Nakanishi and coworkers (1979). Following the signal peptide, there is an N-terminal fragment followed by ACTH (1–39) which is followed by the β-lipotropin (β-LPH) (1–91) sequence. γ-Melanotropin (γ-MSH) refers to the sequence of amino acid residues −55 to −44 (amino acid residues N-terminal to the ACTH sequence are given minus designation) in the N-terminal fragment as determined from the cloned complementary DNA sequence of an mRNA molecule. ACTH (1–39) contains within it the sequences of α-MSH and corticotropin-like intermediate lobe peptide (CLIP). β-Lipotropin contains within it the sequences of γ-lipotropin, β-MSH, and β-endorphin. Note that all the indicated peptide fragments are flanked on both sides by a pair of basic amino acid residues that can potentially be proteolytically cleaved to yield the component peptides.

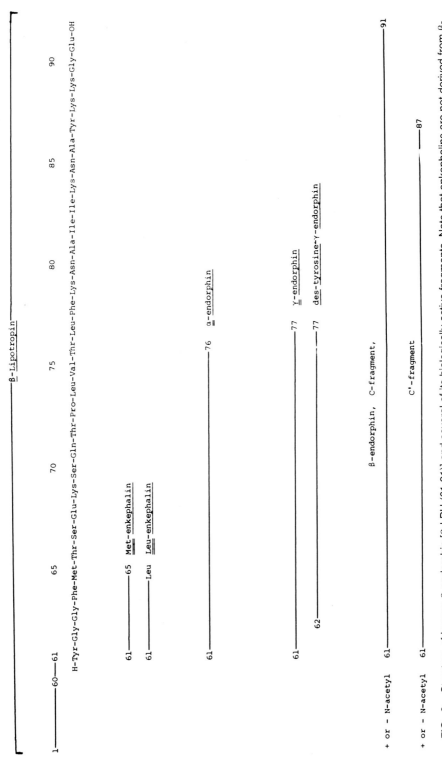

FIG. 2. Structure of human β-endorphin [β-LPH (61-91)] and several of its biologically active fragments. Note that enkephalins are not derived from β-endorphin.

to mean that endogenous opiates are responsible for the clinical syndrome. This has been questioned, however, because naloxone is not necessarily specific in antagonizing opiate effects.

OTHER EFFECTS OF OPIATES

Enkephalin, injected into the thalamus, can cause seizure activity (not analgesia). The systemic application of α or γ-des-tyr-endorphin has effects on learning and memory. In addition, enkephalins and endorphins have been reported to influence the secretion of pituitary hormones. For the most part this last effect is believed to be mediated not by a direct action on the pituitary but by some effect on the neurotransmitter substrate that regulates the releasing factors responsible for the release of the pituitary hormones. There is good evidence that opiates can affect the turnover of catecholamines, dopamine, serotonin, and acetylcholine within the central nervous system.

CHRONOLOGY OF THE ENDORPHINS AND ENKEPHALINS

Li (1964) isolated β-lipotropin from sheep pituitary glands, and it was subsequently isolated and characterized from the pituitary of a variety of species, with minor variations in amino acid sequence (Fig. 1). Its function is still unknown—it has moderate lipolytic activity in several animal systems, and minimal melanophoretic activity. The latter can be ascribed to the presence within its 91-amino acid sequence of the β-melanotropin (β-MSH) sequence (β-LPH 41–58) (Fig. 2).

In a completely different area, investigating mechanisms of narcotic and analgesic action, evidence was presented by Martin (1967) for the heterogeneity of opiate receptors, based on the different syndromes produced by morphine and its congeners in the chronic spinal dog. At that time, no endogenous ligands for these receptors were known. Specific characterization of the brain opiate receptor (Pert and Snyder, 1973; Simon et al., 1973) became possible with the availability of high-specific-activity radiolabeled opiates. Such receptors demonstrated stereospecific binding and ion and cyclic nucleotide sensitivity, and the analgesic potency of various opiates could be correlated with their binding affinities. Subsequent neurophysiological studies and further characterization of the properties of opiate receptor binding suggested the existence of an endogenous ligand for such a receptor. Hughes and coworkers (1975) showed that brain extracts contained morphine-like activity that competed for opiate receptor binding and that was blocked by naloxone (an opiate antagonist). This endogenous substance was isolated from pig brain and characterized as consisting of two pentapeptides—Met-enkephalin and Leu-enkephalin (Tyr-Gly-Gly-Phe<$^{Met}_{Leu}$)—with concentrations of Met-enkephalin four times greater than those of Leu-enkephalin. Soon after, it was realized that the enkephalin sequence was contained within that of β-lipotropin (i.e., β-LPH 61–65). During the same period, a pituitary peptide-like substance also possessing opiate activity was demonstrated, but not characterized (Cox et al., 1975). Attention then turned to other portions of the lipotropin molecule, and it was shown that fragments of β-lipotropin, all incorporating the Met-enkephalin sequence, occurred as intact polypeptides in pituitary (Fig. 3). These include α-endorphin (β-LPH 61–76), γ-endorphin (β-LPH 61–77), C' fragment (β-LPH 61–87), and β-endorphin (β-LPH 61–91), with β-endorphin being the most potent of these fragments in opiate binding assays (Bradbury et al., 1976; Chretien et al., 1976; Cox et al., 1976; Li and Chung, 1976; Ling et al., 1976). Lesser concentrations of β-endorphin were found in brain (Bradbury et al., 1976).

With the development of specific radioimmunoassays for endorphin and enkephalin (in contrast to radioreceptor binding and bioassays in which both peptides are active), it became possible to distinguish between the individual opioid peptides in different tissues. Concentrations

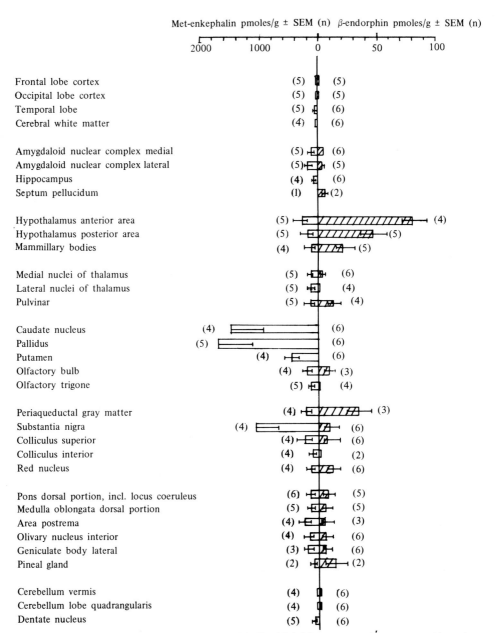

FIG. 3. Met-enkephalin-like **(left)** and β-endorphin-like **(right)** immunoreactivity measured in various parts of human brain. Values represent the mean ± SEM; number of determinations in parentheses. The scale for β-endorphin is expanded 20 times with respect to that for Met-enkephalin. (From Gramsch et al., 1979, with permission.)

of enkephalin were greater than those of β-endorphin in brain, whereas endorphin predominated in the pituitary (intermediate lobe). Pituitary enkephalin was localized mainly to the posterior lobe and was shown to arise from cell bodies in the magnocellular hypothalamic nuclei, i.e., the same area producing vasopressin and oxytocin (Fig. 3 and Table 2).

TABLE 2. *Nature and concentrations of opioid peptides in rat pituitary and comparison of such concentrations to those present in the nervous system and adrenal medulla*[a]

Opioid peptide	Concentration in rat pituitary (pmoles/g)	Comparison with nervous system and adrenal medulla concentrations
Enkephalin		
> Neurointermediate lobe (Leu > Met) (anterior lobe: Met > Leu: ? in growth hormone cells)	1,000–15,000	? 20× > hypothalamus < striatum 10–100× > adrenal
Dynorphin		
> Posterior lobe (none in intermediate) (anterior lobe concentration, c. 1/10th)	1,000–1,200	40× > hypothalamus
Endorphin		
>> Intermediate lobe (anterior lobe concentration, c. 1/50th–1/500th; however, only 10% are ACTH cells)	160,000–550,000	1,000× > hypothalamus

[a]In human pituitary, there is 50–100 times more endorphin (even though only in anterior lobe) than in hypothalamus. Pituitary concentration of opiate receptors is greatest in posterior lobe but concentration is less than in brain. Endorphin has the greatest affinity.

It next became apparent that other peptides, such as adrenocorticotropic hormone (ACTH), were also present in brain as well as in pituitary, with brain concentrations unchanged by hypophysectomy (Krieger et al., 1977). A major finding in assembling these various jigsaw pieces was the demonstration of a common precursor molecule for corticotropin and endorphin (Mains et al., 1977; Roberts and Herbert, 1977) in pituitary. It was subsequently demonstrated that this precursor molecule was also synthesized in brain (Liotta et al., 1979).

Continuing investigation indicated that endorphin and enkephalin were also present outside the pituitary and brain sites. ACTH and β-endorphin were detected in the gastrointestinal tract (Orwoll and Kendall, 1980) and in the human placenta, with demonstration of *in vitro* biosynthesis of the precursor molecule in the latter tissue (Liotta et al., 1979). Immunoreactive enkephalin was detected in fibers within the intestinal tract (Hughes et al., 1977; Miller and Cuatrecasas, 1978) and in cells of the adrenal medulla (Schultzberg et al., 1978; Lewis et al., 1979; Viveros et al., 1979; Yang et al., 1979). The finding of multiple-sized enkephalin forms in the adrenal led to the detection of such forms in brain.

Structural characterization of the pituitary opioid-like peptide, dynorphin, was achieved and its presence was demonstrated in brain (Cox et al., 1976; Goldstein et al., 1979). It contained the Leu-enkephalin sequence followed by a carboxyl-terminal extension (Fig. 4). Dynorphin was several hundredfold more potent than enkephalin in a guinea pig bioassay (A. Goldstein, *this volume*), although less potent than β-endorphin as an analgesic agent.

At this point, a number of questions arose. The observation that endorphin and enkephalin were distributed differently in the brain and that these substances were not found to coexist in individual neurons raised the question of whether enkephalins were derived from endorphin or from another precursor substance. Although soluble and particulate brain endopeptidases have been found that can generate Met-enkephalin from 61–69 β-LPH, the distribution of the soluble enzyme does not coincide with that of enkephalin; the distribution of the particulate form may more closely follow that of the neuropeptides (Knight and Klee, 1979). A 50-kD protein detected in adrenal medulla appears to be an enkephalin precursor and contains seven copies of Met-enkephalin and one of Leu-enkephalin (Lewis et al., 1980; S. Undefriend, *this volume*). A high molecular weight enkephalin precursor has also been detected in brain (Lewis et al., 1978).

PRECURSOR (ADRENAL MEDULLA, STRIATUM)
M.W. ~ 50,000

CONTAINS 7 COPIES OF	1 COPY OF
MET-ENKEPHALIN	LEU-ENKEPHALIN
TYR-GLY-GLY-PHE-MET	TYR-GLY-GLY-PHE-LEU

MET-ENKEPHALIN ARG-PHE*
TYR-GLY-GLY-PHE-MET-ARG-PHE

? α-NEO-ENDORPHIN   TYR-GLY-GLY-PHE-LEU-ARG-LYS-ARG
                                 (PRO,GLY,TYR$_2$,LYS$_2$,ARG)

DYNORPHIN$^+$ TYR-GLY-GLY-PHE-LEU-ARG-ARG-ILE-
                                 ARG-PRO-LYS-LEU-LYS . . . .

*POTENT IN GUINEA PIG ILEUM AND VAS DEFERENS ASSAYS; NALOXONE ANTAGONIZES; APPEARS TO BIND TO δ RECEPTOR PREFERENTIALLY.

$^+$> POTENCY IN GUINEA PIG ILEUM OR VAS DEFERENS ASSAYS THAN OTHER OPIOIDS.

FIG. 4. Some described enkephalin forms.

Concomitant with the complexity of multiple types of opiates was the pharmacological demonstration of multiple types of opiate receptors. There appear to be three different types of receptors, one (δ) with a greater affinity for enkephalin, another with a greater affinity for morphine (μ), and one with apparently equal affinity for these ligands. β-endorphin appears to bind equally to the μ and δ receptors; μ receptors appear to be more involved in analgesic effects, and δ receptors with behavioral effects (Lord et al., 1977; Waterfield et al., 1979). Studies in which the relative numbers of these types of receptors are determined in various brain areas and tissues will perhaps give some insight into the functional significance of the various opiates so far described. It is also possible that for opiates that have relatively little affinity for the already identified receptors, new types of receptors will be found, providing further insights into their function(s).

BIOSYNTHESIS AND PROCESSING OF THE ENDORPHIN PRECURSOR MOLECULE IN DIFFERENT TISSUES

Prior to the identification in 1977 of the common precursor molecule containing ACTH and endorphin, several studies investigating the biosynthesis of ACTH had been under way. When pulse-chase studies were begun in 1975, there had been several reports of the existence of forms of ACTH with molecular weights considerably greater than that of the ACTH 1–39 molecule (4.5 kD). Such high molecular weight forms had been shown not to be artifacts of protein aggregation, protein binding, or disulfide bridge formation. Pulse-chase studies (Mains and Eipper, 1978a; Roberts et al., 1978) demonstrated that the highest molecular weight form present (c. 31 kD) represented an ACTH precursor (pro-ACTH.) This was converted to ACTH 1–39 in two proteolytic cleavage steps, involving the formation of a biosynthetic intermediate form (c. 22 kD) plus an unidentified peptide, and cleavage of this intermediate form to yield ACTH 1–39 plus a second unidentified peptide (Fig. 5). Similar findings were obtained in experiments using a cell-free protein synthesis system in which mRNA from the ACTH-secreting AtT-20/D-16V mouse pituitary tumor cell line was translated. In view of reports of co-release of immunoreactive ACTH and β-LPH, and the demonstration that these peptides occur in the same secretory granules within pituitary corticotrophs, the possibility was investigated that in

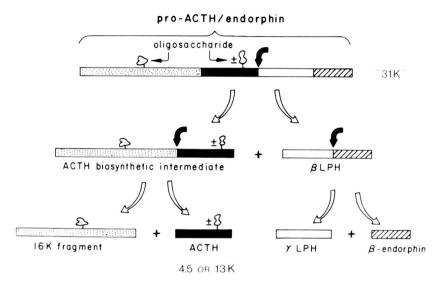

FIG. 5. Model of the glycoprotein common precursor of ACTH and β-endorphin illustrating the peptide backbone and heterogeneity with respect to carbohydrate (oligosaccharide) composition. The *stippled area* is the N-terminal "16K" (MW 16,000) region: the *solid black area* represents the position of the ACTH (1–39) sequence, and the remaining sections represent the β-lipotropin (β-LPH) sequence, the *open area* representing the γ-lipotropin portion and the *hatched area* the β-endorphin portion of the β-lipotropin molecule. The possible points of attachment of carbohydrate moieties are shown. *Black arrows* indicate the sites of posttranslational enzymatic cleavage, and *open arrows* indicate the products formed by such cleavage. As can be seen, the N-terminal "16K" fragment is always glycosylated in both the precursor and the cleaved fragment, and the ACTH region of the precursor may or may not be glycosylated. Hence, on cleavage, unglycosylated ACTH (1–39), referred to as 4.5K (MW 4,500) ACTH, or glycosylated ACTH (1–39), referred to as 13K (MW 13,000) ACTH are possible products, and indeed both have been isolated. (From Mains and Eipper, 1978b, with permission.)

view of its size, β-LPH could represent the first unidentified peptide cleaved from the high molecular weight form. In pursuing these studies, advantage was taken of the then-available information that β-endorphin was contained within the LPH molecule. It was first demonstrated that AtT-20/D-16V cells synthesized material in which β-endorphin activity was present in three peaks with apparent molecular weights of 31 kD (the same size as the already-characterized pro-ACTH), 11.7 kD (similar in size to β-LPH), and 3.5 kD (similar in size to β-endorphin). It was next demonstrated that the 31-kD material contained common antigenic determinants for both β-endorphin and ACTH. Pulse-chase studies confirmed the generation, first of β-LPH and then of endorphin, from the high molecular weight material. Additional studies demonstrated that β-LPH was situated immediately after the carboxyl end of the 22-kD form and that ACTH in turn occupied the carboxyl terminus of the intermediate 22-kD form, leaving an uncharacterized N-terminal portion of this intermediate, which was termed "16K" (see Fig. 5). This arrangement is identical to the protein structure subsequently predicted from the nucleotide sequence of cDNA coding for bovine pituitary intermediate lobe precursor. These studies also identified the presence, as had been demonstrated for other secretory proteins, of a signal peptide preceding the N-terminal sequence (see Fig. 1).

Following the initial characterization of the high molecular weight precursor in AtT-20/D-16V pituitary tumor cells, the presence of a similar high molecular weight form containing both ACTH and endorphin was demonstrated in rodent anterior and intermediate lobes (as well as in those of other species). Similar high molecular weight material has subsequently been characterized and its synthesis demonstrated in rodent and bovine brain and human placenta (Krieger et al., 1980). It has also been characterized in rodent, feline, and human gastrointestinal tract, in human pituitary tumors with clinical manifestations of ACTH hypersecretion (i.e., Nelson's syndrome, Cushing's disease), and in human nonpituitary tumors in which there is "ectopic" ACTH production.

The initial description of the high molecular weight precursor also noted some heterogeneity of its molecular size (Fig. 6). Since it was a glycoprotein, the question arose as to whether such heterogeneity represented differences in the extent of glycosylation of the peptide backbone or whether there was more than one precursor form, which would bespeak the presence of more than one gene coding for the different precursor forms. To date, disparate answers have been provided to this question. Haralson and coworkers (1979) and Loh (1979) characterized material synthesized from toad intermediate lobe or from polysomes or mRNA prepared from cultured AtT-20/D-16V cells, and presented evidence they felt favored the existence of two separate gene products. Other studies were interpreted as being consistent with the existence of one gene product with differential glycosylation. Consistent with this view are the following observations: (1) Roberts and Herbert (1977) demonstrated that translation of the mRNA from AtT-20/D-16V cells yielded a 28-kD product that could be further glycosylated to a larger molecular weight form. (2) Crine and coworkers (1979) demonstrated the presence of two high molecular weight forms (34 and 36 kD) in rat pituitary, but noted that when synthesis took

PRECURSOR MOLECULE PRESENT IN MANY TISSUES

2 TYPES OF SIZE HETEROGENEITY OF PRECURSOR MOLECULE ARE PRESENT
    INTERSPECIES DIFFERENCE: BUT SIMILAR FORMS PRESENT IN DIFFERENT TISSUES WITHIN A SPECIES
    INTRASPECIES DIFFERENCE: SAME HIGH MOLECULAR WEIGHT FORMS PRESENT IN ALL TISSUES WITHIN A SPECIES

RODENT	32 - 36K
TOAD	29.5-32K
BOVINE	36 - 41.5K
HUMAN	34 - 36K

DOES THIS REPRESENT:
    DIFFERENTIAL GLYCOSYLATION:
        mRNA OF AtT-20/D-16V
            YIELDS 28K FORM THAT IS FURTHER GLYCOSYLATED (ROBERTS & HERBERT)
            YIELDS 31K PRE-PROFORM (UNGLYCOSYLATED)
        34K (GLYCOSYLATED)   28K PROFORM (")
        (WITH PANCREAS   (WITH T-RX TUMOR CELL
        MICROSOMAL MEMBRANES) MEMBRANES)
                              (LEIPOLD ET AL)

        IN PRESENCE OF TUNICAMYCIN (T) ONLY 34K FORM SYNTHESIZED BY RAT PITUITARY, WHEREAS NORMALLY ALSO GET 36K (CRINE ET AL)

TWO GENE PRODUCTS: POLYSOMES OR mRNA FROM ONE CLONE OF AtT-20/D-16V DIRECT SYNTHESIS OF 32.5 AND 28K SPECIES, NEITHER OF WHICH IS GLYCOSYLATED; ONE DOES NOT APPEAR TO BE A PROTEOLYTIC PRODUCT OF THE OTHER (HARALSON ET AL)

HYBRIDIZATION OF RAT DNA TO NICK TRANSLATED MOUSE cDNA FRAGMENT YIELDS 2 BANDS OF UNEQUAL INTENSITY ON ELECTROPHORESIS (DROUIN & GOODMAN)

SEQUENCING OF RADIOLABELED PRECURSOR YIELDS TWO DIFFERENT AMINO ACIDS (TRYPTOPHAN AND ARGININE) AT FIRST CYCLE (GOSSARD ET AL)

FIG. 6. Nature of the heterogeneity of the pro-opiocortin precursor molecule and possible factors involved.

place in the presence of tunicamycin (which prevents the formation of N-acetyl glycosamine-lipid intermediates that serve as donors for the synthesis of mannose-rich oligosaccharide units used to glycosylate polypeptide), a single 34-kD product was obtained. (3) Leipold and coworkers (1980) also showed that mRNA from the AtT-20/D-16V mouse pituitary tumor cell line directed the cell-free synthesis of an unglycosylated 31-kD prepro form, which is co- but not posttranslationally processed to a core glycosylated 34-kD and to an unglycosylated 28-kD proform by microsomal membranes from pancreas or tunicamycin-treated tumor cells, respectively. However, studies with cloned rat genes (Drouin and Goodman 1980) and sequence analysis of rat forms (Gossard et al., 1980) indicate that this question has not yet been definitively answered. Should more than one gene be demonstrated, the question will still arise as to the extent to which such genes are expressed in different tissues.

In addition to the precursor molecule serving as a prohormone for ACTH and β-LPH, each of these peptides has potential itself to serve as a prohormone ACTH for α-MSH (acetylated ACTH 1-13-NH$_2$) and CLIP (corticotropin-like intermediate lobe peptide; ACTH 18-39); and β-LPH for β-endorphin, γ-LPH (β-LPH 1-58), and in some species, possibly for β-MSH (β-LPH 41-58) (see Fig. 1). This potential is not expressed in all tissues. In rodent, bovine, and human anterior pituitary, there is apparently no, or very minor, further processing of ACTH, and only a minor amount of β-LPH appears to be processed to β-endorphin. On the other hand, in rat and bovine intermediate lobe, ACTH appears to be further processed to at least two forms of α-MSH and of CLIP, whereas β-LPH is processed almost entirely to β-endorphin, so that virtually no ACTH and LPH are detected in this lobe.

Recently, however, Halmi and coworkers (1981) have described the presence of two distinct cell types in dog intermediate lobe. In only one of these is ACTH apparently further processed to α-MSH. Because of species specificity of LPH antisera, it has not been possible to determine whether β-LPH is also processed differently in the two types of cells. In brain and placenta, processing closely resembles, but is not identical to, that seen in the intermediate lobe (Krieger et al., 1980). In human anterior pituitary adenomas associated with ACTH hypersecretion, there is evidence of a greater extent of processing of β-LPH to β-endorphin (Suda et al., 1979) than is seen in normal anterior pituitary, and in cases of ectopic production of ACTH by tumors, a multiplicity of fragments characteristic of both lobes is seen (see Table 3). (Shibasaki et al., 1981). Evidence has been presented for the presence of two different turnover pools of pro-opiocortin-derived peptides in frog pituitary neurointermediate lobe (Loh and Jenks, 1981). These two pools have different turnover rates, respond differently to dopamine and l-isoproterenol with regard to peptide release, and differ in their relative contents of β-LPH and β-endorphin. This might facilitate preferential release of certain peptides of the ACTH-endorphin

TABLE 3. *Nature of processing of the pro-opiocortin precursor molecule in pituitary and nonpituitary ACTH-producing tumors*

Tumor	Nature of processing
Human pituitary tumors	Increased amounts of precursor molecule in tumor; secretions compared to normal.
	Increased amounts of immunoreactive β-endorphin present in tumor; secretions compared to normal.
	Reported presence in tumor of immunoreactive β-MSH; α-MSH; γ-MSH.
Human ectopic ACTH-producing tumors	Secretion of precursor molecule as well as of ACTH, β-LPH, γ-LPH, β-endorphin.
	? Presence of additional high molecular weight form of ACTH not containing β-LPH.

family during different physiological states of the animal. These observations raise the questions whether such differences in processing are secondary to enzymatic constraints imposed by conformational changes resulting from the expression of different gene products, whether they are due to the differential expression of enzymes in different tissues, or whether they reflect activation of different turnover pools. They also raise the question of whether "disease" may arise in clones of cells with different genetic material from neighboring "normal" cells.

SPECIES DIFFERENCES IN STRUCTURE OF THE PRECURSOR MOLECULE

The determination of possible species differences in the structures of the component peptides of the precursor molecule may, by enabling characterization of conserved portions, help to delineate those portions that may possess significant physiological functions. In addition to the occurrence of several molecular forms of the precursor in the same tissue of a given species and evidence of differential processing of the precursor molecule in different tissues of a given species, there is also evidence of species differences in the molecular sizes of the precursor (but with identity of form in different tissues of a given species) (Krieger et al., 1980). Such reported size heterogeneity varies from approximately 32 to 36 kD in the rodent, 29.5 to 32 kD in the toad (Loh, 1979), 36 to 41.5 kD in bovine, and 34 to 46 kD in humans (Krieger et al., 1980). Another reported species variation is the presence of glycosylated and nonglycosylated forms of ACTH 1–39 in the mouse, rat, and toad, whereas no glycosylated forms are present in the guinea pig, cow, sheep, pig, ostrich, and humans. Structural analysis of ACTH has been performed in the species lacking glycosylated forms, revealing absence of the asparagine residue necessary for attachment of oligosaccharide side chains. In rodent β-MSH, there is a substitution of valine for methionine in LPH residue 47, and there is also evidence for structural dissimilarity between rodent β-LPH and human β-LPH with regard to the total number of residues present. Deletions may also be present in the N-terminal portion of the rodent molecule. There are also differences in two amino acids in the C-terminal portion of the endorphin molecule between these two species. Human γ-LPH has been shown to be shorter by two amino acids than ovine or porcine γ-LPH, its C-terminal 23-amino acid sequence being highly conserved (the segment containing the primary structure of β-MSH) (Li et al., 1980). Differences between this structure proposed for human γ-LPH and those presented by other investigators have suggested that one possible explanation may be the presence of polymorphism in the human precursor gene, similar to that described for the globin gene. Sequencing of the signal peptide has indicated identity in 8 of 12 of the amino acids identified in mouse (Crine et al., 1981) and bovine forms (Gossard et al., 1980). It has also been shown that human 16-kD fragments are not precipitated by murine antibody, indicating structural differences in this portion of the molecule.

STRUCTURAL CHARACTERIZATION OF THE PRECURSOR MOLECULE BY RECOMBINANT DNA

The advent of recombinant DNA techniques has greatly facilitated structural delineation of the precursor molecule, and has provided an explanation for some of the changes noted above. To date, such characterization has been performed either by cloning of the complete bovine neurointermediate lobe cDNA for the precursor molecule (Nakanishi et al., 1979) or of the AtT-20/D-16V cDNA complementary to sequence 44–90 of β-LPH (Roberts et al., 1979), or by isolation from rat (Drouin and Goodman, 1980), bovine (Nakanishi et al., 1980), and human fetal DNA genomic material (Chang et al., 1980) of segments encoding for the precursor molecule. In the bovine neurointermediate lobe studies, the amino acid sequence of the precursor molecule was determined from the nucleotide sequences of the cloned 1,091 base-pair insert. This demonstrated the presence of a signal peptide and gave the first insight into the structure of the N-terminal fragment, indicating that there was a third melanotropin sequence in this portion, termed γ-MSH, that could exist (by virtue of the position of basic residues within the segment) as an 11- or 27-amino acid sequence (i.e., positions 51–61 or 51–77 of the N-terminal sequence), see Fig. 1. It also demonstrated the presence in this N-terminal segment of four cysteine residues plus an additional one within the signal peptide. Subsequently, Nakamura and coworkers (1979) were able to show that the partial amino terminal sequence of the precursor molecule determined from the cell-free translation product of bovine neurointermediate lobe mRNA agreed with the amino acid sequence predicted from the nucleotide sequence of cDNA. Independent studies using degradation of purified 16-kD material from the AtT-20/D-16V cells (Keutmann et al., 1979) indicated sequence homology with regard to the location of cysteine residues between the predicted bovine and AtT-20/D-16V N-terminal sequences; and Gossard and coworkers (1980), by sequence analysis of radiolabeled precursor forms from rat pars intermedia, were able to demonstrate that, of 19 amino acids identified in the N-terminal fragment, 14 were identical to that postulated for the bovine form.

The studies using bovine genomic DNA fragments indicated the presence of an intron of approximately 2.2 Kb existing between the 18th and 19th amino acids of the N-terminal fragment, with no introns occurring between the codon for amino acid 19 and the presumptive poly(A) addition site; another intron was also identified in the region corresponding to the 5' noncoding sequence of the mRNA. Studies using rat genomic DNA presented evidence for similar location of an intervening sequence, whereas studies with fetal genomic DNA demonstrated the presence of an intron at position 14 of the N-terminal fragment. The different location of the human intron from that in rat and the bovine form was secondary to a different number of codons in the human N-terminal fragment.

Studies with rat genomic DNA showed that although the predicted rat β-MSH sequence was essentially similar to that of bovine β-MSH, it was not preceded by a pair of basic amino acids as was the bovine sequence. So, unless a different type of processing was involved, β-MSH as such could not be produced from the rat precursor molecule. Also, the regions preceding the following ACTH have diverged considerably in the rat, bovine, and human sequences, suggesting that rat γ-LPH is shorter than its bovine and human homologues (Mains and Eipper, 1978b). It is of interest that the pre-ACTH sequence divergence ceases just prior to the carboxyl terminus of γ-MSH (indicating conservation of this latter moiety across species) and the post-ACTH sequence divergence stops prior to the amino terminus of β-MSH. Finally, the rat genomic DNA studies demonstrate that when rat cell DNA was hybridized to a nick-translated mouse cDNA fragment, most digests showed hybridization of unequal intensity in two bands when electrophoresis in agarose gels was performed, suggesting the presence of at least two precursor genes in the rat.

The further extension of these techniques to the characterization of the structure of the gene for the precursor molecule in other tissues and in ACTH-producing tumors should begin to answer questions of peptide expression. One area of interest concerns the nature of expression of fetal versus adult genes, since it has been observed (although not confirmed) that human fetal pituitary, but not adult human pituitary, contains considerable amounts of α-MSH-like material (Silman et al., 1978). This observation was considered to be in keeping with the presence of an intermediate lobe-like structure in the human fetus, which disappears prior to parturition. Gene identification would also be of aid in evolutionary studies. The presence of multiple molecular forms of immunoreactive ACTH and β-endorphin has recently been reported (LeRoith et al., 1982) in extracts of *Tetrahymena pyriformis*, a unicellular organism. One form of immunoreactive ACTH migrated similarly to synthetic human ACTH 1–39 in both gel filtration and SDS polyacrylamide gel electrophoretic systems, and exhibited ACTH bioreactivity; the endorphin-like material also exhibited opiate radioreceptor activity. High molecular weight material containing both ACTH and β-endorphin antigenic determinants was demonstrated, suggesting, but not proving, the presence of a precursor molecule that awaits demonstration of its coding gene.

ACKNOWLEDGMENTS

This work was supported by USPHS Grant NB-02893 and by the Lita Annenberg Hazen Charitable Trust.

REFERENCES

Bradbury, A. F., Smyth, D. G., and Snell, C. R. (1976): Lipotropin: Precursor to two biologically active peptides. *Biochem. Biophys. Res. Commun.*, 69:950–956.
Chang, A. C. Y., Cochet, M., and Cohen, S. N. (1980): Structural organization of human genomic DNA encoding the pro-opiomelanocortin peptide. *Proc. Natl. Acad. Sci. USA*, 77:4890–4894.
Chretien, M., Benjannet, S., Dragon, N., Seidah, N. G., and Lis, M. (1976): Isolation of peptides with opiate activity from sheep and human pituitaries. Relation to β-lipotropin. *Biochem. Biophys. Res. Commun.*, 72:472–478.
Cox, B. M., Goldstein, A., and Li, C. H. (1976): Opioid activity of a peptide, β-lipotropin-(16-91), derived from β-lipotropin. *Proc. Natl. Acad. Sci. USA*, 73:1821–1823.
Cox, B. M., Opheim, K. E., Teschemacher, H., and Goldstein, A. (1975): A peptide-like substance from pituitary that acts like morphine. 2. Purification and properties. *Life Sci.*, 16:1777–1782.
Crine, P., Gossard, F., Seidah, N. G., Blanchette, L., Lis, M., and Chretien, M. (1979): Concomitant synthesis of β-endorphin and α-melanotropin from two forms of pro-opiomelanocortin in the rat pars intermedia. *Proc. Natl. Acad. Sci. USA*, 76:5085–5089.
Crine, P., Lemieux, E., Fortin, S., Seidah, N. S., Lis, M., and Chretien, M. (1981): Expression of variant forms of proopiomelanocortin, the common precursor to corticotropin and β-lipotropin in the rat pars intermedia. *Biochemistry*, 20:2475–2481.
Drouin, J., and Goodman, H. M. (1980): Most of the coding region of rat ACTHβ-LPH precursor gene lacks intervening sequences. *Nature*, 288:610–613.
Faden, A. I., and Holaday, J. W. (1979): Opiate antagonists: A role in the treatment of hypovolemic shock. *Science*, 205:317–318.
Gibson, M. J., Liotta, A. S., and Krieger, D. T. (1981): The Zucker fa/fa rat: Absent circadian corticosterone periodicity and elevated β-endorphin concentrations in brain and neurointermediate lobe. *Neuropeptides*, 1:349–362.
Goldstein, A., Tachibana, S., Lowney, L. L., Hunkapiller, M., and Hood, L. (1979): Dynorphin-(1-13), an extraordinarily potent opioid peptide. *Proc. Natl. Acad. Sci. USA*, 76:6666–6670.
Gossard, F., Seidah, N. G., Crine, P., Routhier, R., and Chretien, M. (1980): Partial N-terminal amino acid sequence of pro-opio-melanocortin (ACTH/beta-LPH precursor) from rat pars intermedia. *Biochem. Biophys. Res. Commun.*, 92:1042–1051.
Gramsch, C., Hollt, V., Mehraein, P., Pasi, A., and Herz, A. L. (1979): Regional distribution of methionine-enkephalin- and beta-endorphin-like immunoreactivity in human brain and pituitary. *Brain Res.*, 171:261–270.
Halmi, N. S., Peterson, M. E., Colurso, G. J., Liotta, A. S., and Krieger, D. T. (1981): Dog pituitary intermediate lobe: Dual cell population and high bioactive adrenocorticotropin concentration. *Science*, 211:72–74.
Haralson, M. A., Fairfield, S. J., Nicholson, W. E., Harrison, R. W., and Orth, D. N. (1979): Cell-free synthesis of mouse corticotropin. Evidence for two high molecular weightgene products. *J. Biol. Chem.*, 254:2172–2175.
Hughes, J., Kosterlitz, H. W., and Smith, T. W. (1977): The distribution of methionine-enkephalin and leucine-enkephalin in the brain and peripheral tissues. *Br. J. Pharmacol.*, 61:639–647.
Hughes, J., Smith, T. W., Kosterlitz, H. W., Fothergill, L., Morgan, B. A., and Morris, H. R. (1975): Identification of two related pentapeptides from the brain with potent opiate agonist activity. *Nature*, 258:577–579.
Keutmann, H. T., Eipper, B. A., and Mains, R. E. (1979): Partial characterization of a glycoprotein comprising the NH_2-terminal region of mouse tumor cell pro-adrenocorticotropin hormone/endorphin. *J. Biol. Chem.*, 254:9204–9208.
Knight, M., and Klee, W. A. (1979): Enkephalin generating activity of rat brain endopeptidases. *J. Biol. Chem.*, 254:10426–10430.
Krieger, D. T., Liotta, A. S., and Brownstein, M. J. (1977): Presence of adrenocorticotropin in brain of normal and hypophysectomized rats. *Proc. Natl. Acad. Sci. USA*, 74:648–652.
Krieger, D. T., Liotta, A. S., Brownstein, M. J., and Zimmerman, E. A. (1980): ACTH, β-lipotropin and related peptides in brain, pituitary and blood. *Recent Prog. Horm. Res.*, 36:272–344.

Leipold, B., Schmale, H., and Richter, D. (1980): Processing and core-glycosylation of pre-pro-opiocortin synthesized *in vitro* by RNA from a mouse pituitary tumor cell line. *Biochem. Biophys. Res. Commun.*, 94:1083–1090.

LeRoith, D., Liotta, A. S., Roth, J., Shiloach, J., Lewis, M., Pert, C. G., and Krieger, D. T. (1982): ACTH and β-endorphin-like materials are native to unicellular organisms. *Proc. Natl. Acad. Sci. USA*, 79:2086-2090.

Lewis, R. V., Stein, S., Gerber, L. D., Rubinstein, M., and Udenfriend, S. (1978): High molecular weight opioid-containing proteins in the striatum. *Proc. Natl. Acad. Sci. USA*, 75:4021–4023.

Lewis, R. V., Stern, A. S., Kimura, S., Ross, C. A. J., Stein, S., and Udenfriend, S. (1980): An about 50,000-dalton protein in adrenal medulla: A common precursor of [met]- and [leu]-enkephalin. *Science*, 208:1459–1461.

Lewis, R. V., Stern, A. S., Rossier, J., Stein, S., and Udenfriend, S. (1979): Putative enkephalin precursors in bovine adrenal medulla. *Biochem. Biophys. Res. Commun.*, 89:822–829.

Li, C. H. (1964): Lipotropin, a new active peptide from pituitary glands. *Nature*, 201:924–925.

Li, C. H., and Chung, D. (1976): Isolation, primary structure, and synthesis of α-endorphin and γ-endorphin, two peptides of hypothalamic-hypophysial origin with morphinomimetic activity. *Proc. Natl. Acad. Sci. USA*, 73:3942–3946.

Li, C. H., Chung, D., and Yamashiro, D. (1980): Isolation and amino acid sequence of γ-lipotropin from human pituitary glands. *Proc. Natl. Acad. Sci. USA*, 77:7214–7217.

Ling, N., Burgus, R., and Guillemin, R. (1976): Isolation, primary structure and synthesis of α-endorphin and γ-endorphin, two peptides of hypothalamic-hypophysial origin with morphomimetic activity. *Proc. Natl. Acad. Sci. USA*, 73:3942–3946.

Liotta, A. S., Gildersleeve, D., Brownstein, M. J., and Krieger, D. T. (1979): Biosynthesis *in vitro* of immunoreactive 31,000-dalton corticotropin/β-endorphin-like material by bovine hypothalamus. *Proc. Natl. Acad. Sci. USA*, 76:1448–1452.

Loh, Y. P. (1979): Immunological evidence for two common precursors to corticotropins, endorphins, and melanotropin in the neurointermediate lobe of the toad pituitary. *Proc. Natl. Acad. Sci. USA*, 76:796–800.

Loh, Y. P., and Jenks, B. G. (1981): Evidence for two different turnover pools of adrenocorticotropin, α-melanocyte stimulating hormone, and endorphin-related peptides released by the frog pituitary neurointermediate lobe. *Endocrinology*, 109:54–61.

Lord, J. A. H., Waterfield, A. A., Hughes, J., and Kosterlitz, H. W. (1977): Endogenous opioid peptides: Multiple agonists and receptors. *Nature*, 267:495–500.

Mains, R. E., and Eipper, B. A. (1978a): Coordinate synthesis of corticotropins and endorphins by mouse pituitary tumor cells. *J. Biol. Chem.*, 253:651–655.

Mains, R. E., and Eipper, B. A. (1978b): Studies on the common precursor to ACTH and endorphin. In: *Endorphins*, edited by L. Graf, M. Palkovitz, and A. Z. Ronai, pp. 78–126. Elsevier/North Holland Biomedical Press, Amsterdam.

Mains, R. E., Eipper, B. A., and Ling, N. (1977): Common precursor to corticotropins and endorphins. *Proc. Natl. Acad. Sci. USA*, 74:3014–3018.

Margules, D. L., Moisset, B., Lewis, M. J., Shibuya, H., and Pert, C. B. (1978): Beta-endorphin is associated with overeating in genetically obese mice (ob/ob) and rats (fa/fa). *Science*, 202:988–991.

Martin, W. E. (1967): Opioid antagonists. *Pharmacol. Rev.*, 19:463–521.

Miller, R. J., and Cuatrecasas, P. (1978): Enkephalins and endorphins. *Vitam. Horm.*, 36:297–382.

Nakamura, M., Inoue, A., Nakanishi, S., and Numa, S. (1979): Partial amino-terminal sequence of cell-free translation product encoded by bovine corticotropin-β-lipotropin precursor messenger RNA. *FEBS Lett.*, 105:357–359.

Nakanishi, S., Inoue, A., Kita, T., Nakamura, M., Chang, A. C. Y., Cohen, S. N., and Numa, S. (1979): Nucleotide sequence of cloned cDNA for bovine corticotropin-β-lipotropin precursor. *Nature*, 278:423–427.

Nakanishi, S., Teranishi, Y., Noda, M., Notake, M., Watanabe, Y., Kakidani, H., Jingami, H., and Numa, S. (1980): The protein coding sequence of the bovine ACTH-β-LPH precursor gene is split near the signal peptide region. *Nature*, 287:752–755.

Orwoll, E. S., and Kendall, J. W. (1980): β-endorphin and adrenocorticotropin in extrapituitary sites: Gastrointestinal tract. *Endocrinology*, 107:438–442.

Pert, C. B., and Snyder, S. H. (1973): Opiate receptor: Demonstration in nervous tissue. *Science*, 197:1011–1014.

Roberts, J. L., and Herbert, E. (1977): Characterization of a common precursor to corticotropin and β-lipotropin: Identification of β-lipotropin peptides and their arrangement relative to corticotropin in the precursor synthesized in a cell-free system. *Proc. Natl. Acad. Sci. USA*, 74:5300–5304.

Roberts, J. L., Phillips, M., Rosa, P. A., and Herbert, E. (1978): Steps involved in the processing of common precursor forms of adrenocorticotropin and endorphin in cultures of mouse pituitary cells. *Biochemistry*, 17:3609–3618.

Roberts, J. L., Seeburg, P. H., Shine, J., Herbert, E., Baxter, J. D., and Goodman, H. M. (1979): Corticotropin and β-endorphin: Construction and analysis of recombinant DNA complementary to mRNA for the common precursor. *Proc. Natl. Acad. Sci. USA*, 76:2153–2157.

Rossier, J., Rogers, J., Shibasaki, T., Guillemin, R., and Bloom, F. E. (1979): Opioid peptides and α-melanocyte-stimulating hormone in genetically obese (ob/ob) mice during development. *Proc. Natl. Acad. Sci. USA*, 76:2077–2080.

Schultzberg, M., Lundberg, J. M., Hokfelt, T., Terenius, L., Brandt, J., Elde, R. P., and Goldstein, M. (1978): Enkephalin-like immunoreactivity in gland cells and nerve terminals of the adrenal medulla. *Neuroscience*, 3:1169–1186.

Shibasaki, T., Masui, H., Sato, G., Ling, N., and Guillemin, R. (1981): Secretion pattern of pro-opiomelanocortin-derived peptides by a pituitary adenoma from a patient with Cushing's disease. *J. Clin. Endocrinol. Metab.*, 52:350–353.

Silman, R. E., Holland, D., Chard, T., Lowry, P. J., Hope, J., Robinson, J. S., and Thorburn, G. D. (1978): The ACTH 'family tree' of the rhesus monkey changes with development. *Nature*, 276:526–528.

Simon, E. J., Hiller, J. M., and Edelman, I. (1973): Stereospecific binding of the potent narcotic analgesic ^{3}H etorphine to rat brain homogenate. *Proc. Natl. Acad. Sci. USA*, 70:1947–1949.

Suda, T., Abe, Y., Demura, H., Demura, R., Shizume, K., Tamahashi, N., and Sasano, N. (1979): ACTH, β-LPH and β-endorphin in pituitary adenomas of the patients with Cushing's disease: Activation of β-LPH conversion to β-endorphin. *J. Clin. Endocrinol. Metab.*, 49:475–477.

Terenius, L., Wahlstrom, A., and Johansson, L. (1979): Endorphins in human cerebrospinal fluid and their measurement. In: *Endorphins in Mental Health Research*, edited by E. Usdin, W. E. Bunney, and N. S. Kline, pp. 553–560. Macmillan Press, London and Basingstoke.

Viveros, O. H., Diliberto, E. J., Jr., Hazum, E., and Chang, K. -J. (1979): Opiate-like materials in the adrenal medulla: Evidence for storage and secretion with catecholamines. *Mol. Pharmacol.*, 16:1101–1108.

Waterfield, A. A., Leslie, R. J., Lord, J. A. H., Ling, N., and Kosterlitz, H. W. (1979): Opioid activities of fragments of β-endorphin and of its leucine^{65}-analogue. Comparison of the binding properties of methionine- and leucine-enkephalin. *Eur. J. Pharmacol.*, 58:11–18.

Yang, H.-Y. T., Costa, E., Di Giulio, A. M., Fratta, W., and Hong, J. S. (1979): Met-enkephalin (ME)-like peptides in bovine adrenal gland: Characterization of possible ME precursors. *Fed. Proc.*, 38:364.

Since submission of this chapter, three laboratories have reported the cDNA structure of the enkephalin precursor [Gubler, U., Seeburg, P., Hoffman, B. J., Gage, L. P., and Udenfriend, S. (1982): Molecular cloning established proenkephalin as precursor of enkephalin-containing peptides. *Nature*, 295:206–208; Noda, M., Furutani, Y., Takahashi, H., Toyosato, M., Hirose, T., Inayama, S., Nakanishi, S., and Numa, S. (1982): Cloning and sequence analysis of cDNA for bovine adrenal preproenkephalin. *Nature*, 295:202–206; Comb, M., Seeburg, P. H., Adelman, J., Eiden, L., and Herbert, E. (1982): Primary structure of the human Met- and Leu-enkephalin precursor and its mRNA. *Nature*, 295:663–666]. This contains 6 Met-enkephalin sequences and 1 Leu-enkephalin sequence.

Molecular Genetic Neuroscience, edited by
F.O. Schmitt, S.J. Bird, and F.E. Bloom.
Raven Press, New York © 1982.

Regulation of Expression of Pro-opiomelanocortin and Related Genes in Various Tissues: Use of Cell-free Systems and Hybridization Probes

Edward Herbert, Olivier Civelli, Neal Birnberg,
Patricia Rosa, and Michael Uhler

ABSTRACT

β-Endorphin is derived from a protein, 245 to 265 amino acids in length (depending on the species), that contains the sequences of adrenocorticotropic hormone and other neuropeptides including melanocyte-stimulating hormones. The protein, known as pro-opiomelanocortin, is present in anterior and intermediate lobes of the pituitary where it is processed to different hormones. The processing, both proteolytic and glycosylatory, has been studied in several experimental systems including an anterior pituitary mouse cell line, AtT-20/D16v. Whether ACTH and endorphin peptides present in extrapituitary tissues are derived from a precursor similar to POMC in the pituitary is studied. A cloned cDNA complementary to the β-lipotropin region of mouse POMC mRNA is used to determine both the levels of POMC-like mRNA in various tissues by solution hybridization and the size of the POMC-like mRNA by Northern blot analysis. Finally, Southern blot analysis of genomic DNA of mouse and rat have suggested that there is more than one POMC-related gene in these species.

Small neuroactive peptides such as the enkephalins, endorphins, somatostatin, gastrin, and adrenocorticotropic hormone (ACTH) have invariably been found to be synthesized from much larger proteins. Often these large proteins give rise to more than one biologically active component. The peptides resulting from the cleavage may be related functionally in coordinating some kind of response, but it is not fully understood why nature has built these different peptides into the same precursor protein.

The precursors are themselves inactive; enzymatic cleavage yields the active form. The cleavage is regulated, as is the synthesis of the precursor protein, but the mechanism of this regulation is not yet understood. This chapter concentrates on the biosynthesis and regulation of pro-opiomelanocortin (POMC). Figure 1 shows a detailed structure of POMC (Roberts and Herbert, 1977a,b), which consists of three domains: the β-lipotropin domain, the ACTH domain, and an N-terminal piece. In addition to working together to produce several kinds of responses, each domain appears to have a specific function.

The precursor occurs in both the anterior and intermediate lobes of the pituitary in mice, rats, and several other species. However, different processing of POMC yields different peptides in these sites, as shown in Fig. 2 (Roberts et al., 1978). Sequencing of the precursor has revealed pairs of basic residues surrounding the biologically active domains (Nakanishi et al., 1979; Roberts et al., 1979b). In both the anterior and intermediate lobes, cleavage of POMC yields β-lipotropin, ACTH, and an N-terminal fragment as major products. In the intermediate

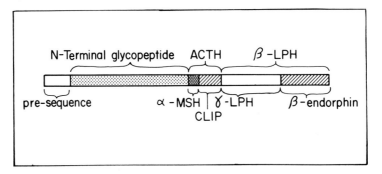

FIG. 1. Model of the structure of pro-opiomelanocortin showing the three principle domains of the molecule.

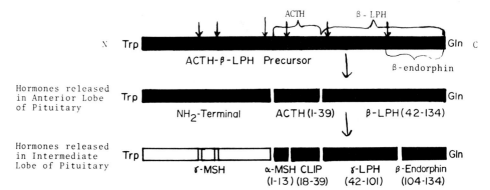

FIG. 2. Principle products of processing of pro-opiomelanocortin in anterior and intermediate lobes of the pituitary. The *arrows* indicate the cleavage sites where pairs of basic amino acids are located.

lobe, ACTH is further processed to give rise to α-melanotropin (α-MSH) and corticotropin-like intermediate lobe peptide (CLIP), whereas β-lipotropin (β-LPH) is cleaved to produce γ-LPH, β-endorphin, and derivatives of β-endorphin (Roberts et al., 1978). The cleavage of ACTH to yield α-MSH is an example of the conversion of one hormone to another with very different biological activities.

Not only does processing of POMC differ in the two lobes of the pituitary, but the regulation of production and release of these peptides also varies (Moriarty et al., 1975). The anterior lobe is a separate structure from the intermediate lobe, and although both are connected to the hypothalamus, this is accomplished in different ways. The hypophysialportal circulatory system delivers releasing factors to the anterior pituitary, including corticotropin-releasing factor (CRF). This small peptide, which has yet to be chemically characterized, is released into the portal circulation by neurons in the hypothalamus. It is then carried to the anterior pituitary where, possibly in conjunction with other peptides, it stimulates the release of ACTH into the bloodstream. This, in turn, stimulates the adrenal cortex to release glucocorticoids. The glucocorticoids feed back on the anterior pituitary, decreasing release of ACTH and other hormones derived from POMC. Glucocorticoids may also have inhibitory effects at other levels, including the hypothalamus (Guillemin and Rosenberg, 1955). The intermediate lobe, on the other hand, receives direct synaptic input from dopaminergic neurons originating in the hypothalamus (Dawson, 1953; Kurosumi et al., 1961). These have a negative influence on the release of α-MSH, the endorphins, and derivatives of β-endorphin. Thus, in addition to different processing in the two lobes, regulation is quite different as well. Most of these control features can also be demonstrated in primary tissue culture systems (Rosa et al., 1980).

Yet unknown is the basis of the observed differences in expression of the POMC molecule in the two lobes of the pituitary. Furthermore, it has been found that the component peptides of the precursor exist in a number of other sites, including various brain regions and the gut. Whether expression of POMC differs in these sites is unknown.

The pituitary tumor cell line, AtT-20-$D_{16}v$, derived from the anterior pituitary of mice, has been used as a model system to study the details of POMC processing. This tumor has been maintained in culture for a number of years, and has been subcloned to derive several lines incorporating favorable features for studies of the synthesis of these components (Herbert et al., 1978). The cells synthesize ACTH, β-LPH, and β-endorphin in large quantities and in the same forms as are synthesized by the anterior lobe of the pituitary. They also respond to the same regulators to which anterior lobe cells are sensitive, including glucocorticoids which inhibit production of ACTH and β-LPH, and partially purified CRF, which stimulates release of these compounds. Messenger RNA has been isolated from these cells and translated in cell-free systems. The translation product contains the immunological determinants of both ACTH and endorphin (Roberts and Herbert, 1977b). This has also been demonstrated using radioactive labeling of AtT-20 cell proteins and specific immunoprecipitation (Mains et al., 1977).

After an overall outline of the molecular structure of POMC had been determined by peptide mapping, its processing in cells was studied. The cells were incubated with radioactive amino acids or sugars, lysed, and the POMC-related proteins specifically immunoprecipitated. The proteins were then separated by gel electrophoresis, and peptide structure analysis was done to identify each of the pieces that were isolated from the gels. An essential requirement of this methodology is illustrated in Fig. 3. Antibodies specific to each of the three domains of the molecule—the ACTH region, the endorphin region, and the N-terminal region (Roberts et al., 1978)—must react with the antigenic determinant in the precursor, the intermediate forms, and the end product. All fragments derived from the three domains of the molecule can thus be

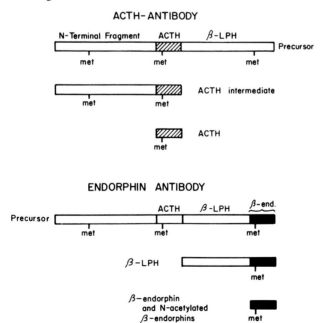

FIG. 3. Fragments of pro-opiomelanocortin recognized by antisera to ACTH, endorphin, and N-terminal regions.

accounted for during label incorporation and in the chase of label from the precursor to end products.

Figure 4 demonstrates the way in which the proteolytic processing of this molecule was followed with ACTH antiserum. The cells were incubated with [^{35}S]methionine for varying periods of time, and then an extract was immunoprecipitated with the ACTH antiserum and the products were electrophoresed on an SDS slab gel. The figure is an autoradiogram of such a gel and shows that the label is first incorporated into a 29,500-MW precursor (Roberts and Herbert, 1977a; Roberts et al., 1978). Higher molecular weight forms are observed with time because of changes in states of glycosylation, and somewhat later a series of lower molecular weight intermediates appears. (These bands are diffuse as a result of heterogeneity of the oligosaccharide side chains.) Finally, the end products of processing of the ACTH portion appear as two forms of ACTH—ACTH$_{1-39}$ plus a glycosylated form of ACTH$_{1-39}$ (Eipper and Mains, 1977).

The precursors and products of POMC processing in the AtT-20 cells can also be labeled with radioactive sugars, as shown in Fig. 5. To identify the sugars in each fragment, the cells were first incubated with [^{3}H]mannose or [^{3}H]glucosamine (Eipper et al., 1976; Roberts et al., 1978; Phillips et al., 1981). The extracts were then immunoprecipitated with anti-ACTH antiserum and subjected to SDS slab gel electrophoresis. From autoradiograms of the gels, it is possible to determine the kinds of sugars present in the oligosaccharides and the time course of addition of these sugars to each component (Roberts et al., 1978; Phillips et al., 1981). These types of experiments can be repeated using antisera specific to each domain of the molecule. A processing scheme can then be deduced. The schemes for two major kinds of

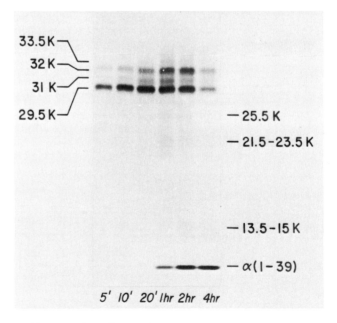

FIG. 4. Time course of incorporation of [^{35}S]methionine into the forms of ACTH in AtT-20 cells. AtT-20 cell cultures were grown in microtest wells, pretreated for 15 min with low-methionine Dulbecco-Vogt minimal essential medium (40 μM methionine) without horse serum, and incubated with 55 μCi of L-[S]methionine dissolved in 50 μl of the same medium for the times indicated in the figure. The cells were extracted with 5N acetic acid containing 1 mM phenylmethyl sulfonylfluoride and 1 mM iodoazetamide, lyophilized, and immunoprecipitated with ACTH antiserum (antiserum Wilma). The immunoprecipitates were analyzed by SDS slab gel electrophoresis and visualized by autoradiography.

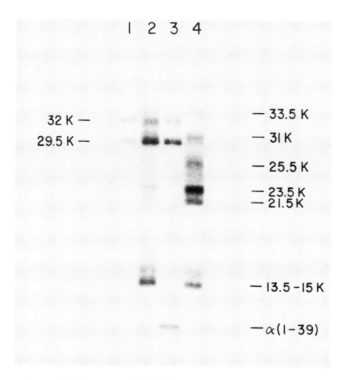

FIG. 5. [³H]glucosamine and [³H]mannose-labeled forms of ACTH in AtT-20 cells and in culture medium from AtT-20 cells. AtT-20 cells were grown in Falcon microtest wells in complete Dulbecco-Vogt minimal essential medium containing 10% horse serum. One hour before the beginning of the labeling, the medium was changed to Dulbecco-Vogt minimal essential medium without horse serum and without glucose to deplete the intracellular sugar pools. Cells were incubated with the following radioactive precursors dissolved in 50 μl of Dulbecco-Vogt minimal essential medium with glucose and without horse serum for 2 hr; lane 1, 100 μCi D-[1-³H]mannose (culture medium); lane 2, 100 μCi of D-[6-³H] glucosamine HC1 (cell extract); lane 3, 100 μCi of L-[4-³H]phenylalanine (cell extract); lane 4, 100 μCi of D-[1-³H]galactose (cell extract). The cells were extracted with 5N acetic acid containing 1 mM phenylmethyl sulfonylfluoride and 1 mM iodoacetamide. The extracts were lyophilized and ACTH proteins precipitated with antiserum Wilma. The immunoprecipitates were analyzed by SDS slab gel electrophoresis and fluorography.

processing are shown in Fig. 6. The most likely course of processing in these cells is the synthesis, on membrane-bound ribosomes, of a 28,000- to 29,000-dalton precursor protein, including a signal sequence of 26 amino acids (Policastro et al., 1981). During synthesis the signal sequence is removed and one or two carbohydrate side chains are added to the N-terminal region of the molecule (two side chains are shown in Fig. 6.) This partially processed molecule then follows alternative routes for further processing. It can either be converted to another form (32,000-dalton form) by the addition of another carbohydrate side chain to the ACTH region of the molecule or it can be processed to lower molecular weight fragments by proteolytic processing. In either case, the first pieces cleaved out are β-LPH (21,000 daltons) and ACTH intermediates (23,000 daltons). Further processing leads to the formation of the end products found in the AtT-20 cell. These include $ACTH_{1-39}$ (4,500 daltons), a glycosylated form of $ACTH_{1-39}$ (14,000 daltons), an N-terminal fragment (18,000 daltons), and β-LPH. The two ACTH end products have typical ACTH bioactivity, and are equivalent in a steroidogenic assay (Gasson, 1979). The work of Pedersen and Brownie (1980) indicates that the N-terminal fragment potentiates the action of ACTH in an *in vitro* steroidogenic assay.

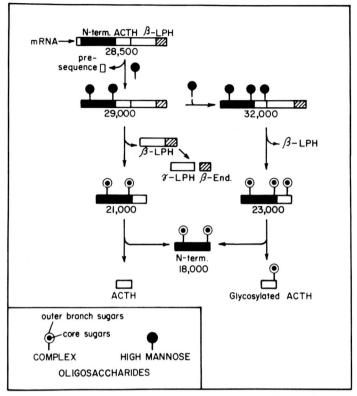

FIG. 6. Processing of pro-opiomelanocortin in AtT-20 cells.

The end products are released together in the AtT-20 cells: ACTH and the glysolated form of ACTH, the N-terminal fragment, and β-LPH. Release of all three peptides can be inhibited by treatment with glucocorticoids; CRF stimulates their release. (Roberts et al., 1979a). The peptides appear to be packaged together in the same granules and released by exocytosis.

The same type of labeling experiments can be carried out with primary cultures of normal anterior and intermediate lobe cells from the pituitary (Crine et al., 1978; Mains and Eipper, 1979; Hinman and Herbert, 1980). The processing events that have been shown to take place in the pituitary tumor cells also occur in normal cells in primary culture. It has been found by pulse label and pulse chase studies that culture of anterior and intermediate lobe cells of rat pituitary synthesize the same kinds of ACTH-labeled proteins (precursors, intermediates, and end products) (Hinman and Herbert 1980; Rosa et al., 1980). The difference in processing between the anterior and intermediate lobe cells is the result of later steps in which ACTH is cleaved further to α-MSH and CLIP, and LPH is cleaved to endorphin and derivatives of endorphin.

Figure 7 illustrates a processing scheme that can be deduced from these kinds of data. As a result of extensive peptide mapping, it is thought that the POMC formed in the anterior lobe is very similar to that in the intermediate lobe of the mouse and the rat pituitary, although these molecules have not yet been sequenced (Rosa et al., 1980). All available data indicate that the first steps, involving glycosylation and proteolytic cleavage, are the same or very similar in the two lobes. Differences arise when the intermediate lobe β-LPH is cleaved to form β-endorphin and γ-LPH; β-endorphin is then acetylated (Zakarian and Smyth, 1979). ACTH is also further cleaved to α-MSH and CLIP. These differences in the final end products of precursor

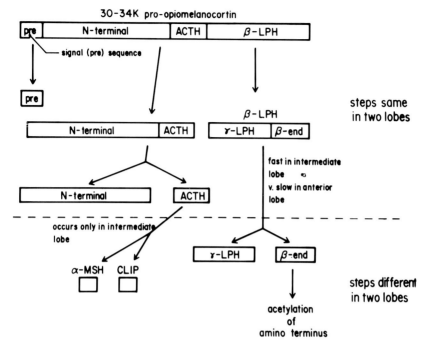

FIG. 7. Similarities and differences in processing of pro-opiomelanocortin in anterior and intermediate lobe cells of rat and mouse pituitary.

cleavage may be due to differences in mRNA translation or posttranslational processing, enzyme processing in the two lobes, or tissue-specific structures in the precursor molecules. Furthermore, peptides that are derived from POMC are also found in a number of other tissues, including the hypothalamus, amygdala, liver, placenta, gut, pancreas, and lung. It is possible that the peptides in these tissues arise from a precursor similar to that found in the pituitary. It has been reported that a precursor similar to the pituitary POMC is present in the bovine hypothalamus and gives rise to immunoreactive ACTH and β-endorphin peptides (Liotta et al., 1979).

To determine whether a common precursor molecule is synthesized in these tissues, radioactive labeling of gene products in tissue explants from various regions was attempted, but insufficient label was incorporated. Therefore, a search was undertaken to find the mRNA that would code for a precursor molecule. A probe for hybridization was made using mRNA purified from AtT-20 cells (Roberts et al., 1979b; J. Baxter et al., *this volume*). The mRNA was reverse-transcribed into DNA, made double-stranded, and inserted into the pBR322 plasmid. The probe is specific for the LPH region of the precursor and was made single-stranded by cloning the insert into the single-stranded phage M-13. This allowed the preparation of single-stranded cDNA, which could then be labeled with [^{32}P] prior to performing hybridizations. Solution hybridization allowed quantification of mRNA in various RNA preparations, and a Northern blot allowed size determination of the mRNA. In addition, the probe could be used to screen for POMC genes in a gene library.

Figure 8 shows the results of using this probe for solution hybridization with RNA isolated from various tissues, including AtT-20 cells, rat pituitary intermediate and anterior lobes, hypothalamus, and amygdala. Plotted are parts per million of hybridizing RNA to total RNA. Clearly the intermediate lobe has by far the highest level of POMC mRNA at about 1,100 parts per million by weight. The AtT-20 cells also have fairly high levels, and the anterior lobe

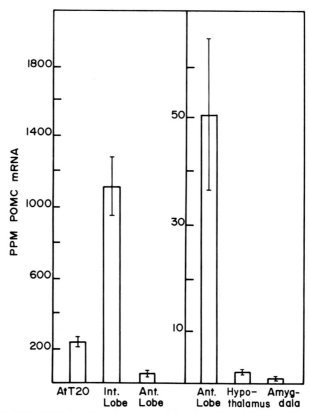

FIG. 8. Levels of POMC mRNA in various tissues as determined by solution hybridization with LPH-cDNA. Total nucleic acid was extracted from tissue by homogenizing each tissue in a standard buffer solution containing sodium dodecylsulfate to inactivate nucleases. Appropriate amounts of RNA were then assayed to a solution hybridization method (N. Birnberg, *unpublished data*) using [^{32}P]cDNA (labeled by the nick translation method). The *error bars* are the standard deviation of three determinations over a fivefold range of nucleic acid in the assay mixture.

contains much less. These numbers do not directly reflect the number of copies of mRNA per cell because the intermediate lobe has many more cells expressing this mRNA than does the anterior lobe. When the scale is expanded, as on the right side of Fig. 8, it can be seen that the anterior lobe has 50 parts per million and the hypothalamus and amygdala have 3.5 and 1.0 parts per million, respectively (sensitivity level is 0.15 ppm). A number of other tissues, such as liver and reticulocytes, were screened and POMC RNA could not be detected by solution hybridization.

However, the Northern blot procedure appears to be more sensitive. RNA was isolated from various tissues and poly(A)-enriched RNA was prepared and fractionated by the Thomas (1980) method. Figure 9 shows that rat pituitary intermediate lobe cells have a single band of mRNA about 1,150 nucleotides long (the same length as in the anterior lobe, not shown). In addition, distinct bands of RNA can be seen in the hypothalamus and amygdala but not in the cerebral cortex, cerebellum, or striatum. The hypothalamic mRNA is the same length as pituitary mRNA, but the amygdala mRNA is slightly smaller (1,100 nucleotides). The reason for the smaller size of amygdala mRNA is not known.

It is possible to estimate the number of copies of precursor mRNA per cell for the different tissues that synthesize POMC. On the basis of total number of cells, there is an average of 5

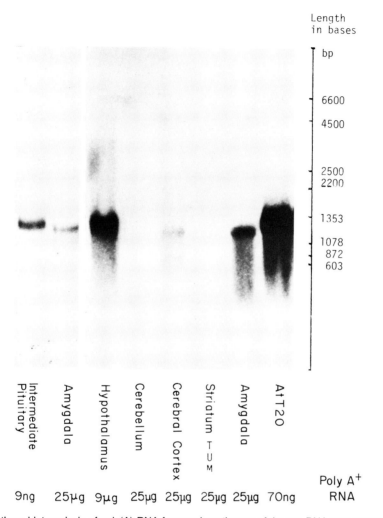

FIG. 9. Northern blot analysis of poly(A) RNA from various tissues of the rat. RNA was extracted from the tissues with guanidinium thiocyanate and poly(A) RNA was prepared by oligo-dTMP-cellulose chromatography. The poly(A) RNA was applied to 1.5% agarose gels in the amounts shown, and electrophoresis was performed. The RNA was transferred to sheets of nitrocellulose and hybridized to [^{32}P] LPH-cDNA by the method of Thomas (1980). Autoradiography was performed for 6 hr **(left)** or for 24 hr **(right)**.

to 10 copies of POMC mRNA per cell in the anterior lobes of the rat pituitary and about 250 copies per cell in the intermediate lobe. However, *in situ* hybridization studies indicate that the percentage of cells that are actually engaged in making POMC/mRNA are about 2 to 4% in the anterior pituitary and 50 to 100% in the intermediate lobe (Hudson et al., 1981; J. Roberts, *personal communication*). Thus, when only the cells that are actually synthesizing precursor mRNA are considered, there are about 500 copies of mRNA per cell in each lobe. Since the percentage of cells expressing the precursor mRNA in other tissues is unknown, this analysis cannot presently be extended to those tissues. *In situ* hybridization methods should allow future determination of this value.

A key question is whether the peptides made in all of these sites are the result of expression of one or more than one POMC gene. Drouin and Goodman (1980) have isolated a rat genomic DNA clone coding for POMC and have found that there is at least one more closely related

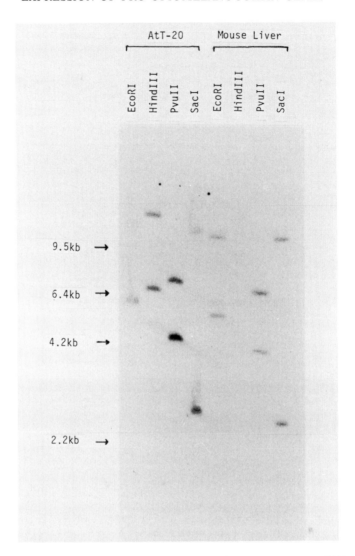

FIG. 10. Southern blot analysis of genomic DNA from mouse liver and AtT-20 cells. DNA from mouse liver tissue and AtT-20 cells was digested with the restriction endonucleases shown in the blot, fractionated by gel electrophoresis on 1.5% agarose gels, and transferred to nitrocellulose by the method of Southern (1975). The DNA bound to the nitrocellulose was hybridized to the [^{32}P]cDNA probe and autoradiography was performed for 4 hr.

POMC gene. However, it is not known whether both genes are expressed. The AtT-20 cell DNA and mouse liver DNA have been studied by transferring restriction enzyme-digested DNA to nitrocellulose filters and then hybridizing with a POMC gene probe. The results are shown in Fig. 10. It is interesting to note that two bands in each lane light up by binding the radioactive probe. In this experiment, which has been done with a large number of different restriction enzymes, at least two hybridizing bands are always found, suggesting that there are at least two genes for this precursor molecule in the mouse as well as in the rat.

DNA probes are also being prepared to different regions of the POMC sequence. The probe discussed here is specific for the LPH region of the molecule. Probes have also been made to the ACTH region, to the N-terminal region, and to the 3' noncoding region of the gene. When various tissues are tested for their ability to express the two different genes, it is expected that the coding regions will be quite similar. However, the 3' noncoding regions may differ, since they diverge in other cases of multigene families. If so, probes could be devised to determine in which tissues the different genes are expressed.

Recombinant DNA research has provided powerful new tools for investigating the role of peptides in the central nervous system. Their application has answered questions that seemed virtually unapproachable a few years ago. The excitement generated by discoveries made with these tools seems justified in terms of the impressive progress made in the few years that these methods have been available.

ACKNOWLEDGMENTS

This work was supported by National Institutes of Health Grants 2 Rol AM 16879 and DA 02736.

REFERENCES

Crine, P., Gianoulakis, C., Seidah, N. G., Gossard, F., Pezalla, P. D., Lis, M., and Chretien, M. (1978): Biosynthesis of β-endorphin from β-lipotropin and a larger molecular weight precursor in rat pars intermedia. *Proc. Natl. Acad. Sci. USA*, 75:4719–4723.

Dawson, A. (1953): Evidence for the termination of neurosecretory fibers within the pars intermedia of the hypophysis of the frog, *Rana pipiens*. *Anat. Rec.*, 115:63–67.

Drouin, J., and Goodman, H. M. (1980): Most of the coding region of rat ACTH β-LPH precursor gene lacks intervening sequences. *Nature*, 288:610–613.

Eipper, B. A., and Mains, R. E. (1977): Peptide analysis of a glycoprotein form of adrenocorticotropic hormone. *J. Biol. Chem.*, 252:8821–8832.

Eipper, B. A., Mains, R. E., and Guenzi, D. (1976): High molecular weight forms of adrenocorticotropic hormone are glycoproteins. *J. Biol. Chem.*, 251:4121–4126.

Gasson, J. C. (1979): Steroidogenic activity of high molecular weight forms of corticotropin. *Biochemistry*, 19:5395–5402.

Guillemin, R., and Rosenberg, B. (1955): Humoral hypothalamic control of anterior pituitary: A study with combined tissue cultures. *Endocrinology*, 57:599–607.

Herbert, E., Allen, R. G., and Paquette, J. P. (1978): Reversal of dexamethasone inhibition of adrenocorticotropin release in a mouse pituitary tumor cell line either by growing cells in the absence of dexamethasone or by addition of hypothalamic extract. *Endocrinology*, 102:218–226.

Hinman, M. B., and Herbert, E. (1980): Processing of the precursor to adrenocorticotrophic hormone and β-lipotropin in monolayer cultures of mouse anterior pituitary. *Biochemistry*, 19:5395–5402.

Hudson, P., Penshow, J., Shine, J., Ryan, G., Niall, H., and Coughlan, J. (1981): Hybridization histochemistry: Use of recombinant DNA as a "homing probe" for tissue localization of specific mRNA populations. *Endocrinology*, 108:353–356.

Kurosumi, K., Matsuzawa, T., and Shibasaki, I. (1961): Electron microscope studies on the fine structures of the pars nervosa and pars intermedia and their morphological interrelation in the normal rat hypophysis. *Gen. Comp. Endocrinol.*, 1:433–452.

Liotta, A., Gildersleeve, D., Brownstein, M. J., and Krieger, D. T. (1979): Biosynthesis *in vitro* of immunoreactive 31,000-dalton corticotropin/β-endorphin-like material by bovine hypothalamus. *Proc. Natl. Acad. Sci. USA*, 76:1448–1452.

Mains, R. E., and Eipper, B. A. (1979): Synthesis and secretion of corticotropins, melanotropins and endorphins by rat intermediate pituitary cells. *J. Biol. Chem.*, 254:7885–7894.

Mains, R. E., Eipper, B. A., and Ling, N. (1977): Common precursor to corticotropin and endorphins. *Proc. Natl. Acad. Sci. USA*, 74:3014–3018.

Moriarty, G. C., Halmi, N. S., and Moriarty, C. M. (1975): The effect of stress on the cytology and immunocytochemistry of pars intermedia cells in the rat pituitary. *Endocrinology*, 96:1426–1436.

Nakanishi, S., Inoue, A., Kita, T., Nakamura, A., Chang, A. C. Y., Cohen, S. N., and Numa, S. (1979): Nucleotide sequence of cloned cDNA for bovine corticotropin-β-lipotropin precursor. *Nature*, 278:423–427.

Pedersen, R. C., and Brownie, A. C. (1980): Adrenocortical response to corticotropin is potentiated by part of the amino terminal region of pro-corticotropin/endorphin. *Proc. Natl. Acad. Sci. USA*, 77:2239–2243.

Phillips, M. A., Budarf, M. L., and Herbert, E. (1981): Glycosylation events in the processing secretion of pro-ACTH-endorphin in mouse pituitary tumor cells. *Biochemistry*, 20:1666–1675.

Policastro, P., Phillips, M., Oates, E., Herbert, E., Roberts, J. L., Seidah, N., and Chrétien, M. (1981): Evidence for a signal sequence at the N-terminus of the common precursor to adrenocorticotropin and β-lipotropin in mouse pituitary cells. *Eur. J. Biochem.*, 116:255–259.

Roberts, J. L., Budarf, M. L., Baxter, J. D., and Herbert, E. (1979a): Selective reduction of pro-adrenocorticotropin/endorphin proteins and messenger ribonucleic acid activity in mouse pituitary tumor cells by glucocorticoids. *Biochemistry*, 18:4907–4915.

Roberts, J., and Herbert, E. (1977a): Characterization of a common precursor to corticotropin and β-lipotropin: Cell-free synthesis of the precursor and identification of corticotropin peptides in the molecule. *Proc. Natl. Acad. Sci. USA*, 74:4826–4830.

Roberts, J., and Herbert, E. (1977b): Characterization of a common precursor to corticotropin and β-lipotropin: Identification of β-lipotropin peptides and their arrangement relative to corticotropin in the precursory synthesized in the cell-free system. *Proc. Natl. Acad. Sci. USA*, 74:5300–5304.

Roberts, J. L., Phillips, M. A., Rosa, P. A., and Herbert, E. (1978): Steps involved in the processing of common precursor forms of adrenocorticotropin and endorphin in cultures of mouse pituitary cells. *Biochemistry*, 17:3609–3618.

Roberts, J. L., Seeburg, P. H., Shine, J., Herbert, E., Baxter, J. D., and Goodman, H. M. (1979b): Corticotropin and β-endorphin: Construction and analysis of recombinant DNA complementary to mRNA for the common precursor. *Proc. Natl. Acad. Sci. USA*, 76:2153–2157.

Rosa, P. A., Policastro, P., and Herbert, E. (1980): A cellular basis for the differences in regulation of synthesis and secretion of ACTH/endorphin peptides in anterior and intermediate lobes of the pituitary. *J. Exp. Biol.*, 89:215–237.

Southern, E. M. (1975): Detection of specific sequences among DNA fragments separated by gel electrophoresis. *J. Mol. Biol.*, 98:503–517.

Thomas, P. S. (1980): Hybridization of denatured RNA and small DNA fragments transferred to nitrocellulose. *Proc. Natl. Acad. Sci. USA*, 77:5201–5205.

Zakarian, S., and Smyth, D. G. (1979): Distribution of active and inactive forms of endorphins in the pituitary and brain of the rat. *Proc. Natl. Acad. Sci. USA*, 76:5975–5979.

Molecular Genetic Neuroscience, edited by
F. O. Schmitt, S. J. Bird, and F. E. Bloom.
Raven Press, New York © 1982.

Structure and Expression of Hormone Genes

John D. Baxter, Peter L. Whitfeld, Peter H. Seeburg,
Andrea Barta, Nancy E. Cooke, Norman L. Eberhardt,
Robert I. Richards, Guy Cathala, Maurice Wegnez,
Joseph A. Martial, and John Shine

ABSTRACT

The structure and expression of the growth hormone set of genes and the pro-opiomelanocortin gene have been examined. Complementary DNA to POMC mRNA has been used to synthesize β-endorphin in bacteria and to isolate the POMC gene. The latter was found to contain two introns, one of which interrupts the 5'-noncoding structure, and the other the codons of the N-terminal fragment. Complementary DNAs to growth hormone, prolactin, and placental lactogen mRNA show substantial homology and have been used to isolate the respective chromosomal genes. All of these have four introns located at identical locations when the sequences are aligned to maximize homology. There are three examples whereby a single pre-mRNA product of these genes is spliced to yield two mRNAs. Growth hormone mRNA in cultured pituitary cells is regulated by glucocorticoid and thyroid hormones.

The studies reported here focus on the structure and expression of the pro-opiomelanocortin (POMC) gene and the growth hormone, prolactin, and placental lactogen set of genes. The last set shows interesting structural contrasts with the POMC gene. It is also of particular interest in terms of its evolution and regulated expression.

STRUCTURE AND EXPRESSION OF THE POMC GENE, THE COMMON PRECURSOR TO CORTICOTROPIN AND β-ENDORPHIN

Complementary DNA (cDNA) has been used to clone the human gene for POMC (Whitefeld et al., 1982) and to localize it on chromosome 2 (Owerbach et al., 1981b). The POMC gene has been found to contain two large intervening sequences (introns). One of these, of approximately 3,500 base pairs, interrupts the coding region of the N-terminal fragment of the common precursor. The second intron (>3,000 base pairs) interrupts the 5'-untranslated portion of the mRNA. Thus, the second exon of the gene contains the sequence for a portion (around 20 nucleotides) of the 5'-untranslated portion of the mRNA, all of the amino acids of the signal-peptide portion, and eight amino acids of the N-terminal fragment. The region of the gene spanning corticotropin (ACTH), α-, β-, and γ-melanocyte-stimulating hormone (MSH), corticotropin-like intermediate lobe peptide (CLIP), and β-lipotropin (β-LPH), including β-endorphin, is not interrupted by introns. Thus, with this gene, many regions coding for different functional domains in proteins are not located on exons separated by introns.

Of additional interest is the finding of striking homology in the amino acid codons in the N-terminal fragment of neuron POMC with those previously detected in the N-terminal fragment for bovine POMC (Nakanishi et al., 1979). This finding suggests that the N-terminal fragment may serve a functional physiologic role. Indeed, several reports suggest that the N-terminal fragment might be involved in aldosterone secretion.

A DNA fragment complementary to a portion of mRNA coding for mouse POMC (Roberts et al., 1979b) was used to study the regulation of ACTH synthesis by cultured mouse pituitary (AtT-20) cells (Roberts et al., 1979c). It was found that glucocorticoids induce a long-term (hours) decrease in mRNA production that paralleled the decrease in ACTH production (Roberts et al., 1979a). These changes were preceded by binding of glucocorticoids in the nucleus and by rapid effects on chromatin structure that were reflected by a decrease in its ability to bind bacterial RNA polymerase. When the cells were stimulated with vasopressin, a faster (5-min) effect of glucocorticoids, inhibiting ACTH release, was observed (Johnson et al., 1980). Thus, glucocorticoids appear to regulate the expression of this gene at posttranscriptional as well as transcriptional levels.

Recombinant DNA technology has also been used successfully to synthesize β-endorphin in bacteria. A fragment of DNA complementary to mouse POMC mRNA was cloned previously in bacteria (Fig. 1). This fragment contained the information for the carboxyl-terminal 15 amino acids of β-MSH and all the information except the C-terminal amino acid of β-endorphin. It was necessary, therefore, to modify it to recreate the codon for the C-terminal amino acid, to insert a stop codon, to link the fragment in place to a bacterial gene, and to devise a method for the release of mature β-endorphin from the hybrid protein (Shine et al., 1980). After recreation of the codon for the last amino acid and the addition of a stop codon, the fragment was ligated to the bacterial β-galactosidase gene (in plasmid pBR322) (Fig. 2). The fusion gene of β-galactosidase, β-MSH, and β-endorphin directed the synthesis of a β-galactosidase/β-MSH/β-endorphin hybrid protein in *E. coli*. To release authentic β-endorphin from the hybrid protein by treatment with trypsin, the lysine side chains present in β-endorphins were protected with citraconic anhydride. Therefore, trypsin could only cleave at the arginine residue in the connecting dipeptide (Fig. 1) between the β-MSH fragment and β-endorphin. The protecting groups were then removed by acid hydrolysis. The released peptide was shown to possess β-endorphin activity by radioimmunoassay and opiate receptor binding assay (Shine et al., 1980).

Human β-endorphin has subsequently been synthesized by bacteria by taking advantage of the fact that the natural gene is not interrupted by intervening sequences in the region spanning β-endorphin. In fact, a Hae III fragment could be obtained that spanned the β-endorphin sequences, and this could be ligated into the β-galactosidase gene in a similar way to create a gene that could direct the synthesis of the fusion protein from which authentic β-endorphin could be derived.

STRUCTURE AND EXPRESSION OF GENES FOR GROWTH HORMONE AND RELATED HORMONES

Growth hormone, a protein of about 22 kD synthesized in the anterior pituitary, is required for preadult growth. The hormone has growth-promoting activity and is also involved in the regulation of carbohydrate metabolism. It is structurally, evolutionarily, and functionally related to two other protein hormones, prolactin (also produced in the anterior pituitary) and placental lactogen (also called chorionic somatomammotropin and produced in the placenta). These three hormones show considerable structural homology, one to the other; in addition, each protein shows four areas of internal homology.

The genes for these proteins are under the control of different hormones. To understand the evolution and regulation of expression of these genes, cDNAs from mRNAs and genomic DNAs

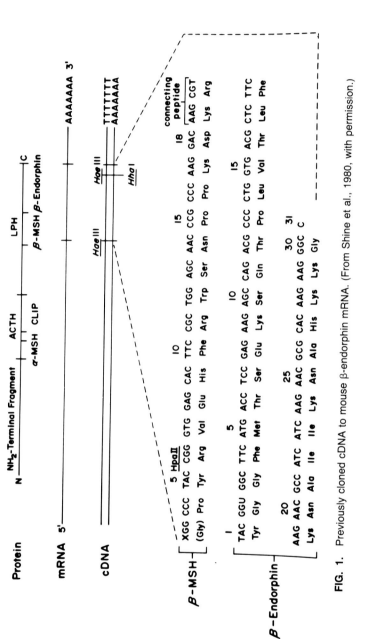

FIG. 1. Previously cloned cDNA to mouse β-endorphin mRNA. (From Shine et al., 1980, with permission.)

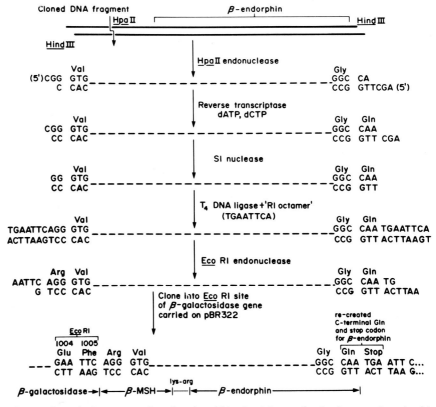

FIG. 2. Steps in the construction of a plasmid to direct the synthesis of mouse β-endorphin.

(gDNAs) from different species (rat, human, and bovine) have been cloned and sequenced (Seeburg et al., 1977; Shine et al., 1977; Martial et al., 1979; Cooke et al., 1980; Miller et al., 1980; Cooke et al., 1981). There is significant nucleic acid homology in all cases. For instance, comparison of rat growth hormone and human placental lactogen genes (two genes for two different proteins in two different species) reveals a substantial degree of homology (Fig. 3).

The genes for human growth hormone and placental lactogen show the greatest homology; the gene for prolactin shows the least homology with the other two. A comparison of the nucleic acid sequence data supports the hypothesis that the genes for these proteins evolved from a common ancestral gene that first duplicated—about 380 million years ago—to form one gene that was the precursor to prolactin and another that later duplicated to yield the precursor gene to growth hormone and placental lactogen. Whereas the genes for the latter proteins are localized on human chromosome 17 (Owerbach et al., 1980), that for prolactin is part of chromosome 6 (Owerbach et al., 1981a).

However, further comparisons of nucleic acid sequences raise questions regarding the extent to which nucleotide substitutions or differences can really serve as a reasonable evolutionary clock. For example, comparison of the genes for human growth hormone and human placental lactogen reveals that they are much more similar to each other than is human growth hormone to rat growth hormone. This might be interpreted to mean that placental lactogen diverged from growth hormone at a time after humans and rats had diverged, which is unlikely because many other species possess placental lactogen. Likewise, a comparison of this family of genes in

		NUMBER	134			137			141	
RGH		AA	ARG	ILE	GLY	GLN	ILE	LEU	LYS	GLN
		mRNA	CGU	AUU	GGG	CAG	AUC	CUC	AAG	CAA
HCS		mRNA	CGG	ACU	GGG	CAG	AUC	CUC	AAG	CAG
		AA	ARG	THR	GLY	GLN	ILE	LEU	LYS	GLN
			142			145			149	
RGH		AA	THR	TYR	ASP	LYS	PHE	ASP	ALA	ASN
		mRNA	ACC	UAU	GAC	AAG	UUU	GAC	GCC	AAC
HCS		mRNA	ACC	UAC	AGC	AAG	UUU	GAC	ACA	AAC
		AA	THR	TYR	SER	LYS	PHE	ASP	THR	ASN

AMINO ACID HOMOLOGY = 80%
NUCLEIC ACID HOMOLOGY = 85%

FIG. 3. Amino acid and nucleic acid sequences in comparable regions of rat growth hormone (RGH) and human chorionic somatomammotropin (HCS; placental lactogen). (From Baxter et al., 1979, with permission.)

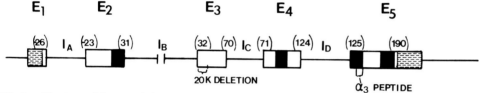

FIG. 4. Structure of the growth hormone gene. *Dashed area*, untranslated gene sequences; *black areas*, homologous sequences. *Number in parentheses* indicates amino acid number. (From Barta et al., 1981, with permission.)

different species reveals that bovine and rat genes are more similar, and bovine and human prolactin genes are more similar (Miller et al., 1981). Thus, this type of analysis may have limited validity. These data are most likely explained by the hypothesis that there have been rearrangements between the growth hormone and placental lactogen genes, of which there are several copies in the human genome, and all are located on the same chromosome. These rearrangements (similar to the ones proposed for some of the globin genes) (see T. Maniatis et al., *this volume*) might have minimized divergence in human genes.

The genomic DNAs for human and rat growth hormone and human and rat prolactin, as well as for human placental lactogen, have been analyzed. Whereas to date there seems to be only one gene for growth hormone and for prolactin in the rat, there are several human growth hormone and placental lactogen genes. In some cases there is a close linkage between the two genes. In each case examined so far, five exons containing the coding regions are separated by four introns. The introns in the prolactin gene, which has about 12,000 bases, are much larger than the introns in growth hormone, which has about 2,000 bases.

Furthermore, when the sequences are aligned to maximize homology, the introns are placed at identical locations, providing additional support for the notion of a common evolutionary origin of these genes as well as of the globin set of genes. As already mentioned, there are several areas of internal homology in growth hormone (Fig. 4): a homologous sequence once on each of exons 2 and 4 and twice on exon 5.

Exon 3, however, does not contain any of the repeated sequences and appears to have a separate functional domain. If one prepares a proteolytic digest of growth hormone, the fragments that contain the amino acids of exon 3 possess the carbohydrate metabolism-regulating properties of growth hormone mentioned earlier. Similarly, a 20,000 MW variant of growth hormone that is devoid in part of amino acids of exon 3 also lacks certain carbohydrate metabolism-regulating activities but still has growth-promoting activity. It is known that the disulfide bridge between amino acids 53 and 165 (Fig. 5) is essential for the growth-promoting activities. The loop also contains exon 3, some of which is essential for, but does not itself have, growth-promoting activity.

These data have led to the hypothesis that an ancestral gene (A) was duplicated four times and into this was ligated a second structure (B) to form the gene that served as the precursor of all members of the set (Fig. 6). It might be that a deletion of an intron may have occurred that led to two homologous sequences in one exon (i.e., exon 5).

Of particular interest is the observation that the DNA in intron B of the rat growth hormone gene is highly repetitive and also contains terminal and internally repeated sequences. Repetitive DNA is also found in the 3'-region flanking the gene and in one intron of the rat prolactin gene. This repetitive DNA has features similar to transposable elements of prokaryotes, which move DNA segments within the genome (Calos and Miller, 1980). Thus it may be involved in some of the rearrangements that have participated in gene evolution.

With the growth hormone set of genes, there are at least three examples where the same pre-mRNA appears to be processed in two different ways. The first example occurs with human growth hormone. It has been known for some time that, in addition to the normal human growth hormone (191 amino acids, 22kD), there exists a variant that lacks amino acids 32 to 46 and has growth-promoting activity but not certain of the carbohydrate metabolism-regulating activity of the hormone (Wallis, 1980). This 20-kD variant represents 15% of the growth hormone made by the pituitary. Apparently, the same pre-mRNA generates two different mRNAs by different processing of the primary transcript (Fig. 7) (Wallis, 1980). In the case of prolactin, sequencing of the cDNA revealed the presence of an additional triplet coding for alanine, which does not occur in the protein structure. This alanine difference can be explained by splicing at either of two different locations, deleting or including the alanine triplets. Under certain experimental conditions (growth conditions in cell cultures, presence of certain hormones), two forms of rat growth hormone mRNA are found in GH_3 cells. The translation products of these

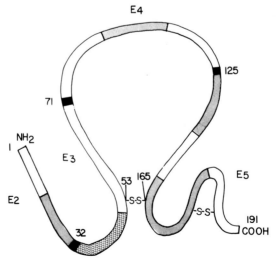

FIG. 5. Growth hormone structure and its exon (E) sources. *Black area*, intron exon junction; *dotted area*, homologous region; *dashed area*, 20 kD deletion.

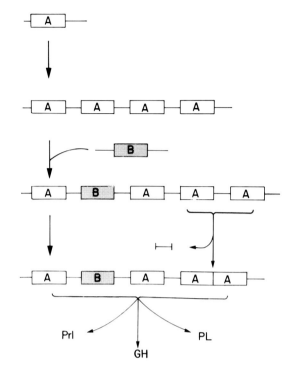

FIG. 6. Evolution of the growth hormones (GH) and related genes. A, ancestral gene; B, second structure; Prl, prolactin; PL, placental lactogen. (From Barta et al., 1981, with permission.)

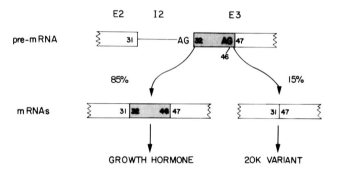

FIG. 7. Processing of growth hormone pre-mRNA to different mRNAs for growth hormone.

mRNAs are indistinguishable from each other, however. Sequencing of the corresponding cDNAs does not reveal differences in the 5' end or the coding region. Thus, the difference may reside in the 3' end, perhaps in the polyadenylation site.

Hormones such as triiodothyronine (T_3) or dexamethasone (a synthetic glucocorticoid) regulate the expression of the growth hormone gene in cells and culture (Martial et al., 1977). Solution hybridization experiments (see W. Hahn, *this volume*) show that both hormones increase the mRNA levels for growth hormone, acting synergistically. Whereas T_3 is normally more potent than dexamethasone in increasing mRNA levels, dexamethasone produces more predominant mRNA precursor bands than does T_3, as seen in Northern blotting experiments. This suggests that there is additional regulation at levels other than that of transcription.

ACKNOWLEDGMENT

Studies reported here were supported in part by NIH Grant AM 19997-05.

REFERENCES

Barta, A., Richards, R. I., Baxter, J. D., and Shine, J. (1981): Primary structure and evolution of rat growth hormone gene. *Proc. Natl. Acad. Sci. USA*, 78:4867–4871.
Baxter, J. D., Seeburg, P. H., Shine, J., Martial, J. A., Ivarie, R. D., Johnson, L. K., Fiddes, J. C., and Goodman, H. M. (1979): Structure of growth hormone gene sequences and their expression in bacteria and in cultured cells. In: *Hormones and Cell Culture*, edited by G. H. Sato and R. Ross, pp. 317–337. Cold Spring Harbor Laboratory, New York.
Calos, M. P., and Miller, J. H. (1980): Transposable elements. *Cell*, 20:579–595.
Cooke, N. E., Coit, D., Shine, J., Baxter, J. D., and Martial, J. A. (1981): Human prolactin: cDNA structural analysis and evolutionary comparisons. *J. Biol. Chem.*, 256:4007–4016.
Cooke, N. E., Coit, D., Weiner, R. I., Baxter, J. D., and Martial, J. A. (1980): Structure of cloned DNA complementary to rat prolactin messenger RNA. *J. Biol. Chem.*, 255:6502–6510.
Johnson, L. K., Eberhardt, N. L., Spindler, S. R., Martial, J. A., Dallman, M. F., Jones, M. T., and Baxter, J. D. (1980): Regulation of the genes for ACTH and growth hormone by glucocorticoid hormones. In: *Endocrinology 1980*, edited by I. A. Cumming, J. W. Funder, and F. A. O. Meldelsolm, pp. 70–73. Australian Academy of Science, Canberra.
Martial, J. A., Baxter, J. D., Goodman, H. M., and Seeburg, P. H. (1977): Regulation of growth hormone messenger RNA by thyroid and glucocorticoid hormones. *Proc. Natl. Acad. Sci. USA*, 74:1816–1820.
Martial, J. A., Hallewell, R. A., Baxter, J. D., and Goodman, H. M. (1979): Human growth hormone: Complementary DNA cloning and expression in bacteria. *Science*, 205:602–607.
Miller, W. L., Coit, D., Baxter, J. D., and Martial, J. A. (1981): Cloning of bovine prolactin cDNA and evolutionary implications of its sequence. *DNA*, 1:37–50.
Miller, W. L., Martial, J. A., and Baxter, J. D. (1980): Molecular cloning of DNA complementary to bovine growth hormone mRNA. *J. Biol. Chem.*, 255:7521–7524.
Nakanishi, S., Inoue, A., Kita, T., Nakamura, M., Chang, A. C. Y., Cohen, S. N., and Numa, S. (1979): Nucleotide sequence of cloned cDNA for bovine corticotropin-β-lipotropin precursor. *Nature*, 278:423–427.
Owerbach, D., Martial, J. A., Baxter, J. D., Rutter, W. J., and Shows, T. B. (1980): Genes for growth hormone, chorionic somatomammotropin and growth hormone-like gene on chromosome 17 in humans. *Science*, 209:289–292.
Owerbach, D., Rutter, W. J., Cooke, N., Martial, J. A., and Shows, T. B. (1981a): The prolactin gene is located on chromosome 6 in humans. *Science*, 212:815–816.
Owerbach, D., Rutter, W. J., Roberts, J. L., Whitfeld, P., Shine, J., Seeburg, P. H., and Shows, T. B. (1981b): The proopiocortin (adrenocorticotropin/β-lipotropin) gene is located on chromosome 2 in humans. *Somatic Cell Genet.*, 7:359–369.
Roberts, J. L., Budarf, M. L., Baxter, J. D., and Herbert, E. (1979a): Selective reduction of proadrenocorticotropin/endorphin proteins and messenger ribonucleic acid activity in mouse pituitary tumor cells by glucocorticoids. *Biochemistry*, 18:4907–4915.
Roberts, J. L., Johnson, L. K., Baxter, J. D., Budarf, M. L., Allen, R. G., and Herbert, E. (1979b): Effect of glucocorticoids on the synthesis and processing of the common precursor to adrenocorticotropin and endorphin in mouse pituitary tumor cells. In: *Hormones and Cell Culture*, edited by H. H. Sato and R. Ross, pp. 827–841. Cold Spring Harbor Laboratory, New York.
Roberts, J. L., Seeburg, P. H., Shine, J., Herbert, E., Baxter, J. D., and Goodman, H. M. (1979c): Corticotropin and β-endorphin: Construction and analysis of recombinant DNA complementary to mRNA for the common precursor. *Proc. Natl. Acad. Sci. USA*, 76:2153–2157.
Seeburg, P. H., Shine, J., Martial, J. A., Baxter, J. D., and Goodman, H. M. (1977): Nucleotide sequence and amplification in bacteria of structural gene for rat growth hormone. *Nature*, 270:486–494.
Shine, J., Fettes, I., Lan, N. C., Roberts, J. L., and Baxter, J. D. (1980): Expression of cloned β-endorphin gene sequences by *E. coli*. *Nature*, 285:456–461.
Shine, J., Seeburg, P. H., Martial, J. A., Baxter, J. D., and Goodman, H. M. (1977): Construction and analysis of recombinant DNA for human chorionic somatomammotropin. *Nature*, 270:494–499.
Wallis, M. (1980): Growth hormone: deletions in the protein and introns in the gene. *Nature*, 284:512.
Whitefeld, P. L., Seeburg, P. H., and Shine, J. (1982): The human pro-opiomelanocortin gene: primary structure of the coding regions. *DNA*, 1:133–143.

Molecular Genetic Neuroscience, edited by
F. O. Schmitt, S. J. Bird, and F. E. Bloom.
Raven Press, New York © 1982.

A Multivalent Proenkephalin and its Processing

Sidney Udenfriend

ABSTRACT

Methodology and instrumentation have been developed to isolate and characterize peptides and proteins in picomole concentrations. Many enkephalin-containing polypeptides (ECPs) have been isolated from beef adrenal medulla and characterized. Some of these same peptides are present in brain and intestine. These ECPs are apparently intermediates in a biosynthetic pathway that starts with a proenkephalin precursor of about 40 to 50 kilodaltons (kD). Proenkephalin mRNA contains about 1,500 bases, consistent with a translation product of about 50 kD. Like several of the smaller and intermediate-sized enkephalin-containing polypeptides, proenkephalin includes many copies of the enkephalin sequence: Met-enkephalin, along with an occasional Leu-enkephalin. Such repetitive (multivalent) polypeptides are unusual and possibly unique. Evidence is presented that ECPs may be physiologic agents in their own right and not just intermediates in the biosynthesis of free enkephalins.

The search for opioid peptides in the adrenal medulla was sparked by the observation of Schultzberg and colleagues (1978) that this tissue contains material that cross-reacts with antibodies against enkephalins. In further studies, Lewis et al. (1980a) revealed the existence in bovine adrenal medulla of many enkephalin-containing polypeptides (ECPs), and a proenkephalin that has a molecular weight in the range of 40 to 50 kD. This precursor also contains many enkephalin sequences, both Met-enkephalin and Leu-enkephalin (in a ratio of approximately 7 to 1) and is therefore a multivalent (or repetitive) precursor. This enkephalin pathway of biosynthesis is clearly different from the proopiocortin pathway because it does not involve β-lipotropin or β-endorphin. Brain and intestine also contain ECPs and a similar multivalent proenkephalin.

IMMUNOREACTIVE ENKEPHALINS IN ACIDIC AND NEURAL EXTRACTS OF BOVINE ADRENAL MEDULLA CHROMAFFIN GRANULES

Sizing of an acid extract from bovine adrenal chromaffin granules (on Sephadex G-100) shows the presence of immunoreactive enkephalins ranging in size from about 1 to 30 kD (Fig. 1). Preincubation of the extract with trypsin increases the amount of immunoreactivity detectable. If the extracted material is also treated with carboxypeptidase B[1], further increases in immunoreactivity can be observed, particularly in the higher molecular weight range. The ECPs were arbitrarily divided into fractions according to size (Table 1).

---

[1] A combination of these two enzymes is thought to be responsible for the enzymatic conversion of prohormones to hormones, i.e., proinsulin to insulin (see D. Steiner, *this volume*).

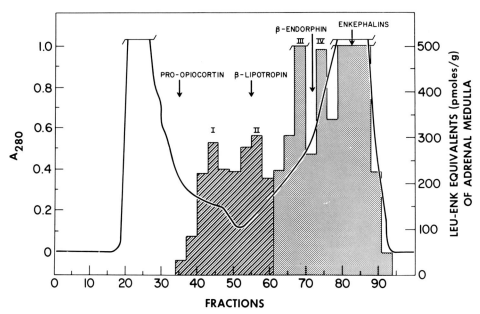

FIG. 1. G-100 chromatography of an acid extract from bovine adrenal medulla chromaffin granules with or without preincubation with trypsin. The pituitary peptides over the arrows are standards used as markers. *Striped area*, with trypsin; *dotted area*, without trypsin. (From Lewis et al., 1979.)

TABLE 1. *Adrenal enkephalin-containing peptide acid extracts*

Fraction	Size (kilodaltons)
I	13–25
II	10–15
III	5–9
IV	1–5
V	<1

TABLE 2. *Tryptic peptides from the 8,000-Dalton ECP*

Peptide	Composition	Yield (nmole)	Yield (%)
T8-I	Lys	1.7	71
T8-II	Ala, Leu, Arg	1.8	75
T8-III	CMCys, Thr_2, Glx, Lys	2.2	92
T8-IV	Ser, Glx_3, Ala, Leu_2, His, Lys	1.9	79
T8-V	Ser, Pro, Leu_2, Lys	1.4	58
T8-VI	Tyr-Gly-Gly-Phe-Met	0.8	33
T8-VII	Lys-Tyr-Gly-Gly-Phe-Met	0.4	17
T8-VIII	Thr, Glx_2, Leu_3, Lys	2.2	37
T8-IX	Asx, Thr, Ser_2, Glx, Pro_3, Ala_2, Leu_3, Lys	1.9	79
T8-X	CMCys, Asx, Thr, Ser, Glx, Pro, Gly, Ala, Leu_2, Lys	1.8	75

PURIFICATION AND CHARACTERIZATION OF ADRENAL ECPS

For the purification and characterization of adrenal ECPs, the following methodology has been developed (for a review, see Stein, 1980): high performance liquid chromatography (HPLC) coupled to fluorescence detection for isolation; fluorescamine amino acid analyzer; fingerprinting and isolation of tryptides; manual sequencing-carboxy and aminopeptidase-dansyl Edman; and automated Edman sequencer. This system offers high sensitivity (in the picomole range), high resolution, and the possibility of working with volatile buffer systems. The procedure(s) used for purification and characterization and the criteria applied to establish the homogeneity of the purified peptides are: single symmetrical peak on final HPLC, single band on SDS gels (proteins only), amino acid analysis, single end group, partial sequencing, and identification and assay of enkephalins and 6-substituted enkephalins in trypsin digests. With this methodology the following small peptides, structurally related to Leu- and Met-enkephalin, were found in fraction V (Kimura et al., 1980): Leu-enkephalin, Met-enkephalin, Met-enkephalin Arg^6, Met-enkephalin Lys^6, Leu-enkephalin Arg^6 or Leu-enkephalin Lys^6, Met-enkephalin Arg^6 Phe^7, and Met-enkephalin Arg^6 Arg^7.

Fraction IV of the same preparation contains at least 10 peptides in the range of 1 to 5 kD (Stern et al., 1981). One of them, peptide F (3,800 kD) contains two copies of the Met-enkephalin sequence, one at the amino-terminal region and the other at the carboxyl-terminal end (Fig. 2).

Figure 3 demonstrates the steps of purification by HPLC of a 14 kD peptide. The crude mixture was chromatographed on an RP 18 or cyanopropyl column (using a pyridine formate/n-propanol system as the mobile phase), and the biological activity was assayed using a radioreceptor assay. Rechromatography of the active fraction (using different stationary phases and elution conditions) yielded a homogeneous product. The 14-kD peptide contains three copies of Met-enkephalin (lewis et al., 1980b). The tryptic map of an 8-kD peptide, purified in essentially the same manner, is schematically illustrated in Fig. 4.

Amino acid analysis and bioassay of the tryptic peptides of this peptide (Table 2) showed it to contain one copy of Met-enkephalin (Lewis et al., 1980b). Adrenal medullary chromaffin cells contain several ECPs ranging in molecular weight from 0.5 to 22 kD (Fig. 5). They all appear to be derived from the processing of a ~50 kD multivalent proenkephalin (polyenkephalin) that contains Met- and Leu-enkephalin sequences in a copy ratio of approximately 7 to 1. Some of the intermediate-sized polypeptides contain both Met- and Leu-enkephalin sequences, and the largest contain several Met-enkephalin sequences. Neither β-endorphin nor its congeners were detected in beef adrenal medulla; if they are present they are in amounts far below those of the ECPs.

The fact that both Met- and Leu-enkephalin are present in the same precursor molecules suggests that it is unlikely that there are distinct Met- and Leu-enkephalinergic cells. In fact, the cellular ratio of these two compounds is determined, in part, by the precursor.

```
        1              5                   10
Tyr-Gly-Gly-Phe-Met-Lys-Lys-Met-Asp-Glu-Leu-Tyr-

             15                  20
Pro-Leu-Glu-Val-Glu-Glu-Glu-Ala-Asn-Gly-Gly-Glu-

 25                  30              34
Val-Leu-Gly-Lys-Arg-Tyr-Gly-Gly-Phe-Met
```

FIG. 2. Primary structure of peptide F. (Adapted from Jones et al., 1980.)

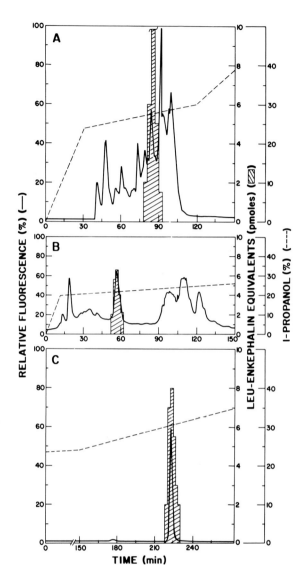

FIG. 3. Purification of a 14-kD peptide in three consecutive HPLC steps. (From Lewis et al., 1980b.)

All of the internal enkephalin sequences that have been characterized thus far in the various ECPs are bracketed by pairs of basic amino acids (arginine and/or lysine) in a manner similar to hormone sequences in prohormones (see D. Steiner, *this volume*). Precursors and intermediates all occur in the chromaffin granules of the adrenal medulla. Brain and intestine also contain a similar proenkephalin. Some of the smaller enkephalin-containing peptides have also been found in these tissues.

BIOLOGICAL ACTIVITY OF ADRENAL OPIOID PEPTIDES

In addition to the enkephalins, there are several other biologically active peptides that are apparently intermediates in the processing of proenkephalin (Inturrisi et al., 1980). Some of the adrenal opioid peptides are far more active than the enkephalins in the guinea pig myenteric plexus assay (Table 3) (see A. Goldstein, *this volume*).

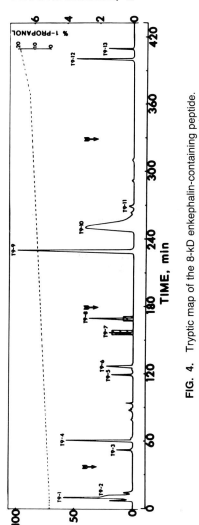

FIG. 4. Tryptic map of the 8-kD enkephalin-containing peptide.

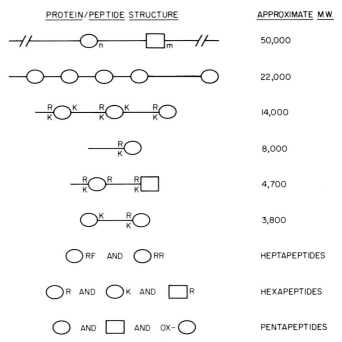

FIG. 5. Enkephalin precursors. *Circle*, Met-enkephalin; *square*; Leu-enkephalin; R, Arg; K, Lys; F, Phe; R/K, Arg or Lys; OX, oxidized Met. (From Lewis et al., 1980a.)

TABLE 3. *Biological activities of enkephalins and naturally occurring enkephalin-containing peptides*[a]

Peptide	IC_{50} (nmole)[b]	
	Guinea pig ileum assay	Receptor binding assay
Met-enkephalin	36	20
Leu-enkephalin	504	—
β-Endorphin	31	—
Dynorphin$_{1-13}$	0.52	15
Adrenal opioid peptides		
Peptide I	100	—
Peptide F	306	100
Peptide B	200	—
Peptide E	0.96	40
Des-Leu^{25}-peptide E[c]	1.5	—
BAM-12P	15.5	—
BAM-20P	2.2	—
BAM-22P	1.3	—

[a]Values reported for both guinea pig ileum and receptor binding activities are from individual experiments. Similar results were found in three other experiments.

[b]Concentration that inhibits maximal response by 50%.

[c]The carboxyl-terminal residue of peptide E, Leu^{25}, was removed by digestion with carboxypeptidase Y (Kilpatrick et al., 1981). Analysis of the digest mixture demonstrated greater than 95% removal of Leu^{25}.

4900 DALTON ADRENAL PEPTIDE I (REF.)
Ser-Pro-Thr-Leu-Glu-Asp-Glu-Leu-Lys-Glu-Leu-Gln-Lys-Arg-Tyr-Gly-Gly-Phe-Met-Arg-Arg-Val-Gly-Arg-Pro-Glu-Trp-Trp-Met-Asp-Tyr-Gln-Lys-Arg-Tyr-Gly-Gly-Phe-Leu
(positions 1, 5, 10, 15, 20, 25, 30, 35, 39)

3200 DALTON ADRENAL PEPTIDE E
Tyr-Gly-Gly-Phe-Met-Arg-Arg-Val-Gly-Arg-Pro-Glu-Trp-Trp-Met-Asp-Tyr-Gln-Lys-Arg-Tyr-Gly-Gly-Phe-Leu
(positions 1, 5, 10, 15, 20, 25)

BAM-22 P (REF.)
Tyr-Gly-Gly-Phe-Met-Arg-Arg-Val-Gly-Arg-Pro-Glu-Trp-Trp-Met-Asp-Tyr-Gln-Lys-Arg-Tyr-Gly
(positions 1, 5, 10, 15, 20, 22)

BAM-20 P (REF.)
Tyr-Gly-Gly-Phe-Met-Arg-Arg-Val-Gly-Arg-Pro-Glu-Trp-Trp-Met-Asp-Tyr-Gln-Lys-Arg
(positions 1, 5, 10, 15, 20)

BAM-12 P (REF. 21)
Tyr-Gly-Gly-Phe-Met-Arg-Arg-Val-Gly-Arg-Pro-Glu
(positions 1, 5, 10, 12)

3800 DALTON ADRENAL PEPTIDE F
Tyr-Gly-Gly-Phe-Met-Lys-Lys-Met-Asp-Glu-Leu-Tyr-Pro-Leu-Glu-Val-Glu-Glu-Ala-Asn-Gly-Gly-Glu-Val-Leu-Gly-Lys-Arg-Tyr-Gly-Gly-Phe-Met
(positions 1, 5, 10, 15, 20, 25, 30, 34)

3600 DALTON ADRENAL PEPTIDE B (REF.)
Phe-Ala-Glu-Pro-Leu-Pro-Ser-Glu-Glu-Glu-Gly-(Ser,Glx,Glx,Glx,Glx,Pro,Val,Met,Tyr,Lys)Lys-Arg-Tyr-Gly-Phe-Met-Arg-Phe
(positions 1, 5, 10, 25, 30)

DYNORPHIN₁₋₁₃
Tyr-Gly-Phe-Leu-Arg-Arg-Ile-Arg-Pro-Lys-Leu-Lys
(positions 1, 5, 10, 13)

α-NEO-ENDORPHIN
Tyr-Gly-Phe-Leu-Arg-Lys-Arg(Pro,Gly,Tyr,Tyr,Lys,Lys,Arg)
(positions 1, 5)

FIG. 6. Adrenal medullary peptides with similar partial structures isolated by different procedures. Comparison is made with other known enkephalin-containing proteins. (From Kilpatrick et al., 1981.)

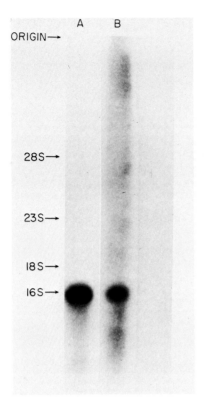

FIG. 7. Hybridization of decahexamer or cDNA to adrenal medullary poly(A)-RNA. 10 mg of poly(A)-RNA were fractionated on parallel lanes on a 1.4% agarose gel in the presence of 10 mM CH_3HgOH for 16 hr at 3 V/cm. The RNA was transferred onto nitrocellulose sheets. Adjacent lanes were hybridized with decahexamer [2 pmole (specific activity = 5.7×10^6 cpm/pmole) in 3 ml of solution A for 17 hr at 30° C] or with cDNA (12,000 cpm for 12 hr at 42° C). Filters were washed, dried, and autoradiographed at $-70°$ C. Size markers were *E. coli* rRNA and bovine rRNA, run in parallel lanes and visualized by staining with ethidium bromide prior to transfer. **Lane A**: decahexamer; **lane B**: cDNA. (From Gübler et al., 1981.)

One of these, peptide E, is far more active than the enkephalins (Kilpatrick et al., 1981) with a potency comparable to dynorphin (see A. Goldstein, *this volume*; Gübler et al., 1981). This peptide is related to a group of highly active peptides, called BAM (bovine adrenal medulla) 12P, 20P, and 22P, which were recently isolated by Mizuno and coworkers (1980) in Japan (Fig. 6). The latter represent shorter and varying extensions of exactly the same sequence as peptide E. It should be noted that these workers did not use washed chromaffin granules as starting material for extraction but rather whole adrenal medulla. It may be that the presence of lysosomal proteases in initial extracts may explain the heterogeneity of the peptides isolated.

The marked biological activity of proenkephalin intermediates may indicate that the enkephalins are not the only physiologically significant peptides included in this precursor.

PROENKEPHALIN mRNA

A large proportion of the amino acid sequences of some regions of the enkephalin precursor have been obtained. As an alternative to sequencing the whole protein, most of the cDNA for proenkephalin has been cloned and sequenced (Gübler et al., 1982). A primer was synthesized based on a sequence of six amino acids present in peptide I: Trp-Trp-Met-Asp-Tyr-GLn.

Using this synthetic primer of 18 nucleotides, five distinct polynucleotide species of bovine adrenal medullary cDNA were obtained by reverse transcription of medullary mRNA. The synthetic primer and cDNAs were then used in hybridization experiments to identify the mRNA that encodes the proenkephalin message. This was found to be a 16S species (approximately 1,500 bases), and may represent as little as 0.03% of the total mRNA in this tissue (Fig. 7)

(Gübler et al., 1981). Work is in progress to obtain the DNA-sequence of the entire gene corresponding to this mRNA.

The findings reported here provide evidence for a biosynthetic pathway that starts with a polyenkephalin translation product, proenkephalin, which yields a large number of enkephalin-containing polypeptides through appropriate processing enzymes. The physiologic significance of these enkephalin-containing polypeptides must be considered as distinct from free enkephalin.

ACKNOWLEDGMENTS

The studies reported here were carried out by many colleagues including Stanley Stein, Jack Shively, Jean Rossier, Larry Brink, Louise Gerber, Kohichi Kojima, Barry Jones, Dan Kilpatrick, Sadao Kimura, Alvin Stern, and Randy Lewis.

REFERENCES

Gübler, U., Seeburg, P. H., Hoffmann, B. J., Gage, L. P., Udenfriend, S. (1982): Molecular cloning establishes proenkephalin as precursor of enkephalin-containing peptides. *Nature*, 295:206–208.

Gübler, U., Kilpatrick, D. L., Seeburg, P. H., Gage, L. P., and Udenfriend, S. (1981): Detection and partial characterization of proenkephalin mRNA. *Proc. Natl. Acad. Sci. USA*, 78:5484–5487.

Inturrisi, C. E., Umans, J. G., Wolff, D., Stern, A. L., Lewis, R. V., Stein, S., and Udenfriend, S. (1980): Analgesic activity of the naturally occurring heptapeptide of [Met]enkephalin-Arg^6-PHe^7. *Proc. Natl. Acad. Sci. USA*, 77:5512–5514.

Jones, B. N., Stern, A. S., Lewis, R. V., Kimura, S., Stein, S., Udenfriend, S., and Shively, J. E. (1980): Structure of two adrenal polypeptides containing multiple enkephalin sequences. *Arch. Biochem. Biophys.*, 204:392–395.

Kilpatrick, D. L., Taniguchi, T., Jones, B. N., Stern, A. S., Shively, J. E., Hullihan, J., Kimura, S., Stein, S., and Udenfriend, S. (1981): A highly potent 3200-dalton adrenal opioid peptide that contains both a [Met]- and [Leu]enkephalin sequence. *Proc. Natl. Acad. Sci. USA*, 78:3265–3268.

Kimura, S., Lewis, R. V., Stern, A. S., Rossier, J., Stein, S., and Udenfriend, S. (1980): Probable precursors of [Leu]- and [Met]enkephalin in the adrenal medulla: Peptides in the range of 3 to 5 kilodaltons. *Proc. Natl. Acad. Sci. USA*, 77:1681–1685.

Lewis, R. V., Stern, A. S., Kimura, S., Rossier, J., Stein, S., and Udenfriend, S. (1980a): An about 50,000-dalton protein in the adrenal medulla: A common precursor of [Met]- and [Leu]-enkephalin. *Science*, 208:1459–1461.

Lewis, R. V., Stern, A. S., Kimura, S., Stein, S., and Udenfriend, S. (1980b): Enkephalin biosynthetic pathway: Proteins of 8,000 and 14,000 daltons in bovine adrenal medulla. *Proc. Natl. Acad. Sci. USA*, 77:5018–5020.

Lewis, R. V., Stern, A. S., Rossier, J., Stein, S., and Udenfriend, S. (1979): Putative enkephalin precursors in bovine adrenal medulla. *Biochem. Biophys. Res. Commun.*, 89:822–829.

Mizuno, K., Minamino, N., Kangawa, K., and Matsuo, H. (1980): A new family of endogenous "big" met-enkephalins from bovine adrenal medulla: Purification and structure of docosa-(Bam-22P) and eicosapeptide (Bam-20P) with very potent opiate activity. *Biochem. Biophys. Res. Commun.*, 97:1283–1290.

Schultzberg, M., Hökfelt, T., Lundberg, J., Terenius, L., Elfirm, L. -G., and Elde, R. (1978): Enkephalin-like immunoreactivity in nerve terminals in sympathetic ganglia and adrenal medulla and in adrenal medullary gland cells. *Acta Physiol. Scand.*, 103:475–477.

Stein, S. (1980): Ultramicro isolation and analysis of peptides and proteins. In: *Polypeptide Hormones*, edited by R. F. Beers and E. G. Bassett, pp. 77–86. Raven Press, New York.

Stern, A. S., Jones, B. N., Shively, J. E., Stein, S., and Udenfriend, S. (1981): Two adrenal opioid polypeptides: Proposed intermediates in the processing of proenkephalin. *Proc. Natl. Acad. Sci. USA*, 78:1962–1966.

Dynorphin and the Dynorphin Receptor: Some Implications of Gene Duplication of the Opioid Message

Avram Goldstein

ABSTRACT

Dynorphin is a recently discovered and highly potent endogenous opioid peptide. Evidence is presented in support of the presence of a highly specific dynorphin receptor in guinea pig myenteric plexus. Since dynorphin is a widely distributed neuropeptide, dynorphin receptors are presumably also found throughout the central and peripheral nervous system. Some speculations are offered concerning the significance of multiple opioid ligands and multiple opioid receptors.

Dynorphin is a widely distributed opioid peptide (found in the pars nervosa of the pituitary gland, the brain, spinal cord, and intestine). Of the 17 amino acid residues, the first 13 are known (Fig. 1). These known residues carry essentially all the bioactivity of the molecule.

The biological activity of dynorphin was first described in 1975 when it was found in crude porcine adrenocorticotropic hormone (ACTH) extracts (Cox et al., 1975). The properties of this opioid peptide were clearly different from those of β-endorphin or any fragment of β-lipotropin (Teschemacher et al., 1975). It was much more basic, its apparent molecular weight was only half as great, its naloxone-reversible inhibiting effect in the guinea pig myenteric plexus preparation could not be washed out as quickly, and its biological activity was more easily destroyed by trypsin. Most important, its biological activity was completely resistant to cyanogen bromide—a reagent that attacks methionine side chains—whereas the activity of every opioid peptide containing Met-enkephalin is destroyed by this reagent. The cyanogen bromide result suggested the possibility that this non-β-endorphin-like "slow reversing endorphin" might contain Leu-enkephalin (Cox et al., 1976). Enough of the new endorphin for partial

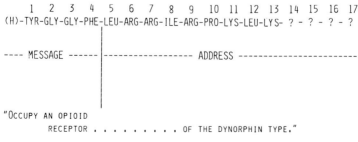

FIG. 1. The amino acid sequence of dynorphin. Residues 14 through 17 have not yet been identified with certainty.

sequencing was not isolated until 1979 when a few nanomoles were obtained in a state suitable for microsequencing. The sequence of the first 13 amino acids was elucidated in collaboration with Michael Hunkapiller and Leroy Hood by a method described by the latter (L. Hood, *this volume*) (Goldstein et al., 1979).

Fortunately, all the potency of the natural peptide was displayed by the synthetic tridecapeptide. The natural peptide was named dynorphin in recognition of its great potency (from the Greek *dynamis*, "power"). In the guinea pig myenteric plexus preparation, for example, it is—on a molar basis—50 times more potent than β-endorphin, 200 times more potent than normophine, and 700 times more potent than Leu-enkephalin. Because dynorphin contains the Leu-enkephalin sequence (Tyr-Gly-Gly-Phe-Leu), followed by two basic residues (Arg-Arg)—a common processing signal for the generation of biologically active peptides from their precursors (see D. Steiner, *this volume*)—it could be argued that this peptide is an intermediate precursor of Leu-enkephalin without a physiological role of its own. Evidence now indicates that dynorphin is a physiologically important entity and that there is in fact a dynorphin receptor, i.e., that the peptide with the dynorphin sequence has a specific receptor in at least one tissue, which is distinctively different from the opiate receptors that interact with enkephalins or morphine.

SOME STRUCTURAL CHARACTERISTICS OF EXOGENOUS OPIATES

Exogenous opiates possess several structural characteristics that led to the initial identification of opiate receptors (Goldstein et al., 1971). Levorphanol (Fig. 2), for instance, which has the full potency of morphine in most systems, has the following structural features: (1) a basic N-atom, that is, it associates a proton, to become cationic at physiological pH; (2) a phenolic hydroxyl group, which adds greatly to the potency, although it is not absolutely essential for

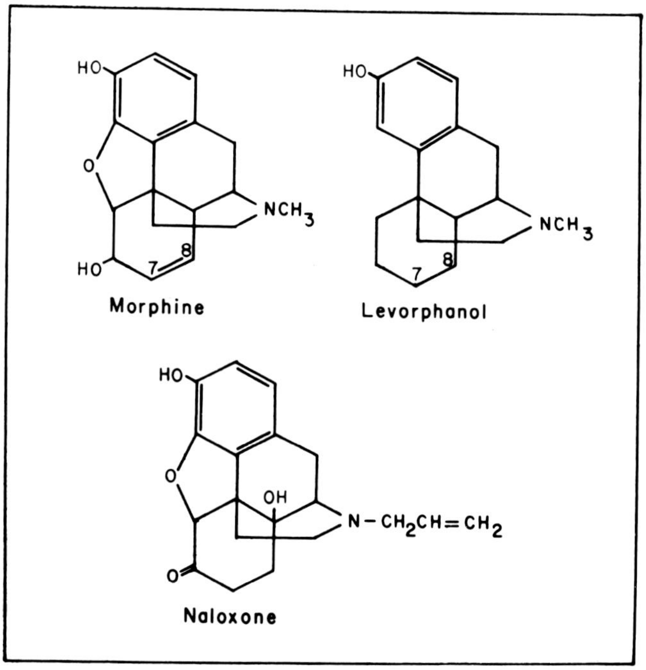

FIG. 2. Opiate alkaloid agonists morphine and levorphanol and antagonist naloxone.

activity; and (3) a hydrophobic three-ring structure in which one ring is perpendicular to the others creating a T-shaped conformation.

Whereas levorphanol is fully active, dextrorphan, its enantiomer, is entirely inactive; this is also true for radioreceptor binding assays, a finding that led to the initial development in the early 1970s of the concept for studying stereospecific binding of opiates and therefore of studying opiate receptors (Goldstein et al., 1971). Naloxone, a specific antagonist of the opiate receptor, played a large role in the search for endogenous opioid ligands (endorphins).

The comparison of a molecular model of an endogenous ligand-like dynorphin (or Leu-enkephalin) with these exogenous ligands shows how the flexible structure of a peptide is capable of matching almost exactly the structure of the opiate series both with respect to the basic nitrogen atom, the distance between the N-atom and the phenolic hydroxyl group, the benzene ring, and the T-shape and angle of that benzene ring to the rest of the ring structure (Fig. 3).

BIOLOGICAL ASSAY SYSTEM FOR OPIATE LIGANDS: THE MYENTERIC PLEXUS ASSAY

Because practically all experimental data presented here are based on this assay, its principle will be described briefly. If one peels off the external longitudinal muscle layer of a guinea pig intestine (Fig. 4) (a few centimeters), the myenteric plexus nerve network comes with it. This layer is mounted in a tissue bath attached to a strain gauge, shocked electrically by field stimulation, and the twitch of the muscle recorded.

It has been known for 25 years that electrical stimulation excites cholinergic neurons in the plexus, which innervate the longitudinal muscle, that this causes the release of acetylcholine,

FIG. 3. Molecular model of dynorphin-(1-13). The NH_2-terminal portion, containing the tetrapeptide Tyr-Gly-Gly-Phe is at extreme left. Phenolic ring of tyrosine is horizontal, with cationic amino group at left, phenolic OH group at right. Angled position of Phe^4 benzene ring is evident. High complexity (information content) of the COOH-terminal extension ("address") is obvious.

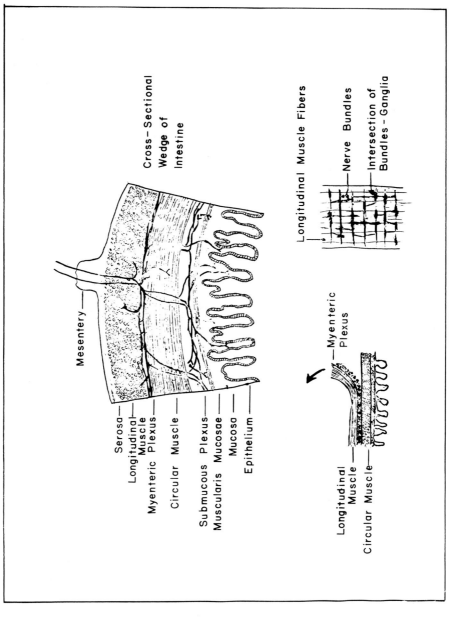

FIG. 4. Microanatomy of the enteric nervous system and method of preparing the myenteric plexus-longitudinal muscle bioassay system. **Left inset** shows how myenteric plexus adheres to longitudinal muscle during stripping operation. **Right inset** depicts ganglia of the myenteric plexus.

and that opiates—in a stereospecific manner and blockable by naloxone—inhibit the release of acetylcholine (Fig. 5). One could therefore use an assay based on the release (or inhibition of release) of acetylcholine, but it is easier to use the twitch of the muscle as an experimental measure.

One must be careful, however, because many factors will inhibit the muscle twitch, such as atropine, which blocks acetylcholine receptors on the muscle, or catecholamines, which alter the strength of the contraction. It is therefore mandatory—if this assay is to be used to detect opioid activity—to use the reversibility of the effect by naloxone as the proper control. Thus, in principle, the muscle is electrically stimulated to cause a twitch, which is blocked by the substance in question, and this inhibition is abolished by naloxone. The concentration needed for a 50% inhibition of the electrically induced muscle twitch—the IC_{50}—can be used to characterize agonist activity. On the other hand, antagonist activity can be characterized by the apparent equilibrium dissociation constant, K_e, as described below.

Figure 6 illustrates the assay. Addition of 10^{-7}M levorphanol leads to a 60% inhibition of the twitch; addition of a 10 times higher concentration of dextrorphan has no effect of its own, and by the subsequent application of levorphanol the same effect as before is obtained. Thus, dextrorphan is absolutely inactive and does not compete with levorphanol for the binding sites. This indicates that the recognition sites of the receptor are stereospecific in admitting only molecules with the correct conformation.

Figure 7 shows the activity of normorphine, pituitary extract containing β-endorphin, and pituitary extract containing dynorphin, the activity of which is completely reversed by 200 nM naloxone. The characteristic slow wash-out of the dynorphin activity, as compared with normorphine or β-endorphin, is evident.

MESSAGE AND ADDRESS IN OPIOID PEPTIDES

Using Schwyzer's concept (Schwyzer et al., 1980) describing the information content of peptides, one can distinguish between a message (a sequence with opioid activity) and an address on the C-terminal extension, which specifies the kind of opioid receptor to be occupied.

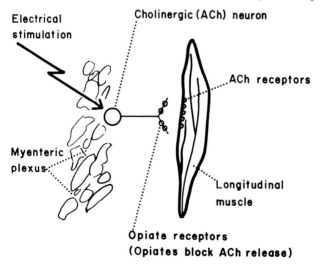

FIG. 5. Principle of the guinea pig myenteric plexus-longitudinal muscle bioassay. Since opioids inhibit release of acetylcholine, the opiate receptors are shown deployed on terminals of the cholinergic neuron. Additional evidence supports such a presynaptic action in other tissues as well.

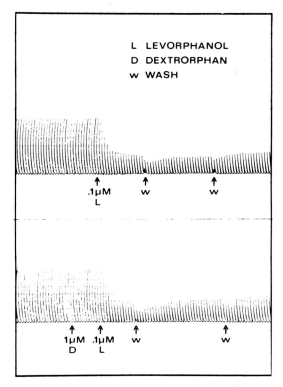

FIG. 6. Activity of levorphanol (L) and its (+) enantiomer dextrorphan (D) in the bioassay. Concentrations indicated are final concentrations in the tissue bath, 100 nM levorphanol, 1,000 nM dextrorphan. Reversal of the levorphanol effect by washing (W) is slow. Electrically stimulated muscle twitches are every 10 sec.

The tetrapeptide Tyr-Gly-Gly-Phe is considered to contain the opioid message because it is the smallest entity with opioid activity. Moreover, it contains the minimum features (phenolic hydroxyl group, basic nitrogen atom and hydrophobic T-shape) that match the characteristic conformation of the opiate alkaloids. It is not potent (concentrations in the micromolar range are required), but one does obtain a typical opioid effect that is blocked by naloxone. The addition of leucine or methionine in the fifth position greatly enhances the potency, but, as in the case of dynorphin, the addition of the next eight residues increases the potency (as compared to Leu-enkephalin) by a factor of 700.

THE ROLE OF DOUBLE BASIC RESIDUES IN PEPTIDES

There is an interesting analogy between dynorphin and an embedded sequence of a 4,900-dalton peptide described by Udenfriend *(this volume)* in which Lys-Arg precedes and follows the opioid structure (Fig. 8). This peptide also possesses two basic residues in positions 6 and 7 (as do dynorphin and α-neo-endorphin). Another peptide, described by this group, has an Arg-Phe assembly in positions 6 and 7.

The basic pair is often a processing signal, a classic example being the conversion of proinsulin to insulin; its presence, however, does not mean that processing to a smaller active unit will necessarily take place. In general, Lys-Arg, Lys-Lys, Arg-Arg, and Arg-Lys cannot be assumed to signal a cleavage. Instances are known in which such pairs signal phosphorylation or specific addition of other functional groups. Moreover, such basic pairs are found in numerous proteins as part of the intrinsic biologically active segment, not intended for processing at all, as for

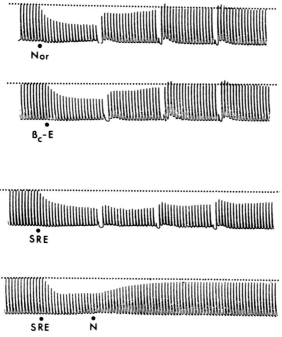

FIG. 7. Typical bioassay records for normorphine (Nor), camel β-endorphin (β$_C$-E), and dynorphin (SRE). "Slow reversing endorphin" was an early name for dynorphin prior to its isolation. Gaps in records indicate replacement of bath fluid by drug-free fluid. Bottom record shows reversal of dynorphin effect by naloxone (N), 200 nM. (From Lowney et al., 1979.)

Embedded sequence in Udenfriend 4.9 kDal peptide:

```
       1         6 7                    18
-KR-   Y G G F M R R V G R P E W W M D Y Q   -KR-
```

Dynorphin:

```
       1         6 7                17
-??-   Y G G F L R R I R P K L K X X X X
```

"Alpha-neo-endorphin":

```
                 6 7 8
??-    Y G G F L R K Y P K
```

Udenfriend heptapeptide:

```
-KR-   Y G G F M R F
```

Beta Endorphin:

```
                 6                17       28       31
-KR-   Y G G F M T S E K S Q T P L V T L .....  K K G E
```

FIG. 8. Various opioid peptides containing paired basic residues. The 18-residue peptide *(top line)* contained within the pro-enkephalin sequence shows considerable homology to dynorphin. As with dynorphin, it is much more potent than its contained enkephalin. Other peptides are shown for comparison. Standard single-letter code is used here; K = lysine, R = arginine.

example in the insulin-like growth factors. It is likely, therefore, that Arg-Arg in dynorphin is not a processing signal, unless, perhaps, it signals the termination of action of this highly potent peptide by cleavage, yielding Leu-enkephalin as a degradation product.

EVIDENCE FOR A SPECIFIC DYNORPHIN RECEPTOR

There are several independent lines of evidence that point to a specific dynorphin receptor. These include (1) structure-activity studies with dynorphin fragments and related peptides in the myenteric plexus assay; (2) measurement of the apparent dissociation constant (K_e) of naloxone as antagonist to the various dynorphin fragments and related peptides; (3) effect of D-Ala^2 substitutions on potencies and naloxone K_e values for the various fragments; (4) selective protection or destruction of the binding affinity of dynorphin; (5) development of selective tolerance in chronically treated mice; (6) selective tolerance to intraventricular application of opioid peptides; and (7) distribution of immunoreactive dynorphin in various tissues. Most of the work to be described was carried out in this laboratory by Charles Chavkin.

Structure Activity Studies with Dynorphin Fragments

Removal of one residue at a time from the carboxy terminal of dynorphin-(1-13) allowed the identification of those residues that are critically important for the potency, using the myenteric plexus preparation as an assay. The free carboxyl group of lysine ^{13} is not required, as the methyl ester is fully active and carboxy terminally extended peptides (like natural dynorphin) retain full potency. There are three critically important basic residues, removal of which causes major loss of potency; these are lysine^{11}, arginine^7, and lysine^{13}. Our present picture of the dynorphin receptor is shown in Fig. 9. The important point is that the tetrapeptide pocket, occupied by Tyr-Gly-Gly-Phe or, presumably, by opiate alkaloids, or even the whole region that interacts with Tyr-Gly-Gly-Phe-Leu, represents only a small part of the extensive recognition site in this opiate receptor. The tyrosine residue at the first position is, as would be expected, absolutely indispensable for activity. In all these fragment studies, to avoid the

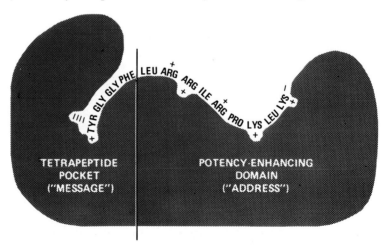

FIG. 9. Hypothetical model of the dynorphin receptor. Tetrapeptide pocket presumably also accommodates opiate alkaloids, interacting through its postulated anionic site and hydrogen bond acceptor. Three additional anionic sites are critically placed to interact with arginine^7, lysine^{11}, and lysine^{13}. (From Chavkin and Goldstein, 1981b.)

confounding effect of malpositioned negatively charged carboxylate as the chain was shortened, each fragment was tested as the methyl ester (Chavkin and Goldstein, 1981b).

Determination of the Apparent Dissociation Constant for Naloxone

In the myenteric plexus bioassay preparation, it is believed that both normorphine and Leu-enkephalin act on the same receptor (μ-receptor), which is very sensitive to naloxone. The reversibility by naloxone of the pharmacological effect of a peptide can be converted to an apparent dissociation constant of naloxone, which is a measure of how much naloxone is required to reverse the effects of the peptide.

For the μ-receptor, the site of interaction of Leu-enkephalin or normorphine in the guinea pig myenteric plexus, the K_e is about 2 nM. In principle, reversal by an antagonist has nothing to do with the potency of the agonist: very potent peptides are used in low concentrations and a certain concentration of naloxone is then needed to reverse their effect. If the peptide is less potent, higher concentrations of the peptide are required to obtain the same biological effect; however, the concentration of naloxone to reverse the effect remains unaltered. Thus, what is measured is the apparent affinity of naloxone itself for the particular receptor site with which the given agonist interacts. If there is a homogeneous population of receptors, one should find by this method of computation the same apparent affinity constant for naloxone no matter what agonist is used. On the contrary, what is actually observed is that a 10 times higher concentration of naloxone is required to reverse the effects of dynorphin-(1-13) than to reverse the effects of either normorphine or Leu-enkephalin. In other words, this indicates the presence of a different receptor for dynorphin in this tissue, one that is less sensitive to naloxone.

Effect of D-Ala^2 Substitutions on Potencies and Naloxone K_e Values

Under the conditions of the guinea pig myenteric plexus assay, Leu-enkephalin with a D-alanine substitution in position 2 is greatly increased in potency. The same substitution in dynorphin causes a very large decrease in potency. This indicates that the tetrapeptide pocket of the dynorphin receptor is different from that of the Leu-enkephalin receptor.

Selective Protection or Destruction of the Receptor Sites

This method is based on selective protection of the receptor against irreversible destruction by the alkylating agent β-chlornaltrexamine (CNA). If this agent is used alone, many receptors are nonselectively destroyed (Fig. 10). If the receptors are protected by concurrent treatment with D-Ala^2-D-Leu^5-enkephalin (DADLE), there is no protective effect of this compound when dynorphin-(1-13) is the test ligand. If, however, the receptor site is protected with dynorphin-(1-13) during the treatment with CNA, there is a significant protection.

Reciprocal experiments with Leu-enkephalin as test ligand show similar results (Fig. 11). Whereas DADLE acts as a protective agent, dynorphin-(1-13) does not.

In summary, one can selectively protect the receptor sites for dynorphin and for Leu-enkephalin, respectively, in this tissue (Chavkin and Goldstein, 1981a).

Development of Selective Tolerance

If mice are treated chronically with DADLE, the IC$_{50}$ for Leu-enkephalin in the vas deferens assay[1] for treated animals is increased to a value of 1,000 nM (the control being 15 nM), i.e.,

---

[1] An assay similar to that previously described for the myenteric plexus. Opioid agonist activity is characterized by the ability of a substance to inhibit electrically induced longitudinal contractions of mouse vas deferens.

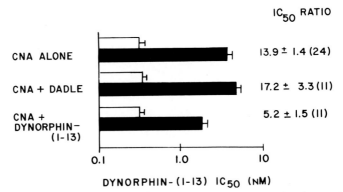

FIG. 10. Selective protection of dynorphin receptors by dynorphin-(1-13). IC_{50} of dynorphin-(1-13) in myenteric plexus-longitudinal muscle bioassay was determined before and after treatment of the muscle strip with CNA, an alkylating agent. *Open bars*, before CNA; *solid bars*, after CNA. **Top line:** CNA alone; **middle line:** DADLE present during CNA treatment; **bottom line:** dynorphin-(1-13) present during CNA treatment. IC_{50} ratio = IC_{50} after CNA divided by IC_{50} before CNA. (From Chavkin and Goldstein, 1981a.)

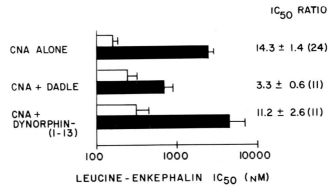

FIG. 11. Selective protection of μ receptors by DADLE. Here Leu-enkephalin was the test ligand. For explanation see Fig. 10. (From Chavkin and Goldstein, 1981a.)

a high degree of tolerance has developed. In the same tissue there is, however, no tolerance at all for dynorphin-(1-13).

Interestingly, in the vas deferens, normorphine and the enkephalins combine with different receptors. Thus, there are at least three different receptors: the δ receptor for Leu- and Met-enkephalin; the μ-receptor, the natural ligand of which is unknown; and the dynorphin receptor, which (according to evidence not presented here) appears to be the receptor heretofore described as κ.

Intraventricular Application of Opioid Peptides

If opioid peptides are applied into the ventricular system of the rat, the animals become cataleptic (Fig. 12). In selective tolerance experiments, Herman and Goldstein (1981) have demonstrated that an animal made tolerant to the cataleptic effect of a typical μ-receptor agonist will not show tolerance to dynorphin-(1-13). Thus, the catalepsy effect is mediated by at least two different receptors, the μ-receptor and the dynorphin receptor.

Distribution of Immunoreactive Dynorphin in Various Tissues

Using a very sensitive and highly specific antiserum (Fig. 13) (Ghazarossian et al., 1980), immunoreactive dynorphin can be detected in the central nervous system (Table 1), in the

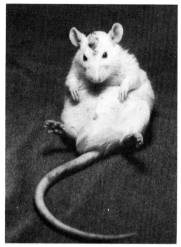

FIG. 12. Catalepsy produced by intracerebroventricular injection of dynorphin-(1-13). **Left:** Typical rigidity allowing rat to maintain an immobile posture stretched between two platforms. **Right:** Typical frozen immobility when placed in a bizarre posture.

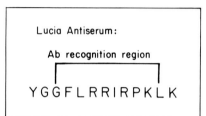

FIG. 13. Specificity of the "Lucia" antiserum. The exact region of the epitope was determined by studies with dynorphin fragments.

TABLE 1. *Distribution of immunoreactive dynorphin in rat brain[a]*

Region	Total immunoreactive dynorphin (pmole)		Density (mg)	
	Mean	SEM	Mean	SEM
Cerebellum	0.36	0.06	1.34	0.23
Medulla-pons	1.88	0.36	8.45	1.62
Striatum	0.40	0.04	4.25	0.40
Midbrain	1.97	0.44	8.45	1.81
Hippocampus	0.93	0.17	5.97	0.81
Cortex C	1.30	0.16	3.08	0.28
Cortex B	1.74	0.23	3.87	0.46
Hypothalamus	0.77	0.15	13.88	2.07
Anterior	0.51	0.12	16.36	4.00
Posterior	0.26	0.10	10.48	2.32
Whole brain	9.36	0.93	4.94	0.54

[a]Data are based on six rats, radioimmunoassays of extracts with "Lucia" antiserum using dynorphin-(1–13) as reference standard. Cortex sections C and B are rostral and caudal, respectively, to optic chiasma.
Data from Goldstein and Ghazarossian, 1980.

peripheral nervous system, and in the gut, as well as in the pituitary gland (pars nervosa) (Goldstein and Ghazarossian, 1980). The specificity of the antiserum is excellent, in that Leu-enkephalin is not recognized even at 10 million times the IC_{50} of dynorphin-(1-13) (Fig. 14). Another opioid peptide that contains the Leu-enkephalin sequence, α-neo-endorphin (Fig. 8), cross-reacts only at about $10^{-4}\%$. This antiserum is also indifferent to tyrosine[1] and lysine[13], and therefore also recognizes both COOH-terminally and NH_2-terminally extended dynorphin peptides. Preliminary gel permeation experiments using this antiserum for detection of dynorphin sequences indicate that rat brain extracts contain peaks of larger molecular weight, which could be precursor peptides ("prodynorphin").

Immunoreactive dynorphin is also present in ganglia of the myenteric plexus where it may have functional importance in the context of intestinal motility.

CONCLUSIONS AND QUESTIONS

With this characterization of a specific dynorphin receptor and the previous differentiation of μ and δ receptor types, the number of clearly defined opiate receptor types comes to three. How are these related to the several endogenous opioid peptides? There are three well-characterized opioid peptides specified by different genes. The enkephalins are contained in multiple copies in a proenkephalin protein (see S. Udenfriend, *this volume*), which does not seem to contain either the β-endorphin or dynorphin sequence. Using pro-enkephalin material kindly furnished by Dr. Udenfriend, we have failed to observe any dynorphin immunoreactivity either in the intact protein or after fragmenting it by means of a glutamyl endopeptidase, which does not affect the immunoreactivity of dynorphin-(1-13). It is not yet clear if this pro-enkephalin contains the α-neo-endorphin sequence. Beta-endorphin, which contains Met-enkephalin at its NH_2-terminus, is specified by the pro-opiomelanocortin gene, the entire sequence of which is

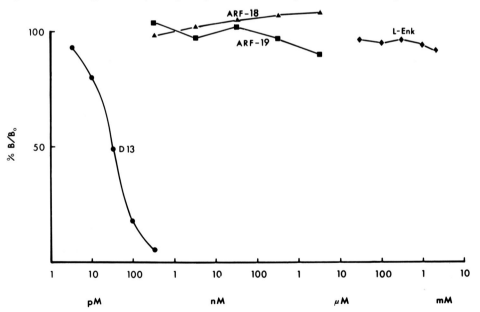

FIG. 14. Radioimmunoassay with "Lucia" antiserum. Displacement of ^{125}I-labeled dynorphin-(1-13) by dynorphin-(1-13) (D13), lack of cross-reactivity of dynorphin-(6-13) (ARF-18), α-N-acetyl-dynorphin-(6-13) (ARF-19), and Leu-enkephalin (L-Enk). Concentrations shown are final concentrations in the radioimmunoassay. (From Ghazarossian et al., 1980.)

known, and which specifies only a single copy of the opioid message. Dynorphin, which contains Leu-enkephalin at its NH_2-terminus, is assumed to be specified by a unique gene; certainly there is evidence of large immunoreactive peptides that could be pro-dynorphins.

Will it turn out that there is a one-to-one relationship between receptor types and the several opioid ligands? Was there first a single opioid message, with complex addresses later added through evolution, to confer selectivity on newly developed receptor types? Could leucine[5] (in Leu-enkephalin) or methionine[5] (in Met-enkephalin) be primitive addresses? Transmitters, acting locally at synapses, may not require much address; they are already at their site of action. However, with the development of neurohumoral agents, which act over a wide distance and for an extended time, the informational specificity has to be contained in the chemical structure. In this sense, the complex addresses in β-endorphin and dynorphin seem quite reasonable.

In some tissues there seems to be a redundancy of opiate receptors. For example, selective tolerance experiments have indicated the presence of at least three such receptors in the mouse vas deferens, where one doubts there are also three different endogenous opioid ligands. It has been suggested that such a redundancy may be a way to provide both a low affinity and a high affinity receptor at the same effector, for a wider range of responsiveness to different ambient ligand concentrations. This could have regulatory significance, related to the phenomenon of receptor desensitization. Occupancy of a high affinity receptor could lead to desensitization, yet the low affinity receptor could still respond to a higher concentration of the same ligand. This would provide a mechanism for raising the threshold of response pari passu with the intensity of stimulation, in a manner analogous to what happens at the gross neural level for many sensory inputs. It would, in essence, be a neurochemical damping mechanism.

Finally, receptor redundancy associated with ligand redundancy (i.e., parallel opioid inputs) could provide nature with a mechanism for circumventing tolerance (desensitization). Tonic inhibitory regulation (as of the release of luteinizing hormone-releasing hormone) could be maintained by an oscillatory transfer of control between two opioid receptor types activated by two different ligands. Could physiological pulsatile hormone release be due to tolerance to one ligand, followed then by activation of the alternate receptor by the alternate ligand?

ACKNOWLEDGMENTS

The work described here was supported by DA-1199 and DA-7063 from the National Institute on Drug Abuse, and by gifts from generous private donors.

REFERENCES

Chavkin, C., and Goldstein, A. (1981a): Demonstration of a specific dynorphin receptor in guinea pig ileum myenteric plexus. *Nature*, 291:591–593.
Chavkin, C., and Goldstein, A. (1981b): Specific receptor for the opioid peptide dynorphin: Structure-activity relationships. *Proc. Natl. Acad. Sci. USA*, 78:6543–6547.
Cox, B. M., Gentleman, S., Su, T. P., and Goldstein, A. (1976): Further characterization of morphine-like peptides (endorphins) from pituitary. *Brain Res.*, 115:285–296.
Cox, B. M., Opheim, K., Teschemacher, H., and Goldstein, A. (1975): A peptide-like substance from pituitary that acts like morphine. 2. Purification and properties. *Life Sci.*, 16:1777–1782.
Ghazarossian, V. E., Chavkin, C., and Goldstein, A. (1980): A specific radioimmunoassay for the novel peptide opioid dynorphin. *Life Sci.*, 26:75–86.
Goldstein, A., and Ghazarossian, V. E. (1980): Immunoreactive dynorphin in pituitary and brain. *Proc. Natl. Acad. Sci. USA*, 77:6207–6210.
Goldstein, A., Lowney, L. I., and Pal, B. K. (1971): Stereospecific and non-specific interactions of the morphine congener levorphanol in subcellular fractions of mouse brain. *Proc. Natl. Acad. Sci.*, 68:1742–1747.
Goldstein, A., Tachibana, S., Lowney, L. I., Hunkapiller, M., and Hood, L. (1979): Dynorphin-(1-13): An extraordinarily potent opioid peptide. *Proc. Natl. Acad. Sci. USA*, 76:6666–6670.
Herman, B., and Goldstein, A. (1981): Dynorphin-(1-13) catalepsy is not mediated by a μ receptor in rat brain. *Soc. Neurosci. Abstr.*, 7:130.

Lowney, L. I., Gentleman, S. B., and Goldstein, A. (1979): A pituitary endorphin with novel properties. *Life Sci.*, 24:2377–2384.
Schwyzer, R., Karlaganis, G., and Lang, U. (1980): Hormone-receptor interactions: A study of the molecular mechanism of receptor stimulation in isolated fat cells by the partial agonist corticotropin-(5-24)-icosapeptide. In: *Frontiers of Bioorganic Chemistry and Molecular Biology*, edited by S. N. Ananchenko, pp. 277–283. Pergamon Press, Oxford and New York.
Teschemacher, H., Opheim, K. E., Cox, B. M., and Goldstein, A. (1975): A peptide-like substance from pituitary that acts like morphine. 1. Isolation. *Life Sci.*, 16:1771–1776.

Section IX

Steroid Hormones and Gene Expression

The fact that neuroactive compounds may have a short-term (hours to days) and/or long-term (days to months) influence over gene expression is of importance to neuroscience. This is fundamentally different from the rapid (milliseconds to seconds) changes that characterize studies of excitability, and it suggests possible mechanisms for modifying behavior and other aspects of "higher" functions of the brain. Alteration in gene expression may affect metabolic processes within neuronal target cells, including synthesis and degradation of their own neurotransmitters, neurohormones, and/or neuromodulators. The reactivity of target neurons to other inputs, whether synaptic, endocrine, or paracrine, may be modified through regulation of a second messenger system or by altering the availability or affinity of receptors for these neuroactive agents.

Neuroactive substances combine with receptors that are either plasmalemmal or cytosolic. The ligand–receptor complex may then be translocated to the nucleus where it may affect gene expression. Alternatively, ligands may interact with their chromosome-bound nuclear receptors.[1] Ligands might also influence gene expression indirectly through second messengers, such as cyclic nucleotides or calcium (see A. M. Martonosi, *this volume*).

Steroids apparently modulate gene expression through an activated cytosolic receptor and therefore provide a model from which some general principles may be deduced regarding the regulation of gene function by neuroactive substances. Various molecular, cellular, and behavioral aspects of this model system are described in this section.

B. S. McEwen presents biochemical evidence localizing neural sites involved in steroid-related behavior. In addition, it seems many estrogen- and progesterone-induced behaviors are dependent on protein synthesis and that some estrogen effects are likely to be related to the induction of synthesis of receptors for progesterone. Preliminary observations indicate that estrogen also alters the level of receptors for specific neurotransmitters on particular neuronal types. This has wide-ranging implications for concepts of the central role of receptors, not only in classic neuronal response to neuroactive agents, but also in the plasticity of that response, and for receptor function in behavior.

The mechanism of action of insect steroid hormones has been examined in impressive detail, including visualization and mapping of the binding of the steroid–receptor complex to specific chromosomal sites. As described by P. Cherbas and colleagues, such binding leads to increased synthesis of specific mRNAs that may encode more than one polypeptide. This represents a possible mechanism for coordinate control of cellular response to the steroid.

As with other neuroactive substances, the effect of a given steroid is dependent on the properties and localization of the receptor with which it interacts. This is complicated by

---

[1]Oppenheimer, J. H. (1979): Thyroid hormone action at the cellular level. *Science*, 203:971–979.

metabolic conversion of one active compound to another. T. O. Fox and associates discuss the role of androgens and androgen metabolites, and their respective receptors in the sexual differentiation of the brain. In the course of their discussion, models for the possible mechanism of action of the steroid-receptor complex, both in the cytoplasm and in the nucleus, are presented.

E. B. Thompson and J. S. Strobl have found that although the presence of an appropriate receptor is a necessary link between hormones and their cellular effects, other factors are also involved in the regulation of gene expression. They discuss efforts designed to elucidate these factors.

Molecular Genetic Neuroscience, edited by
F. O. Schmitt, S. J. Bird, and F. E. Bloom.
Raven Press, New York © 1982.

Steroid Hormone Action in the Brain: Cellular and Behavioral Effects

Bruce S. McEwen

ABSTRACT

Steroid hormone action on brain offers unique opportunities for localizing neural sites of environmentally controlled neural plasticity. By localizing the neuroanatomical sites of hormone action through brain implants and uptake of [^{3}H] steroids and by elucidating the temporal characteristics of hormone effects, it has been possible to formulate some idea of how hormones can regulate the occurrence of species-typical behavior associated with reproduction and stress. Localization of hormone action in brain is also leading to further studies of cellular and molecular events that are the consequences of hormone action. Among such events are changes in gene expression that give rise to altered levels of gene products associated with neural activity and neurotransmission.

The study of hormonal influences on behavior has long been an area of fruitful investigation for animal behaviorists and psychologists. It is now becoming a fertile area for neurobiological investigations of the cellular mechanisms underlying this aspect of neural plasticity. What makes this topic relevant to a volume on molecular genetics and neuroscience is that the regulation of gene expression in brain cells is a major means by which steroid hormones bring about changes in behaviors such as courtship and mating, and trigger neuroendocrine events such as ovulation. Implicit in this statement is the important notion that regulation of the type exerted by steroid hormones on the brain is not on a time scale of milliseconds or seconds, but rather on one of minutes, hours, or days. Two types of steroid actions are recognized: organizational effects, such as those leading to brain sexual differentiation, which are permanent and which occur during a "critical period" of early development; and activational effects, which are reversible effects of gonadal steroids that occur repeatedly during adult life. There is considerable evidence that both kinds of steroid actions are mediated by intracellular receptors of the type discussed by T. Fox and coworkers *(this volume)*. Steroid action on the brain can be discussed in cellular and, in some cases, synaptic terms, and thus one can address the question: What are the changes in neuronal function and connectivity that are required for a hormone to organize or activate a behavior?

One of the keys to answering this question is the localization of hormone action. The studies in this and other laboratories that identified, characterized, and mapped steroid hormone receptor systems in the brains of various vertebrate species provide the starting point for such investigations (for a review, see McEwen et al., 1979). Further progress was made by locally applying tiny amounts of hormone within hormone-sensitive brain regions to localize sites of greatest behavioral sensitivity. In addition, efforts were made to localize hormone-induced changes in neuronal properties by microsampling of the brain or by quantitative histological procedures.

These approaches will be illustrated by using the example of estradiol's and progesterone's activation of feminine sexual behavior in ovariectomized rats. These hormones appear to activate feminine sexual behavior by a mechanism involving stimulation of protein synthesis within target neurons of the ventromedial hypothalamus. The protein products are presumably responsible for changing synaptic functions.

ORGANIZATIONAL EFFECTS

Gonadal steroids play a central role in sexual maturation and sexual differentiation of the brain. During the 1970s, there was considerable progress in defining the hormonal factors and cellular processes involved in these events, as well as in describing the ontogeny of gonadal steroid receptors and other aspects of the steroid-response systems in the brain (for reviews, see McEwen, 1978a, 1980).

Brain sexual differentiation among mammals and birds is determined by gonadal steroids during late fetal and early postnatal life. The presence or absence of testosterone during this critical period specifies whether the animal will display masculine or feminine behavioral and neuroendocrine characteristics when sexually mature (Gorski, 1979; Goy and McEwen, 1980). It seems, however, that testosterone metabolites produced in the brain, rather than testosterone itself, bring about brain sexual differentiation. The hypothalamus of neonatal aminals contains enzymes for producing the two major metabolites of testosterone, estradiol (E_2) and 5-dihydrotestosterone (DHT) (McEwen, 1978a,b). The actions of E_2 and DHT are presumably mediated by cellular cytoplasmic estrogen and androgen receptors, which, in the rat, appear at the beginning of the critical period in brain regions, including the hypothalamus. Gonadal steroid receptors are detectable in the rat brain during the last 5 days of fetal life and increase markedly within the first 2 postnatal weeks (MacLusky et al., 1979; Vito et al., 1979; Lieberburg ct al., 1980; MacLusky and McEwen, 1980).

As shown schematically in Fig. 1A, in neonatal males or testosterone-treated neonatal females, the interaction of E_2 produced from testosterone with estrogen receptors is the primary event responsible for defeminization, i.e., the suppression of the development of feminine characteristics such as lordosis behavior and endocrine control of ovulation (McEwen et al., 1977). Determination of masculine characteristics (masculinization) appears to be more complicated, involving the estrogen pathway, the conversion of testosterone to DHT, and its subsequent interaction with androgen receptors (McEwen, 1982). Progestin receptors become detectable in the rat brain shortly after birth, increasing rapidly during the first 10 days of life. By the end of this period, progestin receptors become inducible in response to estrogens (MacLusky and McEwen, 1980). The function of these receptors in brain at this developmental stage is not presently known.

One of the major advances in the area of brain sexual differentiation in the last 10 years has been the discovery of sexual dimorphism in the structure of particular brain regions, which arises during the critical period (reviewed by Gorski, 1979; Goy and McEwen, 1980; Nottebohm, 1980). Sexual dimorphism in the preoptic area and hypothalamus of rodents and birds consists of localized differences in synapse type and distribution (Raisman and Field, 1973), in the shape of the dendritic tree (Greenough et al., 1977), or in the size of cell groupings (Nottebohm and Arnold, 1976; Gorski et al., 1978). In rodents, the functions of sexual dimorphic brain structures are not yet known. In songbirds, sexually dimorphic nuclei are part of the circuitry for singing, which is an androgen-dependent behavior (Nottebohm, 1980).

One of the key events in producing morphological changes may be the hormonal stimulation of neurite outgrowth. Toran-Allerand (1976) has shown that the addition of testosterone or E_2

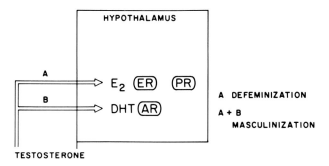

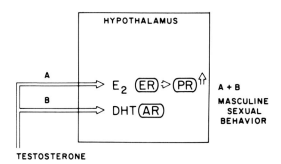

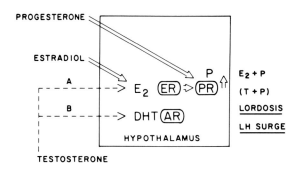

FIG. 1. Overview of gonadal steroid interactions with the hypothalamic region of the rat brain **(A)** in neonatal male or testosterone-treated female rats; **(B)** in adult male or perinatally testosterone-treated female rats; and **(C)** in female or neonatally castrated male rats. E_2, estradiol; DHT, 5α dihydrotestosterone; ER, estrogen receptor; AR, androgen receptor; PR, progestin receptor.

to organ-cultured newborn mouse hypothalamus stimulates a remarkable outgrowth of neurites from distinct regions of the hypothalamus. Outgrowth is blocked by the estrogen antagonist CI-628, and DHT (which cannot be converted to E_2) is not able to induce the response (C. Toran-Allerand, *unpublished data*). These results strongly suggest that the effect is mediated by estrogen receptors and are consistent with the idea that the neurites emanate from cells that contain estrogen receptors. Evidence supporting this suggestion recently has been obtained; groups of cells corresponding to the bed nucleus of the stria terminalis, an area from which the hormone-induced outgrowth appears to emanate, were found to concentrate labeled E_2

(Toran-Allerand et al., 1980). Thus, it seems that this effect is at least in part a direct result of the interaction of estrogen with intracellular estrogen receptors. Whether other factors in the culture medium, such as growth factors, are also required for this effect is not clear.

ACTIVATIONAL EFFECTS

The behavioral and neuroendocrine effects of endogenous or exogenous gonadal steroids in adult animals are reversible and depend on the perinatal hormonal history of the animal. Some of these activational effects are summarized schematically in Fig. 1B and C. In an adult male or in a neonatally androgen-treated female rat, testosterone operates both through aromatization to E_2 and through reduction to DHT to stimulate masculine sexual behavior (Fig. 1B). Progesterone alone or in combination with E_2 will not activate feminine sexual behavior, although progestin receptors are present in the brain of these animals and are inducible by E_2, as they are in normal females (Krey et al., 1981). In a female or in a neonatally castrated male rat, E_2 and progesterone activate feminine sexual behavior and the luteinizing hormone surge, which in a normal female triggers ovulation. In addition, it is possible to activate feminine sexual behavior in these animals with testosterone, presumably through aromatization and interaction of the locally produced E_2 with the estrogen receptor system.

Activation of lordosis behavior in ovariectomized female rats by estrogens and progestins is a sensitive assay for the behavioral effects of these hormones. Figure 2 illustrates a central characteristic of the lordosis response: when a single pulse of tritium-labeled E_2 is administered to an ovariectomized female, the levels of nuclear steroid-occupied receptor in the hypothalamus rise and then fall rapidly, so that after 12 hr very little of the injected hormone remains in the cell nuclei. Lordosis behavior is not elicited unless progesterone is given after E_2. If progesterone is injected at various times after E_2 treatment and lordosis behavior is measured, the behavior does not appear until the nuclear estrogen receptor levels are very low, and it shows a gradual activation so that peak response occurs 24 hr after E_2 treatment (Green et al., 1970; Parsons et al., 1980). Lordosis behavior is measured as the lordosis quotient (LQ), which is calculated as number of lordoses per number of mounts attempted by the male × 100. This time lag between maximal receptor occupancy and maximal response probably reflects various biosynthetic events that occur within the target brain cells.

The temporal requirements of E_2 and the possible involvement of protein synthesis in the activation of lordosis were examined by Rainbow and coworkers (1980a) and Parsons and coworkers (1981). Continuous 24-hr exposure to E_2 induced an LQ of about 60%; the same

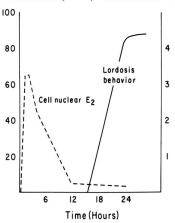

FIG. 2. Time course of [³H]estradiol retention by hypothalamic cell nuclei (---) compared with time course of activation of feminine sexual behavior (———) caused by the [³H]estradiol injection. (Adapted from Green et al., 1970 and McEwen et al., 1975.)

effect can be obtained by a 6-hr, but not by a 3-hr, exposure to the hormone (Fig. 3). However, two 1-hr exposures were able to activate lordosis behavior to a comparable degree with that induced by one 24-hr exposure, provided that they were separated from each other by more than 4 hr and less than 14 hr. Two 30-min exposures could not activate the behavior (Fig. 3). These results indicate that a constant level of estrogens is not necessary to produce lordosis behavior and that more than one phase of cellular events—such as changes in macromolecular synthesis—may be stimulated by estrogens.

Injection of the protein synthesis inhibitor anisomycin shortly before the first or the second 1-hr exposure to E_2, or between the two exposures, blocks lordosis activation (Fig. 4); the drug has no effect if given 2 hr after the second exposure. This supports the idea that, as with other steroid hormones, E_2 action in this system involves new protein synthesis. Earlier studies have shown that actinomycin D, an RNA synthesis inhibitor, can also reversibly block estrogen effects on sexual behavior (Quadagno et al., 1971; Terkel et al., 1973). However, protein synthesis inhibitors were used rather than RNA synthesis inhibitors in in vivo experiments to avoid the complication of nonspecific effects. Anisomycin is not very toxic to rats, nor does it have any effect on sexual behavior.

There is a direct correlation between activation of lordosis behavior and induction of progestin receptors in rat hypothalamus by E_2 (Parsons et al., 1980, 1981). Continuous exposure of an ovariectomized rat to E_2, which will activate sexual behavior, also causes a parallel induction of progestin receptors in the hypothalamus, as shown in Fig. 4. When the source of E_2 is removed, E_2 blood levels decrease rapidly to zero, and yet the behavior can be elicited for more than 24 hr; the levels of hypothalamic progestin receptors decrease with a half-life of about 24 hr (Fig. 4). A similar correlation exists when E_2 is applied discontinuously (Parsons et al., 1981). When animals are exposed to E_2 for 6 hr or for two 1-hr periods separated by at least 4 to 5 hr, cytosolic levels of progestin receptors in the hypothalamus are elevated to

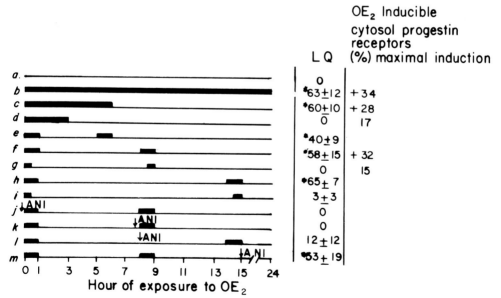

FIG. 3. Effect of estradiol (OE_2) and anisomycin (ANI) on the lordosis quotient (LQ) and on the induction of cytosol progestin receptor in the hypothalamus and preoptic area of ovariectomized female rats. *Solid bars* indicate the duration of an implant of Silastic capsules containing estradiol under the skin. *Significantly different from control group using a Newman-Kvels test after a one-way analysis of variance. (From Parsons et al., 1980, with permission.)

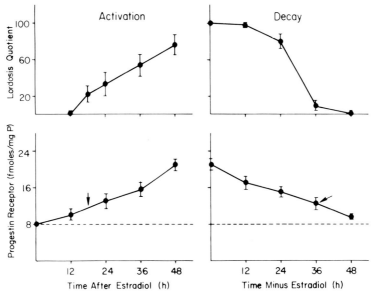

FIG. 4. The appearance and disappearance of feminine reproduction behavior and cytosol progestin receptors in the hypothalamus and preoptic area after estradiol treatment (activation) and removal of estradiol (decay). *Arrow* indicates level of cytosol progestin receptors, which correlate with ability of rat to display lordosis. (From Parsons et al., 1980, with permission.)

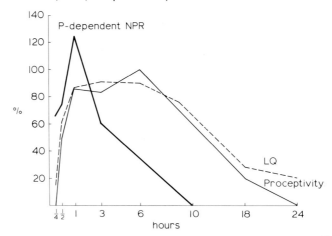

FIG. 5. Time course of nuclear progestin receptor translocation (P-dependent NPR) and feminine sexual behavior (LQ) and hop-darting or ear-wiggling (proceptivity) in estrogen-primed ovariectomized rats after intravenous progesterone administration. (From McGinnis et al., 1981, with permission.)

approximately 30% of maximal progestin receptor induction. This seems to be a threshold level that is necessary for the sexual behavior to be expressed. A 3-hr exposure, a single 1-hr exposure, or two 30-min to 1-hr exposures to E_2 were insufficient to produce the behavior and also failed to raise cytosolic progestin receptor levels to the threshold. It seems, therefore, that progestin receptor induction may be an important part of estrogen action in activating sexual behavior.

Lordosis behavior is fully induced within approximately 1 hr after intravenous injection of progesterone in an estrogen-primed animal, and begins to decline slowly after a few hours (Fig. 5). As with E_2, nuclear translocation of the progesterone-receptor complexes occurs prior to the onset of the behavioral effects, and the behavior outlasts the disappearance of hormone from the cell nucleus (McGinnis et al., 1981). The effects of progesterone can also be blocked by the protein synthesis inhibitor anisomysin (Rainbow et al., 1980a). In nonovariectomized animals, sexual receptivity declines in parallel with progesterone plasma levels without any apparent lag (Fig. 6). In addition to inducing sexual receptivity, progesterone may initiate a more delayed process, also dependent on protein synthesis, which inhibits sexual receptivity (Parsons and McEwen, 1981). Thus, it appears that progesterone action is biphasic.

Localization of hormonal action is another important parameter. It has been shown by electrical stimulation and lesion studies that lordosis behavior in the rat is controlled by the ventral region of the ventromedial nucleus of the hypothalamus (Pfaff and Sakuma, 1979a,b). Other studies have shown that this nucleus is among the hypothalamic structures that contain estrogen receptors and estrogen-inducible progestin receptors (Pfaff and Keiner, 1973; Warembourg, 1978). Implantation experiments in which tiny amounts of steroid are applied locally indicate that gonadal steroids act directly on the ventromedial nucleus to activate lordosis behavior. Implantation of estrogen in this region activates lordosis behavior in ovariectomized rats after progesterone treatment (Davis et al., 1979). Implantation of progesterone into the same area of ovariectomized, estrogen-primed animals also induces sexual behavior (B. Rubin and R. Barfield, *personal communication*), and implantation of anisomycin can block the effects of both E_2 and progesterone (Rainbow et al., 1980a). Identification of the anatomical site at which steroids act to enhance sexual behavior will now make possible an analysis of the cellular and molecular changes caused by steroids in this particular system.

CELLULAR EFFECTS

Steroid hormones have been shown to influence several aspects of neuronal function related to synaptic transmission. Steroid effects on neural tissues are reviewed in McEwen (1981a).

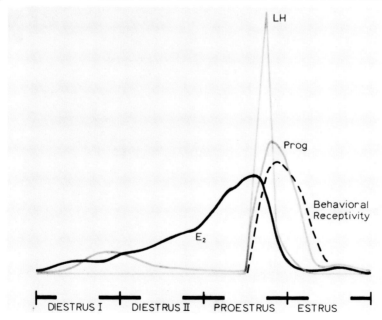

FIG. 6. Estradiol (E_2), progesterone (Prog), and luteinizing hormone (LH) levels and behavioral receptivity in 4-day cycling female rats. (Courtesy of Dr. Marilyn McGinnis.)

It is generally recognized that steroid hormones act on target tissues at the level of the genome via intracellular receptors, but it is also evident that certain steroid effects that occur very rapidly are not mediated by changes in gene expression. This is summarized diagrammatically in Fig. 7. Genomic action of the steroid can lead to changes in the synthesis of specific proteins, which after axonal or dendritic transport, may influence pre- or postsynaptic events. Nongenomic effects may involve direct interaction of the steroid with the pre- or postsynaptic membranes or neurotransmitter receptors.

One of the first and most impressive examples of steroid action on a specific synaptic system is that of the song-controlling circuit of songbirds. It has been shown that a number of structures that are involved in song production concentrate labeled androgens. Among them is a region of the hypoglossal nucleus that innervates the syrinx, the song organ, and the syrinx muscle itself (Nottebohm, 1980). The hypoglossal nucleus and the syrinx muscle contain androgen receptors, and castration results in a decrease in syringeal mass (Luine et al., 1980), in acetylcholine receptor binding (W. Bleisch, *unpublished data*), and in acetylcholinesterase activity (Luine et al., 1980). In addition, castration causes a decrease in choline acetyltransferase and acetylcholinesterase activities of the tracheosyringealis nerve. These effects are reversed by administration of androgens (Luine et al., 1980). It appears, therefore, that major components of the cholinergic transmission pathway of this nerve-muscle system in the songbird are regulated directly by androgens. Protein I, a neuron-specific phosphoprotein associated with neurotransmitter vesicles (Dolphin et al., 1980), is another example of a synaptic protein whose steroid hormone regulation is correlated with the presence of specific receptors: glucocorticoids increase the level of protein I in rat hippocampus, a region rich in glucocorticoid receptors. In brain regions that contain low levels of glucocorticoid receptors, protein I levels do not increase in response to glucocorticoids (Nestler et al., 1981).

However, as in the case of behavioral effects, the presence of steroid receptors may be necessary but not sufficient for some cellular effects of steroids. For instance, it has been shown that estrogen treatment results in induction of choline acetyltransferase in the preoptic area and in the amygdala of the female but not of the male rat, although estrogen receptors exist in these areas in both sexes (Luine et al., 1975). This interesting sex difference is under investigation again to determine whether the lack of induction in the male is due to sexually dimorphic "programming" or to the absence of cholinergic cells in these areas of the male brain.

A sensitive point of control of neuroendocrine functions and behavior is that of neurotransmitter receptors. Certain cholinergic systems of the rat brain appear to be influenced by steroids

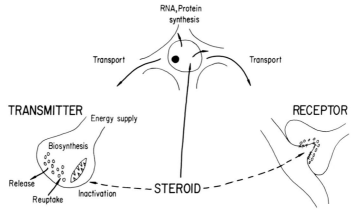

FIG. 7. Direct (---) and indirect (———) (genomic) action of steroids on nerve cells. (From McEwen et al., 1978, with permission.)

at that level, as shown by Rainbow and co-workers (1980b). Estrogen treatment of ovariectomized rats increases the levels of muscarinic receptors in a number of the estrogen-sensitive cell groups of the preoptic area and hypothalamus, but has no effect on muscarinic receptors of brain regions that do not contain estrogen receptors, such as the cerebral cortex and the corpus striatum. Among the hypothalamic structures exhibiting this effect is the lateral region of the ventromedial nucleus, which is implicated in the lordosis response. This result takes on additional significance from the observation that local application of a cholinergic agonist in the hypothalamus potentiates feminine sexual behavior in overiectomized rat primed with low doses of hormones, and that hypothalamic application of a muscarinic antagonist inhibits hormonally activated sexual behavior (Clemens and Dohanich, 1980; Clemens et al., 1980).

Another example of a steroid hormone acting on neurotransmitter receptors is the estrogenic regulation of serotonin binding to putative membrane receptors of the rat brain. Within 48 hr after the initiation of estrogen treatment, an increase in the total number of membrane-associated serotonin-binding sites has been observed in three estrogen-concentrating regions of the rat brain: hypothalamus, amygdala, and preoptic area (Biegon et al., 1981; Fischette et al., 1981; McEwen and Biegon, 1981). A finer anatomical analysis of this effect is currently under way.

Indirect genomic actions of steroids are not the only ones to be considered in the brain. Brief incubation of rat cerebral cortex membranes with an estrogen results in a reduction of the number of available serotonin-binding sites (McEwen and Biegon, 1981). This effect seems too rapid to be mediated by genomic events and is presumably due to some direct action of the steroid. The effectiveness of various estrogens in eliciting this response correlates with their known relative activities; corticosterone, progesterone, and testosterone are completely inactive. In addition, 1 hr after injection of physiological doses of E_2 to ovariectomized females, a brain-wide reduction of serotonin-binding sites, similar to that produced in vitro, is observed (McEwen and Biegon, 1981). It appears, therefore, that estrogens may exert a biphasic effect on brain serotonin receptors: an acute, brain-wide reduction in the number of available receptor sites, followed by a slower (48 to 72 hr later) increase in receptor sites localized in some of the estrogen receptor containing neurons.

By localizing the neuroanatomical sites of hormone action, by specifying the temporal characteristics of the hormone effects, and, more recently, by beginning to unveil some of the cellular events associated with hormone action, it has been possible to formulate a description of how steroid hormones can influence behavior. The techniques described in this volume will help to elucidate some of the gene products under hormonal control in the brain.

ACKNOWLEDGMENTS

Supported by USPHS grant NS 07080 and by institutional grant RF 70095 from Rockefeller Foundation for Research in Reproductive Biology.

REFERENCES

Biegon, A., Rainbow, T. C., and McEwen, B. S. (1981): Serotonin receptor localization by autoradiography: Quantitative analysis and the effect of steroid hormones. *Soc. Neurosci. Abstr.*, 7:9.
Clemens, L. G., and Dohanich, G. P. (1980): Inhibition of lordotic behavior in female rats following intracerebral infusion of anticholinergic agents. *Pharmacol. Biochem. Behav.*, 13:89–95.
Clemens, L. G., Humphrys, R. R., and Dohanich, G. P. (1980): Cholinergic brain mechanisms and the hormonal regulation of female sexual behavior in the rat. *Pharmacol. Biochem. Behav.*, 13:81–88.
Davis, P. G., McEwen, B. S., and Pfaff, D. W. (1979): Localized behavioral effects of tritiated estradiol implants in the ventromedial hypothalamus of female rats. *Endocrinology*, 104:898–903.
Dolphin, A. C., Goelz, S. E., and Greengard, P. (1980): Neuronal protein phosphorylation: Recent studies concerning Protein I, a synapse-specific phosphoprotein. *Pharmacol. Biochem. Behav.*, 13:169–174.

Fischette, C. T., Biegon, A., and McEwen, B. S. (1981): Serotonin receptors in estrogen concentrating brain nuclei and their modulation by estradiol. *Soc. Neurosci. Abstr.*, 7:9.

Gorski, R. A. (1979): Long-term modulation of neuronal structure and function. In: *The Neurosciences: Fourth Study Program*, edited by F. O. Schmitt and F. G. Worden, pp. 969–982. MIT Press, Cambridge.

Gorski, R. A., Gordon, G. H., Shryne, J. E., and Southam, A. M. (1978): Evidence for a morphological sex difference within the medial preoptic area of the rat brain. *Brain Res.*, 148:333–346.

Goy, R. W., and McEwen, B. S., eds. (1980): *Sexual Differentiation of the Brain*. MIT Press, Cambridge.

Green, R., Luttge, W. G., and Whalen, R. E. (1970): Induction of receptivity in ovariectomized female rats by a single intravenous injection of estradiol 17β. *Physiol. Behav.*, 5:137.

Greenough, W. T., Carter, C. S., Steerman, C., and DeVoogd, T. (1977): Sex differences in dendritic patterns in hamster preoptic area. *Brain Res.*, 126:63–72.

Krey, L. C., Lieberburg, I., MacLusky, N. J., and Davis, P. G. (1981): Aromatization in the brain of Stanley-Gumbreck testicular feminized male rats: Implications for testosterone modulation of neuroendocrine activity. *Endocrine Society 63rd Annual Meeting Abstr.* 607:234. Cincinnati.

Lieberburg, I., MacLusky, N. J., and McEwen, B. S. (1980): Androgen receptors in the perinatal rat brain. *Brain Res.*, 196:125–138.

Luine, V. N., Khylchevskaya, R. I., and McEwen, B. S. (1975): Effect of gonadal steroids on activities of monoamine oxidase and choline acetylase in rat brain. *Brain Res.*, 86:293–306.

Luine, V. N., Nottebohm, F., Harding, C., and McEwen, B. S. (1980): Androgen affects cholinergic enzymes in syringeal motor neurons and muscle. *Brain Res.*, 192:89–107.

MacLusky, N. J., Lieberburg, I., and McEwen, B. S. (1979): The development of estrogen receptor systems in the rat brain: Perinatal development. *Brain Res.*, 178:129–142.

MacLusky, N. J., and McEwen, B. S. (1980): Progestin receptors in the developing rat brain and pituitary. *Brain Res.*, 189:262–268.

McEwen, B. S. (1978a): Sexual maturation and differentiation: The role of gonadal steroids. *Prog. Brain Res.*, 48:281–307.

McEwen, B. S. (1978b): Gonadal steroid receptors in neuroendocrine tissues. In: *Hormone Receptors, Vol. 1: Steroid Hormones*, edited by B. O'Maley and L. Birnbaumer, pp. 353–400. Academic Press, New York.

McEwen, B. S. (1980): Gonadal steroids and brain development. *Biol. Reprod.*, 22:43–48.

McEwen, B. S. (1981): Neural gonadal steroid action. *Science*, 211:1301–1311.

McEwen, B. S. (1982): Sexual differentiation of the brain: Gonadal hormone action and current concepts of neuronal differentiation. In: *Molecular Approaches to Neurobiology*, edited by I. Brown. pp. 195–213. Academic Press, New York.

McEwen, B. S., and Biegon, A. (1981): Estrogen modulation of serotonin receptors in the rat brain. *Soc. Neurosci. Abstr.*, 7:818.

McEwen, B. S., Davis, P. G., Parsons, B., and Pfaff, D. W. (1979): The brain as a target for steroid hormone action. In: *Annual Review of Neuroscience*, edited by M. Cowan, pp. 65–112. Annual Reviews, Palo Alto.

McEwen, B. S., Krey, L. C., and Luine, V. N. (1978): Steroid hormone action in the neuroendocrine system: When is the genome involved? In: *The Hypothalamus*, edited by S. Reichlin, R. J. Baldessarini, and J. B. Martin, pp. 255–268. Raven Press, New York.

McEwen, B. S., Lieberburg, I., Chaptal, C., and Krey, L. C. (1977): Aromatization: Important for sexual differentiation of the neonatal rat brain. *Horm. Behav.*, 9:249–263.

McEwen, B. S., Pfaff, D. W., Chaptal, C., and Luine, V. (1975): Brain cell nuclear retention of ^{3}H estradiol doses able to promote lordosis: Temporal and regional aspects. *Brain Res.*, 86:155–161.

McGinnis, M. Y., Parsons, B., Rainbow, T. C., Krey, L. C., and McEwen, B. S. (1981): Temporal relationship between cell nuclear progestin receptor levels and sexual receptivity following intravenous progesterone administration. *Brain Res.*, 218:365–371.

Nestler, E. J., Rainbow, T. C., McEwen, B. S., and Greengard, P. (1981): Corticosterone increases the level of protein I, a neuron-specific protein, in rat hippocampus. *Science*, 212:1162–1164.

Nottebohm, F. (1980): Brain pathways for vocal learning in birds: A review of the first 10 years. In: *Progress in Psychobiology and Physiological Psychology, Vol. 9*, edited by J. M. Sprague and A. N. Epstein, pp. 85–124. Academic Press, New York.

Nottebohm, E., and Arnold, A. P. (1976): Sexual dimorphism in vocal control areas of the songbird brain. *Science*, 194:211–213.

Parsons, B., MacLusky, N., Krey, L. C., Pfaff, D., and McEwen, B. S. (1980): The temporal relationship between estrogen-inducible progestin receptors in the female rat brain and the time course of estrogen activation of mating behavior. *Endocrinology*, 107:774–779.

Parsons, B., and McEwen, B. S. (1981): Sequential inhibition of sexual receptivity by progesterone is prevented by a protein synthesis inhibitor and is not causally related to decreased levels of hypothalamic progestin receptors in the female rat. *J. Neurosci.*, 1:527–531.

Parsons, B., Rainbow, T. C., Pfaff, D. W., and McEwen, B. S. (1981): Oestradiol, sexual receptivity and cytosol progestin receptors in rat hypothalamus. *Nature*, 292:58–59.

Pfaff, D. W., and Keiner, M. (1973): Atlas of estradiol-concentrating cells in the central nervous system of the female rat. *J. Comp. Neurol.*, 151:121–158.

Pfaff, D. W., and Sakuma, Y. (1979a): Facilitation of the lordosis reflex of female rats from the neuromedial nucleus of the hypothalamus. *J. Physiol.*, 288:189–202.
Pfaff, D. W., and Sakuma, Y. (1979b): Deficit in the lordosis reflex of female rats caused by lesions in the ventromedial nucleus of the hypothalamus. *J. Physiol.*, 288:203–210.
Quadagno, D. M., Shryne, J., and Gorski, R. A. (1971): The inhibition of steroid-induced sexual behavior by intrahypothalamic actinomycin D. *Horm. Behav.*, 2:1–10.
Rainbow, T. C., Davis, P. G., and McEwen, B. S. (1980a): Anisomycin inhibits the activation of sexual behavior by estradiol and progesterone. *Brain Res.*, 194:548–555.
Rainbow, T. C., DeGroff, V., Luine, V. N., and McEwen, B. S. (1980b): Estradiol 17β increases the number of muscarinic receptors in hypothalamic nuclei. *Brain Res.*, 198:239–243.
Raisman, G., and Field, P. M. (1973): Sexual dimorphism in the neuropil of the preoptic area of the rat and its dependence on neonatal androgen. *Brain Res.*, 54:1–29.
Terkel, A. S., Shryne, J., and Gorski, R. A. (1973): Inhibition of estrogen facilitation of sexual behavior by the intracerebral infusion of actinomycin-D. *Horm. Behav.*, 4:377–386.
Toran-Allerand, C. (1976): Sex steroids and the development of the newborn mouse hypothalamus and preoptic area *in vitro:* Implications for sexual differentiation. *Brain Res.*, 106:407–412.
Toran-Allerand, C., Gerlach, J. L., and McEwen, B. S. (1980): Autoradiographic localization of 3H estradiol related to steroid responsiveness in cultures of newborn mouse hypothalamus and preoptic area. *Brain Res.*, 184:517–522.
Vito, C. C., Wieland, S. J., and Fox, T. O. (1979): Androgen receptors exist throughout the "critical period" of brain sexual differentiation. *Nature*, 282:308–310.
Warembourg, M. (1978): Radioautographic study of the rat brain, uterus and vagina after ^{3}H R5020 injection. *Mol. Cell. Endocrinol.*, 12:67–79.

Molecular Genetic Neuroscience, edited by
F.O. Schmitt, S.J. Bird, and F.E. Bloom.
Raven Press, New York © 1982.

Steroid-Controlled Gene Expression in a *Drosophila* Cell Line

Peter Cherbas, Charalambos Savakis, Lucy Cherbas, and M. Macy D. Koehler

ABSTRACT

The Kc-H cell line is one of numerous *Drosophila melanogaster* cell lines that are responsive to the steroid molting hormone 20-hydroxyecdysone. In Kc-H cells, the hormone provokes morphological and enzymatic differentiation and proliferative arrest. The earliest biochemical changes so far detected consist of increases in the synthesis of a few ecdysteroid-inducible polypeptides (EIPs). Molecular cloning has shown that 9 of the 10 EIPs are encoded by just two genes. Hormone treatment leads to rapid increases in the transcripts for these EIPs.

How gene activity is regulated has long been a central concern of developmental biology. Intense interest has been focused on two questions that can be approached by the molecular techniques now available: (a) How are genes coordinately regulated in eukaryotic cells, and (2) how is it that different genes are active in different cells and that these apparently epigenetic restrictions are passed from a parental cell to its progeny? We will concern ourselves here with the ways in which the actions of steroid hormones exemplify and may help to illuminate these processes.

To illustrate steroid action it is useful to consider the ecdysteroid-induced[1] development of the salivary glands in the fruitfly *Drosophila melanogaster* (for review, see Ashburner and Berendes, 1978; Berendes and Ashburner, 1978). Throughout the final larval stage, the giant polytene gland cells are engaged in the synthesis of glue polypeptides, which are temporarily sequestered in secretory vesicles. When molting hormone (the steroid 20-hydroxyecdysone, 20-HE) appears in the hemolymph 4 to 8 hr prior to the onset of metamorphosis, it causes the gradual exocytosis of the secretory product (glue) into the gland lumen. These events can be reproduced in organ culture where, as in the animal, hormone addition provokes exocytosis after a 3- to 4-hr lag. Of course, what makes this system informative is that some of the major transcriptional changes that occur can be monitored by observing the puffing pattern of the polytene chromosomes. Puffing at a site is correlated with increased transcriptional activity at that site (Ashburner, 1977); therefore, a record of puffing during the ecdysteroid response is a sampling of transcriptional events. M. Ashburner, H. D. Berendes, and their colleagues have analyzed in detail the effects of ecdysteroids on puffing.

---

[1]Ecdysteroid is the generic name for steroids related to 20-hydroxyecdysone, the insect molting hormone.

When salivary glands from a mature larva are cultured, the addition of 20-HE leads to a sequence of changes in the puffing pattern that is identical to that occurring *in vivo* during metamorphosis (Ashburner, 1972a). Sites whose puffing is altered by 20-HE belong, in general, to one of three classes, examples of which are illustrated in Fig. 1. Prior to hormone addition, only about half a dozen "intermolt puffs" are present; these correspond to the structural genes for the glue polypeptides. Ecdysteroid treatment leads to the rapid regression of these puffs and, within minutes, to the appearance of a small number of "early puffs" at new locations. The early puffs reach maximal size within 4 hr; they then regress and are replaced by a very large number (of the order of 100) of "late puffs," which appear and regress on individual schedules during the next 10 to 20 hr (Ashburner, 1974).

There is persuasive evidence that at least some of the early puffs are primary responses, whereas most late puffs are secondary. Aside from timing, early puffs are distinguishable from late puffs by three criteria: (1) the sizes of early puffs show a graded response to increasing hormone dose; late puffs exhibit an all or nothing dependence on ecdysteroid concentration (Ashburner, 1973); (2) early puffs regress rapidly after hormone withdrawal; after 5 hr late puffing becomes hormone-independent (Ashburner and Richards, 1976); and (3) protein synthesis inhibitors do not prevent the induction of early puffs but block the eventual regression of the early puffs and the induction of the late puffs (Ashburner, 1973). These observations, among others, have led to the conclusion that ecdysteroids induce transcription at a small group of primary-responsive loci, including the early puff sites, and inhibit transcription of a second set of primary loci, including the intermolt puff sites. Induction of transcription at secondary—late—puff sites and inhibition of transcription at the early puff sites are caused by products of the primary induced loci (Ashburner et al., 1973).

The outline of steroid hormone action implied by such systems as the *Drosophila* salivary gland suggests an experimental approach to the question about gene activation with which we began. These questions, restated in terms of steroid hormones, become: (1) how do steroid–receptor complexes recognize the primary responsive genes in a given cell, and (2) if the set

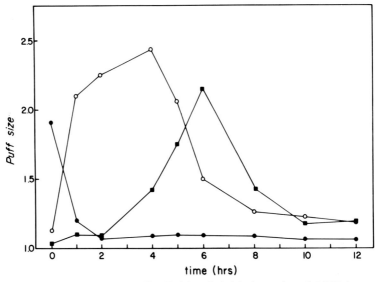

FIG. 1. Time courses of the intermolt puff 25B *(closed circle)*, the early puff 74EF *(open circle)*, and the late puff 78D *(square)*. Puff size is expressed as the ratio of puff diameter to the diameter of a nearby constant region. (Data from Ashburner, 1972a; figure republished from Cherbas and Cherbas, 1981.)

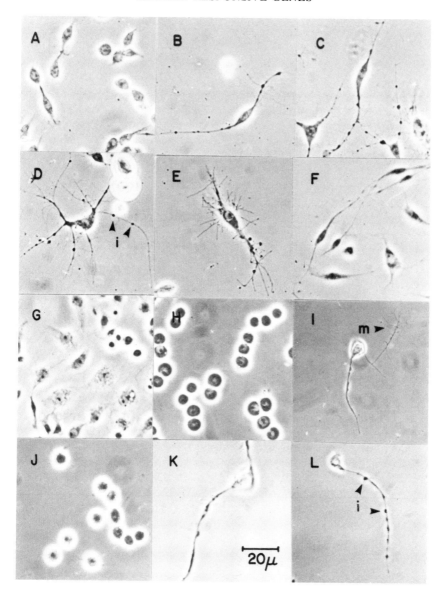

FIG. 2. Kc-H cells before and after treatment with 20-HE. **A** to **G** are cells growing on glass surfaces; **H** to **L** are cells growing in suspension. Cells in **A, H,** and **J** are untreated controls. Those in **G** were treated for 3 days with 10^{-8} M 20-HE. All others were exposed for at least 3 days to concentrations of hormone sufficient to elicit a maximal response ($10^{-7} - 10^{-5}$ M 20-HE). Examples of inclusions *(i)* and membranous extensions *(m)* are indicated by *arrows*. (Figure republished from Cherbas et al., 1980a.)

of primary responsive genes is tissue-specific, what is the molecular basis of this specificity and of its maintenance through cell division?

ECDYSTEROID RECEPTORS IN *DROSOPHILA*

The lack, until recently, of a suitable radiolabeled ecdysteroid hormone has delayed progress in the biochemical study of ecdysteroid receptors. Recently, by use of the extremely active analog ponasterone A, radiolabeled at a high specific radioactivity, ecdysteroid-binding proteins were detected and characterized in cells of the *Drosophila* cell line Kc (Maroy et al., 1978) and in *Drosophila* imaginal disks (Yund et al., 1978). The receptors from the two cell types have properties similar to those of vertebrate steroid receptors. The idea that ecdysteroid-receptor complexes participate directly in puff induction is supported by the experiments of Gronemeyer and Pongs (1980), which show that in salivary glands ecdysteroids become concentrated at distinct chromosomal sites, including the early puffs. Thus, one site of ecdysteroid action appears to be the chromosome. [This does not, of course, exclude the possibility that the same hormone also acts at other sites, such as the cell or nuclear membrane (for a discussion of models for ecdysteroid action on the membrane, see Kroeger and Lezzi, 1966; Ashburner and Cherbas, 1976).]

ECDYSTEROID EFFECTS ON THE *DROSOPHILA* CELL LINE Kc

Efforts have been made to identify and isolate primary responsive genes from a *Drosophila* cell line. Dozens of independent cell lines have been established from dissociated embryos of *D. melanogaster*, and many of these lines exhibit some sensitivity to ecdysteroid hormones (for recent reviews, see Cherbas and Cherbas, 1981; Sang, 1981). In general, these lines are near-diploid, have stable phenotypic and karyotypic properties, and are relatively easy to grow. They are probably derived from one or more imaginal cell types; imaginal cells in *Drosophila* are undifferentiated, dividing diploid cells in the larva, and are differentiated during metamorphosis to form the adult.

The effects of ecdysteroids have been studied most extensively on the Kc line and its subline Kc-H. When these cells are exposed to ecdysteroids, they differentiate morphologically and enzymatically and cease to proliferate. The morphological response becomes visible within several hours after addition of ecdysteroid and consists of the extension of long processes that often contain multiple inclusions (Fig. 2) (Cherbas et al., 1980). Acetylcholinesterase activity, which is not detectable in untreated Kc-H cells, appears in these cells after 24 hr of continuous exposure to ecdysteroids (Cherbas et al., 1977). In flies, AChE activity is apparently confined to the nervous system (Hall and Kankel, 1976); this, along with the morphology of the treated cells, suggests that the Kc line may be a neural derivative.

These effects of ecdysteroids on Kc-H cells are reproducible, synchronous, and specific to active ecdysteroids. Two lines of evidence suggest that these effects are mediated by the ecdysteroid receptor that has been detected in Kc cells. First, the relative affinities for the receptor of a number of ecdysteroids are indistinguishable from their relative activities in inducing the various responses (Beckers et al., 1980; Cherbas et al., 1980a; Cherbas et al., 1980b). Second, two ecdysteroid-resistant clones that were selected for growth in the presence of 20-HE and failed to show any response to ecdysteroid are deficient in ecdysteroid-binding activity (Cherbas et al., 1980b); J. D. O'Connor and L. Cherbas, *unpublished observations*).

FIG. 3. Two dimensional separations of pulse-labeled Kc-H cells and of translation products of their RNAs. **Left:** Control cells (*C*) or cells treated 4 hr with 10^{-6} M 20-HE (*E*) were labeled 20 min with ^{3}H-leucine, immediately solubilized in detergent-urea, and separated according to the procedures of O'Farrell (1975). The polypeptides were detected by autofluorography. **Right:** Polyadenylated RNAs from untreated cells (*C*) and from cells treated 4 hr with 10^{-6} M 20-HE (*E*) were translated in the RNA-dependent rabbit reticulocyte lysate (Pelham and Jackson, 1976). The [^{3}H]-labeled polypeptides were separated and detected as in the left panel. Isofocusing was in the horizontal dimension with the anode (acidic) at the right; the subsequent SDS-polyacrylamide separation ran from top to bottom. *Arrows* indicate spots discussed in the text. The names of the individual spots are given in the keys below. "A" designates the two spots corresponding to actin. (Figure republished from Cherbas et al., 1981.)

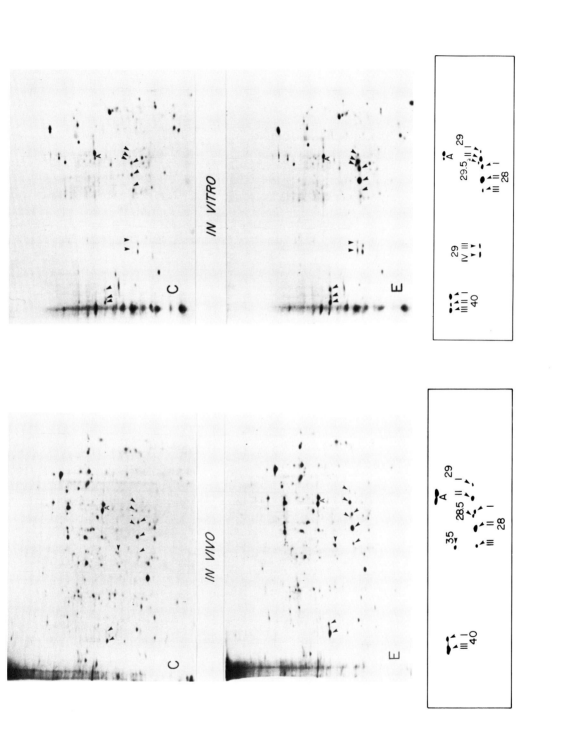

ECDYSTEROID-RESPONSIVE GENES

The pattern of protein synthesis in a clone of Kc-H cells was examined by pulse-labeling with [^{3}H]leucine and separation of the labeled polypeptides on SDS-polyacrylamide gels or on two-dimensional gels (O'Farrell, 1975). Within 1 hr of hormone addition, increased relative synthesis can be detected of polypeptides that fall into three molecular weight classes: 40, 29, and 28 kD (Savakis et al., 1980). We call these the ecdysteroid-inducible polypeptides (EIPs) and distinguish EIP 40, EIP 29, and EIP 28. In fact, as shown in Fig. 3A, two-dimensional electrophoresis resolves each of the EIPs into more than one form in the isofocusing dimension, and the catalog of changes compromises eight major spots—EIP 40 (I, II, III), EIP 29 (I, II), and EIP 28 (I, II, III)—where the most acidic form of each molecular weight class is designated I (Cherbas et al., 1981; C. Savakis, M. M. D. Koehler, and P. Cherbas, *unpublished data*). Two-dimensional electrophoresis resolves an additional major ecdysteroid-inducible polypeptide of approximately 35 kD (Fig. 3A). Except for EIP 28 III, all EIPs are stable during prolonged chases. Increased synthesis of these polypeptides becomes detectable within 1 hr, reaches a maximum within 4 to 8 hr, and continues for 2 to 3 days. During this time, few additional ecdysteroid-provoked changes, and none of comparable magnitude, are detectable using the two-dimensional separations (M. M. D. Koehler and P. Cherbas, *unpublished data*). The induction of the EIPs is ecdysteroid-specific, increases with hormone concentration over at least a 10-fold range of concentrations, and does not occur in ecdysteroid-resistant cell clones (Savakis et al., 1980).

In vitro translations of Kc-H cell RNAs have demonstrated that the EIP mRNAs are predominantly polyadenylated, and that the levels of translatable EIP mRNAs increase essentially in parallel with the *in vivo* synthesis of these polypeptides (Fig. 3B) (Savakis, 1981). All eight forms of the EIPs are represented in two-dimensional separations of translation products in approximately the same relative ratios as in *in vivo* synthesized polypeptides (see Fig. 3). This, combined with the results of the *in vivo* pulse-chase experiments, suggested that the various EIP forms represent independent products of translation rather than products of posttranslational modifications. The 35-kD ecdysteroid-inducible polypeptide was not detectable among the translation products of polyadenylated RNAs. Therefore, for technical reasons, efforts were concentrated on EIPs 40, 29, and 28.

Since the translation results indicated that the EIP mRNAs are among the 10 most abundant polyadenylated mRNA species in hormone-treated cells and that the titers of these mRNAs are elevated in hormone-treated cells severalfold relative to untreated cells, molecular clones of the EIP mRNA sequences were isolated by the following method (for details, see Savakis, 1981). Polyadenylated RNA from ecdysteroid-treated cells was used as template for the synthesis of double-stranded copy DNA (cDNA). This was used to prepare a cDNA library by inserting the cDNA at the Pst I site of plasmid pBR322 and transforming *E. coli* cells with the hybrid plasmid molecules. The resulting tetracycline-resistant clones were screened by comparing their patterns of hybridization to radioactive probes representing a control cell polyadenylated RNA and (b) 4-hr ecdysteroid-treated cell polyadenylated RNA. In a screen of around 5,000 bacterial colonies, 22 that showed reproducibly greater hybridization to probe (b) were recovered (8). The plasmids from these bacterial colonies were tested for the presence of EIP cDNA sequences by hybrid selection and translation: plasmid DNA immobilized on nitrocellulose filters was challenged with polyadenylated RNA from Kc-H cells and the hybridized RNA was eluted from the filter and translated *in vitro*.

Three EIP-related plasmids were recovered, and two-dimensional patterns of the translation products of the corresponding mRNAs were produced (Fig. 4). Plasmid pKc441 selected RNAs

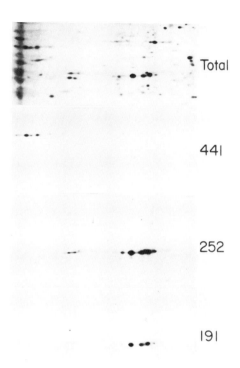

FIG. 4. Two-dimensional separations of hybrid-selected translation products. Polyadenylated RNA from 20-HE-treated Kc-H cells was translated without treatment (4 hr 20-HE) or after selection by hybridization to pKc252, pKc441, or pKc191. The translation products were displayed as in Fig. 3.

TABLE 1. *Properties of two plasmids containing EIP cDNA inserts*

Plasmid	EIPs selected	Insert length (nucleotides)	Chromosome site
pKc252	28 (I,II,III) 29 (I,II)	850	71C3,4-D1,2
pKc441	40 (I,II)	450	55B-D

encoding all the EIPs 40. Plasmids pKc252 and pKc191 are homologous to RNAs encoding all the EIPs 29 and 28. Plasmid pKc191 contained a cDNA insert homologous to that of pKc252, but considerably shorter. In addition to the five EIP 28/29 polypeptides, mRNA hybridizing to pKc252 and pKc191 directed the synthesis of two other 29-kD polypeptides with isoelectric points intermediate between the EIP 28/29 and EIP 40 clusters. These polypeptide spots are detectable among the *in vitro* translated, but not among the *in vivo* synthesized, polypeptides (see Fig. 3).

Properties of the prototype EIP-specific plasmids pKc252 and pKc441 are summarized in Table 1. Each of these plasmids is homologous to mRNAs encoding a family of polypeptides. It was initially hypothesized that two gene families exist, one homologous to pKc252, encoding EIPs 28/29, and another homologous to pKc441, encoding EIPs 40. However, each haploid genome has been shown to contain only one gene homologous to pKc252 and one gene homologous to pKc441, mapping at chromosomal sites 71C3.4-D1.2 and 55B-D, respectively. The situation with the gene encoding EIPs 28/29 is summarized in Fig. 5A. Labeled plasmid pKc252 DNA was used as a probe to isolate clones containing the EIP 28/29 genes from a library of *D. melanogaster* genomic DNA fragments cloned in a bacteriophage λ vector (L. Cherbas, *unpublished observations*). All clones recovered from this screen were overlapping, and the restriction map of the genomic region covered by these clones is shown at the top of Fig. 5A. When Kc-H cell DNA, digested with various restriction endonucleases, was subjected to electrophoresis on agarose gels, and then blotted onto nitrocellulose filters and challenged with labeled pKc252 DNA, hybridization patterns such as those shown in Fig. 5B were obtained. In this experiment, DNA from a genomic clone was restricted with the same enzyme and run in parallel with the genomic DNA. The amounts of genomic and phage DNA loaded in each lane were adjusted so that approximately equal levels of hybridization should be obtained if one copy of pKc252-specific sequences existed per haploid genome. The absence of heterogeneity in the restriction patterns of the cloned genomic sequences (Fig. 5A) and in the Kc

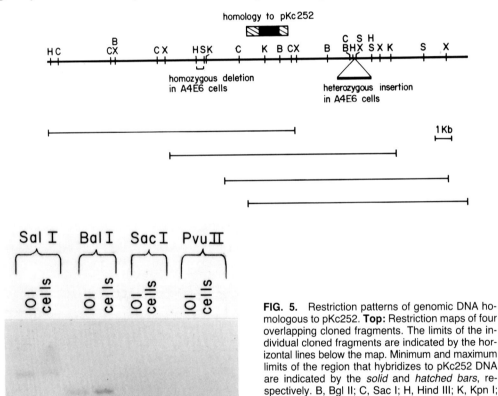

FIG. 5. Restriction patterns of genomic DNA homologous to pKc252. **Top:** Restriction maps of four overlapping cloned fragments. The limits of the individual cloned fragments are indicated by the horizontal lines below the map. Minimum and maximum limits of the region that hybridizes to pKc252 DNA are indicated by the *solid* and *hatched* bars, respectively. B, Bgl II; C, Sac I; H, Hind III; K, Kpn I; S, Sal I; X, Xho I. **Left:** DNA extracted from Kc-H cells and DNA of the clone λ (Charon 4)252-101 (the second of the four clones above) were digested with the restriction endonucleases Sal I, Bgl I, Sac I, and Pvu II. The fragments were separated by electrophoresis on an agarose gel, blotted onto nitrocellulose, and hybridized to [^{32}P]-labeled pKc252.

cell DNA (Fig. 5B) imply that there is, per haploid genome, only one gene encoding the five EIP 28/29 polypeptides. Similar results have been obtained for the EIP 40 gene (J. Rebers and L. Cherbas, *unpublished observations*).

The apparent paradox of one gene encoding a family of polypeptides is under investigation. We believe that the multiple EIP forms are not due to artifactual modifications such as those described by O'Farrell (1975), because the two-dimensional pattern is very reproducible and is seen in gels where other polypeptides do not exhibit heterogeneity in the isofocusing dimension (see Fig. 5). Possible explanations for the EIP heterogeneity include: (1) alternative mRNA splicing events; (2) alternative translation modes of the same message; and (3) some kind of posttranslational modification—such as phosphorylation or autophosphorylation—that occurs in the reticulocyte lysate as well as in intact Kc cells, and that gives rise very rapidly to a terminal equilibrium pattern.

Using plasmids pKc252 and pKc441 as probes, the EIP mRNAs were quantified during the early phase of the ecdysteroid response (Savakis, 1981). The titers of mRNAs encoding the EIPs are approximately 50% higher than control levels 30 min after hormone addition. Preliminary experiments indicate that the presence of cycloheximide administration during this period, at concentrations sufficient to inhibit about 97% of the total protein synthetic activity, does not prevent the EIP mRNA induction. Cycloheximide alone appears to cause increases in the EIP mRNA titers, by an unknown mechanism; the effects of cycloheximide and ecdysteroid are additive (Savakis, 1981).

In general, the properties of EIP induction are very similar to those of early puff induction in salivary glands. The induction is rapid; it occurs in the presence of cycloheximide, and the extent of induction increases with ecdysteroid concentration over at least a 10-fold range of concentrations. Therefore, the EIP genes, like the salivary gland early puff sites, are probably primary ecdysteroid-responsive loci.

TISSUE-SPECIFICITY OF THE PRIMARY RESPONSIVE GENES

Early ecdysteroid-responsive genes have been identified in larval tissues (salivary glands, prothoracic glands, and fat body) by analysis of puffing patterns, and in imaginal disks by *in situ* hybridization of newly synthesized RNA (Ashburner, 1972a; Bonner and Pardue, 1976; Holden and Ashburner, 1978; Richards, 1980). Puffs are induced at regions 74EF and 75B in the three larval tissues. Other regions puff in one tissue but not in another; 71CD, for example, puffs rapidly in salivary glands but not in the prothoracic gland. (Of course, one cannot tell from such studies whether apparently identical early puff sites are identical loci or are merely closely linked. Resolution of this problem will require the use of cloned sequences.) Early induced RNAs in imaginal disks do not hybridize to any of the regions that give early puffs in salivary glands. The EIP 40 gene maps to 55B-D, a site that is not among the early induced regions in the three larval tissues or in imaginal disks. The EIP 28/29 gene maps to 71C3.4-D1.2, very close to a salivary gland early puff site, but none of the EIP genes appears to be expressed in ecdysteroid-treated salivary glands (W. Koerwer and C. Savakis, *unpublished data*).

The region 71C-E is of particular interest (see Fig. 6). In salivary glands, this region is the site of a large complex puff that has been analyzed by Ashburner and his colleagues (Ashburner, 1972b; Velissariou, 1980). At 71CD there is a salivary gland early puff; immediately next to it is an intermolt puff corresponding to a glue polypeptide gene. The site of the largest late salivary gland puff is contained in 71E. The structural gene for EIPs 28/29 is located between the early and intermolt puffs, as demonstrated by *in situ* hybridization of pKc252. Thus, three

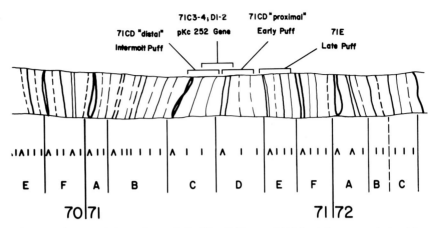

FIG. 6. Cytogenetic map of the region 71C-E. (After Bridges, 1941.) Loci that are ecdysteroid-responsive in salivary glands or in Kc cells are indicated.

ecdysteroid-controlled genes, two of which appear to be tissue-specific primary responsive genes, are located within a four to five band region. It is intriguing to speculate about possible evolutionary or functional significance for a clustering of tissue-specific hormone-responsive genes; in any case, such a clustering will be useful in the isolation of tissue-specific ecdysteroid-responsive genes.

PROSPECTS

The availability of cloned EIP sequences will enable researchers to map the pattern of EIP mRNA in larval and adult tissues and to examine the regulation of EIP mRNA during development. This type of analysis may give some clues about the origin of Kc cells and may indicate whether their response to ecdysteroids has anything to do with neural development. More important, cloned EIP genes can be used in functional assays of gene control using *in vitro* reconstituted transcription systems and DNA-mediated transformation. Such experimental techniques should enable us to begin to answer basic questions about the control of primary responsive genes by a steroid-receptor complex and the basis for tissue specificity in the steroid-responsiveness of individual genes.

DISCUSSION

W. Hahn asked whether, in Kc cells, there are any translatable mRNA species that lack poly(A) and are distinct from the polyadenylated species. P. Cherbas answered that there are many mRNAs in the so-called poly(A)$^-$ fraction (i.e., RNAs not retained by oligo(dT)-cellulose) that are translated into discrete products that are not detectable among the translation products of oligo(dT)-binding RNAs. On this basis, there seems to exist a discrete population of poly(A)$^-$ mRNAs in Kc-H cells. It is now known, however, whether these mRNAs lack poly(A) completely or have very short poly(A) tracts.

It was also asked whether juvenile hormone blocks any of the ecdysteroid effects in Kc cells. P. Cherbas answered that juvenile hormone analog ZR515, at 10^{-9} M, blocks many aspects of the ecdysteroid response including the ability of short pulses of ecdysteroids to reduce the cloning efficiency of Kc-H cells. It has no detectable effect on EIP induction (L. Cherbas, M. M. D. Koehler, and P. Cherbas, *unpublished data*).

ACKNOWLEDGMENTS

Previously unpublished work was supported by grants from the National Science Foundation (PCM-80-03931 and PCM-78-07614) and the American Cancer Society (VC316 and CD107A).

REFERENCES

Ashburner, M. (1972a): Patterns of puffing activity in the salivary gland chromosomes of *Drosophila*. VI. Induction of ecdysone in salivary glands of *D. melanogaster* cultured in vitro. *Chromasoma*, 38:255–281.
Ashburner, M. (1972b): Puffing patterns in *Drosophila melanogaster*. In: *Developmental Studies on Giant Chromosomes*, edited by W. Beermann, pp. 101–151. Springer-Verlag, New York.
Ashburner, M. (1973): Sequential gene activation by ecdysone in polytene chromosomes of *Drosophila melanogaster*. I. dependence upon ecdysone concentration. *Dev. Biol.*, 35:47–61.
Ashburner, M. (1974): Sequential gene activation by ecdysone in polytene chromosomes of *Drosophila melanogaster*. II. The effects of inhibitors of protein synthesis. *Dev. Biol.*, 39:141–157.
Ashburner, M. (1977): Happy birthday-puffs! In: *Chromosomes Today*, edited by A. de la Chapelle and M. Sorsa, pp. 213–222. Elsevier, Amsterdam.
Ashburner, M., and Berendes, H. D. (1978): Puffing of polytene chromosomes. In: *The Genetics and Biology of Drosophila*, Vol. 2b, edited by M. Ashburner and T. R. F. Wright, pp. 315–395. Academic Press, New York.
Ashburner, M., and Cherbas, P. (1976): The control of puffing by ions—The Kroeger hypothesis: A critical review. *Mol. Cell. Endocrinol.*, 5:89–107.
Ashburner, M., Chihara, C., Meltzer, P., and Richards, G. (1973): Temporal control of puffing activity in polytene chromosomes. *Cold Spring Harbor Symp. Quant. Biol.*, 38:655–662.
Ashburner, M., and Richards, G. (1976): Sequential gene activation in polytene chromosomes of *Drosophila melanogaster*. III. Consequences of ecdysone withdrawal. *Dev. Biol.*, 54:241–255.
Beckers, C., Maroy, P., Dennis, R., O'Connor, J. D., and Emmerich, H. (1980): Uptake and release of ^{3}H-ponasterone A by the Kc cell line of *Drosophila melanogaster*. In: *Progress in Ecdysone Research*, edited by J. A. Hoffmann, pp. 335–347. Elsevier, Amsterdam.
Berendes, H. D., and Ashburner, M. (1978): The salivary glands. In: *The Genetics and Biology of Drosophila*, Vol. 2b, edited by M. Ashburner and T. R. F. Wright, pp. 453–498. Academic Press, New York.
Bonner, J. J., and Pardue, M. L. (1976): Ecdysone-stimulated RNA synthesis in imaginal discs of *Drosophila melanogaster*. Assay by in situ hybridization. *Chromosoma*, 58:87–99.
Bridges, P. N. (1941): A revised map of the left limb of the third chromosome of *Drosophila melanogaster*. *J. Hered.*, 32:64–65.
Cherbas, L., and Cherbas, P. (1981): The effects of ecdysteroid hormones on *Drosophila melanogaster* cell lines. *Adv. Cell Culture*, 1:91–124.
Cherbas, L., Yonge, C. D., Cherbas, P., and Williams, C. M. (1980a): The morphological response of Kc-H cells to ecdysteroids: Hormonal specificity. *Wilhelm Roux Arch.*, 189:1–15.
Cherbas, P., Cherbas, L., Demetri, G., Manteuffel-Cymborowska, M. Savakis, C., Yonge, C. D., and Williams, C. M. (1980b): Ecdysteroid hormone effects on a *Drosophila* cell line. In: *Gene Regulation by Steroid Hormones*, edited by A. K. Roy and J. H. Clark, pp. 278–305. Springer-Verlag, New York.
Cherbas, P., Cherbas, L., Savakis, C., and Koehler, M. M. D. (1981): Ecdysteroid-responsive genes in a *Drosophila* cell line. *Am. Zool.*, 21:743–750.
Cherbas, P., Cherbas, L., and Williams, C. (1977): Induction of acetylcholinesterase activity by β-ecdysone in a *Drosophila* cell line. *Science*, 197:275–277.
Gronemeyer, H., and Pongs, O. (1980): Localization of ecdysterone on polytene chromosomes of *Drosophila melanogaster*. *Proc. Natl. Acad. Sci. USA*, 77:2108–2112.
Hall, J. C., and Kankel, D. R. (1976): Genetics of acetylcholinesterase in *Drosophila melanogaster*. *Genetics*, 83:517–535.
Holden, J. J., and Ashburner, M. (1978): Patterns of puffing activity in the salivary gland chromosomes of *Drosophila*. IX. The salivary and prothoracic gland chromosomes of a dominant temperature sensitive lethal of *D. melanogaster*. *Chromosoma*, 68:205–227.
Kroeger, H., and Lezzi, M. (1966): Regulation of gene action in insect development. *Annu. Rev. Entomol.*, 11:1–22.
Maroy, P., Dennis, R., Beckers, C., Sage, B. A., and O'Connor, J. D. (1978): Demonstration of an ecdysteroid receptor in a cultured cell line of *Drosophila melanogaster*. *Proc. Natl. Acad. Sci. USA*, 75:6035–6038.
O'Farrell, P. H. (1975): High resolution two-dimensional electrophoresis of proteins. *J. Biol. Chem.*, 250:4007–4021.
Pelham, H. R. B., and Jackson, R. J. (1976): An efficient mRNA-dependent translation system from reticulocyte lysates. *Eur. J. Biochem.*, 67:247–257.
Richards, G. (1980): Ecdysteroids and puffing in *Drosophila melanogaster*. In: *Progress in Ecdysone Research*, edited by J. Hoffmann, pp. 363–378. Elsevier, Amsterdam.
Sang, J. H. (1981): *Drosophila* cells and cell lines. *Adv. Cell Culture*, 1:125–177.
Savakis, C. (1981): Studies on Ecdysteroid-Inducible Polypeptides and their mRNAs in a *Drosophila melanogaster* Cell Line. Ph.D. thesis, Harvard University.

Savakis, C., Demetri, G., and Cherbas, P. (1980): Ecdysteroid-inducible polypeptides in a *Drosophila* cell line. *Cell*, 22:665–674.
Velissariou, V. (1980): The Cytogenetics of the Salivary Gland Glue Proteins of *Drosophila melanogaster*. Ph.D. thesis, University of Cambridge.
Yund, M. A., King, D. S., and Fristrom, J. W. (1978): Ecdysteroid receptors in imaginal discs of *Drosophila melanogaster*. *Proc. Natl. Acad. Sci. USA*, 75:6039–6043.

Putative Steroid Receptors: Genetics and Development

Thomas O. Fox, Kathie L. Olsen, Christine C. Vito, and Steven J. Wieland

ABSTRACT

Receptors for steroid hormones may exhibit diverse modes of action, depending on the specific function, tissue, and agonist. In developing brains of mice and rats, receptors for androgens and estrogens exist in the hypothalamus and in other neural regions and may partially determine the timing of the critical period of brain sexual differentiation. The patterns of appearance of these receptor proteins in normal animals and in mutants that display androgen-resistance are correlated with the physiological effectiveness of androgens and their metabolites. By combining these developmental and genetic comparisons, hypotheses emerge for differential roles of androgens and estrogens in sexual differentiation.

In steroid-responsive tissues, including neural tissues, the analysis of putative receptors for steroids serves in two ways to further our understanding of the mechanisms of steroid hormone action. First, it provides models for correlating specific responses and definable hormonal interactions with cells. Second, it provides clues regarding the possible ligands that may be agonistic for a given steroid type. For example, the actions of androgens in the developing brain appear to be mediated by a set of androgens and androgen metabolites, most notably testosterone, dihydrotestosterone (DHT), and estradiol (E_2). Knowing which putative steroid receptors—with specificities for which ligands—are present in developing neural tissues, and knowing when during development these binding proteins appear, help in suggesting and testing possible mechanisms for the actions of androgens in sexual differentiation of male and female brains.

The most sensitive, or "critical," period of sexual differentiation of the brain is believed to begin several days before birth and to extend into the first postnatal week in mice and rats. Androgens and androgen metabolites influence development, leading to phenotypes of adult neuroendocrine controls and behaviors, including those of reproduction (Goy and McEwen, 1980). It was widely believed that most, if not all, of the effects of androgens during this critical period were mediated by aromatization of androgens to E_2, and binding of E_2 to the estrogen receptor (McEwen et al., 1975; Naftolin et al., 1975; MacLusky et al., 1976; Plapinger and McEwen, 1978). This conclusion was based primarily on studies of defeminization, such as sterilization effects on the pituitary, elimination of cyclic behavior, and reduction of lordosis behavior in female rats (Barraclough, 1967, 1973; Goldman, 1978; Gorski, 1979). As other aspects of behavior were examined with increasing subtlety, it became clear that there are multiple ways to distinguish between defeminization (suppression of development of female-typical behavior and estrous cyclicity) and masculinization (enhancement of potential to display male-typical responses) (Whalen, 1974). There is evidence that defeminization and masculinization can differ (see Goy, 1970; Beach, 1971; Baum, 1979) in their dose-response characteristics (e.g., DeBold and Whalen, 1975), hormone specificities (e.g., Goy, 1978; Baum, 1979), and timing (e.g., Goy et al., 1964; Noble, 1973). Thus there is reason to believe that multiple sites of hormone action and possibly more than one receptor type may be involved in androgen action in brain sexual differentiation.

Early work on sexual differentiation of brain focused on defeminization. The direction of research in this field was influenced greatly by observations that, in mice and rats, those androgens that appear to be most effective during the critical period tend to be those that can be metabolized by "aromatization" to estrogens. Androgens such as DHT, which are not aromatizable, generally are less active in rats and mice (Whalen and Rezek, 1974), although they can be effective during the perinatal period in guinea pigs, monkeys, hamsters, and rats (see Vito et al., 1979). The tendency for aromatizable androgens to be active, and the difficulty of detecting androgen receptors in brain in several earlier experiments, were compelling reasons for proposing aromatization as the mechanism for mediation of androgen effects in male and female brains (Naftolin et al., 1971). However, existence of an aromatization mechanism does not preclude parallel involvement of androgen receptors in the process of brain sexual differentiation. Evidence for androgen receptor involvement is supported by investigations of androgen receptors in developing brain and from studies on the behavioral and biochemical phenotypes of androgen-resistant mutants. Perhaps both androgen and estrogen receptors act during brain differentiation. With demonstration of androgen receptors in neonatal (Fox, 1975a,b; Attardi and Ohno, 1976; Fox et al., 1978) and prenatal mouse and rat brains (Fox et al., 1978; Vito et al., 1979; Lieberburg et al., 1980; Vito and Fox, 1982), it is possible to examine two other alternatives to the aromatization hypothesis: aromatizable androgens might be more active than nonaromatizable androgens because they are able to accumulate in neuronal targets in greater concentrations, or they are more agonistic ligands for androgen receptors in developing brain tissues.

MECHANISMS OF STEROID RECEPTOR ACTION

Steroid receptors can redistribute within cells, and their locations are influenced by occupation with hormones. Although most notable models invoke action at the nucleus, the sites of action within the cell are not fully defined. It is possible that different actions, mediated by the same binding proteins, occur at alternate sites in the cell.

Membrane-Interacting Steroid Receptors

Generally, models of steroid receptors depict steroid hormone molecules as entering the cell readily, probably facilitated by their lipoidal character, and binding inside the cell to cytoplasmic receptors. Receptor–steroid complexes then accumulate in the nucleus, their site of function. There are, however, reports of changes in electrical activity of neurons occurring in some cases within seconds or minutes in response to nanomolar or subnanomolar concentrations of steroid hormones (Kelly et al., 1977, 1978, 1980; Kendrick and Drewett, 1980; Teyler et al., 1980). Such effects are not easily explained by the mechanisms of steroid action that involve nuclear accumulation of receptor and subsequent effects on genetic expression; these actions would require a longer time course.

Two features of the classic nuclear model might be viewed differently. First, the lipoidal nature of steroids may result in their partitioning into the membrane, where they may accumulate rather than diffusing through the membrane to the cytoplasm. Second, some of the receptors may loosely associate with the inner surface of the membrane and there form complexes with the steroid (Fig. 1). A consequence of this mechanism might be that migration of the hormone-receptor complex from the membrane would cause a perturbation at the membrane that could account for the reported fast membrane effect. Although this effect may be possible for many tissues, it may have evolved especially in neurons. In essence, this hypothesis emphasizes how redistribution of steroid receptors may be used for actions that precede those in the nucleus. Much remains to be learned about how, in neurons, the receptor-steroid complex acts in the nucleus, and to what extent receptor action in the nucleus explains observed neural responses to steroids.

The model depicted in Fig. 1 suggests how conclusions about conventional steroid receptors, which can be found in solubilized forms in tissue homogenates, might account for both nuclear and extranuclear activities, without invoking a specific class of membrane receptors. This picture could represent an extreme that is physiologically functional in neurons and is also demonstrable in other tissues. For example, the ability of dissociated uterine, hepatic, and intestinal cells to bind to estrogen that was covalently immobilized on albumin-derivatized nylon fibers (Pietras and Szego, 1977) could occur if these ligands entered the plasma membrane and there bound the "soluble" intracellular estrogen receptors. Experiments demonstrating estradiol receptors tightly associated with membrane fragments (Pietras and Szego, 1979; Pietras et al., 1981) were performed with buffers containing Ca^{2+}, with the intent of preserving certain properties that are believed to be typically membrane associated. Although this atypical component for steroid receptor analysis may have preserved authentic membrane architecture (Pietras and Szego, 1979), a significant literature on Ca^{2+}-activated proteases that cleave steroid receptors (Andre and Rochefort, 1973; Puca et al., 1977; Sherman et al., 1978; Vedeckis et al., 1980) raises the possibility that attachment of cleavage products to membrane fragments might have occurred. Figure 1 permits both a mechanism for steroid action at the membrane and a substrate to which such an attachment might occur.

In another analysis of steroid receptors (Barrack and Coffey, 1980) that also included the divalent cation Mg^{2+} in certain buffers, another atypical and unusually tight localization was obtained. Clearly the specific ionic components of buffers used for extraction are critical for extrapolations to the distribution for intact cells. With amphibian oocytes, progesterone and other agents may act at the surface to activate meiotic division (Baulieu et al., 1978; Brachet, 1980), possibly with direct action on membrane-associated cyclic adenosine monophosphate (cAMP) activities (Maller and Krebs, 1977; Finidori-Lepicard et al., 1981). The questions raised by the model in Fig. 1 may apply to several steroid hormones.

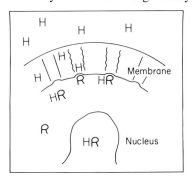

FIG. 1. Model for possible interaction of soluble steroid receptors (R) with steroid hormones (H) in the cellular membrane.

Interactions of Steroid Receptors with DNA

In many steroid-responsive tissues and cell types, accumulation of receptor-steroid complexes in the nucleus is followed by accumulation of specific mRNAs in the cytoplasm (Palmiter and Smith, 1973; Ringold et al., 1975), and there is convincing evidence that mRNA accumulation is the result of increased transcription of specific genes (McKnight, 1978; McKnight and Palmiter, 1979). It has been hypothesized that receptor-steroid complexes function in the nucleus in a way analogous to prokaryotic transcriptional effectors, e.g., the *lac* repressor, by binding specific DNA sequences (for reviews on steroid receptor action, see Gorski and Gannon, 1976; Yamamoto and Alberts, 1976). Receptor-steroid complexes do bind to DNA but usually with low affinity, without apparent sequence specificity; only recently, with relatively short and specific DNA sequences, have specific high-affinity interactions between steroid receptors and DNA been compellingly reported (Payvar et al., 1981). It is probable that receptor-steroid complexes also bind nonspecifically to DNA in the nucleus, and that this nonspecific binding directly or indirectly modulates their function. In addition, receptor-steroid complexes may bind to other sites in the nucleus, independent of their binding to DNA. These might include proteinaceous sites, and these interactions also might influence transcriptional activity.

Yamamoto and Alberts (1975), building on considerations (von Hippel et al., 1974; Lin and Riggs, 1975) of the *lac* repressor model, hypothesized that high-affinity binding of receptor to specific DNA sites could be masked *in vitro* by low-affinity nonspecific receptor binding to DNA. In addition to its high affinity for *lac* operator DNA, *lac* repressor exhibits a low affinity for nonoperator DNA (Lin and Riggs, 1972, 1975; von Hippel et al., 1974) roughly comparable in magnitude to the general affinity of steroid receptors for DNA. Because of this low-affinity nonspecific binding, specific repressor-operator interactions cannot be detected unless the DNA used in binding assays is enriched in operator sites, a condition that is achieved when phage DNA containing the *lac* operon is used (Gilbert and Muller-Hill, 1967) and when homologous phage operators and repressors are examined (Ptashne, 1967).

Even with steroid receptor binding to operator-like specific DNA sequences, nonspecific binding to DNA in the cell nucleus may influence receptor function, because it could compete with binding of receptor to specific functional nuclear sites. This possibility is demonstrated in the *lac* system by a class of repressor mutants (Chamness and Willson, 1970) that exhibit increased affinity for both operator and nonoperator DNA (Pfahl, 1976). In the wild type, addition of inducer causes derepression of the *lac* operon (Fig. 2A, top). *Escherichia coli* cells bearing the X86 mutation of the *lac* repressor (Pfahl, 1976) exhibit the phenotype represented by Fig. 2A, bottom. In the absence of inducer, the *lac* operon is incompletely repressed; intermediate concentrations of inducer elicit repression, whereas induction appears at higher inducer concentrations.

The X86 repressor has an approximately 50-fold higher affinity for *lac* operator DNA than does the wild-type repressor; it also has a higher affinity for nonoperator DNA (Pfahl, 1976). Based on this, Pfahl (1976) proposed the following model to explain the X86 phenotype. In a normal, dividing *E. coli* cell, about 90% of repressor molecules are bound to nonoperator DNA; the remainder are free or, in the absence of inducer, bound to the operator site causing repression (Fig. 2B, top). In the mutant, the higher affinity of the repressor for all DNA sequences results in a much slower dissociation rate from nonoperator DNA. Thus, as the cell divides and new operator sites become available, many of these sites are not occupied by repressor. Consequently, under normal growing conditions, there is only partial repression in the mutant of the operon (Fig. 2B). As inducer is added, it binds to repressor and the affinity of the mutant repressor for DNA generally decreases. More repressor becomes available for operator binding, and repression occurs.

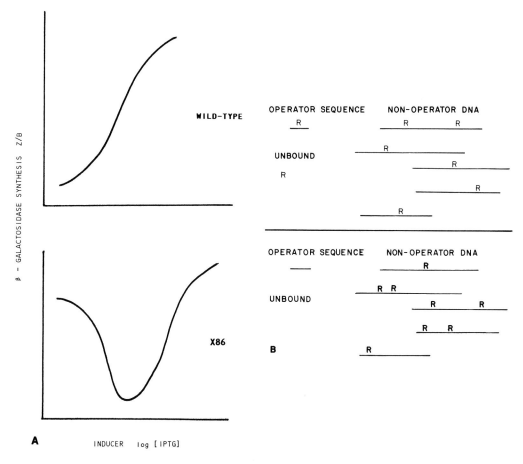

FIG. 2. Diagrammatic representations of wild-type and X86 lactose *(lac)* operon repressor mutant. **A:** Idealized enzyme induction curves for wild-type **(top)** and X86 mutant **(bottom)**. **B:** Hypothetical distributions of wild-type *lac* repressor (light R) and tight-binding X86 repressor (dark R) in growing wild-type **(top)** and mutant **(bottom)** *E. coli* bacteria. IPTG, isopropylthiogalactoside. (Based on data derived from Pfahl, 1976.)

The *lac* repressor mutants of the X86 type (Pfahl, 1976; Barkley and Bourgeois, 1978; Miller, 1978; Pfahl, 1981) provide a model to explain the phenotype of a class of glucocorticoid receptor variants with increased affinity for DNA. Two interesting classes of receptor variants that have been characterized in selected glucocorticoid-resistant mouse lymphoma cells (Sibley and Tomkins, 1974a,b) have receptors that show decreased or increased nuclear accumulation (designated nt^- and nt^i, respectively). When these receptors were examined for DNA binding by DNA-cellulose chromatography, a correlation was found between nuclear accumulation and apparent affinity for DNA. Both nt^- and nt^i receptors bind to DNA, but compared to normal receptor, nt^- receptor shows a more rapid elution and nt^i receptor shows a retarded elution from DNA-cellulose with increasing ionic strength (Yamamoto et al., 1974). This correlation between apparent affinity for DNA, nuclear accumulation, and biological activity is evidence that the binding of receptor-steroid complex to DNA is biologically significant. The X86 *lac* repressor model provides a plausible explanation of how increased affinity for DNA can result in loss of function for the nt^i variant receptors, although other explanations for these particular steroid-response mutations are possible. Other steroid receptor variants might prove to exhibit this proposed phenotype.

ANDROGEN AND ESTROGEN RECEPTORS IN DEVELOPING RAT AND MOUSE BRAIN

A powerful analytical method, DNA-cellulose chromatography, takes advantage of nonspecific low-affinity binding of receptor–steroid complexes to DNA. This technique was used to detect and analyze putative androgen and estrogen receptors in extracts from rat and mouse hypothalamus (Fig. 3A).

Previously there had been difficulties in demonstrating androgen receptors in hypothalamus. The conditions required for effective DNA-cellulose chromatography, such as sufficient ionic strength (Fig. 3B) and the presence of glycerol (Fig. 3C) also facilitate the detectability of androgen receptors. Furthermore, since androgen receptors are partially purified by DNA-cellulose chromatography, it has become possible to examine them in tissues such as the hypothalamus and cerebellum (Fox, 1977), which contain very low levels of receptor. Testosterone and DHT bind to putative androgen receptors from mouse and rat hypothalamus with comparable affinities, and androgen binding can be blocked with E_2 (Bullock and Bardin, 1974), although it is approximately 10-fold less active as a competitor than testosterone or DHT (Fox, 1975a; Attardi et al., 1976; Chamness et al., 1979).

Detection of estrogen receptors in tissue extracts of neonatal mouse or rat hypothalamus was initially prevented by the presence in these animals of α-fetoprotein, which binds E_2 strongly and competes with the receptor for added radioactively labeled E_2 (Plapinger et al., 1973). DNA-cellulose chromatography was used to separate α-fetoprotein from hypothalamic estrogen receptor, which selectively bound to the column (Fox, 1975b). Because α-fetoprotein does not adhere to DNA-cellulose, extracts can be applied to a column to remove α-fetoprotein (Fox et al., 1978; Vito and Fox, 1979). Bound receptor can then be labeled with hormone and eluted from the column (Fox et al., 1978; Vito and Fox, 1979, 1982). An alternative approach to the same problem took advantage of estrogen analogs that have relatively low affinity for α-fetoprotein (McEwen et al., 1975).

Androgen and estrogen receptors can be distinguished by ligand specificity and by their differential elution profiles in DNA-cellulose chromatography (Fig. 3A). Androgen receptors elute at a lower salt concentration than estrogen receptors. This allows separation of androgen and estrogen receptors from the same extract in a single experiment.

The presence of macromolecules with properties of putative androgen and estrogen receptors was demonstrated in extracts of late embryonic, neonatal, and prepubertal mouse and rat hypothalamus (Plapinger and McEwen, 1973, Barley et al., 1974; Kato et al., 1974; Fox, 1975a,b; Attardi and Ohno, 1976; Kato, 1976; MacLusky et al., 1976, 1979a,b,c; Westley et al., 1976; Westley and Salaman, 1976, 1977; Fox et al., 1978; Lieberburg et al., 1979, 1980; Vito et al., 1979; Vito and Fox, 1979, 1982; White et al., 1979). These receptors are qualitatively similar to androgen and estrogen receptors from adult brain by virtue of their hormone-binding characteristics, their ability to bind to DNA, their differential elution in DNA-cellulose chromatography, and their velocity sedimentation properties. Figure 4 summarizes the developmental profiles of androgen and estrogen receptors in rat and mouse hypothalamus. In both rats and mice, androgen receptors are detectable in the hypothalamus as early as 5 to 7 days before birth, and their levels increase steadily after birth. The developmental profiles of estrogen receptors differ in the two species. In mouse, appreciable levels of receptor can be detected as early as 6 days before birth, and these levels remain unchanged until the end of the first postnatal week. In rat, the levels of estrogen receptor are initially low, and rise dramatically during the last embryonic week.

These results indicate that both androgen and estrogen receptors are present in hypothalamus during the critical period of brain sexual differentiation in mice and rats. The presence of these

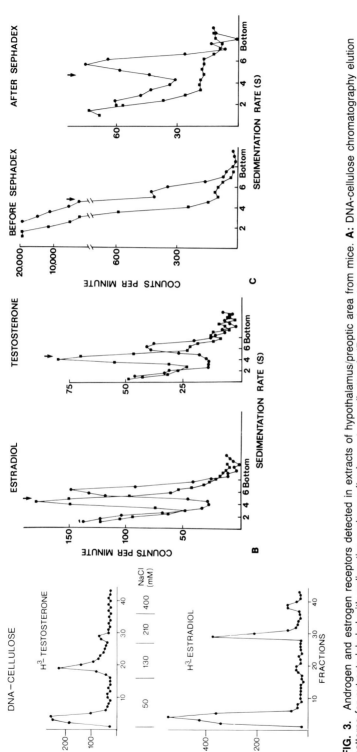

FIG. 3. Androgen and estrogen receptors detected in extracts of hypothalamus/preoptic area from mice. **A:** DNA-cellulose chromatography elution patterns for extracts labeled with radioactive androgen **(top)** or estrogen **(bottom)**. (Reproduced from Fox et al., 1978.) **B:** Sucrose density gradient sedimentation of receptors in buffer containing no added NaCl (*circles*) or containing 0.15 M NaCl (*squares*). The inclusion of NaCl increased the yield of detectable androgen receptor. This result and the methods used are reported in Fox, 1975a. **C:** Effect of 10% (vol/vol) glycerol (*circles*) and no glycerol (*squares*) on detectability of androgen receptors labeled with [³H] testosterone and detected by sedimentation in sucrose density gradients before or after Sephadex G-25 chromatography to remove unbound radioactivity. *Arrows* indicate internal albumin marker for sedimentation. These results and the methods used are reported in Fox, 1975a.

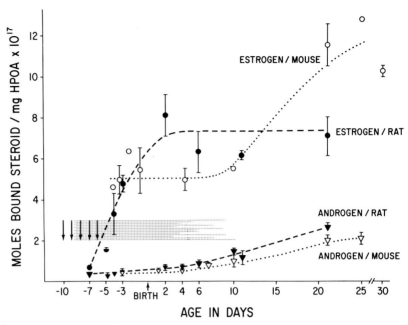

FIG. 4. Comparison of androgen and estrogen receptor ontogeny in hypothalamus/preoptic area (HPOA) of rats and mice. The *arrows* indicate the dates of final neuronal division for most hypothalamic neurons, as described in the text. *Stippled area* represents critical period of brain sexual differentiation. (From Vito and Fox, 1982.)

receptors in hypothalamus during the last embryonic week suggests a mechanism for initiation of events in the critical period. The critical period of sexual differentiation might simply be a result of hypothalamic neurons maturing at a time when there are sex differences in circulating gonadal steroids. Some events concurrent at this stage are as follows: Most of the neurons in the hypothalamus and related area cease dividing (Ifft, 1972; Shimada and Nakamura, 1973; Altman and Beyer, 1978a,b; Anderson, 1978) in mice and rats near the beginning of the last embryonic week (arrows in Fig. 4), and therefore are differentiating during the critical period. Another group of neurons ceases to divide late in the final embryonic week (Jacobson and Gorski, 1981). Therefore, the sexual dimorphism that has been observed within this nucleus might be a consequence of their late proliferation at a time when sex steroids differ between male and female embryos, and estrogen and androgen receptors are present (Vito and Fox, 1979, 1982; Vito et al., 1979). Evidence for prenatal differences in the levels of circulating androgens between females and males exists for both rats and mice (vom Saal and Bronson, 1980; Weisz and Ward, 1980; vom Saal, 1981). A series of experiments indicates that sexual differentiation of the brain does occur prenatally. In female rats (Clemens, 1974; Clemens et al., 1978; Meisel and Ward, 1981) and mice (Gandleman et al., 1977; vom Saal and Bronson, 1978), individual differences in the degree of masculinization are correlated with the in utero position. Female mice situated in utero between two males show higher levels of androgens before birth and a higher degree of masculinization in adult life than do females situated between females (vom Saal and Bronson, 1980).

The presence of androgen receptors in hypothalamus during the critical period is consistent with the hypothesis that androgens may act directly in brain sexual differentiation, as well as indirectly via conversion to estrogens (Vito et al., 1979; Vito and Fox, 1982).

ANDROGEN RECEPTORS IN ANDROGEN-RESISTANT MUTANTS

As indicated by the studies on glucocorticoid receptor variants, mutations affecting receptors can help to evaluate the biological significance of biochemical data. Spontaneous mutations affecting androgen action are well characterized in several mammals, including mice, rats, and humans (see Bardin and Catterall, 1981). Although the androgen-resistant mammalian mutants are less amenable to biochemical and genetic manipulations than glucocorticoid-resistant cell variants, they allow investigations of events that are mediated directly by androgens. Androgen-resistant mutants have been used to define which androgens or androgen metabolites are active during brain differentiation and in adult endocrine and behavioral functions (Naess et al., 1976; Beach and Buehler, 1977; Olsen, 1979a,b; Shapiro et al., 1980; Olsen and Whalen, 1981). As a necessary beginning for such studies, androgen-binding activity in androgen-resistant mice and rats was investigated.

In mice, the androgen-resistant mutation, testicular feminization *(Tfm)* (Lyon and Hawkes, 1970), is X-linked and causes X/Y individuals to develop a phenotype that is externally "feminized" even though they possess unmatured testes that produce androgens. Presumably, failure to respond to their own androgens results in lack of androgen-dependent, "masculinized" traits in these mutants. It has been shown that the kidney, normally an androgen-responsive tissue, is deficient in androgen-binding activity for *Tfm*/Y mice (Attardi and Ohno, 1974; Bullock and Bardin, 1974; Gehring and Tomkins, 1974). This result provides a genetic correlation in support of the hypothesis that these steroid-binding proteins are receptors.

Low levels of androgen-binding activity have been detected in tissues of *Tfm*/Y mice (Fox, 1975a). When cytosolic extracts from mutant hypothalamus are labeled with $[^3H]$-testosterone or $[^3H]$-DHT and subjected to sucrose gradient sedimentation, low levels of binding can be detected at a region of the gradient where the normal androgen receptor migrates (Fig. 5A). The residual androgen-binding activity from *Tfm* hypothalamus adheres reversibly to DNA-cellulose but elutes with a higher salt concentration than the predominant wild-type receptor (Fig. 5B). Mouse hypothalamic extracts contain a minor androgen-binding entity, which has an elution profile similar to the residual *Tfm* form (compare Fig. 3A to Fig. 5B). In contrast, estrogen receptors were not detectably changed in the mutant (Fig. 5C). For kidney, androgen receptor is resolved by DNA-cellulose chromatography into two major components with elution characteristics similar to the two components from hypothalamus (Fig. 6). If the two kidney receptor components are separated and reapplied to DNA-cellulose, each elutes at its characteristic position, indicating that the observed elution pattern is not caused by multiple binding sites in the column (Wieland, 1979). As in hypothalamus, the residual androgen-binding activity in *Tfm* kidney elutes from DNA-cellulose together with the higher salt eluting component (Fig. 6).

When extracts from wild-type and *Tfm* kidney are mixed and subjected to DNA-cellulose chromatography, the elution pattern resembles that expected from the addition of the individual patterns; the same result is obtained with hypothalamic extracts (Fig. 7). This result reduces the possibility that a cystolic component present in *Tfm* animals modifies the DNA-binding behavior of androgen receptors, and suggests that the elution profile of the residual *Tfm* receptor is an intrinsic property of this entity. These data do not resolve whether the residual androgen-binding activity from *Tfm* mice is a higher salt eluting form of normal androgen receptors or represents an aberrant form of receptor.

The *tfm* mutation of the rat (Stanley and Gumbreck, 1964) results in androgen resistance and external female appearance (Stanley et al., 1973). It also is deficient in the level of androgen

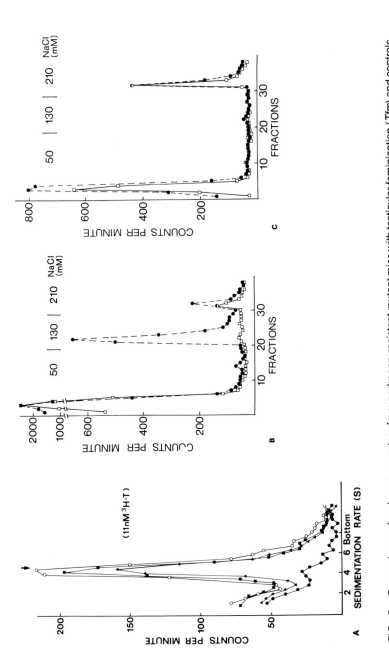

FIG. 5. Comparisons of androgen receptors from androgen-resistant mutant mice with testicular feminization (*Tfm*) and controls. **A:** Sucrose density gradient sedimentation; the *Tfm* mutant (*solid circles*) is compared to males and females that are wild type (*solid squares*, B6C3H F₁ hybrid male; *open circles* Ta/Y male) or heterozygous (*triangles*, *Tfm*/+ female) for the *Tfm* locus. (From Fox, 1975a). DNA-cellulose chromatography of androgen (**B**, dihydrotestosterone) and estrogen (**C**, estradiol) radioactively labeled receptors from *Tfm* mutants (*open squares*) or wild-type controls (*solid circles*). (From Wieland et al., 1978.)

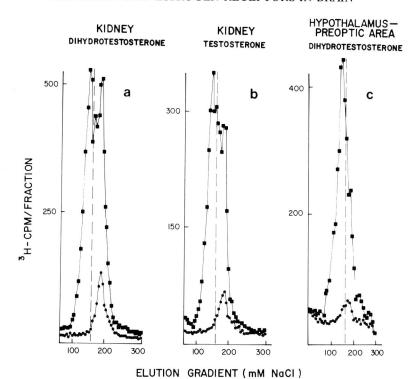

FIG. 6. Variations of DNA-cellulose chromatography of androgen receptors from mice. Two tissue homogenates, kidney (**a,b**) and hypothalamus/preoptic area (**c**), two ligands, DHT (**a** and **c**) and testosterone (**b**), and Tfm/Y mutant (*circles*) and wild type (*squares*) were compared. *Dashed line* at 160 mM NaCl assists in distinguishing lower and higher salt eluting receptors. (From Wieland and Fox, 1979.)

receptors (Bullock and Bardin, 1972), but *tfm* rats appear to be somewhat more responsive to androgen than *Tfm* mice (Bardin and Catterall, 1981). Examination of androgen binding in the *tfm* rat reveals a low level of receptor (approximately 10 to 30% of normal), which, in contrast to mouse *Tfm*, appears to be identical to the wild-type receptor by all the available criteria, including chromatographic behavior on DNA-cellulose (Fig. 8). As with the *Tfm* mouse, rat estrogen binding in brain (Olsen and Whalen, 1982) and estrogen receptors appear to be unaffected by the mutation (T. O. Fox, K. L. Olsen, and S. J. Wieland, *unpublished results*).

The *tfm* rat and *Tfm* mouse mutants differ in the display of masculine sexual behavior (Table 1) (Olsen and Fox, 1981). Whereas the gonadally intact *tfm* rats mounted receptive females, and some even exhibited the ejaculatory pattern, none of the gonadally intact *Tfm* mice showed any masculine mating responses. Female sexual behaviors were not readily displayed by the mutants or by the male wild-type mice and rats.

These data suggest that these two mutations differentially affect androgen receptors, and this could cause differential androgen responsiveness. Alternately, even if the mutations were the same, the androgen receptor defects might be expressed differently in rats and mice or between these particular stocks of animals. In either case, such comparisons provide tools for examining other genetic determinants of steroid hormone action in behavior.

CONCLUSION

Evidence from many lines of research indicate that testosterone can enter the cell and accumulate in the nucleus, either without being metabolized or after being converted in cells to

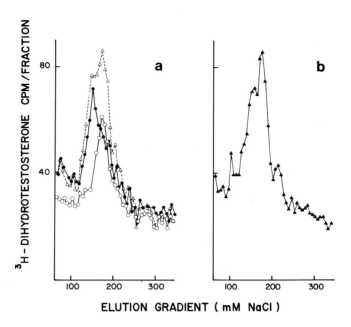

FIG. 7. DNA-cellulose chromatography of mixed hypothalamus/preoptic area cytosols labeled with [³H]-DHT from wild type (*solid circles*) and nine times that amount of tissue from *Tfm*/Y mutants (*open circles*). The summation of the wild-type and *Tfm*/Y patterns, corrected first for background, is shown (**a**, *open triangles*) and can be compared to the actual elution pattern for both extracts labeled and chromatographed together (**b**, *solid triangles*). (From Wieland and Fox, 1979.)

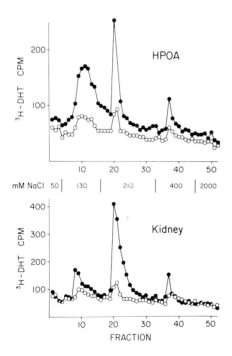

FIG. 8. DNA-cellulose chromatography of hypothalamus/preoptic area (HPOA) androgen receptors and kidney androgen receptors from *tfm* mutant rats (*open circles*) and wild-type controls (*solid circles*). (From Wieland and Fox, 1981.)

TABLE 1. *Sexual behavior*

	Feminine[a]	Masculine[b]
Mice		
Tfm +/Y	—	—
+ *Ta*/Y (wild-type male)	—	+++
Rats		
tfm/Y	—	++
+/Y (wild-type male)	—	++++

— indicates that the animals rarely exhibited the behavior; + + + + indicates the display of the full copulatory response.
[a]The lordosis quotient and rejection quotient were measured to determine the degree of feminine behavior.
[b]Masculine behavior was assessed by recording the latency and frequency of mounts, mounts with intromissions, and ejaculation.

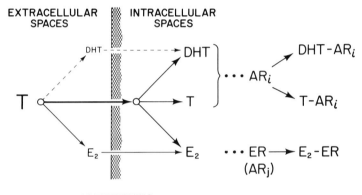

CONVERSIONS

FIG. 9. Model for androgen reception mechanism involving multiple androgen ligands and metabolites (T, DHT, E_2) and multiple binding proteins (AR_i). The estrogen receptor (ER) can be considered to be one of several receptors for androgen metabolites, (AR_i). (From Wieland and Fox, 1979).

DHT or to E_2. The extent to which single neurons contain both types of receptors is not clear. Testosterone and both metabolites, DHT and E_2, were detected in cell nuclear fractions from brains of rats injected with radioactive testosterone (Lieberburg and McEwen, 1977). Presumably, testosterone and DHT can both bind to androgen receptors, and E_2 can bind to the estradiol receptor in brain. The ability of selective treatments of cultured hypothalamic cells with norepinephrine (Vaccaro et al., 1980) and kainic acid (Livingstone et al., 1980) preferentially to reduce aromatase activity versus 5α-reductase suggests that these metabolic pathways may be under different controls and may be differentially located among neurons and other neural cells. Because androgens are, to some extent, converted to estrogens, which in turn mediate some of the effects of androgens during development, estrogen and androgen receptors can be considered to be components of a single receptor system as opposed to two separate receptor systems. These relationships are shown schematically in Fig. 9.

It is plausible that estrogens functioned as more primitive hormones (Callard et al., 1977), and that androgens were only anabolic precursors (Heard et al., 1955) for these hormones. Later, receptors for the precursors (testosterone) and a second group of their metabolites, including DHT (Farnsworth and Brown, 1963), may have evolved to permit changes in sex steroid regulation. An opposite relationship is also possible (Wieland and Fox, 1979). The significance here is that many interrelationships exist between androgens and estrogens, and a better understanding of these factors in metabolism and receptor mechanisms may help to elucidate the variable ways in which these hormones are used for sexual differentiation of the brain.

ACKNOWLEDGMENTS

Portions of this research were supported generously by the March of Dimes Birth Defects Foundation, the National Institutes of Health, and the Mental Retardation Research Center at Children's Hospital Medical Center.

REFERENCES

Altman, J., and Bayer, S. A. (1978a): Development of the diencephalon in the rat. I. Autoradiographic study of the time of origin and settling patterns of neurons of the hypothalamus. *J. Comp. Neurol.*, 182:945–972.
Altman, J., and Bayer, S. A. (1978b): Development of the diencephalon in the rat. II. Correlation of the embryonic development of the hypothalamus with the time of origin of its neurons. *J. Comp. Neurol.*, 182:973–994.
Anderson, C. H. (1978): Time of neuron origin in the anterior hypothalamus of the rat. *Brain Res.*, 154:119–122.
Andre, J., and Rochefort, H. (1973): Estrogen receptor: Loss of DNA binding ability following trypsin or Ca^{2+} treatment. *FEBS Lett.*, 32:330–334.
Attardi B., Geller, L. N., and Ohno, S. (1976): Androgen and estrogen receptors in brain cytosol from male, female and testicular feminized (*tfm/y*) mice. *Endocrinology*, 98:864–874.
Attardi B., and Ohno S. (1974): Cytosol androgen receptor from kidney of normal and testicular feminized (*Tfm*) mice. *Cell*, 2:205–212.
Attardi, B. and Ohno, S. (1976): Androgen and estrogen receptors in the developing mouse brain. *Endocrinology*, 99:1279–1290.
Bardin, C. W., and Catterall, J. F. (1981): Testosterone: A major determinant of extragenital sexual dimorphism. *Science*, 211:1285–1294.
Barkley, M. D., and Bourgeois, S. (1978): Repressor recognition of operator and effectors. In: *The Operon*, edited by J. H. Miller and W. S. Reznikoff, pp. 177–220. Cold Spring Harbor Laboratory, New York.
Barley, J., Ginsburg, M., Greenstein, B. D., MacLusky, N. J., and Thomas, P. J. (1974): A receptor mediating sexual differentiation? *Nature*, 252:259–260.
Barrack, E. R., and Coffey, D. S. (1980): The specific binding of estrogens and androgens to the nuclear matrix of sex hormones responsive tissues. *J. Biol. Chem.*, 255:7265–7275.
Barraclough, C. A. (1967): Modifications in reproductive function after exposure to hormones during the prenatal and early postnatal period. In: *Neuroendocrinology, Vol. II*, edited by L. Martini, pp. 62–100. Academic Press, New York.
Barraclough, C. A. (1973): Sex steroid regulation of reproductive neuroendocrine processes. In: *Handbook of Physiology, Sec. 7, Endocrinology, Vol. 2, Female Reproductive System (Part I)*, edited by R. O. Greep and E. B. Astwood, pp. 29–56. American Physiological Society, Washington, D.C.
Baulieu, E.-E., Godeau, F., Schorderet, M., and Schorderet-Slatkine, S. (1978): Steroid-induced meiotic division in *Xenopus laevis* oocytes: Surface and calcium. *Nature*, 275:593–598.
Baum, M. J. (1979): Differentiation of coital behavior in mammals: A comparative analysis. *Neurosci. Biobehav. Rev.*, 3:265–284.
Beach, F. A. (1971): Hormonal factors controlling the differentiation, development and display of copulatory behavior in the ramstergig and related species. In: *The Biopsychology of Development*, edited by E. Tobach, L. Aronson, and E. Shaw, pp. 249–296. Academic Press, New York.
Beach, F. A., and Buehler, M. G. (1977): Male rats with inherited insensitivity to androgen show reduced sexual behavior. *Endocrinology*, 100:197–200.
Brachet, J. (1980): Induction of maturation in full-grown and vitellogenic amphibian oocytes by steroids and protein factors. In: *Steroids and Their Mechanism of Action in Nonmammalian Vertebrates*, edited by G. Delrio and J. Brachet, pp. 75–83. Raven Press, New York.
Bullock, L. P., and Bardin, C. W. (1972): Androgen receptors in testicular feminization. *J. Cell Endocrinol. Metab.*, 35:935–937.
Bullock, L. P., and Bardin, C. W. (1974): Androgen receptors in mouse kidney: A study of male, female and androgen-insensitive (*tfm/y*) mice. *Endocrinology*, 94:746–756.
Callard, G. V., Petro, Z., and Ryan, K. J. (1977): Identification of aromatase in the reptilian brain. *Endocrinology*, 100:1214–1218.
Chamness, G. C., King, T. W., and Sheridan, P. J. (1979): Androgen receptor in the rat brain—assays and properties. *Brain Res.*, 161:267–276.
Chamness, G. C., and Willson, C. D. (1970): An unusual *lac* repressor mutant. *J. Mol. Biol.*, 53:561–565.

Clemens, L. G. (1974): Neurohormonal control of male sexual behavior. In: *Reproductive Behavior*, edited by W. Montagna and W. A. Sadler, pp. 23–53. Plenum Press, New York.

Clemens, L. G., Gladue, B. A., and Coniglio, L. P. (1978): Prenatal endogenous androgenic influences on masculine sexual behavior and genital morphology in male and female rats. *Horm. Behav.*, 10:40–53.

DeBold, J. F., and Whalen, R. E. (1975): Differential sensitivity of mounting and lordosis control systems to early androgen treatment in male and female hamsters. *Horm. Behav.*, 6:197–209.

Farnsworth, W. E., and Brown, J. R. (1963): Metabolism of testosterone by the human prostate. *JAMA*, 183:436–439.

Finidori-Lepicard, J., Schorderet-Slatkine, S., Hanoune, J., and Baulieu, E.-E. (1981): Progesterone inhibits membrane-bound adenylate cyclase in *Xenopus laevis* oocytes. *Nature*, 292:255–257.

Fox, T. O. (1975a): Androgen and estrogen binding macromolecules in developing mouse brain: Biochemical and genetic evidence. *Proc. Natl. Acad. Sci. USA*, 72:4303–4307.

Fox, T. O. (1975b): Oestradiol receptor of neonatal mouse brain. *Nature*, 258:441–444.

Fox, T. O. (1977): Estradiol and testosterone binding in normal and mutant mouse cerebellum: Biochemical and cellular specificity. *Brain Res.*, 128:263–273.

Fox, T. O., Vito, C. C., and Wieland, S. J. (1978): Estrogen and androgen receptor proteins in embryonic and neonatal mouse brain: Hypotheses for roles in sexual differentiation and behavior. *Am. Zool.*, 18:525–537.

Gandelman, R., vom Saal, F. S., and Reinisch, J. M. (1977): Contiguity to male foetuses affects morphology and behavior of female rats. *Nature*, 266:722–724.

Gehring, U., and Tomkins, G. M. (1974): A new mechanism for steroid unresponsiveness: Loss of nuclear binding activity of a steroid hormone receptor. *Cell*, 3:301–306.

Gilbert, W., and Muller-Hill, B. (1967): The lac operator is DNA. *Proc. Natl. Acad. Sci. USA*, 58:2415–2421.

Goldman, B. D. (1978): Developmental influences of hormones on neuroendocrine mechanisms of sexual behavior: Comparisons with other sexually dimorphic behaviors. In: *Biological Determinants of Sexual Behavior*, edited by J. B. Hutchison, pp. 127–152. John Wiley and Sons, New York.

Gorski, J., and Gannon, F. (1976): Current models of steroid hormone action: A critique. *Annu. Rev. Physiol.*, 38:425–450.

Gorski, R. A. (1979): The neuroendocrinology of reproduction: An overview. *Biol. Reprod.*, 20:111–127.

Goy, R. W. (1970): Early hormonal influences on the development of sexual and sex-related behavior. In: *The Neurosciences: Second Study Program*, edited by F. O. Schmitt, pp. 196–207. Rockefeller University Press, New York.

Goy, R. W. (1978): Development of play and mounting behavior in female rhesus virilized prenatally with esters of testosterone or dihydrotestosterone. In: *Recent Advances in Primatology, Volume 1, Behavior*, edited by D. J. Chivers and J. Herbert, pp. 449–462. Academic Press, New York.

Goy, R. W., Bridson, W. E., and Young, W. C. (1964): Period of maximal susceptibility of the prenatal female guinea pig to masculinizing actions of testosterone propionate. *J. Comp. Physiol. Psychol.*, 57:166–174.

Goy, R. W., and McEwen, B. S. (1980): *Sexual Differentiation of the Brain*, MIT Press, Cambridge, Mass.

Heard, R. D. H., Jellinck, P. H., and O'Donnell, V. J. (1955): Biogenesis of the estrogens: The conversion of testosterone-4-C^{14} to estrone in the pregnant mare. *Endocrinology*, 57:200–204.

Ifft, J. D. (1972): An autoradiographic study of the time of final division of neurons in rat hypothalamic nuclei. *J. Comp. Neurol.*, 144:193–204.

Jacobson, C. D., and Gorski, R. A. (1981): Neurogenesis of the sexually dimorphic nucleus of the preoptic area of the rat. *J. Comp. Neurol.*, 196:519–529.

Kato, J. (1976): Cytosol and nuclear receptors for 5α-dihydrotestosterone and testosterone in the hypothalamus and hypophysis, and testosterone receptors isolated from neonatal female rat hypothalamus. *J. Steroid Biochem.*, 7:1179–1187.

Kato, J., Atsumi, Y., and Inaba, M. (1974): Estradiol receptors in female rat hypothalamus in the developmental stages and during pubescence. *Endocrinology*, 94:309–317.

Kelly, M. J., Kuhnt, U., and Wuttke, W. (1980): Hyperpolarization of hypothalamic parvocellular neurons by 17β-estradiol and their identification through intracellular staining with procion yellow. *Exp. Brain Res.*, 40:440–447.

Kelly, M. J., Moss, R. L., and Dudley, C. A. (1977): The effects of microelectrophoretically applied estrogen, cortisol and acetylcholine on medial preoptic-septal unit activity throughout the estrous cycle of the female rat. *Exp. Brain Res.*, 30:53–64.

Kelly, M. J., Moss, R. L., and Dudley, C. A. (1978): The effects of ovariectomy on the responsiveness of preoptic-septal neurons to microelectrophoresed estrogen. *Neuroendocrinology*, 25:204–211.

Kendrick, K. M., and Drewett, R. F. (1980): Testosterone-sensitive neurones respond to oestradiol but not to dihydrotestosterone. *Nature*, 286:67–68.

Lieberburg, I., Krey, L. C., and McEwen, B. S. (1979): Sex differences in serum testosterone and in exchangeable brain cell nuclear estradiol during the neonatal period in rats. *Brain Res.*, 178:207–212.

Lieberburg, I., MacLusky, N., and McEwen, B. S. (1980): Androgen receptors in the perinatal rat brain. *Brain Res.*, 196:125–138.

Lieberburg, I., and McEwen, B. S. (1977): Brain cell nuclear retention of testosterone metabolites, 5α-dihydrotestosterone and estradiol-17β, in adult rats. *Endocrinology*, 100:588–597.

Lin, S., and Riggs, A. D. (1972): *Lac* repressor binding to non-operator DNA: detailed studies and a comparison of equilibrium and rate competition methods. *J. Mol. Biol.*, 72:671–690.

Lin, S., and Riggs, A. D. (1975): The general affinity of *lac* repressor for E. coli DNA: Implications for gene regulation in procaryotes and eucaryotes. *Cell*, 4:107–111.

Livingston, E. G., Canick, J. A., Vaccaro, D. E., Ryan, K. J., and Fox, T. O. (1980): The effect of kainic acid on aromatization and 5α-reduction in embryonic rat hypothalamic cultures. *Society for Neuroscience, 10th Annual Meeting, Abstr.*, 6:457.

Lyon, M. F., and Hawkes, S. G. (1970): X-linked gene for testicular feminization in the mouse. *Nature*, 227:1217–1219.

MacLusky, N. J., Chaptal, C., Lieberburg, I., and McEwen, B. S. (1976): Properties and subcellular inter-relationships of presumptive estrogen receptor macromolecules in the brain of neonatal and prepubertal female rats. *Brain Res.*, 114:158–165.

MacLusky, N. J., Chaptal, C., and McEwen, B. S. (1979a): The development of estrogen receptor systems in the rat brain and pituitary: Postnatal development. *Brain Res.*, 178:143–160.

MacLusky, N. J., Lieberburg, I., and McEwen, B. S. (1979b): The development of estrogen receptor systems in the rat brain: Perinatal development. *Brain Res.*, 178:129–142.

MacLusky, N. J., Lieberburg, I., and McEwen, B. S. (1979c): Development of steroid receptor systems in the rodent brain. In: *Ontogeny of Receptors and Reproductive Hormone Action*, edited by T. H. Hamilton, J. H. Clark, and W. A. Sadler, pp. 393–402. Raven Press, New York.

Maller, J. L., and Krebs, E. G. (1977): Progesterone-stimulated meiotic cell division in *Xenopus* oocytes. *J. Biol. Chem.*, 252:1712–1718.

McEwen, B. S., Plapinger, L., Chaptal, C., Gerlach, J., and Wallach, G. (1975): Role of fetoneonatal estrogen binding proteins in the association of estrogen with neonatal brain cell nuclear receptors. *Brain Res.*, 96:400–406.

McKnight, G. S. (1978): The induction of ovalbumin and conalbumin mRNA by estrogen and progesterone in chick oviduct explant cultures. *Cell*, 14:403–413.

McKnight, G. S., and Palmiter, R. D. (1979): Transcriptional regulation of the ovalbumin and conalbumin genes by steroid hormones in chick oviduct. *J. Biol. Chem.*, 254:9050–9058.

Meisel, R. L., and Ward, I. L. (1981): Fetal female rats are masculinized by male littermates located caudally in the uterus. *Science*, 213:239–242.

Miller, J. H. (1978): The *lacI* gene: Its role in lac operon control and its use as a genetic system. In: *The Operon*, edited by J. H. Miller and W. S. Reznikoff, pp. 31–88. Cold Spring Harbor Laboratory, New York.

Naess, O., Haug, E., Attramadal, A., Aakvaag, A., Hansson, V., and French, F. (1976): Androgen receptors in the anterior pituitary and central nervous system of the androgen "insensitive" *(Tfm)* rat: Correlation between receptor binding and effects of androgens on gonadotropin secretion. *Endocrinology*, 99:1295–1303.

Naftolin, F., Ryan, K. J., Davies, I. J., Reddy, V. V., Flores, F., Petro, Z., Kuhn, M., White, R. J., Takoaka, Y., and Wolin, L. (1975): The formation of estrogens by central neuroendocrine tissues. *Recent Prog. Horm. Res.*, 31:295–319.

Naftolin, F., Ryan, K. J., and Petro, Z. (1971): Aromatization of androstenedione by the diencephalon. *J. Clin. Endocrinol. Metab.*, 33:368–370.

Noble, R. G. (1973): The effects of castration at different intervals after birth on the copulatory behavior of male hamsters *(Mesocricetus auratus)*. *Horm. Behav.*, 4:45–52.

Olsen, K. L. (1979a): Androgen-insensitive rats are defeminised by their testes. *Nature*, 279:238–239.

Olsen, K. L. (1979b): Induction of male mating behavior in androgen-insensitive (tfm) and normal (King-Holtzman) male rats: Effects of testosterone propionate, estradiol benzoate, and dihydrotestosterone. *Horm. Behav.*, 13:66–84.

Olsen, K. L., and Fox, T. O. (1981): Differences between androgen-resistant rat and mouse mutants. *Society of Neuroscience, 11th Annual Meeting, Abstr.*, 7:219.

Olsen, K. L., and Whalen, R. E. (1982): Estrogen binds to hypothalamic nuclei of androgen-insensitive *(tfm)* rats. *Experientia*, 38:139–140.

Olsen, K. L., and Whalen, R. E. (1981): Hormonal control of the development of sexual behavior in androgen-insensitive *(tfm)* rats. *Physiol. Behav.*, 27:883–886.

Palmiter, R. D., and Smith, L. T. (1973): Synergistic effects of oestrogen and progesterone on ovomucoid and conalbumin mRNA synthesis in chick oviduct. *Nature New Biol.*, 246:74–76.

Payvar, F., Wrange, O., Carlstedt-Duke, J., Okret, S., Gustafsson, J.-A., and Yamamoto, K. R. (1981): Purified glucocorticoid receptors bind selectively in vitro to a cloned DNA fragment whose transcription is regulated by glucocorticoids in vivo. *Proc. Natl. Acad. Sci. USA*, 78:6628–6632.

Pfahl, M. (1976): *Lac* repressor-operator interaction. Analysis of the X86 repressor mutant. *J. Mol. Biol.*, 106:857–869.

Pfahl, M. (1981): Characteristics of tight binding repressors of the *lac* operon. *J. Mol. Biol.*, 147:1–10.

Pietras, R. J., and Szego, C. M. (1977): Specific binding sites for oestrogen at the outer surfaces of isolated endometrial cells. *Nature*, 265:69–72.

Pietras, R. J., and Szego, C. M. (1979): Estrogen receptors in uterine plasma membrane. *J. Steroid Biochem.*, 11:1471–1483.

Pietras, R. J., Szego, C. M., and Seeler, B. J. (1981): Immunologic inhibition of estrogen binding and action in preputial-gland cells and their subcellular fractions. *J. Steroid Biochem.*, 14:679–691.
Plapinger, L., and McEwen, B. S. (1973): Ontogeny of estradiol-binding sites in rat brain. I. Appearance of presumptive adult receptors in cytosol and nuclei. *Endocrinology*, 93:1119–1128.
Plapinger, L., and McEwen, B. S. (1978): Gonadal steroid-brain interactions in sexual differentiation. In: *Biological Determinants of Sexual Behavior*, edited by J. B. Hutchison, pp. 153–218. John Wiley and Sons, New York.
Plapinger, L., McEwen, B. S., and Clemens, L. E. (1973): Ontogeny of estradiol-binding sites in rat brain. II. Characteristics of a neonatal binding macromolecule. *Endocrinology*, 93:1129–1140.
Ptashne, M. (1967): Specific binding of the λ phage repressor to λ DNA. *Nature*, 214:232–234.
Puca, G. A., Nola, E., Sica, V., and Bresciani, F. (1977): Estrogen binding proteins of calf uterus. *J. Biol. Chem.*, 252:1358–1366.
Ringold, G. M., Yamamoto, K. R., Tomkins, G. M., Bishop, J. M., and Varmus, H. E. (1975): Dexamethasone-mediated induction of mouse mammary tumor virus RNA: A system for studying glucocorticoid action. *Cell*, 6:299–305.
Shapiro, B. H., Levine, D. C., and Adler, N. T. (1980): The testicular feminized rat: A naturally occurring model of androgen independent brain masculinization. *Science*, 209:418–420.
Sherman, M. R., Pickering, L. A., Rollwagen, F. M., and Miller, L. K. (1978): Mero-receptors: Proteolytic fragments of receptors containing the steroid-binding site. *Fed. Proc.*, 37:167–173.
Shimada, M., and Nakamura, T. (1973): Time of neuron origin in mouse hypothalamic nuclei. *Exp. Neurol.*, 41:163–173.
Sibley, C. H., and Tomkins, G. M. (1974a): Isolation of lymphoma cell variants resistant to killing by glucocorticoids. *Cell*, 2:213–220.
Sibley, C. H., and Tomkins, G. M. (1974b): Mechanisms of steroid resistance. *Cell*, 2:221–227.
Stanley, A. J., and Gumbreck, L. G. (1964): Male pseudohermaphroditism with feminizing testes in the male rat: A sex-linked recessive character. *The Endocrine Society, Abst.*, 40.
Stanley, A. J., Gumbreck, L. G., Allison, J. E., and Easley, R. B. (1973): Part I. Male pseudohermaphroditism in the laboratory Norway rat. *Rec. Prog. Horm. Res.*, 29:43–64.
Teyler, T. J., Vardaris, R. M., Lewis, D., and Rawitch, A. B. (1980): Gonadal steroids: Effects on excitability of hippocampal pyramidal cells. *Science*, 209:1017–1019.
Vaccaro, D. E., Canick, J. A., Livingston, E. G., Fox, T. O., Ryan, K. J., and Leeman, S. E. (1980): Possible effectors of aromatization and 5-alpha reduction in hypothalamic cell cultures. *The Endocrine Society, Abst.*, 106.
Vedeckis, W. V., Freeman, M. R., Schrader, W. T., and O'Malley, B. W. (1980): Progesterone-binding components of chick oviduct: Partial purification and characterization of a calcium-activated protease which hydrolyzes the progesterone receptor. *Biochemistry*, 19:335–343.
Vito, C. C., and Fox, T. O. (1979): Embryonic rodent brain contains estrogen receptors. *Science*, 204:517–519.
Vito, C. C., and Fox, T. O. (1982): Androgen and estrogen receptors in embryonic and neonatal rat brain. *Brain Res.*, 254 (Dev. Brain Res., vol. 2):97-110.
Vito, C. C., Wieland, S. J., and Fox, T. O. (1979): Androgen receptors exist throughout the "critical period" of brain sexual differentiation. *Nature*, 282:308–310.
vom Saal, F. S. (1981): Variation in phenotype due to random intrauterine positioning of male and female fetuses in rodents. *J. Reprod. Fertl.*, 62:633–650.
vom Saal, F. S., and Bronson, F. H. (1978): In utero proximity of female mouse fetuses to males: Effect on reproductive performance during later life. *Biol. Reprod.*, 19:842–853.
vom Saal, F. S., and Bronson, F. H. (1980): Sexual characteristics of adult female mice are correlated with their blood testosterone levels during prenatal development. *Science*, 208:597–599.
von Hippel, P. H., Revzin, A., Gross, C. A., and Wang, A. C. (1974): Nonspecific DNA binding of genome regulating proteins as a biological control mechanism. I. The *lac* operon: Equilibrium aspects. *Proc. Natl. Acad. Sci. USA*, 71:4808–4812.
Weisz, J., and Ward, I. L. (1980): Plasma testosterone and progesterone titers of pregnant rats, their male and female fetuses, and neonatal offspring. *Endocrinology*, 106:306–316.
Westley, B. R., and Salaman, D. F. (1976): Role of oestrogen receptor in androgen-related sexual differentiation of the brain. *Nature*, 262:407–408.
Westley, B. R., and Salaman, D. F. (1977): Nuclear binding of the oestrogen receptor of neonatal rat brain after injection of oestrogens and androgens; localization and sex differences. *Brain Res.*, 119:375–388.
Westley, B. R., Thomas, P. J., Salaman, D. F., Knight, A., and Barley, J. (1976): Properties and partial purification of an oestrogen receptor from neonatal rat brain. *Brain Res.*, 113:441–447.
Whalen, R. E. (1974): Sexual differentiation: Models, methods and mechanisms. In: *Sex Differences in Behavior*, edited by R. C. Friedman, R. M. Richart, and R. L. Vande Wiele, pp. 467–481. John Wiley and Sons, New York.
Whalen, R., and Rezek, D. L. (1974): Inhibition of lordosis in female rats by subcutaneous implants of testosterone, androstenedione or dihydrotestosterone in infancy. *Horm. Behav.*, 5:125–128.
White, J. O., Hall, C., and Lim, L. (1979): Developmental changes in the content of oestrogen receptors in the hypothalamus of the female rat. *Biochem. J.*, 184:465–468.

Wieland, S. J. (1979): Analysis of putative androgen receptors in wild-type and testicular-feminized *(Tfm)* mice. Ph.D. Dissertation, Harvard University.

Wieland, S. J., and Fox, T. O. (1979): Putative androgen receptors distinguished in wild-type and testicular feminized *(Tfm)* mice. *Cell*, 17:781–787.

Wieland, S. J., and Fox, T. O. (1981): Androgen receptors from rat kidney and brain: DNA-binding properties of wild-type and *tfm* mutant. *J. Steroid Biochem.*, 14:409–414.

Wieland, S. J., Fox, T. O., and Savakis, C. (1978): DNA-binding of androgen and estrogen receptors from mouse brain: Behavior of residual androgen receptor from *Tfm* mutant. *Brain Res.*, 140:159–164.

Yamamoto, K. R., and Alberts, B. (1975): The interaction of estradiol-receptor protein with the genome: An argument for the existence of undetected specific sites. *Cell*, 4:301–310.

Yamamoto, K. R., and Alberts, B. M. (1976): Steroid receptors: Elements for modulation of eukaryotic transcription. *Annu. Rev. Biochem.*, 45:721–746.

Yamamoto, K. R., Stampfer, M. R., and Tomkins, G. M. (1974): Receptors from glucocorticoid-sensitive lymphoma cells and two classes of insensitive clones: Physical and DNA-binding properties. *Proc. Natl. Acad. Sci. USA*, 71:3901–3905.

Molecular Genetic Neuroscience, edited by
F. O. Schmitt, S. J. Bird, and F. E. Bloom.
Raven Press, New York © 1982.

Regulation of Hormonally Controlled mRNA Synthesis and Gene Expression in Variant Somatic Cells and Cell Hybrids

E. Brad Thompson and Jeanine S. Strobl

ABSTRACT

Clones of variant cells and somatic cell hybrids have been used to study the nature of cellular controls over specific, hormonally induced processes. Noninducible variants of the hepatoma tissue culture (HTC) cell line have been isolated. The noncoordinate control of glucocorticoid-induced events in these variants points up the diversity and specificity of the cellular control mechanisms involved. With respect to steroid-controlled genes, somatic cell hybrids between inducible and noninducible cell lines usually show extinction of the controlled product. Through studies on the growth hormone and prolactin genes, it is apparent that this negative control is not the result of gross DNA loss or rearrangement, but of failure of the hybrid cells to produce measurable quantities of transcripts. Therefore these results also point to the great specificity of cell controls over hormonally affected genes.

STEROID HORMONE ACTION AND STEROID RECEPTORS

In many tissues and cells, steroid hormones have been shown to cause accumulation of specific mRNAs. These have been measured either by *in vitro* translation assays or by quantification of the mRNA itself. Despite efforts in many laboratories, the precise biophysical and biochemical mechanism by which steroids produce these effects is still unknown. In some cases, the rapidity of onset of mRNA accumulation following addition of hormone, other kinetic arguments, and RNA polymerase loading experiments have led to the proposal that these steroids are acting "at the level of transcription." However, the molecular mechanisms implicit in this phrase are not entirely clear.

Some puzzles remain from experiments of the last decade that are not solved by the simple idea that steroids act by increasing the rate of initiation of transcriptional events at specific genes. Tsai and coworkers (1975) showed that when *E. coli* RNA polymerase is added to chromatin prepared from oviducts of estrogen-treated chicks, and in vitro transcription is allowed to proceed, the number of sites at which the heterologous polymerase begins transcription is higher than in controls. This was originally interpreted to mean that the estrogen increased the number of specific initiation sites (Schwartz et al., 1975). It is now known that this is not true (Tsai et al., 1976); rather, by an unknown mechanism, the steroid receptor complex profoundly alters the chromatin, causing an increase in the number of sites available to *E. coli* RNA polymerase. Similar effects on chromatin have been reported in various cell lines responding to glucocorticoids (Johnson et al., 1979). Perhaps the resolution of these problems will become possible only when more is known about the organization of eukaryotic chromatin and about the rules governing its transcription. Despite these puzzles, we believe that a major effect of steroids is to cause an increase, somehow, in the transcription of certain genes (Mullinix et al., 1979; Swaneck et al., 1979).

Steroids do not act exclusively by increasing transcription of specific genes; there is evidence that they also decrease transcription of certain genes (Johnson et al., 1979); that they can increase or decrease the general activity of RNA polymerases (Borthwick and Bell, 1975; Courvalin et al., 1976; Mohla et al., 1981); and that they can affect the rate of translation and stability of mRNA (Rosen et al., 1980). Furthermore, in some cases, steroids cause direct membrane effects (Thompson, 1979). In addition, estrogens and androgens can act as trophic hormones and thus exert global effects on cell viability. Finally, and this is particularly important to neuroendocrinologists, steroids may act on tissues or cell types indirectly. For example, it has been shown recently that glucocorticoids not only destroy immature thymus-derived lymphocytes (T cells) but also prevent the production of T-cell growth factor (Gillis et al., 1979a,b). Thus, glucocorticoids destroy lymphoid cells by two presumably independent pathways—one direct (i.e., via interaction with the steroid receptors of the affected cells) and one indirect. It is nearly inevitable that similar indirect steroid effects will occur between cells in the brain.

Strong evidence indicates that in most cases steroids act via specific cytoplasmic receptors. This evidence comes from biochemical studies involving ligand binding and dose-response relationships, and by studies on steroid-resistant variants. In general, the magnitude of steroid responses is proportional to receptor occupancy by steroid, and the potency of agonists and antagonists can usually be predicted by their receptor affinity (King and Mainwaring, 1974; Munck, 1976; Baxter and Rousseau, 1979). Studies of mouse lymphoid cell variants selected for resistance to glucocorticoids have shown that such variants contain altered receptors or no receptors at all (Pfahl et al., 1978; T. Fox et al., *this volume*). In addition, an activation-labile receptor that appears to be the predominant phenotype of spontaneous glucocorticoid-resistant variants of human T cells has recently been described in a human T-cell-derived leukemic cell line (Schmidt et al., 1980; Harmon and Thompson, 1981). This new phenotype is evidence that receptor activation really occurs and is of biological significance. Furthermore, in this system and in other cell lines lacking glucocorticoid receptors, glutamine synthetase activity, which is normally inducible by glucocorticoids, can no longer be induced. Thus in the same cells killed by glucocorticoids, an inducible function also appears to be dependent on glucocorticoid receptors (Harmon and Thompson, 1982). These findings clearly provide evidence that receptors are required for glucocorticoid action in these systems.

Steroid receptors are cytoplasmic proteins and generally require stabilizing functional sulfhydryl groups. There is evidence that phosphorylation is required for activation (Nielsen et al., 1977a) and that activation leads to nuclear association (Parchman and Litwack, 1977; Sakaue and Thompson, 1977; Higgins et al., 1979; Munck and Foley, 1980). In addition, there are several reports that small molecules in the cytoplasm (under 1,000 daltons) are responsible for the stabilization as well as the prevention of activation of glucocorticoid or other receptors (Cake et al., 1976; Bailly et al., 1977; Nielson et al., 1977b; Sando et al., 1977; Sato et al., 1980). The glucocorticoid receptor is probably a monomer of elliptical shape, with a molecular weight of about 90 kilodaltons (Kd); it is prone to specific proteolysis that leads to the appearance of very clear size classes, detectable by SDS-polyacrylamide gel electrophoresis (Rousseau and Baxter, 1979; Wrange et al., 1979; Westphal and Beato, 1980).

Glucocorticoid receptors are ubiquitous; with the exception of certain brain areas and gastric mucosa, virtually every tissue appears to contain them. However, not all cells respond in the same way to glucocorticoids. Tissue and cell specificity in glucocorticoid action must therefore be determined either at the level of receptor—each cell type containing a characteristic set of multiple receptor species—or at some other level distal to the receptor, e.g., chromatin structure or diffusable modulators. In trying to probe the nature of tissue specificity by using somatic cell genetics, coordinate control of multiple, hormone-regulated processes has been studied both in cultured variant somatic cells and in cell hybrids.

CONTROL OF MULTIPLE GLUCOCORTICOID-INDUCED PROCESSES IN HEPATOMA TISSUE CULTURE CELL VARIANTS

The hepatoma tissue culture (HTC) cells are a line of cultured cells derived from a rat hepatoma (Thompson et al., 1966). Glucocorticoids (up to 10^{-5} M dexamethasone) do not affect viability or growth in the original HTC cell strain, but do induce several specific cell processes and inhibit others. Among the glucocorticoid responses of HTC cells is the induction of tyrosine aminotransferase (TAT) activity. Clones of HTC cells can be obtained that vary widely in TAT inducibility, as shown in Fig. 1 (Aviv and Thompson, 1972). When a low- and a high-inducing clone were subcloned, the resulting subclones exhibited overlapping, but distinguishable induction spectra (Fig. 1). Prolonged growth of these high- and low-inducers led to the production of cell lines with stable inducibility. These results encouraged attempts

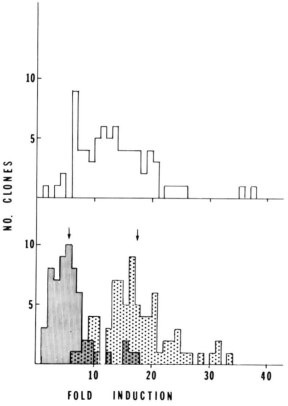

FIG. 1. Clonal variation of tyrosine aminotransferase inducibility. **Top:** Various clones of HTC cells, obtained without selective pressure, were screened for the ability of glucocorticoid (dexamethasone phosphate, 2×10^{-7} M) to induce enzyme. **Bottom:** A pair of clones, relative low- or high-inducers *(arrows)* was recloned and retested. (From Aviv and Thompson, 1972.)

to establish stable variants. By repeated subcloning of low-inducers, cell lines were obtained that were noninducible for TAT (Thompson et al., 1977).

Typical results from glucocorticoid treatment of noninducible cells are shown in Fig. 2. In contrast to wild-type cells (Fig. 2A), levels of TAT activity in three variant clones (Figs. 2B-D) show no increase 8 hr after addition of 10^{-6} M dexamethasone in the culture medium. Exposure to hormone for up to 48 hr did not lead to any detectable TAT induction in these clones. This noninducible phenotype has been stable for several years and is independent of culture conditions. The noninducible variants studied so far have the following properties: (1) basal TAT levels are similar to, or slightly lower than, wild-type cells; (2) TAT is degraded at a normal rate and has normal heat stability and antigenicity; (3) TAT is not induced or is very weakly induced by up to 10^{-4} M dexamethasone; (4) the growth rate of variant cells is indistinguishable from that of wild-type cells and is unaffected by glucocorticoids; (5) variants do not degrade inducing glucocorticoids; and (6) levels of glucocorticoid receptor are normal and capable of nuclear translocation (Thompson et al., 1977). It appears, therefore, that nothing known about the receptors or the enzyme can explain the lack of inducibility in these cells.

The HTC variants present an excellent system for the study of interrelations between multiple steroid-controlled functions in subclones of a single cell. The domain of glucocorticoid-altered polypeptides in HTC cells has been defined and is limited to 20 or so (Ivarie and O'Farrell, 1978). Several of these functions are easily measurable and have been examined (Table 1) (Thompson et al., 1979 a,b). The inducible functions studied were glutamine synthetase (Crook et al., 1978), glucocorticoid induction of TAT, and both steroid-dependent and independent cyclic AMP induction of TAT. Functions inhibited by glucocorticoids were cyclic AMP phosphodiesterase (Manganiello and Vaugham, 1972), uptake of α-aminoisobutyric acid (McDonald and Gelehrter, 1977), and plasminogen activator production and release (Carlson and Gelehrter, 1977). The responses of three TAT-noninducible variants (268E, M714, and 719C) and one

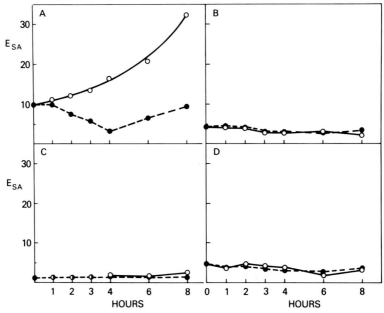

FIG. 2. Time course of induction of tyrosine aminotransferase in wild-type **(A)** and three clones of stably noninducible **(B-D)** HTC cells. Enzyme specific activity (E_{SA}) in cells induced with 10^{-6} M dexamethasone phosphate added at time zero (○); control cells (●). (From Thompson et al., 1977).

TABLE 1. Responses to glucocorticoids by variant HTC cells[a]

Cell	TAT basal	TAT +steroid	TAT +dbcAMP	TAT +steroid +dbcAMP	GS basal	GS +steroid	PA basal	PA +steroid	PDE basal	PDE +steroid	AIB uptake +steroid
Wild type	+	↑	±	↑	+	↑	+	↓	+	↓	↓
Variants											
268E	+	no ↑	no ↑	no ↑	+	←	+	↓↓	+	no ↓	↓↓↓
M714	+	no ↑	no ↑	no ↑	+	←	+	↓↓	+	no ↓	↓↓↓
719C	+	no ↑	no ↑	no ↑	+	←	+	↓	+	no ↓	↓↓↓
921	+	↑	↑	↑	+	←	+	no ↓	n.t.	n.t.	

[a]TAT, tyrosine aminotransferase; GS, glutamine synthetase; PA, plasminogen activator; PDE, phosphodiesterase; AIB, α-aminoisobutyric acid; n.t., not tested.
From Thompson et al., 1979b.

TAT-inducible variant (921) were compared to those of wild-type cells. All three TAT-noninducible variants exhibit the same phenotype. All other responses, except phosphodiesterase suppression, are the same in both wild type and variants. In all three clones the phosphodiesterase response appears to be linked with the TAT response in that they are both lost. Clone 921 is a wild-type variant with respect to TAT but is constitutive with respect to plasminogen activator; in this case there is dissociation in the "opposite direction." In addition, with G. Hager, clones were infected with mouse mammary tumor virus. The normal glucocorticoid inducibility of the virus was not affected in the TAT-phosphodiesterase nonresponding clones.

The simultaneous loss of two apparently unrelated responses (TAT induction and phosphodiesterase suppression) suggests that the alteration, or alterations, responsible for these variants was not limited to the TAT gene. Thus, these experiments show that the interaction of steroid with receptor does not necessarily lead to all subsequent steroid-specific responses. Different responses, or groups of responses, must be controlled through different pathways. Furthermore, because phosphodiesterase suppression and TAT induction were lost whereas other induction and suppression responses were preserved, specific controls in HTC cells are not limited to two types of responses, one suppressive and one inductive, each controlled by a separate pathway.

There are several levels at which subdivision of control mechanisms might occur. The first is at the level of receptor. According to the classic model for steroid hormone action, in each cell there is a single class of receptor molecule for each type of steroid. This conclusion, however, is based on ligand binding experiments that are consistent with a single class of binding site for each type of steroid (King and Mainwaring, 1974, Munck, 1976); such experiments do not necessarily distinguish heterogeneity in receptor molecules. It is possible that heterogeneity exists not in the steroid binding sites of the receptor, but in other sites on the receptor that bind to DNA or that interact with chromatin or other nuclear components. In that case, glucocorticoid receptors might consist of a family of receptor species, all with similar molecular weights and identical affinity for steroid, but differing in the regions of the molecule that are involved in specific interaction and regulation. Immunoglobin production (L. Hood, *this volume*) offers an example of how such heterogeneity can be generated. This possibility can now be directly tested for steroid receptors by studying the organization of their genes through recombinant DNA techniques.

Alternatively, receptors may indeed be what they appear to be, that is homogeneous, and their action may be modulated by specific adapters (acceptors), each corresponding to a specific response, or a subgroup of responses. Furthermore, the sensitivity of specific genes to steroid control could be determined by the availability of regulatory DNA sites for the steroid receptor, or for the steroid-receptor-acceptor complex. This availability presumably would be determined by nuclear proteins or other factors affecting chromatin structure that expose or mask the appropriate regulatory sites. Clearly any or all of these levels of control could exist in the HTC variants discussed here. Phosphodiesterase and TAT, for example, could be controlled by a common subspecies of receptor, or by a common acceptor or masking element. A schematic outline of these possibilities is shown in Fig. 3. The possibility that negative control elements are involved in the determination of tissue specificity in steroid responsiveness is suggested by experiments using somatic cell hybrids.

REGULATION OF TAT, PROLACTIN, AND GROWTH HORMONE IN SOMATIC CELL HYBRIDS

Stable hybrids produced by fusion of somatic cells have been of great value for monoclonal antibody production or for chromosome mapping (V. McKusick, *this volume*). Another classic

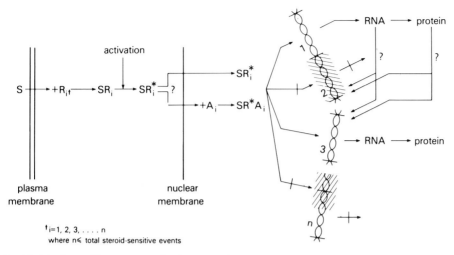

FIG. 3. Model of steroid-receptor and subsequent interactions showing multiple sites at which cellular specificity of response to each class of steroids may occur. *, activated steroid-receptor (SR) complex; A, acceptor. (From Thompson et al., 1979b.)

use of somatic cell hybrids is dominance-recessiveness tests. By fusing cells that display a given function with cells that do not, one can ask the question: Is the function present in the resulting hybrids? The fate of several glucocorticoid-inducible enzymes or peptides has been studied by this test in a variety of somatic cell hybrids. The results of these studies have been summarized and discussed in detail (Gehring and Thompson, 1979). Generally, the rule appears to be that inducible glucocorticoid responses are recessive in somatic cell hybrids; induction is lost in hybrids containing a full set of chromosomes from each parent.

The most widely studied response is TAT induction, which has been examined in several laboratories using a variety of cell types (Table 2). In most cases the enzyme became noninducible. Loss of TAT inducibility has been shown to occur in heterokaryons as early as 24 hr after fusion, when both parental nuclei are still present in the hybrid cells (Thompson and Gelehrter, 1971). This is an important experiment because it suggests that an element from one nucleus or cytoplasm is transactive on the other nucleus or cytoplasm. In other studies, fully selected hybrids with single fused nuclei also have been shown to lack TAT. Similar results have been obtained for alanine aminotransferase, tryptophan oxygenase, growth hormone, and glycerol phosphate dehydrogenase (Table 2). The earlier observation of extinction of growth hormone in hybrids has been confirmed and prolactin loss observed (Thompson et al., 1980). The combination of this cell system and new advances in the molecular biochemistry of the growth hormone and prolactin genes made this an attractive area for further research.

Full-length cDNAs for both rat prolactin and growth hormone have been cloned and sequenced (Seeburg et al., 1977; Gubbins et al., 1980). With a partially purified cDNA to prolactin prepared from estrogen-stimulated rat pituitary, several cloned DNA sequences coding for parts of the prolactin gene, and together covering the entire gene, were obtained from a gene library. In addition, a single clone that appeared to contain the entire growth hormone gene was recovered. By a combination of restriction endonuclease and heteroduplex mapping, both genes were shown to comprise at least five structural and four intervening sequences (Chien and Thompson, 1980). The prolactin gene is much larger, due to larger intervening sequences, and overall is about 10 Kb. The growth hormone gene is about 2.1 Kb; recently, the DNA sequence of an independent isolate from the same library of what appears to be the identical clone has

TABLE 2. Expression of glucocorticoid-inducible proteins in somatic cell hybrids

Inducible protein	Parental cells		Hybrids[b]	Reexpression reported
	Inducible	Noninducible		
Tyrosine aminotransferase (TAT)	HTC[a] (Buffalo rat hepatoma)	BRL-62 (Buffalo rat diploid liver epithelial)	noninducible	no
	FU5 (rat hepatoma)	3T3 (mouse embryo)	noninducible	no
	FU5 (rat)	BRL-1 (Buffalo rat liver epithelial)	noninducible	yes
	HTC (rat)	3T3 (mouse)	noninducible	no
	HTC (rat)	L (mouse transformed fibroblast)	noninducible	no
	FU5AH (rat)	KOP (human fibroblast)	noninducible	yes
	Faza 967 (rat hepatoma)	DON (Chinese hamster)	noninducible	yes
	FU5AH (rat)	Cl. 1D (mouse fibroblast)	noninducible	no
	Faza (rat)	Lc (mouse lymph)	inducible at reduced level	no
	HTCTG-30	YAC (mouse lymphoma) MM (mouse macrophages) RM (rat macrophages) MLy (mouse lymphocytes) RLy (rat lymphocytes) MF (mouse embryo fibroblasts)	(1) rat × mouse mostly non- or borderline inducible. Occasionally inducible. (2) rat × rat often inducible (see text)	no
Glycerol-e-phosphate dehydrogenase	RG6A (rat glial)	3T3-4(E) (mouse embryo fibroblast) LM(TK-) cl. 1D (mouse transformed fibroblast)	noninducible	
Alanine aminotransferase	Faza 967 (rat)	DON (Chinese hamster)	noninducible	yes
Tryptophan oxygenase	FLC (mouse)	LMTK- (mouse transformed fibroblast)	noninducible	no
Growth hormone	GH, 2c, (rat pituitary adenoma)	LM(TK-) cl. 1D	noninducible (no basal)	no

[a] All cell lines containing HTC stem from single line. FU5, FU5-5, FU5AH, Faza 967, and Faza all stem from another independent line, the Reuber H-35 rat hepatoma.
[b] Heterokaryons.
From Gehring and Thompson, 1979.

been obtained (Barta et al., 1982; Page et al., 1981). The partial sequence of this clone appears virtually identical (P. Earl and E. B. Thompson, *unpublished results*). Given these data and molecular reagents, one can hope to begin an exploration of the hybrid extinction of hormone-induced genes at a more detailed level than was previously possible. The cells employed are GH_3, a clone derived from a rat pituitary adenoma that produces prolactin and growth hormone (Tashjian et al., 1968). Figure 4 shows schematically the relevant properties of GH_3 cells and mouse L cells. Thyrotropin-releasing hormone (TRH), acting through cell-surface receptors, and estradiol, acting through cytoplasmic receptors, induce prolactin production in GH_3 cells. L cells lack TRH receptors and, although estrogen and thyroid hormone (T_3) receptors are present, no response of L cells to either estrogen or T_3 is known. Other inducers include glucocorticoids and thyroid hormone, which act through their respective intracellular receptors to induce growth hormone in GH_3 cells. Glucocorticoids also induce increased glutamine synthetase activity in GH_3 as well as L cells.

GH_3 and L cells were hybridized with standard techniques and hybrid clones were examined for prolactin and growth hormone production (Thompson et al., 1980). As shown in Table 3, with the exception of clone GL16, these peptides were not produced in any of the hybrids, even in the presence of the corresponding inducers. Thus, the usual result of extinction was obtained. Assay of the hybrids for receptors of the inducers revealed little or no TRH receptor (Table 4). The level of steroid receptors appeared to be normal or even elevated relative to the parental cells. However, without an appropriate test, it cannot be determined whether these receptors are functional. Induction of glutamine synthetase, a response to glucocorticoids in both parental cell types, can be used as such a test for glucocorticoid receptor. All hybrids examined showed induction of this enzyme activity in response to dexamethasone in degrees comparable with those of the parental cells (Table 5). Therefore, the loss of growth hormone inducibility in the hybrid cells is not due to any gross alteration or loss of functional glucocorticoid receptors. At present, there is no equivalent test of function for estrogen or T_3 receptors.

Using *in vitro* translation as an assay, it was shown that hybrids that do not produce prolactin and growth hormone are lacking functional mRNAs for these peptides (Thompson et al., 1980; Strobl, et al., 1982). Figure 5 shows some of the results for prolactin. The left panel is an electropherogram of the total translation products of poly(A)-containing message from GH_3 cells and from two hybrid clones, GL14 and GL12; the right panel shows the corresponding immunoprecipitated translation products. Estradiol and TRH cause detectable increases in the level of prolactin mRNA in GH_3 cells (data not shown). Hybrid clones GL12 and GL14 exhibit extinction of prolactin and do not contain functional prolactin mRNA. Similar results were

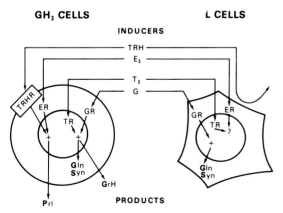

FIG. 4. Hormone responses of GH_3 cells and L cells.

TABLE 3. Extinction of prolactin and growth hormone in new GL hybrids[a]

Cell	Prolactin (μg/ml medium)			Growth hormone (μg/mg protein/4 days)	
	Control	+E$_2$	+TRH	Control	+HC
Parent					
GH3	30 ± 5	58 ± 3	65	0–8	12
LB82		ND[b]		ND	
Hybrids					
Fusion 1					
GL1					
GL4		ND		ND	
GL5					
GL8					
Fusion 2					
GL11					
GL12		ND		ND	
GL14					
GL15					
GL16		ND		4.5[c]	

[a] +E$_2$, 5 × 10^{-9} M 17β estradiol for 7 days; +TRH, 56 × 10^{-9} M thyrolibrin 7 days; +HC, 5 × 10^{-6} M hydrocortisone.
[b] None detectable.
[c] Only after treatment with 10^{-6} M dexamethasone. Otherwise none detectable. Data from Thompson et al., 1980.

TABLE 4. Hormone receptors in GL hybrids

Cells	TRH (fmole/mg)	E$_2$ (fmole/mg)	Dexamethasone		T$_3$ (nuclear sites/ cells × 10^{-3})
			(sites/cell × 10^{-3})	Kd (× 10^8 M)	
Parents					
GH3	266	22	23–32	1.5	6–10
LB82	ND[a]	38	11–87	6.7	2–6
Hybrids					
GL1	ND	25	158	2.9	—
GL2	ND	45	87	8.8	—
GL3	5	62	33	1.1	—
GL4	ND	40	14	1.5	—
GL5	ND	—[b]	204	2.8	—
GL8	ND	15	827	1.6	—
		(sites/cell × 10^{-3})			
GL11	—	106	—	—	—
GL12	—	133	144	—	4.5
GL14	—	265	208	—	2
GL16	—	44	190	—	2.5

[a] ND, none detectable.
[b] Test not performed.
TRH, thytrotropin-releasing hormone, thyrolibrin; E$_2$, 17 β-estradiol.
Steroid receptors reported as fmole/mg were assayed by the cytosol assay; those reported as sites/cell were assayed by a whole cell binding technique.
From Thompson et al., 1980, and Strobl et al., 1981.

TABLE 5. *Glutamine synthetase in GL hybrids*

Cell	Glutamine synthetase (μg product/min/mg cell protein) $\times 10^2$		
	Basal	+ Dexamethasone	Fold
Parents			
GH3	1.0	2.2	2.2
LB82	0.7	3.2	4.6
Hybrids			
GL1	0.64	3.5	5.5
GL2	0.64	1.4	2.2
GL3	0.74	1.3	1.7
GL4 (early)	0.46	2.3	5.0
GL4 (late)	0.83	4.3	5.2
GL5	0.92	6.4	6.9
GL8	0.64	5.3	8.3

From Thompson et al., 1980.

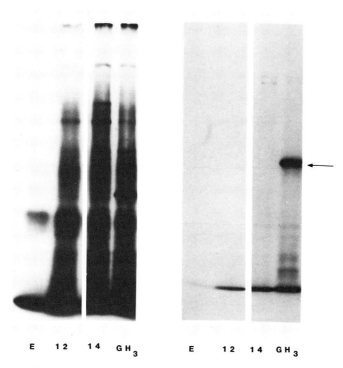

FIG. 5. Cell-free translation of mRNAs from GH_3 cells and $GH_3 \times$ L cell hybrids. Equal amounts of poly(A)-enriched RNAs from GH_3 cells and nonprolactin-producing hybrids GL12 and GL14 were translated in a wheat-germ preparation, with [^{35}S]methionine as the label. The products were electrophoresed on SDS-acrylamide gels and radioautographed. **Left:** Total translation products. **Right:** Products of the same reactions immunoprecipitated with anti-prolactin anti-serum prior to electrophoresis. Lanes L to R: Endogenous translation (E), GL12 hybrid (12), GL14 hybrid (14), GH_3 cells (GH_3). The prominent band in the immunoprecipitated GH_3 lane *(arrow)* is at the location predicted by authentic prolactin (For details, see Thompson et al., 1980).

obtained for growth hormone (Strobl et al., 1982). Furthermore, the presence of nontranslated growth hormone mRNA has been ruled out by DNA:RNA hybridization analysis (Strobl et al., 1982).

TABLE 6. *Reexpression of prolactin by hybrid segregant GL C5[a]*

	Prolactin	
	μg/mg cell protein/24 hr	% control
Control	60 ± 4	100
10 nM E_2	76 ± 8	127
1 μM TM	65 ± 10	109
0.5 μM Dex	38 ± 6	63
0.5 μM TRH	95 ± 10	159

[a] E_2, 17-β-estradiol; TM, tamoxifen; Dex, dexamethasone; TRH, thyrolibrin.

Reexpression occurs in certain of the hybrid clones. Table 6 shows data from an exceptional segregant that has reexpressed prolactin. The amounts of the peptide produced by this clone are very high, about 20 times higher than the GH_3 parent, and are nearly constitutive. A detailed study of the arrangement of the prolactin gene in this hybrid may reveal why the gene is "turned on" at such high constitutive levels.

The use of molecular probes for the gene for growth hormone has shown that the gene exists in hybrids that exhibit extinction. Thus, in several hybrids that do not produce growth hormone, there is no loss or gross rearrangement of the growth hormone gene contributed by the GH_3 parent (Strobl et al., 1982). Before such probes became available, the conclusion that the gene was present in the hybrid was based perforce on either karyotype analysis with its inherent uncertainties or subsequent reexpression in cell segregants. Similar experiments using molecular probes for the prolactin gene are in progress.

Clearly, we have reached an intriguing point in the analysis of these hybrids, which already allows a number of interesting conclusions to be drawn. We can confirm that such hybrids shut off growth hormone expression, a glucocorticoid- and T_3-induced peptide. They also shut off prolactin expression, which is induced by estrogen or TRH. This is the first example studied of the negative control over a sex steroid-induced function. The probable reason for lack of induction by TRH is the extinction of TRH receptors, but the hybrids still display receptors for the other three hormones. Their failure to induce the appropriate peptides therefore remains the nub of the system. Data show that the genes for growth hormone from both inducible and noninducible parent cells are present, grossly intact, in the hybrids. The prolactin genes have not yet been analyzed in this way. The hybrids contain no translatable mRNA for either peptide and little or no growth hormone mRNA in any form capable of annealing with its cDNA. These data, taken together, suggest that existence of a rather specific, negative regulating system acting to prevent the transcription of these genes. The further analysis of these hybrids may be a useful way to study certain cellular regulatory systems. To this end, *in vitro* transcription systems are being developed. In addition, microinjection techniques (see F. Ruddle, *this volume*) are being used in two kinds of experiments: first, injection of large amounts of the relevant DNA to see whether the negative control in the L cell or in the hybrid can be overcome; and second, injection of extracts of the dominant L cells to the wild-type rat cells to determine whether the repressive trans-effect can be reproduced.

ACKNOWLEDGMENTS

We would like to acknowledge the gift of cDNA probes for Southern blots: for growth hormone from John D. Baxter and for prolactin from Rick A. Maurer. We would also like to thank Dr. Stephanie J. Bird for her editorial assistance.

REFERENCES

Aviv, D., and Thompson, E. B. (1972): Variation in tyrosine aminotransferase induction in HTC cell clones. *Science*, 177:1201–1203.
Bailly, A., Sallas, N., and Milgrom, E. (1977): A low molecular weight inhibitor of steroid receptor activation. *J. Biol. Chem.*, 252:858–863.
Barta, A., Richards, R. I., Baxter, J., and Shine, J. (1982): Primary structure and evolution of the rat growth hormone gene. *Proc. Natl. Acad. J. USA*, 78:4867–4871.
Baxter, J. D., and Rousseau, G. G., eds. (1979): *Glucocorticoid Hormone Action*, Springer-Verlag, Berlin.
Borthwick, N. M., and Bell, P. A. (1975): Early glucocorticoid-dependent stimulation of RNA polymerase B in rat thymus cells. *FEBS Lett.*, 60:396–399.
Cake, M. H., Goidl, J. A., Parchman, L. G., and Litwack, G. (1976): Involvement of a low molecular weight component(s) in the mechanism of action of the glucocorticoid receptor. *Biochem. Biophys. Res. Commun.*, 71:45–52.
Carlson, S. A., and Gelehrter, T. D. (1977): Hormonal regulation of membrane phenotype. *J. Supramol. Struct.*, 6:325–331.
Chien, Y., and Thompson, E. B. (1980): Genomic organization of rat prolactin and growth hormone genes. *Proc. Natl. Acad. Sci. USA*, 77:4583–4587.
Courvalin, J.-C., Bouton, M.-M., Baulieu, E.-E., Nuret, P., and Chambon, P. (1976): Effect of estradiol on rat uterus DNA-dependent RNA polymerases. Studies on solubilized enzymes. *J. Biol. Chem.*, 251:4843–4849.
Crook, R. B., Louie, M., Deuel, T. F., and Tomkins, G. M. (1978): Regulation of glutamine synthetase by dexamethasone in hepatoma tissue culture cells. *J. Biol. Chem.*, 253:6125–6131.
Gehring, U., and Thompson, E. B. (1979): Somatic cell fusion in the study of glucocorticoid action. In: *Glucocorticoid Hormone Action*, edited by J. D. Baxter and G. G. Rousseau, pp. 399–421. Springer-Verlag, Berlin.
Gillis, S., Crabtree, G. R., and Smith, K. A. (1979a): Glucocorticoid-induced inhibition of T-cell growth factor production I. The effect on mitogen-induced lymphocyte proliferation. *J. Immunol.*, 123:1624–1631.
Gillis, S., Crabtree, G. R., and Smith, K. A. (1979b): Glucocorticoid-induced inhibition of T-cell growth factor production II. The effect on the *in vitro* generation of cytolytic T-cells. *J. Immunol.*, 123:1632–1638.
Gubbins, E. J., Maurer, R. A., Lagrimini, M., Erwin, C. R., and Donelson, J. E. (1980): Structure of the rat prolactin gene. *J. Biol. Chem.*, 255:8655–8662.
Harmon, J. M., and Thompson, E. B. (1982): Glutamine synthetase induction by glucocorticoids in the glucocorticoid-sensitive human leukemic cell line CEM-C7. *J. Cell. Physiol.*, 110:155–160.
Harmon, J. M., and Thompson, E. B. (1981): Isolation and characterization of dexamethasone-resistant mutants from human lymphoid cell line CEM-C7. *Mol. Cell. Biol.*, 1:512–521.
Higgins, S. J., Baxter, J. D., and Rousseau, G. G. (1979): Nuclear binding of glucocorticoid receptors. In: *Glucocorticoid Hormone Action*, edited by J. D. Baxter and G. G. Rousseau, pp. 136–160. Springer-Verlag, Berlin.
Ivarie, R. D., and O'Farrell, P. H. (1978): The glucocorticoid domain: Steroid-mediated changes in the rate of synthesis of rat hepatoma proteins. *Cell*, 13:41–55.
Johnson, L. K., Lan, N. C., and Baxter, J. D. (1979): Stimulation and inhibition of cellular functions by glucocorticoids. Correlations with rapid influences on chromatin structure. *J. Biol. Chem.*, 254:7785–7794.
King, R. J. B., and Mainwaring, W. I. P. (1974): *Steroid-cell Interactions*, University Park Press, Baltimore.
Manganiello, V., and Vaughan, M. (1972): An effect of dexamethasone on adenosine 3', 5'-monophosphate content and adenosine 3'-5'-phosphodiesterase activity of cultured hepatoma cells. *J. Clin. Invest.*, 51:2763–2767.
McDonald, R. A., and Gelehrter, T. D. (1977): Glucocorticoid inhibition of amino acid transport in rat hepatoma cells. *Biochem. Biophys. Res. Commun.*, 78:1304–1310.
Mohla, S., Clem-Jackson, N., and Hunter, J. B. (1981): Estrogen receptors and estrogen-induced gene expression in the rat mammary glands and uteri during pregnancy and lactation: Changes in progesterone receptor and RNA polymerase activity. *J. Steroid Biochem.*, 14:501–508.
Mullinix, K. P., Myers, M. B., Christmann, J. L., Deeley, R. G., Gordon, J. I., and Goldberger, R. F. (1979): Specific transcription in chicken liver chromatin by endogenous RNA polymerase II. *J. Biol. Chem.*, 254:9860–9866.
Munck, A. (1976): General aspects of steroid hormone-receptor interactions. In: *Receptors and Mechanisms of Action of Steroid Hormones, Part I*, edited by J. R. Pasqualini, pp. 1–40. Marcel Dekker, New York.
Munck, A., and Foley, R. (1980): Activated and non-activated glucocorticoid-receptor complexes in rat thymus cells: Kinetics of formation and relation to steroid structure. *J. Steroid Biochem.*, 12:225–230.
Nielsen, G. J., Sando, J. J., and Pratt, W. B. (1977a): Evidence that dephosphorylation inactivates glucocorticoid receptors. *Proc. Natl. Acad. Sci. USA*, 74:1398–1402.
Nielsen, C. J., Sando, J. J., Vogel, W. M., and Pratt, W. B. (1977b): Glucocorticoid receptor inactivation under cell-free conditions. *J. Biol. Chem.*, 252:7568–7578.
Page, G. S., Smith, S., and Goodman, H. M. (1981): DNA sequence of the rat growth hormone gene: Location of the 5' terminus of the growth hormone mRNA and identification of an internal transposon-like element. *Nucleic Acids Res.*, 9:2087–2104.
Parchman, L. G., and Litwack, G. (1977): Resolution of activated and unactivated forms of the glucocorticoid receptor from rat liver. *Arch. Biochem. Biophys.*, 183:374–382.

Pfahl, M., Kelleker, R. J., Jr., and Bourgeois, S. (1978): General features of steroid resistance in mouse lymphoma cell lines. *Mol. Cell. Endocrinol.*, 10:193–207.

Rosen, J. M., Richards, D. A., Guyetto, W., and Matusik, R. J. (1980): Steroid-hormone modulation of prolactin action in the rat mammary gland. In: *Gene Regulation by Steroid Hormones*, edited by A. K. Roy and J. H. Clark, pp. 58–77. Springer-Verlag, New York.

Rousseau, G. G., and Baxter, J. D. (1979): Glucocorticoid receptors. In: *Glucocorticoid Hormone Action*, edited by J. D. Baxter and G. G. Rousseau, pp. 50–77. Springer-Verlag, Berlin.

Sakaue, Y., and Thompson, E. B. (1977): Characterization of two forms of glucocorticoid hormone-receptor complex separated by DEAE-cellulose column chromatography. *Biochem. Biophys. Res. Commun.*, 77:533–541.

Sando, J. J., Nielsen, C. J., and Pratt, W. B. (1977): Reactivation of thymocyte glucocorticoid receptors in a cell-free system. *J. Biol. Chem.*, 252:7579–7582.

Sato, B., Noma, K., Nishizawa, Y., Nakao, K., Matsumoto, K., and Yamamura, Y. (1980): Mechanism of activation of steroid receptors: Involvement of low molecular weight inhibitor in activation of androgen, glucocorticoid, and estrogen receptor systems. *Endocrinology*, 106:1142–1148.

Schmidt, T. J., Harmon, J. M., and Thompson, E. B. (1980): Activation-labile glucocorticoid-receptor complexes of a steroid resistant variant of CEM-C7 human lymphoid cells. *Nature*, 286:507–510.

Schwartz, R. J., Tsai, M.-J., Tsai, S. Y., and O'Malley, B. W. (1975): Effect of estrogen on gene expression in the chick oviduct v. changes in the number of RNA polymerase binding and initiation sites in chromatin. *J. Biol. Chem.*, 250:5175–5188.

Seeburg, P. H., Shine, J., Martial, J. A., Baxter, J. D., and Goodman, H. M. (1977): Nucleotide sequence and amplification in bacteria of structural gene for rat growth hormone. *Nature*, 270:486–494.

Strobl, J. S., Dannies, P. S., and Thompson, E. B. (1982): Somatic cell hybridization of growth hormone (GH) producing rat pituitary cells and mouse fibroblasts results in extinction of GH expression via a defect in GH RNA production. *J. Biol. Chem. (in press).*

Swaneck, G. E., Nordstrom, J. L., Kreuzaler, F., Tsai, M., and O'Malley, B. W. (1979): Effect of estrogen on gene expression in chicken oviduct: Evidence for transcriptional control of ovalbumin gene. *Proc. Natl. Acad. Sci. USA*, 76:1049–1053.

Tashjian, A. H., Jr., Yosumura, Y., Levine, L., Sats, G. H., and Parker, M. L. (1968): Establishment of cloned strains of rat pituitary tumor cells that secrete growth hormone. *Endocrinology*, 82:342–352.

Thompson, E. B. (1979): Glucocorticoids and lysosomes. In: *Mechanism of Action of Glucocorticoid Hormones*, edited by J. D. Baxter and G. G. Rousseau, pp. 575–581. Springer-Verlag, Berlin.

Thompson, E. B., Aviv, D., and Lippman, M. E. (1977): Variants of HTC cells with low tyrosine aminotransferase inducibility and apparently normal glucocorticoid receptors. *Endocrinology*, 100:406–419.

Thompson, E. B., Dannies, P. S., Buckler, C. E., and Tashjian, A. H., Jr. (1980): Hormonal control of tyrosine aminotransferase, prolactin, and growth hormone induction in somatic cell hybrids. *J. Steroid Biochem.*, 12:193–210.

Thompson, E. B., and Gelehrter, T. D. (1971): Expression of tyrosine aminotransferase activity in somatic-cell heterokaryons: Evidence for negative control of enzyme expression. *Proc. Natl. Acad. Sci. USA*, 68:2589–2593.

Thompson, E. B., Granner, D. K., Gelehrter, T. D., Erickson, J., and Hager, G. L. (1979a): Unlinked control of multiple glucocorticoid-induced processes in HTC cells. *Mol. Cell. Endocrinol.*, 15:135–150.

Thompson, E. B., Granner, D. K., Gelehrter, T. D., and Hager, G. L. (1979b): Unlinked control of multiple glucocorticoid-sensitive processes in spontaneous cell variants. In: *Hormones and Cell Culture*, Vol. 6, edited by R. Ross and G. Sato, pp. 339–360. Cold Spring Harbor Laboratory, New York.

Thompson, E. B., Tomkins, G. M., and Curran, J. C. (1966): Induction of tyrosine α-ketoglutarate transaminase by steroid hormones in a newly established tissue culture cell line. *Proc. Natl. Acad. Sci. USA*, 56:296–303.

Tsai, M.-J., Schwartz, R. J., Tsai, S. Y., and O'Malley, B. W. (1975): Effects of estrogen on gene expression in the chick oviduct IV: Initiation of RNA synthesis on DNA and chromatin. *J. Biol. Chem.*, 250:5165–5174.

Tsai, M.-J., Towle, H. C., Harris, S. E., and O'Malley, B. W. (1976): Effect of estrogen on gene expression in the chick oviduct. Comparative aspects of RNA chain initiation in chromatin using homologous versus Escherichia coli RNA polymerase. *J. Biol. Chem.*, 251:1960–1968.

Westphal, H. M., and Beato, M. (1980): The activated glucocorticoid receptor of rat liver. Purification and physical characterization. *Eur. J. Biochem.*, 106:395–403.

Wrange, O., Carlstedt-Duke, J., and Gustafsson, J. A. (1979): Purification of the glucocorticoid receptor from rat liver cytosol. *J. Biol. Chem.*, 254:9284–9290.

Section X

Genetic Expression in the Nervous System

For many reasons we might expect brain cells to contain many kinds of proteins and conjugated proteins. Neurons display great heterogeneity of structural types—dendritic, somatic, axonal, and synaptic—many of which are highly specific in different brain regions, as well as proteins involved in the synthesis of components of the fibrous cytoskeleton, on which neuronal morphology depends (see F. Gros et al.). Moreover, the number of different kinds of receptor molecules, usually assumed to be proteinaceous in composition, must be very large indeed to provide specific and high-affinity binding of the many classes of neuroactive and other information-processing ligands. Many of these proteins are present in amounts too small to detect using conventional neurochemical methods.

However, the sheer number of different proteins found in mammalian brain cells by W. E. Hahn and colleagues is staggering. Even more surprising are the large numbers of these proteins that are brain-specific. Many years of neurochemical research have revealed but a relative handful of brain-specific proteins. Now the work by Hahn and his colleagues indicates that they are present in many thousands! Because of the central role of receptors in transferring information in the central nervous system, Hahn et al. suggest that oligonucleotide probes would greatly facilitate identification of receptor-encoding genes and thus open up a vast and very rewarding field of research in applying molecular genetic concepts and technology to problems central to neuroscience.

At the other end of the scale of concentration of receptor proteins, the acetylcholine receptor (AChR) ionophore found in the nerve-muscle junction is present in high concentration in the electroplax of certain fish. This receptor has been studied in detail by D. J. Anderson and associates. The characterization of the four types of subunits in the pentameric complex that contains both the ACh binding site and the cation conductance channel was accomplished by using antisera specific for each subunit type. This led to the demonstration that each subunit is generated by a specific mRNA. Experience gained in these experiments from the application of recombinant DNA techniques may prove helpful in the study of receptors present in concentrations orders of magnitude less than that of the ACh receptor in the electroplax.

Another exciting facet of genetic expression in the central nervous system is the apparent parallel occurrence of more than 30 peptides, both in the endocrine system and as neuronal paracrine activators. As hormones serve endocrine functions, neuropeptides are thought to serve paracrine function, i.e., they are released into synaptic and non-synaptic interneuronal space and, presumably by combining with specific receptors, influence the function of local cells (see J. F. Habener and colleagues). This is a pivotal concept in the more general putative principle that some neuroactive substances function in the paracrine mode and mediate gene expression of target neurons, particularly those in the cerebral cortex, by binding to specific receptors and entraining different pathways to the genome (see F. Schmitt, *this volume*). Such a view places additional emphasis on the wide variety of receptors needed and the fact that the entire surface (dendrites, somata, axons, and axon terminals) may be studied with many types of receptors. The mechanisms of synthesis and release of paracrine agents, as well as the chemical and evolutionary relationship between neurohormones, endocrine hormones, and other neuroactive substances need to be examined more fully.

Evidence concerning these questions is provided in this section by employing as models the somatostatin (STN) family of small peptides derived from large precursors that function both in an endocrine and a paracrine mode. Habener and associates investigated the organization and expression of STN genes, primarily at the transcriptional level, through the use of recombinant DNA technology to illuminate the chemical and biological relationship between members of the STN family as endocrine hormones and as paracrine neuropeptides. S. Reichlin deals with posttranslational processing of the STNs, their fast transport in the axon, and release from axon terminals. The study of neuropeptides, and STN in particular, is a valuable model system for integrating the studies of molecular genetics and brain function.

Molecular Genetic Neuroscience, edited by
F. O. Schmitt, S. J. Bird, and F. E. Bloom.
Raven Press, New York © 1982.

Overview of the Molecular Genetics of Mouse Brain

William E. Hahn, Jeffrey Van Ness, and Nirupa Chaudhari

ABSTRACT

An overview of the complexity of the mammalian (mouse) brain in terms of molecular genetics is presented. From nucleic acid hybridization studies, it is estimated that approximately 75,000 different polyadenylated—poly(A)—mRNAs are present in the brain. Many of these mRNAs are encoded discontinuously in the genome (i.e., intervening sequences, or introns, are present between coding sequences). About 70,000 different nonpolyadenylated—poly(A)$^-$—mRNAs, separate from poly(A)mRNAs, are also present in the brain. About half the different poly(A)mRNAs are shared with other organs, whereas the complex population of poly(A)$^-$mRNAs appears to be specific to brain. The majority of poly(A)mRNAs, characteristic of the adult brain, are present in the brain at birth. In contrast, poly(A)$^-$mRNAs are absent in the brain at birth. Genes specifying these mRNAs are activated gradually during the course of several weeks of postnatal development.

The mammalian brain is the most complex biological structure to have evolved on this planet. Not only is its cellular population highly complex, but individual neuronal and glial cells are morphologically complex. Although it is apparent that many thousands of proteins are required in the embryonic and postnatal development and function of the brain, little is known about brain-specific proteins and their respective genes. This chapter presents an overview of genetic expression in the mammalian (mouse) brain. Emphasis is placed on analysis of the sequence complexity of messenger RNA populations from the brain in an attempt to gain an estimate of the number of structural genes that are expressed in the brain. Also, comparisons with other organs are made in an attempt to delineate mRNAs specific to brain tissue.

For purposes of this discussion, those genes that code for mRNAs are defined as structural genes. Most of the different species of mRNA appear to be encoded by sequences in DNA that are present once or in a few copies per haploid set of chromosomes (egg or sperm). Estimates of the number of different functional, structural genes range mostly between 30,000 and 100,000 (Ohno, 1971; Ohta and Kimura, 1971; Krone and Wolf, 1978). The average mammalian genome contains sufficient sequence complexity to code for $\sim 10^6$ different mRNA species of average size assuming a static, linear arrangement of the nucleotide sequences. However, we know that far fewer structural genes are present, because a large portion of the DNA codes for RNA, which is not conserved as mRNA molecules (Lai et al., 1978; Soreq et al., 1979; MacDonald et al., 1980; Schafer et al., 1980).

Pertinent to developing perspective on the potential coding capacity of the mammalian genome is the recombination of sequences in the DNA known to occur within some families of genes. For example, the gene families (about 500–600 different sequences) that specify immunoglobulins have the potential of coding for 10^7 different immunoglobulins owing to re-combinatorial events (see L. Hood, *this volume*). It is possible that recombinatorial events are widespread throughout the DNA, such that during the generation of various cell lineages, exons (mRNA coding regions) are mixed in different combinations to yield various families of proteins. This could well be the case when selection of functional domains within proteins has occurred during the course of evolution, such that a given primary sequence is present in several related proteins. Thus, the number of different proteins that are produced during the mammalian life cycle is difficult to estimate. But whether there are 50,000 or 1 million genes, conceptually the problems pertaining to the molecular genetics of brain development and function remain basically the same.

It is proposed here that a large portion of the total available genetic information is used in the development and functioning of the brain. Perhaps well over half of the structural genes that have come down the evolutionary turnpike code for proteins specific for the mammalian brain. These genes probably specify: (1) the general developmental organization of the cell population, (2) the production of a vast array of regulatory compounds and receptors, (3) synapse formation and refinement, and (4) RNA or proteins specific for information storage and retrieval processes.

METHODOLOGY

An estimate of the complexity of the mRNA molecules (i.e., gene products) present in whole brain of the mouse has been obtained from measurements based on nucleic acid hybridization. Two hybridization techniques have been used to estimate the sequence complexity of RNA, namely the saturation and kinetic methods. In the saturation method, tracer amounts of labeled single-copy fragments of genomic DNA are hybridized in mixtures containing high concentrations of RNA. If the sequence complexity of the tracer is known, then the percentage of the tracer hybridized at saturation with RNA is a direct estimate of the sequence complexity of the RNA. Complexity is conveniently expressed in units of 1,000 nucleotides (1 kilobase).

The kinetic method requires cDNA (copy DNA) obtained by reverse transcription of the RNA population in question. Reverse transcription can be efficient to the extent of obtaining cDNA of sequence complexity equal to that of the template RNA (Van Ness and Hahn, 1980). The rate at which cDNA hybridizes when driven by the respective template RNA population is inversely proportional to the sequence complexity of the RNA. Thus, estimates of the sequence complexity of an unknown RNA population can be made with reference to the hybridization kinetics of a known complexity standard. The complexity standards are usually simple systems, such as a single mRNA and its respective cDNA, or an RNA virus of defined complexity and its respective cDNA.

In the case of a diverse RNA population obtained from mammalian cells or organs, some mRNA species are present in relatively high abundance and others are relatively rare. The presence of copy frequency classes of mRNA is evident in curves depicting the kinetics of hybridization. The sequence complexity of each class can be estimated from such multicomponent curves. In most instances mRNA taken from mammalian cells and organs contains a low abundance class that makes up 30 to 50% of the mass but constitutes the vast majority of the sequence complexity (Williams and Penman, 1975; Hastie and Bishop, 1976; Young et al., 1976).

Both kinetic and saturation hybridization methods harbor a number of potential pitfalls and shortcomings, especially with regard to high sequence complexity. Both over- or underestimates of the complexity of an RNA population may be obtained by saturation hybridization of genomic DNA fragments. Overestimates result if the actual sequence complexity of the genome has itself been overestimated, or if there is repetition of the DNA sequences that code for RNA, which are assumed to be single copy. The sequence complexity of various mammalian genomes has been estimated by reassociation kinetics of denatured DNA fragments 300 to 400 nucleotides in length. Such studies have shown that about 70% of the mass of the mouse genomic DNA appears to be single copy relative to bacterial DNA standards (i.e., sequences in the DNA present once per haploid genome). If the fraction of the genome that appears to be single copy by reassociation kinetics is in fact composed mostly of such sequences, then the complexity of the mouse genome is approximately 3.2×10^9 nucleotides. However, it is probable that many sequences that code for mRNA may be repeated more than once. If this is the case for many or most coding regions, then saturation hybridization would yield a false high value of the complexity of the mRNA. Previous renaturation studies of single-copy DNA recovered from poly(A)mRNA-DNA hybrids indicated that most of the coding sequences were single copy relative to total single-copy DNA and a bacterial DNA standard (Bantle and Hahn, 1976; Kaplan et al., 1978). However, recent work indicates that many mRNA coding sequences may be somewhat repetitious (2 to 5 copies per unit genome). These results do not necessarily indicate that the repeated coding sequences are identical, but they are enough alike to form stable duplexes under stringent conditions.

Underestimates may result if true saturation is not achieved, owing to the presence of rare mRNA species in an mRNA mixture prepared from a heterogeneous population of cells. Also, the method of quantitation of the hybrids is very important, since in most cases only a small fraction of the DNA is hybridized. Therefore, the method of analysis must have a low "noise" level.

Because of uncertainties pertaining to the actual sequence complexity of the mammalian genome and the copy frequency of coding sequences, estimates of mRNA complexity obtained by the kinetic method are probably more reliable. In the kinetic method, the cDNA must be

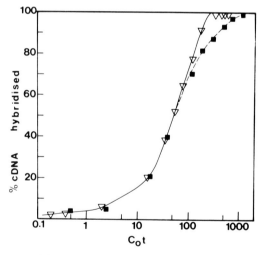

FIG. 1. Effect of continuous mixing on hybridization of cDNA. The kinetics of liver poly(A)mRNA-driven hybridization of respective cDNA under conditions of continuous mixing *(triangle)*, or in the absence of mixing *(square)*, is shown. The cDNA used was the fraction representing the complex class poly(A)mRNAs from mouse liver. Reaction mixtures were contained in sealed capillary tubes. Continuous mixing eliminates the skew that often occurs after about 70% completion of hybridization. Most of cDNA representing the complex class poly(A)mRNA hybridizes, when continuously mixed, as single pseudo first-order component. The rate constant for hybridization can now be more accurately determined and compared to simple complexity standards used in estimating sequence complexity of the mRNA. $C_o t$, concentration of nucleotides times time.

fully hybridizable and the method of assay highly quantitative. Assay of hybridization is conveniently obtained using single-strand specific S_1 nuclease and binding of hybrids to DEAE-cellulose filters (Maxwell et al., 1978). For the kinetic method to be reliable, accurate measurement of the rate of cDNA hybridization relative to the complexity standard must be obtained. A parameter that influences complexity estimates is diffusion of molecules. RNA-driven hybridization involving systems of high sequence complexity becomes increasingly diffusion dependent as the reaction proceeds under conditions standardly applied (i.e., reaction mixtures sealed in capillary tubes, or a single vessel sampled periodically). This phenomenon is not observed in the case of nucleic acid systems of low sequence complexity under routine conditions. However, when continuous mixing of the reactant molecules does not occur during the course of the hybridization, the rate of hybridization of the complex component is retarded and may not be directly comparable to the vastly more simple complexity standards. The complex component of the cDNA is observed to hybridize more rapidly for the last one-third of the reaction if the system is mixed continuously during hybridization (Fig. 1). Also, essentially 100% of the input cDNA is hybridized (resistant to S_1 nuclease) under these conditions. This further decreases the uncertainty of the kinetic method.

Poly(A)mRNAs in the Mouse Brain: Complexity, Biogenesis, Specificity to Brain, and Presence Relative to Development

Messenger RNAs are defined as those RNAs that are encoded by the few or single-copy fraction of the mammalian genome and are components of polyribosomes.[1] Messenger RNAs are released from the polysomal structure as an RNA-protein complex by chelation of magnesium. Messenger RNA molecules may contain a 3' poly(A) tract and a 5' cap structure. In purified form, they serve as templates for polypeptide synthesis in cell-free translation systems. There is no direct evidence that all such RNAs are in fact translated in the cells from which they have been obtained, but the following facts point to the probable and logical conclusion that these molecules function as mRNAs: (1) this RNA population is of high sequence complexity, (2) they are a component of polyribosomes, and (3) they have features mentioned above.

The complexity of poly(A)mRNA from whole brain is given in Table 1. Since the average size of poly(A) mRNA from whole brain is approximately 1,500 nucleotides, the 110,000-Kb complexity given in the table indicates the presence of about 73,000 different, average-sized poly(A)mRNAs. Estimates of sequence complexity obtained by the kinetic and saturation methods are in good agreement, but this could be somewhat fortuitous considering the different potential problems in obtaining reliable measurements with either of these techniques. Some previous estimates of the complexity of brain poly(A) mRNA based on the kinetic method are three- to fourfold less than reported here (Hastie and Bishop, 1976). This discrepancy may be due to the use of cDNA, in the earlier studies, which was not fully hybridizable with respective template mRNA (70 to 75%). Probes approaching 100% hybridizability are required to obtain a reliable rate constant for the last, or complex, component in mRNA-cDNA hybridization systems.

Poly(A)mRNAs are mostly derived from much larger heterogeneous nuclear RNA (hnRNA) molecules of greater sequence complexity. Portions of almost all of the different large poly-

---

[1] Brain polyribosomes, essentially free of contamination by nuclear RNA, can be prepared (Van Ness et al., 1979). Complexity estimates of mRNA via the kinetic method are relatively insensitive to minor contamination of polysomes by nuclear ribonucleoproteins since little of the cDNA mass will represent such contaminant molecules.

TABLE 1. *Sequence complexity of polyadenylated mRNA and heterogeneous nuclear RNA from whole mouse brain*

RNA type	Estimate of sequence complexity in Kb[a]	
	Saturation method[b]	cDNA method[c]
Poly(A) mRNA		
Adult	120,000	110,000
Newborn	110,000	110,000[d]
17-day fetus	—	85,000[d]
Poly(A) hnRNA		
Adult	400,000	350,000
Newborn	400,000	—

[a] 1 Kb = 1,000 nucleotides.
[b] In the saturation method, labeled (~5 × 10⁶ cpm/μg) single-copy genomic DNA fragments are hybridized in the presence of high concentration (3–10 μg/μl) of RNA until saturation is achieved. A complexity value of 3.2×10^9 nucleotides for mouse single-copy DNA was applied in computation of the complexity estimates.
[c] cDNA was obtained by reverse transcription of poly(A) mRNA from adult brain. Complexity was estimated by reference to the EMC virus RNA-driven hybridization of respective cDNA.
[d] Complexity estimate based on maximum percentage hybridization of the cDNA fraction representing the complex class of adult brain poly(A) mRNA.

adenylated hnRNA molecules in the mouse brain are conserved qualitatively in the production of mRNA (Hahn et al., 1978). It is now known for several gene products that large polyadenylated hnRNA molecules undergo a sequence of cutting and splicing to yield functional mRNA molecules, which are usually much smaller (2 to 10 times) than the initial precursor. (For more discussion on this point, see J. Darnell and M. Wilson, and P. Sharp and M. Wilson, *this volume*). As implied by the terms "cutting and splicing," the coding or mRNA sequences contained in large hnRNA molecules are discontinuous. Since large hnRNA molecules are formed from the continuous transcription of the respective region in the DNA, many of the genes specifying poly(A)mRNA in mouse brain are organized such that the coding sequences or exons are interrupted frequently by the presence of intervening sequences, or introns, i.e., they are split genes, a condition well established for a number of specific genes (Breathnach et al., 1977; Jeffreys and Flavell, 1977; Tilghman et al., 1978; Nunberg et al., 1980).

The generality of code sequence discontinuity has been examined in genes expressed via poly(A)mRNA in the mouse brain. The experimental strategy is depicted in Fig. 2A. This method (Maxwell et al., 1980) showed that a large number of genes specifying poly(A)mRNA in the mouse brain are split. Full-length cDNAs (total cDNA as well as the fractions of the cDNA representing the low abundance, complex fraction of the poly(A)mRNA) either annealed with large fragments of genomic DNA (Fig. 2B) or hybridized with large nuclear RNA molecules (Fig. 2C), were cut extensively by S_1 nuclease. These results indicate the frequent presence of introns in genomic DNA, and show that the introns are transcribed, being represented in the large nuclear precursor molecules. Since complementary sequence continuity exists between mRNA and respective cDNA, cDNA hybridized to mRNA was not cut by S_1 nuclease (Fig. 2D).

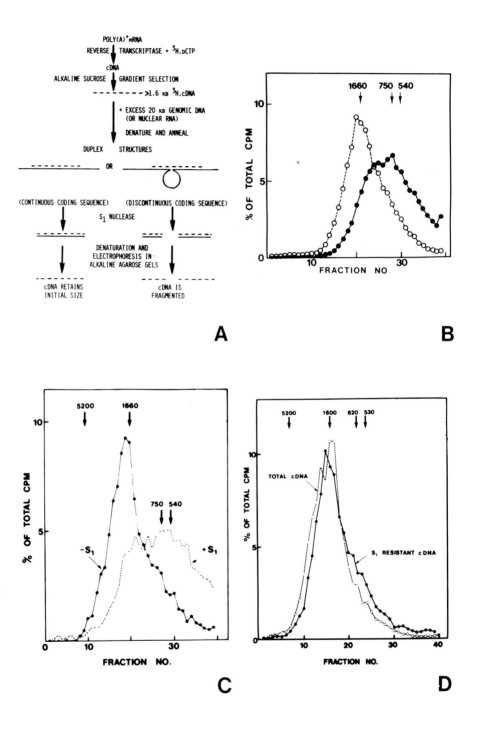

With cDNA probes representing complex class poly(A)mRNA from whole mouse brain prepared by fractionating total cDNA (see Maxwell et al., 1980, for example), it has been shown that a large number of the different poly(A)mRNAs from brain are also present in the polyribosomes from cells of other complex organs such as the liver and kidney. About half the sequence complexity contained in the poly(A)mRNA of mouse brain is shared with these two complex mammalian organs. The remaining 40 to 50% of the total sequence complexity appears, at least relative to liver and kidney, to be unique to the brain. Curiously, about 30 to 40% of the brain specific mRNA sequences (sequences not present in liver or kidney polysomes) can be found in the nuclear RNA of these organs. Thus, these genes are transcribed, but the respective transcripts are not processed and transported to the cytoplasm (posttranscriptional control). An explanation for this lack of tight transcriptional control of some genes remains to be discerned. The majority of the genes specifying poly(A)mRNAs specific to the brain appear not to be transcribed in other organs (J. Van Ness and W. E. Hahn, *unpublished data*). This is in apparent contrast with certain other eukaryotic systems, such as the sea urchin, in which regulation of mRNA populations in different organs appears to be, at least qualitatively, dependent on posttranscriptional mechanisms (Kleene and Humphreys, 1977; Wold et al., 1978; Davidson and Britten, 1979).

Most of the brain poly(A)mRNA species present in the young adult are present in the brain at birth (Table 1). The cDNA probes that were used do not allow determination of whether all of the poly(A)mRNAs of the adult are present in the newborn, but the resolution is probably dependable within about 2,000 different messenger sequences. Consistent with this finding is the observation that complexities of poly(A)hnRNA, precursors of poly(A)mRNA, are also similar in the adult and newborn (Table 1). The brain of the 16- to 17-day mouse fetus (approximately 4 days prior to birth) lacks about 15,000 poly(A)mRNA species that are present in the neonatal and young adult brain. Initially it was speculated that these genes, expressed late in fetal development, might specify poly(A)mRNAs, which are mostly unique to brain. However, about the same proportion of these mRNAs are specific to the brain as in the case for total poly(A) mRNA from adult brain. Genes specifying these messengers appear to be regulated transcriptionally since these sequences are also not found in the nuclear RNA of the 16- to 17-day fetus.

Genes Expressed Via Poly(A)⁻mRNA: Uniqueness in Brain and Postnatal Activation

In most cultured mammalian cells and other organs, virtually all of the sequence complexity contained in the polysomal RNA, or mRNA obtained from polysomes, is represented in polyadenylated RNA molecules, despite the fact that 10 to 30% of the mRNA is nonpolyadenylated

---

FIG. 2. A: Experimental strategy for determining the general occurrence of introns in structural genes. Copy DNA hybridizes only with mRNA coding regions in the DNA. Introns form single-strand loops. Copy DNA is sensitive to cutting by S₁ nuclease at sites opposite intron loops. Hence the presence of introns is indicated if the length of annealed cDNA is decreased after exposure to S₁ nuclease. **B:** Copy DNA representing mostly the complex, low-abundance mRNAs was annealed with large (> 20 Kb) fragments of genomic DNA. Size of the annealed cDNA was reduced from an average length of about 1,660 nucleotides untreated hybridized cDNA *(open circles)* to about 750 nucleotides after treatment with S₁ nuclease *(closed circles)*, thus indicating lack of colinearity between genomic DNA and mRNA (as represented by cDNA). *Arrows* indicate position in the electrophoretic gels of polynucleotides of defined size (in nucleotides) used as length markers. **C:** Experiment here same as depicted in **B** except cDNA was hybridized to high-molecular-weight hnRNA. Reduction in the length of cDNA following treatment with S₁ nuclease indicates intron sequences in DNA are represented in hnRNA. **D:** Copy DNA was hybridized to respective poly(A)mRNA. Following treatment with S₁ nuclease the cDNA modal size remained at about 1,600 nucleotides. This is expected since the cDNA is colinear with respective mRNA, and hence S₁ nuclease-sensitive sites are not present.

(Milcarek et al., 1974; J. Van Ness and W. E. Hahn, *unpublished data*). Even though it has been generally assumed that almost all messengers are polyadenylated (Brawerman, 1976; Darnell, 1979), poly(A)⁻mRNAs that specify histones, actins, and a few other proteins are known to exist (Adesnik and Darnell, 1972; Nemer et al., 1974; Gedamu et al., 1977). Furthermore, poly(A)⁻ and poly(A) forms of apparently the same message species exist (Hunter and Garrels, 1977). Therefore, it was surprising to find that a complex population of poly(A)⁻ mRNAs is present in the mouse brain (Van Ness et al., 1979). Chikaraishi (1979) has made a similar observation on polysomal RNA from the rat brain. A complex population of poly(A)⁻mRNAs appears to be unique to the brain. Poly(A)⁻mRNA can be purified from polysomal RNA to the extent that over 98% of the rRNA can be removed by benzoylated cellulose chromatography. Copy DNA complementary to poly(A)⁻mRNA was prepared using random oligonucleotides as primers (Dudley et al., 1978; Van Ness and Hahn, 1980), and after removal of abundant cDNA sequences representing about 50% of the RNA mass, a cDNA probe was obtained which represents the bulk of the sequence complexity of the poly(A)⁻ mRNA. This probe was shown to have little complementarity with the poly(A)mRNA (see Van Ness et al., 1979). The lack of sequence homology between most of the poly(A)⁻ and poly(A)mRNA species is expected since the sequence complexity of total polysomal RNA is about twice that of either poly(A) or poly(A)⁻mRNA (Tables 1 and 2).

Thus, the sequence complexity of total mRNA from whole mouse brain is equal to about 140,000 different mRNAs of an average length of 1,500 nucleotides. Since more than one functional polypeptide might be specified from some of these mRNAs, as is the case in the production of polypeptide hormones (see E. Herbert et al., and D. Krieger, *this volume*), more than 140,000 different polypeptides are probably present in the brain. This sequence complexity is probably a reflection of an enormously complicated cell population, rather than the result of greater mRNA complexity in individual neural cells than in nonneural cells. However, there is some evidence that in cell lines of neural origin, a larger portion of the genome may be transcribed than in nonneural cells (see review by Kaplan and Finch, 1982).

Copy DNA representing mostly the complex, low copy frequency class of brain poly(A)⁻ mRNAs has been used to probe polysomal RNAs from cultured cells and complex organs such

TABLE 2. *Complexity of nonpolyadenylated mRNA and total nuclear RNA of adult and neonatal mouse brain*[a]

	Estimate of sequence complexity (Kb)	
RNA type	Saturation method	cDNA method
Poly(A)⁻mRNA		
Adult	110,000	100,000
Newborn	—	10,000[b]
Total polysomal RNA		
Adult	230,000	—
Newborn	135,000	—
Total nuclear RNA		
Adult	580,000	—
Newborn	430,000	—

[a]Most of the poly(A)⁻mRNAs characteristic of the adult brain are absent in the brain of the newborn. This is reflected in differences between the complexities of polysomal and nuclear RNAs from adult and newborn.

[b]Complexity estimate based on maximum percentage hybridization of the cDNA fraction representing the complex class of adult brain poly(A) mRNA.

as liver and kidney. These measurements show the bulk of these messengers appear to be specific to the brain. Presence of the poly(A)⁻ mRNAs appears to be mostly regulated transcriptionally, since few of these sequences are present in the nuclear RNAs of organs other than brain (Table 3).

Little is known about the biogenic history of the poly(A)⁻ mRNAs. These mRNAs are not processed from polyadenylated nuclear precursors since only poly(A)⁻ nuclear RNA hybridized respective cDNA with the expected kinetics (J. Van Ness and W. E. Hahn, *unpublished data*). The ratio of sequence complexity of nonpolyadenylated nuclear RNA (equal to the difference between the complexity of nuclear RNA and poly(A) hnRNA) to poly(A)⁻ mRNA from brain is roughly 1.5 to 1. In contrast, this ratio is about 3.5 to 1 for poly(A)hnRNA/poly(A)mRNA. These ratios suggest important differences in the biogenic histories of poly(A)⁻ and poly(A)mRNAs.

Genes specifying poly(A)⁻mRNAs in the mouse brain appear to be mostly activated after birth. The majority of the poly(A)⁻mRNAs characteristic of the young adult brain are absent both in the polysomes and nuclear RNA in the newborn (Table 2). The difference in complexity of total nuclear RNA of adult and newborn also reflects the fact most genes specifying poly(A)⁻mRNAs are inactive in the newborn brain, or at least respective transcripts do not accumulate to levels measurable in hybridization experiments. The time course of the appearance of these mRNAs during postnatal development is being examined and preliminary results indicate a slow progressive appearance of these mRNAs extending beyond 20 days after birth. Since the mouse reaches young adulthood within about 50 days, the general time course of expression of these genes should be easy to determine. Some of these genes are apparently activated soon before or following birth, since about 20% of the sequence complexity of the poly(A)⁻mRNA population can be detected in the nuclear RNA from brains of newborn mice. Because the genes specifying poly(A)⁻mRNAs are largely activated after birth and appear to code for messengers that are putatively specific to the brain, these genes probably specify proteins that function in a manner unique to the properties of the brain. Many of these proteins may be necessary for synaptogenesis and "programming" of membrane surfaces in a manner required for information storage and retrieval processes.

TABLE 3. *Brain specificity of poly(A)⁻ mRNA*

RNA type	Percentage hybridization of cDNA[a] complementary to brain poly(A)⁻mRNA
Brain	
Polysomal	100
Nuclear	100
Liver	
Polysomal	5
Nuclear	12
Kidney	
Polysomal	6
Nuclear	10

[a]The cDNA probe primarily represented complex class, infrequent mRNAs of brain. However, about 5 to 10% of the mass of the cDNA was complementary to moderately abundant mRNAs, which collectively make up little of total sequence complexity. Hybridization of this fraction of the cDNA may constitute most of the observed hybridization with liver and kidney RNAs. Thus, these percentages do not indicate the presence of a moderately complex population of poly(A)⁻mRNAs in these organs.

CONCLUDING REMARKS AND PROSPECTS

These data indicate the presence of about 140,000 different mRNAs in the whole brain of the young adult mouse. Owing to recombinatorial possibilities of sequence in the genome mentioned earlier, and considering that more than one functional polypeptide may be encoded in an mRNA molecule, the number of both mRNAs and proteins could be greater than 140,000.

Although more must be learned about the general anatomical distribution of this complex mRNA population, recombinant DNA technology provides a means for studying individual genes. For example, it should be possible to begin to study genes or gene families that may be expressed only in specific regions of the brain and that specify receptor proteins. Receptors are one mechanism of recognition and transferral of information in the nervous system. Provided receptor proteins can be purified in sufficient quantity for ascertaining a partial amino acid sequence using new equipment functional in the picomolar range, it should be possible to obtain amino acid sequence data needed to make oligonucleotide probes sufficient to identify respective genes in genomic libraries.

Application of genetic engineering technology should provide a basis for generating large quantities of brain-specific protein using recombinant organisms. Most brain proteins are scarce on a whole brain basis, and many are likely to be components of membranes, making their isolation in intact form difficult or impossible in many instances. However, respective mRNAs are soluble and can be used to produce cDNA. Copy DNA "genes" in turn can be expressed in plasmid-bacterial systems to produce large amounts of respective brain proteins, at least in terms of primary structure. Proteins made by recombinant systems may be useful in generating specific monoclonal antibodies against various sites on the protein, which can in turn be used as probes to study the distribution and orientation of specific proteins in the brain. Cloned cDNAs could also be used as probes in *in situ* hybridization experiments to localize specific mRNAs within brain tissue sections. These kinds of approaches may lead to insight, at the macromolecular level, as to how the vast connective network, vital to the functioning of the mature brain, is developed.

Are there specific proteins that delineate specific neural connections? It is likely that there is a large number of proteins involved in the cell-cell communication processes in the brain, for which this organ is specialized. It remains to be seen whether events occur in the genome during the course of differentiation of the nervous system, such that the scrambling of a variety of different exons (mRNA coding domains) results in an enormous array of related proteins required for development of the various neural pathways and fine tuning of the nervous system. These latter developmental processes may occur for years after birth or throughout much of the life span of higher mammals.

Various model systems should be examined in attempts to link molecular genetics with behavior. Genetically programmed behaviors such as web building by spiders and nest building by birds might be worthwhile model systems for genetic approaches to behavior capabilities. Imprint behaviors may also be useful models in which to explore the relationships between environmental input, genetic response, and development of a specific behavioral mode (Oshima et al., 1969, Hahn, 1976; Davison, 1977). Since imprinted behavioral modes have obvious survival advantages, these are old forms of behavior for which perhaps it will be the easiest to elaborate the underlying genetics.

ACKNOWLEDGMENT

This research was supported by grants from NIH.

REFERENCES

Adesnik, M., and Darnell, J. E. (1972): Biogenesis and characterization of histone mRNA in HeLa cells. *J. Mol. Biol.*, 67:397–406.

Bantle, J. A., and Hahn, W. E. (1976): Complexity and characterization of polyadenylated RNA in the mouse brain. *Cell*, 8:139–150.
Brawerman, G. (1976): Characteristics and significance of the polyadenylate sequence in mammalian messenger RNA. *Prog. Nucleic Acid Res. Mol. Biol.*, 17:148.
Breathnach, R., Mandel, J. L., and Chambon, P. (1977): Ovalbumin gene is split in chicken DNA. *Nature*, 270:314–319.
Chikaraishi, D. M. (1979): Complexity of cytoplasmic polyadenylated and nonpolyadenylated rat brain ribonucleic acids. *Biochemistry*, 18:3249–3256.
Darnell, J. E., Jr. (1979): Transcription units for mRNA production in eukaryotic cells and their DNA viruses. *Prog. Nucleic Acid Res. Mol. Biol.*, 22:327–353.
Davidson, E. H., and Britten, R. J. (1979): Regulation of gene expression: Possible role of repetitive sequences. *Science*, 204:1052–1059.
Davison, A. N., ed. (1977): *Biochemical Correlates of Brain Structure and Function*. Academic Press, New York.
Dudley, J. P., Butel, J. S., Socher, S. H., and Rosen, J. M. (1978): Detection of mouse mammary tumour virus RNA in BALB/c tumour cell lines of non-viral etiologies. *J. Virol.*, 28:743–752.
Gedamu, L., Iatrou, K., and Dixon, G. H. (1977): Identification and isolation of protamine mRNP particles from rainbow trout testis. *Biochemistry*, 16:1383–1390.
Hahn, W. E. (1976): Electroencephalography of the olfactory bulb in relation to prespawn homing. *Experientia*, 32:1095–1097.
Hahn, W. E., Van Ness, J., and Maxwell, I. H. (1978): Complex population of mRNA sequences in large polyadenylated nuclear RNA molecules. *Proc. Natl. Acad. Sci. USA*, 75:5544–5547.
Hastie, N. D., and Bishop, J. O. (1976): The expression of three abundance classes of mRNA in mouse tissues. *Cell*, 9:761–774.
Hunter, T., and Garrels, J. I. (1977): Characterization of the mRNAs for α, β and γ actin. *Cell*, 12:767–781.
Jeffreys, A. J., and Flavell, R. A. (1977): The rabbit β-globin gene contains a large insert in the coding sequence. *Cell*, 12:1097–1108.
Kaplan, B. B., and Finch, C. E. (1982): The sequence complexity of brain ribonucleic acids. In: *Molecular Approaches to Neurobiology*, edited by I. R. Brown. Academic Press, New York. *(in press)*.
Kaplan, B. B., Schachter, B. S., Osterburg, H. H., De Vellis, J. S., and Finch, C. E. (1978): Sequence complexity of polyadenylated RNA obtained from rat brain regions and cultured rat cells of neural origin. *Biochemistry*, 17:5516–5524.
Kleene, K. C., and Humphreys, T. (1977): Similarity of hnRNA sequences in blastula and pluteus stage sea urchin embryos. *Cell*, 12:143–155.
Krone, W., and Wolf, U. (1978): Chromosomes and protein variation. In: *The Biochemical Genetics of Man*, edited by D. J. H. Brock and O. Mayo, pp. 93–154. Academic Press, New York.
Lai, E. C., Woo, S. L. C., Dugaiczyk, A., Catterall, J. F., and O'Malley, B. W. (1978): The ovalbumin gene: Structural sequences in native chicken DNA are not contiguous. *Proc. Natl. Acad. Sci. USA*, 5:2205–2209.
MacDonald, R. J., Crerar, M. W., Swain, W. F., Pictet, R. L., Thomas, G., and Rutter, W. J. (1980): Structure of a family of rat amylase genes. *Nature*, 287:117–122.
Maxwell, I. H., Maxwell, F., and Hahn, W. E. (1980): General occurrence and transcription of intervening sequences in mouse genes expressed via polyadenylated mRNA. *Nucleic Acids Res.*, 8:5875–5894.
Maxwell, I. H., Van Ness, J., and Hahn, W. E. (1978): Assay of DNA-RNA hybrids of S_1 nuclease digestion and adsorption to DEAE-cellulose filters. *Nuceic Acids Res.*, 5:2033–2038.
Milcarek, C., Price, R., and Penman, S. (1974): The metabolism of a poly(A) minus mRNA fraction in Hela cells. *Cell*, 3:1–10.
Nemer, M., Graham, M., and Dubroff, L. M. (1974): Coexistence of nonhistone mRNA species lacking and containing poly(A) in sea urchin embryos. *J. Mol. Biol.*, 89:435–454.
Nunberg, J. H., Kaufman, R. J., Chang, A. C. Y., Cohen, S. N., and Schimke, R. T. (1980): Structure and genomic organization of the mouse dihydrofolate reductase gene. *Cell*, 19:355–364.
Ohno, S. (1971): Simplicity of mammalian regulatory systems inferred by single gene determinations of sex phenotypes. *Nature*, 234:134–137.
Ohta, T., and Kimura, M. (1971): Functional organization of genetic material as a product of molecular evolution. *Nature*, 233:118–119.
Oshima, K., Hahn, W. E., and Gorbman, A. (1969): Electroencephalographic olfactory response in adult salmon to waters traversed in the homing migration. *J. Fish. Res. Board Can.*, 26:2123–2133.
Schafer, M. P., Boyd, C. D., Tolstoshev, P., and Crystal, R. G. (1980): Structural organization of a 17 kb segment of the $\alpha 2$ collagen gene and evaluation by R loop mapping. *Nucleic Acids Res.*, 8:2241–2253.
Soreq, H., Harpold, M. M., Evans, R., Darnell, J. E., and Bancroft, F. C. (1979): Rat growth hormone gene: Intervening sequences separate the mRNA regions. *Nucleic Acids Res.*, 6:2471–2482.
Tilghman, S. M., Curtis, P. J., Tiemeier, D. C., Leder, P., and Weissmann, C. (1978): The intervening sequence of a mouse β-globin gene is transcribed within the 15S β-globin mRNA precursor. *Proc. Natl. Acad. Sci. USA*, 75:1309–1313.
Van Ness, J., and Hahn, W. E. (1980): Sequence complexity of cDNA transcribed from a diverse mRNA population. *Nucleic Acids Res.*, 8:4259–4269.

Van Ness, J., Maxwell, I. H., and Hahn, W. E. (1979): Complex population of nonpolyadenylated messenger RNA in mouse brain. *Cell*, 18:1341–1349.
Williams, J. G., and Penman, S. (1975): The messenger RNA sequences in growing and resting mouse fibroblasts. *Cell*, 6:197–206.
Wold, B. J., Klein, W. H., Hough-Evans, B. R., Britten, R. J., and Davidson, E. H. (1978): Sea urchin embryo mRNA expressed in the nuclear RNA of adult tissues. *Cell*, 14:941–950.
Young, B. D., Birnie, G. D., and Paul, J. (1976): Complexity and specificity of polysomal poly(A)$^+$RNA in mouse tissues. *Biochemistry*, 15:2823–2829

Molecular Genetic Neuroscience, edited by
F. O. Schmitt, S. J. Bird, and F. E. Bloom.
Raven Press, New York © 1982.

The Regulation of Gene Expression During Terminal Neurogenesis

François Gros, Bernard Croizat, Marie-Madeleine Portier, Francis Berthelot, and Armando Felsani

ABSTRACT

Neuroblastoma cells constitute a convenient model for studying gene expression during neuronal differentiation and for screening pharmacological drugs acting on the central nervous system. This chapter reports on: (1) polysomal poly(A) RNA complexity studies during neuroblastoma differentiation; and (2) the effect of 1-methylcyclohexane carboxylic acid (CCA), a new inducer of neuroblastoma morphogenesis, on the cytoskeleton components and on some neuronal markers.

Cultured cell lines, derived from mouse neuroblastoma tumors of the C-1300 type, can aid in the study of neuronal differentiation. Many mouse neuroblastoma lines or related somatic cell hybrids have been established in Marshall Nirenberg's group as well as other laboratories over the last 10 to 12 years. These lines, when properly induced, can exhibit, *in vitro*, certain of the characteristic phenotypes of mature neurons. Although they harbor some abnormalities—e.g., they are usually aneuploid, lack nerve growth factor (NGF) receptors, and are often unable to form synapses—they have proved useful in the study of neurite formation, the appearance of neurotransmitter-forming enzymes, and the acquisition of membrane excitability.

A variety of approaches used in the study of various aspects of neuroblastoma differentiation involve both conventional and more specific technologies. The first type of approach involves the attempt to find causal or temporal relationships among changes in mRNA sequences, the rate of synthesis of some classes of proteins, and/or some surface antigens, and the physiological or morphological expression of the neurogenic program. More specific techniques include cloning neuron-specific or semi-specific markers such as tyrosine hydroxylase (TH), β-isotubulin, or the 14-3-2 protein, the γ-subunit of neuron-specific enolase.

MESSENGER RNA STUDIES

Prior to the development of specific cDNA or genomic probes, which will lead to a greater understanding of the organization and control of determinants coding for neuron-specific markers, techniques such as poly(A$^+$) mRNA-cDNA hybridization kinetics were employed to investigate gene-controlling mechanisms during neurogenesis. This technique affords a comparison of mRNA sequences present in undifferentiated and in differentiated cells in culture using either wild-type neuroblastoma cell lines or differentiation-deficient variants.

When cells of the N1E-115 cell line (an adrenergic clone derived from mouse neuroblastoma C-1300) are grown in suspension with 7.5% fetal calf serum, they resemble rounded immature neuroblasts. By a variety of means, these cells can be induced to extend neurites and to acquire enzymatic properties characteristic of mature neurons. Induction techniques include withdrawing serum and transferring cells to a culture dish where they can attach, or adding several substances to a normal serum-containing medium. Among the most commonly used inducers are dimethyl sulfoxide (DMSO), dibutyryl cyclic AMP (Bt_2-cAMP), or 1-methylcyclohexane carboxylic acid (CCA), a novel inducer discussed below.

POLYSOMAL POLY(A) RNA COMPLEXITY IN NEUROBLASTOMA CELLS

In a series of studies, mRNA from cells grown in suspension ("S cells") and used in logarithmic growth phase were compared to mRNA from process-forming, differentiated monolayer cells ("P cells") 3 days after serum deprivation. Cells pretreated with emetin and washed were lysed in NP_{40} plus heparin. The lysate was centrifuged, and a 12,000-g supernatant was processed for sucrose gradient fractionation, which separates polysomes from monosomes. Polysomal RNA was extracted by CsCl density gradient centrifugation, and purified on an oligo-dT cellulose column. After elution the poly(A) RNA was ethanol-precipitated in the presence of 0.15 M LiCl. Its concentration and size distribution could be monitored either by prelabeling with an RNA precursor or by hybridization with [^{3}H]poly U; its average nucleotide length can be calculated from sucrose gradient distribution. This length corresponds to about 1,750 nucleotides, the poly(A) tract representing about 6% of the total length. The amount of polysomal RNA remains stable after differentiation, whereas the poly(A)-containing RNA content is reduced by about 25%. The figures of 520,000 and 390,000 mRNA molecules for undifferentiated and differentiated cells, respectively, are in close agreement with the values later determined by Grouse and colleagues (1980) for another neuroblastoma cell line and with data reported for other mammalian cells (Ryffel and McCarthy, 1975).

Poly(A)-containing polysomal RNA from both cell stages (designated as "S" and "P") was transcribed into cDNA with AMV reverse transcriptase. The size of the resultant cDNA was between 350 to 400 nucleotides (sedimentation constant determined on alkaline sucrose gradients was 5 to 5.5 S), which corresponds to about 20% of the mean length of template RNA.

Under conditions of vast RNA excess, the kinetic constant of hybridization of poly(A) RNA population to its cognate cDNA is inversely proportional to the RNA sequence complexity (Bishop et al., 1974). Therefore, knowing the hybridization kinetics of the RNA population, it is possible to determine its sequence complexity by comparison with the hybridization kinetics of a standard RNA of known complexity. Here globin mRNA was used as a reference. Figure 1 illustrates the kinetics of the two homologous hybridization systems. It will be seen that: (1) a Rot value (i.e., concentration of RNA in moles of nucleotide per liter multiplied by the time in seconds) of 100 moles $\times$ 1^{-1} sec yields complete hybridization; (2) the two curves are superimposable; and (3) in both cases the plateaus reach the same high level (93%). Finally, it is clear that the hybridization curves are heterogeneous, the kinetics extending over several Rot decades and the reaction being a composite of several reactions. Each of these reactions corresponds to an abundance class inside of which all RNA species are present at the same concentration.

The numerical analysis in Table 1 indicates that a good fit is obtained using an RNA frequency distribution in three classes representing 21, 41, and 38% of the poly(A) RNA population with complexities of 5.45×10^4, 1.6×10^6, and 1.09×10^7 nucleotides for the abundant, intermediate, and rare classes, respectively.

The total number of different sequences for both "S" and "P" messenger populations is about 7,200, assuming that the mRNA number average size is 1,750 nucleotides. This indicates

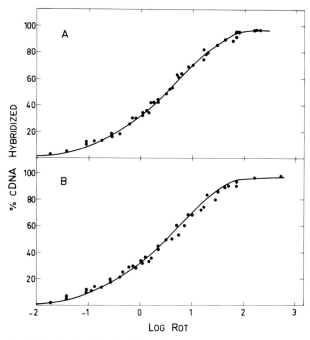

FIG. 1. Homologous hybridization kinetics. Undifferentiated (A) and differentiated (B) cDNA were hybridized with their respective polysomal poly(A)-containing RNA template. For Rot values up to 1 mole × 1⁻¹ sec, hybridization reaction contained RNA at the concentration of 60 μg/ml; for Rot values between 1 and 70 mole × 1⁻¹ sec, RNA at the concentration of 300 μg/ml, and for greater Rot values, the RNA concentration was 1 mg/ml. The points represent the results obtained from five independent experiments. Rot is the product of RNA concentration (mole of nucleotide 1⁻¹) and time (sec). The *solid lines* through the data points represent the same theoretical first-order curve calculated using the parameters determined by the method of Jacquet et al. (1978). (From Felsani et al., 1978.)

clearly that poly(A) RNA populations from both developmental stages display an essentially identical complexity. It should be noted that this does not exclude some minor differences that would be beyond the precision of this approach, and this does not necessarily imply that both populations include the same species. To explore this aspect further, cross-hybridizations have been performed (Fig. 2).

In Fig. 2A, P-cDNA was cross-hybridized to S-poly(A) RNA (dots) and, for comparison, to P-poly(A) RNA (solid line). The two curves reach the same plateau, indicating that undifferentiated cells (S) contain, to a best approximation, all the sequences that are found in the differentiated (P) cells (assuming that reverse transcription of a poly(A)-containing RNA population is random). There is, however, a certain demarcation in the kinetics indicating that transition from the S to the P stage involves a decrease in the concentration of some messengers belonging to the intermediate class.

As seen in Fig. 2B, the reverse is not true: using S-cDNA, poly(A) RNA from P cells does not drive the reaction to a plateau level that is as high as that obtained when S poly(A) is used. Hence, P cells lack approximately 15% of the sequences present in S cells. Because the heterologous kinetics starts to diverge from the homologous at a low Rot value (and continue to do so until Rot equals 10, at which point the divergence remains constant), the missing sequences in differentiated RNA should be relatively abundant in undifferentiated RNA.

To strengthen the relevance of the conclusion regarding the loss of some RNA sequences during the S to P transition, the thermal stability of both heterologous and homologous cDNA-

TABLE 1. Numerical analysis of homologous hybridization kinetics: NIE 115[a]

Class of abundance	Fraction of hybridizable cDNA (α)	Rot 1/2 (mole l^{-1} sec)		Base sequence		Number of copies/cell	
		Observed	Corrected	Complexity in nucleotides (NT)	Number of different mRNA species	Undifferentiated	Differentiated
Abundant	0.21	0.145	0.030	5.45×10^4	31	3,522	2,622
Intermediate	0.41	2.14	0.88	1.60×10^6	914	233	174
Rare	0.38	15.75	5.99	1.09×10^7	6,229	32	24
Total	1.00			1.26×10^7	7,174		

[a]The analysis of the homologous hybridization curve has been performed as described in Fig. 1. (α) represents the fraction of hybridizable cDNA reacting in each abundance class. Rot 1/2 values are expressed as mole l^{-1} sec. Corrected Rot 1/2 is given by Rot 1/2 $\times$ α. The effect of this operation is to correct the observed Rot 1/2 to the value it would have if the RNA of considered abundance class were pure and comprised 100% of the total reaction. The sequence complexity of each abundance class, NT, is estimated as follows: NT = 1,200 (corrected Rot 1/2) 6.6×10^{-4}, where 6.6×10^{-4} corresponds to the Rot 1/2 of the hybridization reaction between α + β globin mRNA (sequence complexity = 1,200 nucleotides) and its cDNA. The number of messenger RNA species is calculated considering an average size of 1,750 nucleotides. The number of copies per cell is given by: [poly(A)-containing RNA/cell (g)]α 6×10^{23}/330 NT. From Felsani et al., 1978.

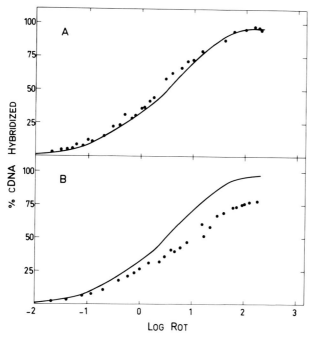

FIG. 2. Kinetics of heterologous hybridization. **(A):** The heterologous hybridization between differentiated cDNA and undifferentiated polysomal poly(A)-containing RNA and for comparison the theoretical homologous curve from Fig. 1 *(solid line)*. **B:** The reciprocal cross-hybridization between undifferentiated cDNA and differentiated polysomal poly(A)-containing RNA and for comparison the theoretical homologous curve from Fig. 1 *(solid line)*. The data are expressed as in Fig. 1. (From Felsani et al., 1978.)

RNA hybrids was determined. Both the homologous and heterologous hybrids exhibited a t_m around 89° C, implying that the observed difference is not related to matching between heterologous molecules.

There is little difference between heterologous and homologous kinetics (Fig. 2) for the intermediate class, and mRNAs from differentiated cells lack some mRNA sequences present in undifferentiated cells. The simplest interpretation of this is that the S to P transition is paralleled by the disappearance of a small number of messengers belonging to the intermediate class. Given the fact that this class represents 41% of the hybridizable cDNA and only 13% of the total complexity, the number of missing sequences should be around 150. This is a minimum estimate since the P cell population is probably contaminated with S cells.

At this stage, at least two possible interpretations could be considered. Either the loss of sequence was related to the expression of the neurogenic program that is induced during the S to P transition, or these changes might reflect the disappearance of mRNA species involved in the mitotic state of the cells since the comparison was between logarithmically grown and postmitotic cells. To answer this question partly, a neuroblastoma variant, N1A 103, was employed. This variant stops dividing when transferred to a serum-deprived medium on culture dishes and can readily attach to the substratum. It partly expresses its biochemical program but fails to develop neurites and excitability (A cells).

Data shown in Table 2 give the computed hybridization parameters for homologous hybridizations. The total complexity of mRNA is similar in both S and A cells (close to 6,300 average-size sequences). Yet, the frequency distribution is not the same; the largest discrepancy is found in the "abundant" class (44 and 29% of hybridizable cDNA in S and A cells, respectively).

TABLE 2. Numerical analysis of homologous hybridization kinetics: N1A 103[a]

Class of abundance	Fraction of hybridizable cDNA (α)	Rot 1/2 (mole 1^{-1} sec)		Base sequence			
		Observed	Corrected	Complexity in nucleotides (NT)	Different mRNA species	Number of copies/cell	mRNA molecules/cell
S cells							
Abundant	0.44	0.089	0.037	6.73×10^4	38	1,486	56,468
Intermediate	0.35	2.11	0.712	1.29×10^6	737	62	45,694
Rare	0.21	26.2	5.29	9.64×10^6	5,510	5	27,550
Total	1.00				6,285		129,712
A cells							
Abundant	0.29	0.165	0.089	1.62×10^5	92	407	37,444
Intermediate	0.40	1.57	0.597	1.09×10^6	621	84	52,164
Rare	0.31	18.14	5.34	9.71×10^6	5,551	7.2	39,967
Total	1.00				6,264		129,575

[a]Numerical analysis was performed as described in Table 1.
From Berthelot el al., 1980.

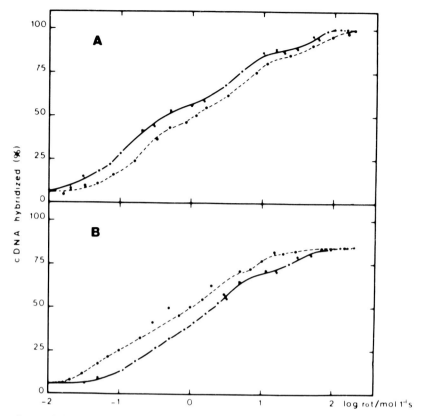

FIG. 3. Heterologous hybridization kinetics [N1A 103]. **A:** cDNA from S cells was hybridized with poly(A) RNA from A cells (●—●). **B:** cDNA from A cells was hybridized with poly(A) RNA from S cells (⋆—⋆). In both cases, the homologous kinetics corresponding to the cDNA was plotted for comparison *(solid line)*. (From Berthelot et al., 1980.)

Figure 3 illustrates the heterologous kinetics. It can be seen that the poly(A)-containing RNAs are able to hybridize to heterologous cDNAs with the same efficiency as that which would correspond to homologous cDNA. Both RNA populations thus comprise the same sequences as evidenced by the same plateau levels. Homologous and heterologous kinetics, however, are different. Hence, the RNA sequences are not at the same concentration in both populations. More specifically, there is a reduced concentration of abundant class messengers during the S to A transition.

It can be concluded that the S to A transition is not paralleled by a disappearance of mRNA species. Since no sequences are lost in the absence of induced morphogenesis, it is tempting to conclude that the disappearance of 150 sequences observed during S to P transition in the N1E-115 cells is related to terminal morphogenesis and not to changes in the state of replication or to cell adhesion, two phenomena that occur normally in the neuroblastoma variant, N1A 103. The modulation in the number of species from the abundant class during S to A transition could either reflect different replication states, metabolic alterations, or partial expression of the biochemical neurogenic program.

Gradual restriction of mRNA complexity, as is observed in the neuroblastoma transition, has already been reported in other developmental systems as differentiation proceeds (Levy and Dixon, 1977). In most cases, new tissue-specific mRNAs also appear in terminally differentiated cells, as a result of either metabolic stabilization or the switching on of new genes, giving rise to new transcripts. In this study the appearance of new sequences was not detected, presumably because the cDNA probe used was not sufficiently sensitive or because *in vitro* differentiation accompanying serum withdrawal was not complete. Using a different neuroblastoma cell line, the cholinergic strain NS20Y, and a better differentiation paradigm, Grouse and coworkers (1980) have confirmed the finding that the S to P transition (as elicited by Bt_2-cAMP) involves the disappearance of old mRNA sequences. In addition, utilizing recycled cDNA probes that are specific either for the S or the P cell mRNAs only (such probes being purified by hydroxyapatite fractionation, which removes hybrids of cDNA and heterologous poly(A) RNA), Grouse and coworkers (1980) were able to demonstrate the appearance of new poly(A) mRNA sequences as well as the disappearance of some old ones. The data are compatible with the conclusion that differentiation results in the disappearance of three abundant poly(A) RNA sequences and about 250 less abundant messengers, and with the appearance of three abundant messengers and 320 less abundant poly(A) mRNAs (Fig. 4).

Taken together, these data make it attractive to speculate that both the loss and gain of mRNA species are related causally to the phenotypic events accompanying neuronal differentiation. This hypothesis is being studied using cloned cDNA probes specific for individual differentiation-specific mRNAs.

MESSENGER RNA COMPLEXITY IN BRAIN TISSUE

It may be useful to comment on the problem of mRNA complexity in brain tissue. With exactly the same technique used in the investigation of mouse neuroblastoma cells, the complexity and frequency class distribution of poly(A) mRNA in whole mouse brain and cortex have been investigated. A summary of the data is given in Table 3. In whole brain tissue there are approximately 12,200 species of the same average size as in neuroblastoma, the best fit being obtained using RNA frequency distribution in three classes. In the cortex, a very similar complexity could be calculated. Heterologous hybridization confirmed this point but showed

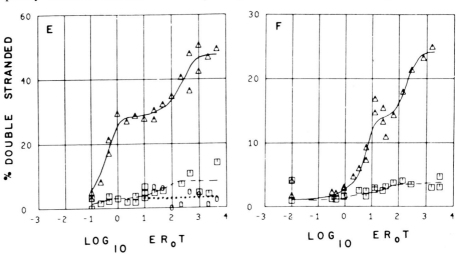

FIG. 4. Complementary DNA reactions with NS20Y poly(A) mRNA. E, a complementary DNA probe specific to sequences present only in the P-cell poly(A) mRNA, was reacted with P-cell poly(A) mRNA (△—△), with S-cell poly(A) mRNA (□---□), or with yeast RNA (○····○). F, a complementary DNA probe enriched in sequences present only in the S-cell poly(A) mRNA, was reacted with S-cell poly(A) mRNA (△—△) or with P-cell poly(A) mRNA (□---□). (From Grouse et al., 1980.)

TABLE 3. Numerical analysis of homologous kinetics: mouse brain[a]

Class of abundance	Fraction of hybridizable cDNA (α)	Rot 1/2		Base sequence	
		Observed	Corrected	Complexity in nucleotides (NT)	Number of different mRNA species
Abundant	0.26	0.31	0.08	1.47×10^5	84
Intermediate	0.49	5.1	2.5	4.54×10^6	2,592
Rare	0.25	36.8	9.2	1.67×10^7	9,562
Total	1.00				12,238

[a]Numerical analysis was performed as described in Table 1.
From Berthelot et al., 1980.

some differences between cortex and whole brain at Rot values characteristic of the intermediate class (smaller frequency for the mRNA of this class in the cortex).

The complexity determined by the cDNA technique in mouse neuroblastoma cells is of the same order as that in mouse myoblasts (Affara et al., 1977; Jacquet et al., 1978). Furthermore, this same value, between 6,000 and 8,000 average size species, has also been found for many mouse cell populations. However, the complexity determined for the brain tissue is significantly greater (1.5- to 2-fold) than that observed in a homogeneous neuronal population. This is consistent with the observation of a higher RNA complexity for mouse brain than for other mouse tissue reported by W. Hahn *(this volume)*. One explanation for this could be that the high complexity register in brain reflects a similarly high heterogeneity of cell types. Alternatively, it could be due to the fact that neuronal cells per se, because of their very large potential for establishing homotypic and heterotypic contacts, express more genes than cells from many other tissues. A neuroblastoma cell, even fully differentiated, would not completely express the neuronal program. There is a large difference between the values reported by W. Hahn *(this volume)* and those reported here regarding the total complexity of brain poly(A) mRNA. This can probably be accounted for by the fact that the technique of saturation of single-copy DNA is far more sensitive for the detection of rare species than is the cDNA-mRNA hybridization technique (Bantle and Hahn, 1976).

In the future, by using developmental stage-specific cDNAs, mercurated and bound to sulfhydryl-Sepharose columns, it should be possible to determine when the synthesis of specific mRNAs is switched off or on during neuroblastoma induction. This has been accomplished for myotube specific cDNA (Affara et al., 1980).

NEUROGENESIS INDUCTION

Another topic worthy of brief discussion is the choice of inducers that maximize differentiation. In the course of some pharmacological studies dealing with antianoxic and anticonvulsive drugs, new inducers of neuroblastoma differentiation have been discovered (Croizat et al., 1979), some of which might be useful for further molecular analysis of gene regulation. One of them, cyclohexane carboscylic acid, CCA, is particularly interesting. N1E-115 cells transferred to a culture dish can extend neurites provided that the serum is withdrawn. Morphological differentiation can also be achieved in the presence of serum, provided that 2% DMSO or 0.1% CCA is added. After three days, CCA-treated cells extend very long and thick processes. In independent experiments these extensions have been shown to be true neurites based on the presence of typical neurofilament proteins.

According to previous reports by Littauer (Gozes et al., 1979), Chan and Baxter (1979), Dahl and Wiebel (1979), and Shelanski and Liem (1979), *in vitro* differentiation of neuroblastoma cultures or brain maturation is accompanied by modulations in the rates of synthesis of isotubulins, cytoplasmic actins, and neurofilaments. Therefore, changes in cytoskeletal proteins during treatment of neuroblastoma, with CCA and other inducing agents, are of interest.

Figure 5 summarizes the effect of various CCA-inducing conditions on relative methionine incorporation into cytoskeletal proteins as well as on proteins Z and Y. The following conclusions can be reached:

1. When CCA is added to cells maintained in a serum-rich medium and in monolayer, there is a reduction in the methionine incorporation into α_1 and β tubulin as well as into actin, whereas, incorporation into Z, Y, and vimentin is largely increased. Interestingly, these modifications do not require actual morphogenesis because the variant line N1A-103 that does not extend neurites responds in the same way as wild-type N1E-115.

2. The presence of serum is required for the stimulation of vimentin and of Z. The effects on tubulin subunits and actin remain unchanged or are greater in the presence of serum.

3. Stimulation of incorporation into vimentin as well as the into Z and Y, but not the modulation of tubulin and actin, requires the capacity for the cells to adhere to the surface of the dish.

4. The effect of CCA on vimentin, Z, and Y is irreversible.

5. The effects of other inducing conditions (such as removal of serum or addition of DMSO) are qualitatively the same with regard to tubulin and actin modulation as those obtained with CCA, but none of them cause any enhancement in the incorporation into vimentin, Z, or Y.

It is tempting to conclude that modulation in the incorporation into tubulins and actin by CCA is related to the arrest of cell division (an hypothesis strengthened by observations with

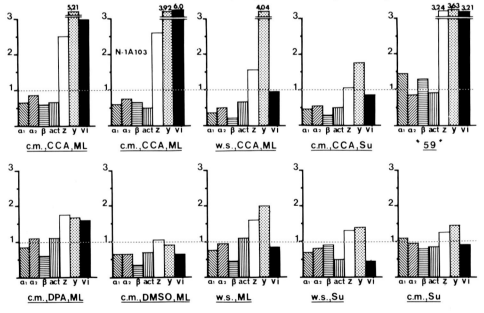

FIG. 5. The effect of various inducing conditions on methionine incorporation into cytoskeletal proteins as well as of proteins Z and Y. The data are expressed relative to the values found in control culture (cells grown in complete medium and maintained in monolayer conditions).

other inducing agents), whereas changes concerning vimentin, and Z and Y proteins could be related to cell adhesion (in the presence of serum). Furthermore, it can be postulated that CCA causes some important modifications of the neural surface since it ensures better adhesion to the substratum than serum deprivation or DMSO in the presence of serum. Concomitant with this, neurite extension proceeds more efficiently, neurites are more stable than after induction by other substances, and the synthesis of many membrane-bound proteins is increased. The sequence of events outlined in Fig. 6 might then be postulated.

Other parameters of neurogenesis, such as the capacity to synthesize neurotransmitters, have been examined in CCA-induced cells (B. Croizat *unpublished data*). The synthesis of TH is stimulated significantly more by CCA than it is by other inducing conditions or inducing agents. Similarly, CCA elicits the appearance of the scorpion toxin receptor to a greater extent than does serum withdrawal (Y. Netter, *personal communication*).

Finally, in view of the CCA effect on membranes and of its antianoxic effect, the effect of CCA on membrane permeability to metabolites has been examined. Figure 7 illustrates some of the results. Using 2-deoxy d-^{14}C-glucose accumulation as an index for glucose utilization, it was found that CCA addition (but neither DMSO nor serum deprivation) promoted a marked

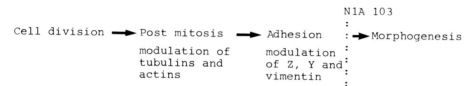

FIG. 6. Sequence of events accompanying terminal morphogenesis.

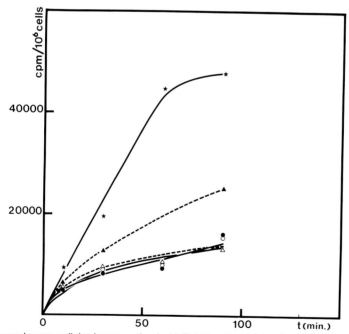

FIG. 7. ^{14}C deoxyglucose cellular incorporation in N1E 115 cells. Total radioactivity was measured in the 12,000-g supernatant and normalized to 10^6 cells. CCA treated cells (★——★); serum-free cultures (○——○); DMSO treated cells (●——●); growing cells in logarithmic phase (△--△); confluent cells in stationary phase (▲--▲).

stimulation of glucose uptake in induced cells compared to noninduced cells, and a significant increase in the accumulation of the deoxyglucose-isomerase complex (not shown). This effect on the cellular energetics may parallel modifications of the ultrastructure in CCA-treated cells, i.e., cytoplasm from CCA-treated cells appears strikingly crowded with mitochondria, compared to the cytoplasm of control cells maintained in the same medium without CCA or induced in another fashion. This effect is somewhat reminiscent of what NGF does to pheochromocytoma (Greene and Rein, 1978).

Whether CCA-induced cells possess a greater capacity for synaptogenesis and display higher action potentials than do cells induced in other ways is currently under investigation. Similarly, CCA induction of the enolase γ subunit is being examined since this protein represents a specific index for highly matured neurons (Pickel et al., 1976). In addition, the status of mRNA modulations using hybridization techniques will also be revisited.

CONCLUSION

The use of neuroblastoma cell lines or of their hybrids, although suffering from many pitfalls due to the neoplastic character of the cells, constitutes an interesting model both for the fine-detailed screening of pharmacological drugs acting on the central nervous system, and for the study of some aspects of normal neurogenesis. Of course, all of the observed effects have to be viewed with caution until the results are matched with those obtained in more natural systems.

Little is known about the histogenic sequence of events underlying different stages of terminal neurogenic development. The choice of an appropriate inducer is not an easy task, but might be crucial for the kind of molecular analysis to be performed.

With regard to gene regulation, terminal differentiation involves some rather modest changes in the level of mRNAs, including both the disappearance of old, and the appearance of new, sequences. Developmentally specific global cDNA probes may be very useful in specifying the nature of the genes, and gene products involved in late commitment to differentiation.

More will certainly be learned about gene regulation during neural differentiation from the use of cloned cDNA or genomic probes corresponding to specific coding sequences. Such work is in progress.

GENERAL DISCUSSION

When asked about the number of activity transcribed genes in neuroblastomas compared to Hahn's estimates of RNA complexity in total brain, Gros estimated that there are 7,000 messengers in these cells, depending on the culture conditions. To answer how many of those messages depend on changes of tissue culture conditions and how many are representative of different cells of origin of the tumor (sympathetic ganglion cells) or other similar classes of cells, data on sympathetic ganglion neurons will be needed. A first experiment would be to cross-hybridize neuroblastoma mRNA and brain cDNA or the reverse. Gros's brain cDNA experiments came out to about 12,000 to 13,000 species, which although slightly more than neuroblastoma, is far less than Hahn's values.

Hahn commented that early estimates of active genes in the liver indicated 8,000 to 10,000 poly(A) messenger sequences. But the value jumped to 35,000 with the use of such techniques as kinetics. Hahn's measurements of myoblasts gave complexity values of RNA populations considerably higher than the 12,000 reported by Gros for the whole brain. If the complexity is attributed to the population of cells, we would have to think that a relatively homogeneous cell population has basically the same complexity of messengers as brain, with at least 10^4 different kinds of cells. Furthermore, Gros's numbers indicate that mouse brain is only four times more complex genetically than *E. coli*. Hahn found this to be rather unlikely. Gros pointed out that the cDNA or the poly(A) RNA studies or techniques should not be discarded too readily, provided the same hybridization conditions are used with the two populations of RNA, since these techniques do yield comparative values.

Asked whether the increase in 2-deoxyglucose transport found with differentiation reflected increased transport per unit of membrane surface, or only an increase in the surface area, Gros replied that the numbers had been expressed as a function of the total protein of the cell. Therefore, it is the amount that is picked up by a given cell mass, and the increase in 2-deoxyglucose transport represents a change in the surface-to-volume ratio, with the numbers representing rates of penetration. Gros and his colleagues have also measured the formation of the isomerase complex, which is influenced to a smaller extent.

ACKNOWLEDGMENTS

This work was supported by the Collège de France as well as by grants from the Centre National de la Recherche Scientifique and the Délégation à la Recherche Scientifique et Technique. We wish also to acknowledge support for the CCA studies from the SANOFI Cy.

REFERENCES

Affara, N. A., Daubas, P., Weydert, A., and Gros, F. (1980): Changes in gene expression during myogenic differentiation. *J. Mol. Biol.*, 140:459–470.
Affara, N. A., Jacquet, M., Jakob, H., Jacob, F., and Gros, F. (1977): Comparison of polysomal polyadenylated RNA from embryonal carcinoma and committed myogenic and erythropoietic cell lines. *Cell*, 12:509–520.
Bantle, J. A., and Hahn, W. E. (1976): Complexity and characterization of polyadenylated RNA in mouse brain. *Cell*, 8:139–150.
Berthelot, F., Gros, F., and Croizat, B. (1980): Complexity of polysomal poly(A) RNA in different developmental stages of a non-differentiating neuroblastoma clone. *FEBS Lett.*, 122:109–112.
Bishop, J. O., Morton, J. G., Rosbash, M., and Richardson, M. (1974): Three abundance classes in Hela cell messenger RNA. *Nature*, 250:199–203.
Chan, V. and Baxter, C. (1979): Compartments of tubulin and tubulin-like proteins in differentiating neuroblastoma cells. *Brain Res.*, 174:135–152.
Croizat, B., Berthelot, F., Ferrandes, B., Eymard, P., and Sahuguillo, C. (1979): Différenciation morphologique du neuroblastome par l'acide 1 methyl cyclohexane carboxylique (CCA) et certains dérivés en C1 ([1], [2]). *C.R. Acad. Sci. Paris*, 289:1283–1287.
Dahl, J. L., and Weibel, V. J. (1979): Changes in tubulin heterogeneity during postnatal development of rat brain. *Biochem. Biophys. Res. Commun.*, 86:822–828.
Felsani, A., Berthelot, F., Gros, F., and Croizat, B. (1978): Complexity of polysomal polyadenylated RNA in undifferentiated and differentiated neuroblastoma cells. *Eur. J. Biochem.*, 92:569–577.
Gozes, I., Saya, D., and Littauer, U. Z. (1979): Tubulin microheterogeneity in neuroblastoma and glioma cell lines differs from that of the brain. *Brain Res.*, 171:171–175.
Greene, L. A., and Rein, G. (1978): Short-term regulation of catecholamine biosynthesis in a nerve growth factor responsive clonal line of rat pheochromocytoma cells. *J. Neurochem.*, 30:549–555.
Grouse, L. D., Schrier, B. K., Letendre, C. H., Zubairi, M. Y., and Nelson, P. G. (1980): Neuroblastoma differentiation involves both the disappearance of old and the appearance of new poly(A)+ messenger RNA sequences in polyribosomes. *J. Biol. Chem.*, 255:3871–3877.
Jacquet, M., Affara, N. A., Robert, B., Jakob, H., Jacob, K., and Gros, F. (1978): Complexity of nuclear and polysomal polyadenylated RNA in pluripotent embryonal carcinoma cell line. *Biochemistry*, 17:69–79.
Levy, B., and Dixon, G. H. (1977): Changes in the sequence diversity of polyadenylated cytoplasmic RNA during testis differentiation in rainbow trout *(Salmo gairdnerii)*. *Eur. J. Biochem.*, 74:61–67.
Pickel, V. M., Reis, D. J., Marangos, P. J., and Zomzely-Neurath, C. (1976): Immunocytochemical localization of nervous system specific proteins (NSP-R) in rat brain. *Brain Res.*, 105:184–187.
Ryffel, G. U., and McCarthy, B. J. (1975): Complexity of cytoplasmic RNA in different mouse tissues measured by hybridization of polyadenylated RNA to complementary DNA. *Biochemistry*, 14:1379–1385.
Shelanski, M. L., and Liem, R. K. H. (1979): Neurofilaments. *J. Neurochem.*, 33:5–13.

Somatostatin, Glucagon, and Calcitonin: A Molecular Approach to Biosynthesis in Peripheral and Neural Tissues

Joel F. Habener, Richard H. Goodman, John W. Jacobs, and P. Kay Lund

ABSTRACT

Recombinant DNA technology provides a potential approach to the biosynthesis of neuropeptides. Cloned cDNAs encoding peptide hormones antigenically related to the neural peptides somatostatin, calcitonin, and glucagon are prepared from endocrine tissues. The cloned cDNAs can then be used to select by specific hybridization mRNAs encoding the neuropeptides. In turn, the mRNA can be used to prepare and clone recombinant cDNAs encoding neuropeptides, and the structures of the neuropeptides can be deduced from the nucleotide sequences of the cDNAs.

The nervous system contains peptide-hormone-like substances immunologically similar, if not identical to, endocrine hormones. This suggests that similar gene products may serve functions both as paracrine neurotransmitters and as endocrine hormones. With the use of radioimmunoassays and immunocytochemical localization techniques, more than 30 different polypeptide-hormone-like activities have now been detected in various regions of the central nervous system. It is of great interest to ascertain the specific functions of these substances: how they are synthesized, how their synthesis and release are regulated, and how they are related to the endocrine hormones.

Among the possible explanations of the relationship between the neuronal and the endocrine peptide hormone substances is the view that these peptides may have served in evolution first as neurotransmitters before the appearance of hormonal functions; the peptides may have evolved along with the growing complexity of the organism and its requirement for a nervous system. Later, as a result of genetic diversification and the need for more highly regulated metabolic systems related to growth and development, similar gene products may have been used as hormones.

Such diversification might have been achieved by: (1) gene duplication (or multiplication) followed by random point mutations and codon deletions or mutations; (2) the use of alternate splicing patterns during the formation of mature mRNA by cleavage from larger RNA precursors resulting in partial differences in protein coding sequences; or (3) by variability in the co- and posttranslation processing of particular gene products. Processing may include proteolytic cleavages, glycosylation, phosphorylation, acetylation, amidation, and formation of intra- and interchain disulfide bonds, all of which may have occurred individually or in various combinations. One might expect, therefore, that when the structures of the paracrine neuropeptides have been determined, they will be found to differ from those of hormones either in the genetic variability of the amino acid sequences or in their processing from precursors.

Little is known about the exact mechanism of neuropeptide biosynthesis. The information available has been largely derived from radioimmunoassay measurements of hormone-like activities in extracts prepared from nerve tissues. Since the amounts of these hormone-like substances in nerve tissue are relatively low compared to the concentrations of the hormones in endocrine organs, study of the biosynthesis of these substances has been difficult. Likewise, the paucity of mRNAs coding for these compounds in nerve tissue has precluded identification of specific cell-free translation products—an approach that has been highly successful in the identification of hormone precursors using mRNAs prepared from endocrine glands.

In view of these technical difficulties, an alternative approach, using the well-established methods of recombinant DNA technology, was adopted to identify mRNA (Fig. 1). Another approach will be needed to identify gene nucleotide sequences coding for the neuropeptides somatostatin, glucagon, and calcitonin.

DNA molecules complementary to mRNA coding for specific hormones (cDNA) can be prepared using the large number of mRNAs obtained from endocrine glands. The cDNAs may then be greatly increased in number by their introduction into, and replication as part of, plasmid vectors such as PBR322 within a bacterial host *(E. coli)*. These more numerous cDNAs may then be used to select, by RNA-DNA hybridization techniques, the scarce mRNAs in nerve tissue that are genetically related to the mRNAs coding for the hormones. In turn, cDNAs complementary to the neural mRNAs can be amplified by molecular cloning of recombinant molecules, and their respective nucleic acid sequences can be analyzed, thus determining the amino acid sequences based on the genetic code.

ANGLERFISH AND RAT SOMATOSTATIN

Current interest in somatostatin arose as a result of its widespread distribution and its importance in modulating functions in many different tissues. Because of the low amounts of

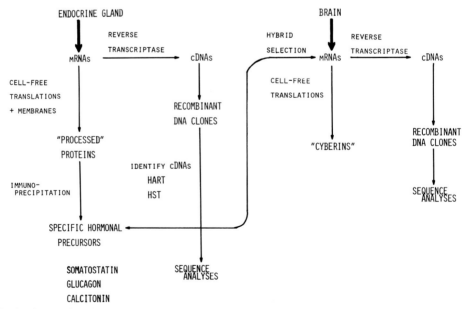

FIG. 1. Approach to biosynthesis of peptides in the nervous system using recombinant DNA technology. HART = hybrid-arrested cell-free translations, HST = hybrid selection and cell-free translations.

somatostatin in the brain, other tissues such as the pancreatic islets and the intestine of the Atlantic anglerfish and the rat medullary carcinoma (a tumor of the thyroid) are used in the isolation, labeling, and structural analysis of this peptide. Cells of these tissues are known to synthesize somatostatin in relatively large amounts. Evidence from other laboratories has shown somatostatin to be synthesized in the form of a large precursor (Patton et al., 1977; Patzelt et al., 1980). Multiple immunoreactive forms of somatostatin larger than the tetradecapeptide have been found in extracts of hypothalamus, extrahypothalamic regions of the brain, spinal cord, intestine, and pancreatic islets. A peptide of 28 amino acids, containing the tetradecapeptide sequence of somatostatin with an N-terminal extension of 14 amino acids, has been isolated from porcine intestine (Pradayrol et al., 1980) and from ovine and porcine hypothalamus (Esch et al., 1980; Schally et al., 1980). Recently, a peptide of 22 amino acids was isolated from the islets of the channel catfish (Oyama et al., 1980b).

Messenger RNAs encoding large precursors of the somatostatins from islets of the anglerfish (Goodman et al., 1980b), extracts of rat hypothalamus (Joseph-Bravo et al., 1980), islets of the channel catfish (Oyama et al., 1980a; Shields, 1980), and a rat medullary thyroid carcinoma (Goodman et al., 1982) have been prepared. Characterizations of the products of translations of these mRNAs in heterologous cell-free systems derived from wheat germ or reticulocyte lysate revealed precursor forms of somatostatin (preprosomatostatins) ranging in size from 14,000 to 16,000 daltons.

Copy DNA libraries have been prepared from the poly(A)-containing mRNA preparations of anglerfish islets (Goodman et al., 1980a; Hobart et al., 1980). Hybrid-arrested translation was used to identify the cDNA molecules containing somatostatin nucleotide sequence. Subsequently, a cDNA encoding a somatostatin precursor from a rat medullary thyroid carcinoma was cloned (Goodman et al., 1982).[1]

The nucleotide sequences of both the islet and carcinoma somatostatin cDNAs are now known (Goodman et al., 1980a; 1982). By decoding the sequences based on the genetic code, Goodman and coworkers (1981) deduced the complete primary structures of one of the two anglerfish islet preprosomatostatins (Fig. 2), as well as a partial sequence of the preprosomatostatin of the rat medullary thyroid carcinoma. A number of interesting features common to both precursors can be distinguished. First, the sequences of the somatostatin-14 (tetradecapeptide) and somatostatin-28 (octacosopeptide) reside at the C-terminus of the large precursors and are preceded by rather long N-terminal extensions. Second, at their amino terminals, the extensions begin with amino acid sequences that contain consecutive stretches of 10 to 12 hydrophobic residues characteristic of "signal" or "leader" sequences found in other globular, secreted proteins. These lipophilic leader sequences are believed to function in the early stages of protein biosynthesis by ensuring the attachment of the nascent polypeptide to the membrane bilayer of the endoplasmic reticulum and by facilitating the unidirectional transport of the nascent polypeptide to the membrane-limited channels of the secretory pathway of the cell (Blobel, 1980). Third, in both cases the somatostatin-14 peptides are connected to the N-proximal peptide extensions via arginyl-lysyl dipeptides. Di- or tripeptides consisting of basic amino acids are typically found in hormone precursors (prohormones) at the sites that are cleaved, during posttranslational processing, to yield the secreted forms of the hormones (Habener and Potts, 1978; Steiner et al., 1980; D. Steiner, *this volume*).

---

[1]In hybridization arrest experiments, mRNA translation is assayed in a cell-free system in the presence of a cloned cDNA. If the cDNA tested is complementary to one of the mRNA molecules present in the reaction mixture, these two molecules form a stable RNA-DNA hybrid that prevents the mRNA from being translated. Thus, an absence of a particular polypeptide indicates that the cDNA contains nucleotide sequences coding for this polypeptide.

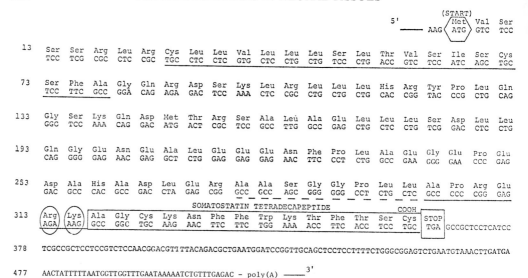

FIG. 2. Preprosomatostatin cDNA. Nucleotide sequence and corresponding amino acid sequence of a cloned cDNA encoding a 14,000-dalton precursor of anglerfish islet somatostatin. Double-stranded cDNAs prepared from islet poly(A) RNA were inserted into the plasmid PBR322 and cloned in *E. coli* following the guidelines established by the National Institutes of Health. The cDNAs containing somatostatin coding sequences were selected by hybridization arrest and cell-free translations of islet poly(A) RNA.

The amino acid sequence of the anglerfish somatostatin-14 that resides in the islet preprosomatostatin I was compared to the primary structures of the somatostatin-14 from mammalian sources and was found to be identical to them (Fig. 3). This is surprising because fish and mammals diverged in evolution approximately 400 million years ago. Such conservation of the primary structure indicates that there has been a strong selective pressure to maintain the exact amino acid sequence of somatostatin-14 peptides during the course of evolution, suggesting that the precise sequence of somatostatin-14 is essential for the expression of the biological activities of this hormone. The high degree of homology becomes even more intriguing in light of the fact that the amino acid sequence of the major form of somatostatin of the catfish, a species much more closely related to the anglerfish than to mammals, bears only partial resemblance to the sequences of the anglerfish, rat, and porcine somatostatin-14 (Fig. 3). It should be noted, however, that the somatostatin-14 contained in the second, i.e., smaller (14,000 daltons), preprosomatostatin of anglerfish islets differs from the other somatostatin-14 peptides in two amino acid residues (Hobart et al., 1980).

The amino acid sequence of the rat carcinoma somatostatin-28 is identical to those of the corresponding somatostatin-28 peptides isolated from porcine intestine and porcine and ovine hypothalamus. However, as shown in Fig. 4, the eight amino acids at the N-terminus of somatostatin-28 contained in the anglerfish preprosomatostatin amino acid sequence differ considerably from those found in the mammalian (rat and porcine) somatostatin-28 peptides. The sequence Glu-Ars and Gln-Ars flanking the N-terminus of the anglerfish and rat sequences, respectively, have not been found previously at sites of proteolytic cleavages of prohormones other than the prosomatostatins. This raises questions regarding the specificity of endopeptidases responsible for prohormone processing and their relevance to possible differences in biological function. Thus, one can hypothesize that the prosomatostatins in anglerfish islets and the C-cells of the thyroid (i.e., cells from which the rat medullary carcinoma originated) are processed to form somatostatin-14 and not somatostatin-28. Contrary to this supposition, however, a recent preliminary report describes the 28-peptide as the major form of somatostatin isolated

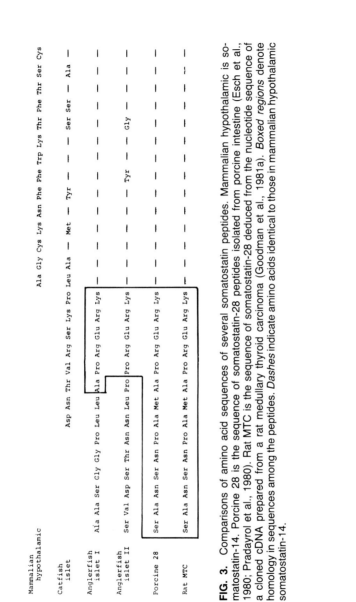

FIG. 3. Comparisons of amino acid sequences of several somatostatin peptides. Mammalian hypothalamic is somatostatin-14. Porcine 28 is the sequence of somatostatin-28 peptides isolated from porcine intestine (Esch et al., 1980; Pradayrol et al., 1980). Rat MTC is the sequence of somatostatin-28 deduced from the nucleotide sequence of a cloned cDNA prepared from a rat medullary thyroid carcinoma (Goodman et al., 1981a). *Boxed regions* denote homology in sequences among the peptides. *Dashes* indicate amino acids identical to those in mammalian hypothalamic somatostatin-14.

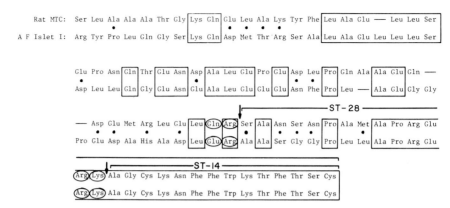

FIG. 4. Sequence homologies between preprosomatostatins of the rat medullary thyroid carcinoma (MTC) and anglerfish (AF). Amino acid sequences were deduced by decoding the nucleotide sequences of cloned recombinant cDNAs. *Boxed sequences* indicate homologous amino acids. *Dots* indicate conservative amino acid substitutions. *Dashes* indicate amino acid (codon) deletions. *Circled residues* indicate putative regions of cleavages *(arrows)* of basic residues in the cellular formation of somatostatin-28 (ST-28) or somatostatin-14 (ST-14).

from an anaplastic rat medullary carcinoma of the thyroid (Benoit et al., 1981). If indeed somatostatin-28 is the predominant peptide in the medullary carcinoma of the rat (whereas somatostatin-14 predominates in the islets of the anglerfish, and perhaps also in the hypothalamus of mammals), then the site of the proteolytic cleavage of the prosomatostatin may vary from one tissue, or species, to the next. As a result, biological activities of the somatostatin peptides so generated may differ.

The biological function of the peptides that lie between the N-terminal leader sequences and the C-terminal somatostatin-28 sequences of the preprosomatostatin molecules (i.e., so-called cryptic peptides) is unknown. These peptides would arise *in vivo* as a result of the cotranslational cleavage of the leader sequence from the nascent preprosomatostatin and posttranslational proteolysis of the prosomatostatin during its processing to form the somatostatin-28 or -14 peptides. It is conceivable that the prosomatostatins are "polyproteins" analogous to the pro-opiomelanocortins and propressophysins (see H. Gainer, *this volume*) in which multiple biologically active peptides are cleaved from a single large precursor. That is, the cryptic peptide(s) may have biological activities other than those of the somatostatins. On the other hand, the amino acid sequence of this peptide may simply function as a protein spacer whose length is required in the processes that are involved in the attachment of the polysome to the membrane of the endoplasmic reticulum and in the transport of the prosomatostatin across the membrane into the secretory pathway. Because the somatostatins are relatively small peptides and lie at the C-terminus of the precursor polypeptides, it may be necessary to have an additional sequence of approximately 50 amino acids interposed between the leader sequence and the somatostatin sequence to allow the nascent polypeptide to span the distance across both the large ribosomal subunit and the membrane bilayer, to the cisternal space of the endoplasmic reticulum. However, in view of the considerable homology between the anglerfish and rat preprosomatostatin amino acid sequences, such an explanation seems unlikely. Despite the fact that fish and mammals diverged in evolution approximately 400 million years ago, considerable similarity has been maintained in the primary structure of these two molecules, suggesting that the precursor peptides have some biological function other than that of a protein spacer. The homology seen within the coding regions is much greater than might be expected if nucleotide changes were random.

The putative biological function of the cryptic peptides is experimentally testable. Now that the primary structures of two of the preprosomatostatins have been deduced from the nucleotide sequences of the corresponding cNDAs, it is possible to synthesize the precursor extensions chemically by the Merrifield solid-phase method and to prepare antisera to the synthetic peptides. The antisera can then be used to monitor the isolation of the cryptic peptides from extracts of islets, intestine, and brain. When isolated, these peptides may be tested for specific biological activities.

INTESTINAL SOMATOSTATIN

The availability of cDNAs containing coding sequences for somatostatin has enabled us to identify somatostatin mRNAs in intestine extracts (Goodman et al., 1981). RNA preparations from these tissues were resolved by agarose gel electrophoresis and transferred, by blotting, onto diazotized paper. Radiolabeled cDNA was then hybridized to this RNA, which had been covalently linked to the paper by a diazo-bond during the blotting process. An autoradiogram from the paper blot identifies the mRNA species that hybridizes specifically with the cDNA (Fig. 5). The somatostatin mRNA of anglerfish intestine identified in this way appears to be 600 nucleotides long, approximately 30 nucleotides smaller than the mRNAs encoding the somatostatin precursors in the anglerfish islets. However, immunoprecipitation of the cell-free translation products of the intestinal mRNA preparation showed a single preprosomatostatin of a molecular weight 16,000 daltons, i.e., identical to the larger of the two islet preprosoma-

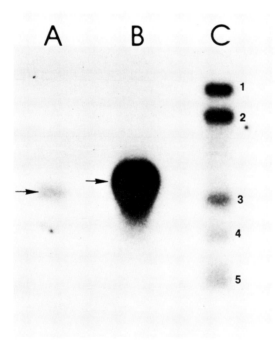

FIG. 5. Reverse Southern blot of mRNAs from anglerfish intestine **(A)** and islets **(B)**. RNAs were separated by electrophoresis on a 2% agarose gel, transferred and covalently linked to DBM paper, and hybridized with ^{32}P-labeled recombinant cDNA encoding anglerfish islet preprosomatostatin I. An autoradiogram was prepared from the paper. **C** is a Taq I restriction endonuclease digest of a recombinant cDNA-plasmid PBR322 for markers of molecular sizes: (1) 1,307 bases, (2) 1,000 bases, (3) 616 bases, (4) 470 bases, and (5) 312 bases. *Arrows* indicate mRNAs of **(A)** 600 and **(B)** 630 bases.

tostatins. Thus, the preprosomatostatin-encoding mRNAs of anglerfish islet and intestine differ in length despite the fact that the polypeptide products appear to be similar in size. There are several possible explanations of this phenomenon. For example, the two preprosomatostatins may be products of two separate genes that are differentially expressed by tissue and that differ in their noncoding regions but code for the same end product. Alternatively, the mRNA precursors may be transcribed from the same gene in both tissues but may be processed differently, presumably by alternative splicing patterns resulting in changes in the precursor outside the area coding for the end product.

CALCITONIN

In addition to preprosomatostatin mRNAs, rat medullary carcinoma of the thyroid contains large amounts of procalcitonin-specific RNA molecules. Cell-free translation of poly(A)-containing RNA fraction isolated from these cells shows a polypeptide of a molecular weight 16,000 daltons, precipitable by anticalcitonin antibodies (Jacobs et al., 1979). When translated in the presence of crude membrane preparations, these polypeptides are cotranslationally processed and glycosylated (Jacobs et al., 1981b). The addition of carbohydrate is accompanied by an increase in size, sensitivity to glycosidases (the N-glycosidase H in particular), and affinity for concanavalin A, a lectin known to bind specifically to mannose residues. As with somatostatin, the calcitonin-specific DNA molecules in the medullary thyroid carcinoma cDNA libraries were identified by hybrid-arrested translation (Jacobs et al., 1981a). The nucleotide sequence that encodes preprocalcitonin has been determined and is 136 codons long (Fig. 6).

	START				-80		
5'-(G)$_{24}$ - TCGCTCACCAGGTGAGCCCTGAGGTTCCTGCTCAGGGAGGCATC	ATG	GGC	TTT	CTG	AAG	TTC	TCC
	Met	Gly	Phe	Leu	Lys	Phe	Ser

							-70										-60	
CCT	TTC	CTG	GTT	GTC	AGC	ATC	TTG	CTC	CTG	TAC	CAG	GCA	TGC	GGC	CTC	CAG	GCA	GTT
Pro	Phe	Leu	Val	Val	Ser	Ile	Leu	Leu	Leu	Tyr	Gln	Ala	Cys	Gly	Leu	Gln	Ala	Val

LEADER SEQUENCE

								-50										-40
CCT	TTG	AGG	TCA	ACC	TTA	GAA	AGC	AGC	CCA	GGC	ATG	GCC	ACT	CTC	AGT	GAA	GAA	GAA
Pro	Leu	Arg	Ser	Thr	Leu	Glu	Ser	Ser	Pro	Gly	Met	Ala	Thr	Leu	Ser	Glu	Glu	Glu

									-30									
GCT	CGC	CTA	CTG	GCT	GCA	CTG	GTG	CAG	AAC	TAT	ATG	CAG	ATG	AAA	GTC	AGG	GAG	CTG
Ala	Arg	Leu	Leu	Ala	Ala	Leu	Val	Gln	Asn	Tyr	Met	Gln	Met	Lys	Val	Arg	Glu	Leu

-20										-10								
GAG	CAG	GAG	GAG	GAA	CAG	GAG	GCT	GAG	GGC	TCT	AGC	TTG	GAC	AGC	CCC	AGA	TCT	AAG
Glu	Gln	Glu	Glu	Glu	Gln	Glu	Ala	Glu	Gly	Ser	Ser	Leu	Asp	Ser	Pro	Arg	Ser	Lys

-1	1									10								
CGG	TGT	GGG	AAT	CTG	AGT	ACC	TGC	ATG	CTG	GGC	ACG	TAC	ACA	CAA	GAC	CTC	AAC	AAG
Arg	Cys	Gly	Asn	Leu	Ser	Thr	Cys	Met	Leu	Gly	Thr	Tyr	Thr	Gln	Asp	Leu	Asn	Lys

CALCITONIN

	20										30							
TTT	CAC	ACC	TTC	CCC	CAA	ACT	TCA	ATT	GGG	GTT	GGA	GCA	CCT	GGC	AAG	AAA	AGG	GAT
Phe	His	Thr	Phe	Pro	Gln	Thr	Ser	Ile	Gly	Val	Gly	Ala	Pro	Gly	Lys	Lys	Arg	Asp

		40										50				
ATG	GCC	AAG	GAC	TTG	GAG	ACA	AAC	CAC	CAC	CCC	TAT	TTT	GGC	AAC	TAG	GTCCCTCCTCT
Met	Ala	Lys	Asp	Leu	Glu	Thr	Asn	His	His	Pro	Tyr	Phe	Gly	Asn	STOP	

CCTTTCCAGTTTCCATCTTGCTTTCTTCCTATAACTTGATGCATGTAGTTCCTCTCTGGCTGTTCTCTGGCTATTAT - (C)$_{14}$ - 3'

FIG. 6. Preprocalcitonin DNA. Nucleotide and corresponding amino acid sequence of a cloned cDNA encoding a precursor of calcitonin (Jacobs et al., 1981a). The cDNA was selected from a cloned cDNA library prepared from poly(A) RNA from a rat medullary carcinoma of the thyroid. *Boxed sequence* is a 32-amino-acid calcitonin. *Circles* denote basic residues at sites of cleavages of the precursor during the posttranslational processing of procalcitonin. *Underlined sequence* is the NH_2-terminal leader or signal sequence removed from the precursor during the cellular translation of the mRNA. Glycine enclosed in *hexagon* is involved in amidation of adjacent proline during posttranslational processing of procalcitonin.

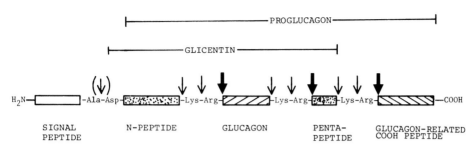

FIG. 7. Diagram of structure of preproglucagon determined from sequencing of a cloned cDNA prepared from pancreatic islets of the anglerfish (Lund et al., 1981). The mRNA encoding this precursor contains two glucagon-related coding sequences arranged in tandem, linked by codons for basic residues and a pentapeptide. Arrows indicate sites of cleavages during the posttranslational processing of the precursor. *Heavy arrows* indicate sites of tryptic-like cleavage; *light arrows*, carboxypeptidase B-like cleavage. *Arrow in parentheses* represents proposed site of cleavage of NH_2-terminal signal (leader) sequence during cellular translation of the mRNA.

The primary structure of the preprocalcitonin consists of several clearly defined regions: (1) the lipophilic leader sequence, (2) the cryptic propeptide extension followed by the Lys-Arg dipeptide, (3) the 32-amino-acid calcitonin sequence, and (4) the sequence Gly-Lys-Lys-Arg. Calcitonin is known to contain a proline amide on its C-terminus. The nucleotide sequence of calcitonin, however, ends with codons for proline and glycine. The presence of glycine followed by basic amino acids at the C-terminal end may be common to peptides with C-terminal amides since it has also been noted for both α-melanotropin and mellitin (Dayhoff, 1972).

GLUCAGON

Glucagon, a 29-amino-acid peptide, is found in the pancreas and intestine. By virtue of its primary structure and biological actions glucagon belongs to a family of vasoactive and inhibitory peptides that includes secretin, gastric inhibitory peptide, vasoactive intestinal peptide, and cholecystokinin. Glucagon-specific cDNA was isolated from the cDNA library of anglerfish islets, cloned, and then sequenced using the same methodological protocol described for somatostatin (Lund et al., 1982). The nucleotide sequence of the glucagon cDNA codes for an N-terminal methionine residue followed by a leader sequence and 29 amino acid residues of the glucagon polypeptide (Fig. 7). The glucagon amino acid sequence is preceded by the Lys-Arg dipeptide; at its C-terminus, it is followed by the residues Lys-Arg-Ser-Gly-Val-Ala-Glu-Lys-Arg, following which there are 43 additional amino acid codons before the stop codon is reached. The most remarkable feature of this C-terminal extension is its pronounced homology to the glucagon amino acid sequence. Does this C-terminal peptide extension possess a glucagon-like activity in the anglerfish islets or is it simply an evolutionary relic of an ancient gene duplication? Chemical synthesis of this polypeptide extension is currently in progress and should provide unambiguous answers to these questions.

Clearly, a great deal of work remains to be done to decipher the complexity of the structures of the neuropeptides before it will be possible to understand the physiological complexities of their actions and regulation. Future investigations will involve analyzing the structure, organization, and complexity of the genomic DNA encoding the polypeptide precursors, determining the ways in which the precursors are processed within the different tissues, and attempting to correlate the structures and processing of the polypeptides with their specific biological functions.

ACKNOWLEDGMENTS

We thank W. W. Chin, P. C. Dee, and G. Heinrich for help in these studies, and J. Sullivan for preparing the manuscript. These studies were supported in part by grants from the NIH and NSF. P. K. Lund was supported by the British Science Research Council.

REFERENCES

Benoit, R., Bohlen, P., Ling, N., Brazeau, P., Shibasaki, T., and Guillemin, R. (1981): A somatostatin-28-like peptide in rat medullary thyroid carcinoma. *Fed. Proc.*, 40:457.
Blobel, G. (1980): Intracellular protein topogenesis. *Proc. Natl. Acad. Sci. USA*, 77:1496–1500.
Dayhoff, M. O. (1972): *Atlas of Protein Sequence and Structure*. National Biomedical Research Foundation, Washington, D.C.
Esch, F., Bohlen, P., Ling, N., Benoit, R., Brazeau, P., and Guillemin, R. (1980): *Proc. Natl. Acad. Sci. USA*, 77:6827–6831.
Goodman, R. H., Jacobs, J. W., Chin, W. W., Lund, P. K., Dee, P. C., and Habener, J. F. (1980a): Nucleotide sequence of a cloned structural gene coding for a precursor of pancreatic somatostatin. *Proc. Natl. Acad. Sci. USA*, 77:5869–5873.
Goodman, R. H., Jacobs, J. W., Dee, P. C., and Habener, J. F. (1982): Somatostatin-28 encoded in a cloned cDNA obtained from a rat medullary thyroid carcinoma. *J. Biol. Chem.*, 257:1156–1159.
Goodman, R. H., Lund, P. K., Barnett, F. H., and Habener, J. F. (1981): Intestinal pre-prosomatostatin: Identification of mRNA coding for a precursor by cell-free translations and hybridization with a cloned islet cDNA. *J. Biol. Chem.*, 256:1499–1501.
Goodman, R. H., Lund, P. K., Jacobs, J. W., and Habener, J. F. (1980b): Preprosomatostatins: Products of cell-free translations of messenger RNAs from anglerfish islets. *J. Biol. Chem.*, 255:6549–6552.
Habener, J. F., and Potts, J. T., Jr. (1978): Biosynthesis of parathyroid hormone. *N. Engl. J. Med.*, 299:580–585, 635–644.
Hobart, P., Crawford, R., Shen, L., Pictet, R., and Rutter, W. J. (1980): Cloning and sequence analysis of cDNAs encoding two distinct somatostatin precursors found in the endocrine pancreas of anglerfish. *Nature*, 288:137–141.
Jacobs, J. W., Goodman, R. H., Chin, W. W., Dee, P. C., Bell, N. H., Potts, J. T., Jr., and Habener, J. F. (1981a): Calcitonin messenger RNA encodes multiple polypeptides in a single precursor. *Science*, 213:457–459.
Jacobs, J. W., Lund, P. K., Potts, J. T., Jr., Bell, N. H., and Habener, J. F. (1981b): Procalcitonin is a glycoprotein. *J. Biol. Chem.*, 256:2803–2807.
Jacobs, J. W., Potts, J. T., Jr., Bell, N. H., and Habener, J. F. (1979): Calcitonin precursor identified by cell-free translation of mRNA. *J. Biol. Chem.*, 254:10600–10603.
Joseph-Bravo, P., Charli, J. L., Sherman, T., Boyer, H., Bolivar, F., and McKelvy, J. F. (1980): Identification of a putative hypothalamic mRNA coding for somatostatin and its product in cell-free translation. *Biochem. Biophys. Res. Commun.*, 64:1004–1012.
Lund, P. K., Goodman, R. H., Dee, P. C., and Habener, J. F. (1982): Preproglucagon cDNA contains two glucagon-related coding sequences arranged in tandem. *Proc. Natl. Acad. Sci. USA*, 79:345–349.
Oyama, H., Bradshaw, R. A., Bates, O. J., and Permutt, A. (1980a): Amino acid sequences of catfish: Pancreatic somatostatin. *J. Biol. Chem.*, 255:2251–2254.
Oyama, H., O'Connell, K., and Permutt, A. (1980b): Cell-free synthesis of somatostatin. *Endocrinology*, 107:845–847.
Patton, G. S., Ipp, E., Dobbs, R. E., Orci, L., Vale, W., and Unger, R. H. (1977): Pancreatic immunoreactive somatostatin release. *Proc. Natl. Acad. Sci. USA*, 74:2140–2143.
Patzelt, C., Tager, H. S., Carroll, R. J., and Steiner, D. F. (1980): Identification of prosomatostatin in pancreatic islets. *Proc. Natl. Acad. Sci. USA*, 77:2410–2414.
Pradayrol, L., Jornvall, H., Mutt, V., and Ribet, A. (1980): N-terminally extended somatostatin: The primary structure of somatostatin-28. *FEBS Lett.*, 109:55–58.
Schally, A. O., Huang, W. Y., Chang, R. C. C., Arimwa, A., Redding, T. W., Millor, R. P., Hunkapillor, M. W., and Hood, L. E. (1980): Isolation and structure of prosomatostatin: A putative somatostatin precursor from pig hypothalamus. *Proc. Natl. Acad. Sci. USA*, 77:4489–4493.
Shields, D. (1980): In vitro biosynthesis of fish islet pre-prosomatostatin: Evidence of processing and segregation of a high molecular weight precursor. *Proc. Natl. Acad. Sci. USA*, 77:4074–4078.
Steiner, D. F., Quinn, P. S., Chin, S. J., Marsh, J., and Tager, H. F. (1980): Processing mechanisms in the biosynthesis of proteins. *Ann. N.Y. Acad. Sci.*, 343:1.

Somatostatin in the Nervous System

Seymour Reichlin

ABSTRACT

Somatostatin is one of the best characterized of the neuropeptides. It is synthesized in neurons of the tuberoinfundibular system and released into the hypophysial portal vessels as a hypophysiotropic factor regulating thyroid-stimulating hormone and growth hormone secretion. It is also synthesized in interneurons of many other neural sites such as the cortex, hippocampus, and amygdala, and it occurs in the brainstem and spinal cord both as a specific neurosecretion of a class of sensory neurons and in spinal interneurons. Somatostatin is synthesized as part of a prohormone whose sequence has been elucidated in nonneural tissues. Two forms occur, a 14-amino-acid sequence, "classic somatostatin," and an N-extended form that contains 28 amino acids (somatostatin-28). These smaller forms are processed within neurons by endoplasmic reticulum and transported to nerve endings in secretory granules conveyed at a rate in excess of 400 mm per 24 hr. Both somatostatin-14 and -28 are released in response to depolarizing stimuli or activation of Na+ channels. Release is dependent on Ca^{2+} entry into cells and is regulated by several neurotransmitters (activated by acetylcholine muscarinic receptors, inhibited by γ-aminobutyric acid). Although the effects of somatostatin on synaptic activity, spontaneous single neuron activity, and on behavior have been demonstrated, and it reacts with receptors on nerve cell membranes, its role as a neuroregulator is still obscure. Major problems to be elucidated are the factors that control synthesis of somatostatin at the molecular level, the mechanism of somatostatin action on neurons, and its role in nervous system function. Although many gaps in our knowledge exist, neuropeptides are excellent models for studies integrating molecular biology and brain function.

Somatostatin is a 14-amino-acid peptide that has been shown by immunohistochemical staining and radioimmunoassay to be present in the central and peripheral nervous system, intestine, and pancreatic islets. This peptide, first identified as a constituent of hypothalamic extracts, inhibits secretion of both growth hormone (Brazeau et al., 1973) and thyroid-stimulating hormone (Vale et al., 1977). In this situation, somatostatin acts as a true neurohormone, literally defined as a factor released from neuronal endings into the circulation to influence function at a remote site. Studies of portal vessel somatostatin (Abe et al., 1978), selective ablation of somatostatinergic pathways (Epelbaum et al., 1977), and administration of antisera to somatostatin (Arimura and Schally, 1976; Terry et al., 1976) have shown that this peptide is a hypophysiotropic factor, involved in both tonic and phasic control of anterior pituitary secretion. In contrast to other somatostatinergic systems, the hypophysiotropic system has a well-defined function. Interest in the neural function of somatostatin has been increased by the recognition that this substance is but one representative of a class of peptide neuromodulators that include the gut-brain peptides and the endorphins (Hökfelt et al., 1980). The study of these substances has provided a crucial link, combining current methods of elucidating gene structure and expression with knowledge of synthesis and processing of proteins, which together can be applied to the understanding of brain function and disease.

HYPOTHALAMIC SOMATOSTATIN

Within the hypothalamus, the highest concentration of somatostatin is found in the median eminence (Hökfelt et al., 1974; Brownstein et al., 1975; Alpert et al., 1976; Patel and Reichlin, 1978). Figure 1 shows a sagittal section of the rat hypothalamus that has been stained by immunohistochemical methods to demonstrate the somatostatinergic system. The white arrows show the distribution of somatostatin immunoreactive material in tracts from the anterior hypothalamus directed toward the stalk median eminence (SME).

The tuberoinfundibular somatostatinergic system arises mainly in the periventricular nuclei of the anterior hypothalamus. Figure 2 shows a dense plexus of bipolar somatostatin-staining cells immediately under the ependymal lining. One of the important features of this anatomical system is that the fibers arc laterally through the substance of the hypothalamus and converge near the midline to enter the median eminence (Fig. 3). That this process involves axonal transport is shown by experiments in which colchicine is introduced into the third ventricle. This substance inhibits transport by interfering with microtubules, and leads to accumulation of somatostatin in cells of origin (R. Lechan, *unpublished data*).

A second group of neurons projecting to the median eminence and involved in pituitary control originates in the paraventricular nucleus. This nucleus is complex, containing a number of neuropeptide tracts in a fairly restricted distribution. It is, in fact, six to eight intermixed nuclei, of which somatostatin-containing neurons constitute but one component. It is not known whether the paraventricular somatostatin projection is functionally important in pituitary regulation or how its function is integrated with that of the periventricular system.

The two components of the tuberoinfundibular somatostatinergic system, periventricular and paraventricular, correspond in anatomical distribution to cells that can be demonstrated by the techniques of retrograde transport of markers following injection into the substance of the median eminence. Lechan and collaborators (1980) initially used horseradish peroxidase as

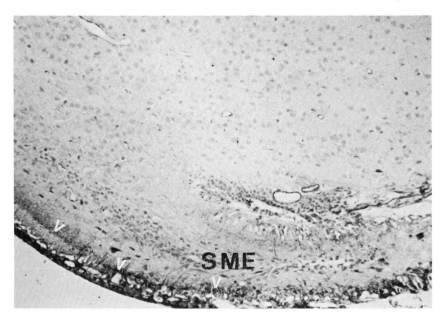

FIG. 1. Sagittal section of the rat hypothalamus stained to show somatostatinergic fibers as they project from anterior hypothalamus to the base of the stalk median eminence (SME). (Photograph courtesy of Dr. Lesley Colton Alpert.)

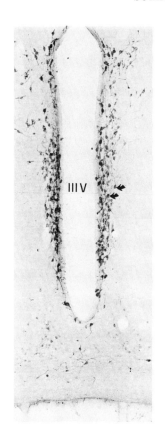

FIG. 2. Most of the somatostatinergic fibers in the hypothalamus arise in a narrow periventricular zone, here illustrated in the anterior hypothalamus of the rat by the PAP technique. *Arrows* point to individual cells. (Photograph courtesy of Dr. Ronald Lechan.)

tracer, but more recently (R. Lechan, J. Nestler, and S. Jacobson, *unpublished data*) have found that wheat germ agglutinin, demonstrable with the use of antiwheat germ agglutinin antibody, is an even more sensitive tracer. Both sets of somatostatin-containing cells can be shown to have an anatomical distribution corresponding to that of the tuberoinfundibular system.

EXTRAHYPOTHALAMIC NEURAL SOMATOSTATIN

The widespread distribution of somatostatin in the nervous system outside the hypothalamus was recognized soon after its discovery as a hypothalamic hypophysiotropic hormone (Hökfelt et al., 1974; Brownstein et al., 1975; Vale et al., 1975, 1977; Kronheim et al., 1976; Patel and Reichlin, 1978; Pelletier, 1980). Immunoreactive somatostatin-containing cells in Fig. 4 are shown to be mainly in sensory cortex layers II and III. The inset shows a single neuron. That extrahypothalamic somatostatin must be important as a neuromodulator or neurotransmitter beyond its function as a pituitary regulating factor is strongly suggested by the findings that (1) the peptide is localized in synaptic nerve endings (Petrusz et al., 1977; Pelletier, 1980); (2) it is released from neural tissues by chemical or electrical depolarization (Reichlin et al., 1982); (3) it exerts neurophysiological changes in neurons when applied locally (Renaud et al., 1975); (4) it causes striking behavioral responses following intracerebral application (Havlicek and Friesen, 1979); (5) it is degraded by enzymes found in brain extracts (Marks and Stern, 1975); and (6) it binds to specific brain receptors (Strikant and Patel, 1981); and (7) it is abnormally low in at least one important human brain disease, Alzheimer's disease (Davies et al., 1980).

FIG. 3. The somatostatin immunoreactive fibers arising in the periventricular region project laterally through the hypothalamus, ultimately to converge posteriorly to enter the median eminence. *Arrows* point to two of the fibers. (Photograph of rat brain courtesy of Dr. Ronald Lechan.)

The hypophysiotropic pathway is only a fraction of the neural somatostatinergic system. As summarized by Krisch (1978), periventricular somatostatin systems project directly to several important hypothalamic nuclei including the preoptic, suprachiasmatic, ventromedial, arcuate, and ventral premammillary, to the subfornical organ, the organum vasculosum of the lamina terminalis, to the corticomedial amygdala, and to the olfactory tubercle (Fig. 5). These structures are in a position to influence major parts of the limbic system that integrate sensory inputs, complex behavioral patterns, drive states, and the determination of circadian rhythm.

Extrahypothalamic somatostatin-containing pathways arise in many other parts of the neuraxis. These include intrinsic fibers, presumably interneurons in the cerebral cortex (Krisch, 1980), the hippocampus, and interneurons in the brainstem and spinal cord (Forssmann, 1978). Somatostatinergic fibers also form a subgroup of sensory afferent neurons whose cell bodies are located in the dorsal root ganglia (type B sensory neurons) (Hökfelt et al., 1975), whose central projections terminate in the dorsal horn of the spinal cord (Fig. 6 shows the Lissauer zone of the dorsal horn of rat). Other sensory fibers arise in the nodose ganglia of the vagus (Lundberg et al., 1978), innervate the gut and pancreas peripherally, and presumably terminate centrally in the solitary nucleus of the vagus, an important visceral-regulating region. The vagus is not the only part of the autonomic nervous system that contains somatostatin. A subpopulation of noradrenergic neurons in sympathetic ganglia also contains this peptide in addition to noradrenergic secretory granules (Lundberg et al., 1978; Hökfelt et al., 1980). Somatostatin is

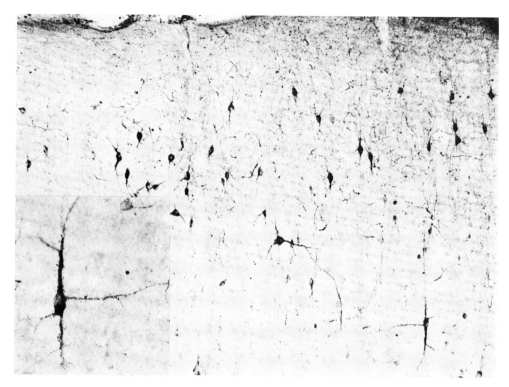

FIG. 4. Somatostatin-immunoreactive fibers in the cerebral cortex of the rat, presumed to be part of an interneuronal system. **Inset:** a single bipolar cell. (Photograph courtesy of Dr. Ronald Lechan.)

also found in amacrine cells of the retina, as determined by radioimmunoassay and by immunohistochemistry (Rorstad et al., 1979a; Yamada et al., 1980).

If the immunoreactivity of the spinal cord is examined by quantitative methods, it can be shown that the dorsal horn contains the highest concentration and the dorsal root a moderate amount (Rasool et al., 1981). Despite the fact that somatostatin is synthesized in the cell body, the sensory ganglia contains relatively small amounts of the peptide, a reflection of the fact that most of the peptide is stored in nerve terminals, either proximally or distally. Somatostatin-containing cells are also found in the ventral horn and elsewhere in the cord where they are presumed to form an interneuronal local control system (Forssmann, 1978). Tessler and coworkers (1980) have recently shown by immunohistochemical means that a population of these cells persists in cat spinal cord even when a segment has been isolated by spinal root section and cord knife cuts. Measurements of the somatostatin content of spinal fragments from Tessler's preparations confirm that the peptide persists even in isolated segments (S. Reichlin, *unpublished observation*). Projections to the Lissauer zone of the dorsal horn come both from sensory afferents and the intrinsic segmental cord system. An interesting problem for the neurophysiologist is how the two systems interact for the processing of sensory coding from the periphery.

One can get an estimate of the quantitative aspects of the somatostatin afferent system by relatively simple methods, including sciatic nerve ligation. It has been shown that there is a progressive accumulation of somatostatin proximal to a ligature (Fig. 7) (Rasool et al., 1981). The amounts accumulated are substantial and confirm the view that somatostatin in the sciatic nerve arises in the cell body and is transported peripherally as well as centrally to the dorsal horn. Quantification of somatostatin synthesis and turnover can be done using standard methods

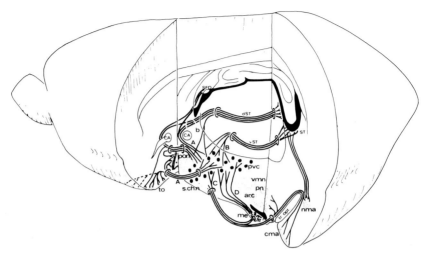

FIG. 5. Schematic drawing of the hypothalamus and extrahypothalamic somatostatin-containing fiber system. A trapezoid piece of the left hemisphere is dissected out up to the subependymal neuropil of the third ventricle. CA, anterior commissure; pvc, paraventricular somatostatin-producing cells; pon, preoptic nucleus; s.ch.n., suprachiasmatic nucleus; vmn, ventromedial nucleus; arc, arcuate nucleus; SFO subfornical organ; b, interstitial (bed) nucleus of the stria terminalis; *, organum vasculosum laminae terminalis; to, olfactory tubercle; A, B, ascending and descending projections; C, caudal projection to the median eminence (me); dST, dorsal component of the stria terminalis; vST, ventral component of the stria terminalis; tr. opt., optic tract; ST stria terminalis at the inferior angle of the posterior horn of the lateral ventricle; nma, nucleus medialis amygdalae; cma, corticomedial amygdala. (Reprinted from Krisch, 1978, with permission.)

for calculating the parameters of transport within the axon (Rasool, 1981). Data from double ligature experiments indicate that only 19.2% of the somatostatin in the sciatic nerve is in the mobile fraction. The calculated absolute transport rate is 415 mm in 24 hr, which is the standard rate for "fast" transport.

As shown in Table 1, if the amount of somatostatin that accumulates proximal to the ligature over a 24-hr period is divided by the somatostatin content of the sciatic nerve, it can be estimated that the content of somatostatin in the sciatic ganglion turns over approximately 9 times in 24 hr. This number would be larger if we knew the amount of somatostatin that was directed centrally. This number gives an idea of the rate of activity of somatostatin synthesis and transport; it does not include the internal breakdown or recycling of somatostatin within the neuron itself, a value that cannot be estimated. Even so, the results suggest a very high rate of protein synthesis. It may be important to know to what extent this reflects the level of activity of somatostatin fibers in the brain or of "peptide fibers" in general. The transport rate is comparable to the range reported for fast transport in several other systems by Wilson and Stone (1979).

Delfs and coworkers (1980) are studying the function of the cerebral cortical somatostatin neurons using dispersed cells in culture grown from 17-day-old fetal rats. These cells are very vigorous, and although there is a variable latent period, it is possible to show the accumulation of somatostatin. Within a week the amount of somatostatin in the cells rises rapidly (Fig. 8). There is apparently a much less rapid rise of somatostatin found in the medium, but this may be accounted for by rapid degradation of the peptide by horse serum. Nevertheless, the preparation is ideal for studying the regulation of somatostatin secretion *in vitro*. This can be illustrated with data from two experiments.

Figure 9 shows the exposure of dispersed cells to high potassium media. It can be seen that depolarization brought about by addition of K^+ stimulates Ca^{2+}-dependent somatostatin release.

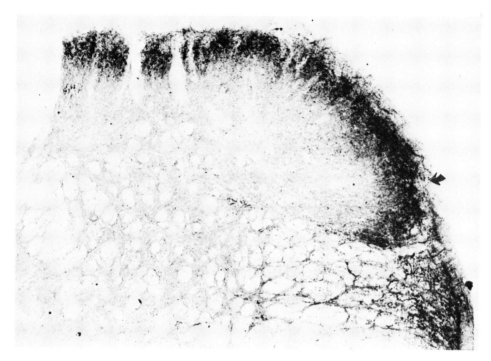

FIG. 6. The dorsal horn of rat spinal cord illustrates the rich innervation by somatostatin immunoreactive fibers. *Arrow* points to a selected neuron. (Photograph courtesy of Dr. Ronald Lechan.)

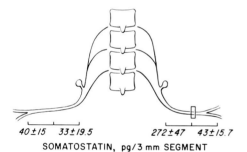

SOMATOSTATIN, pg/3 mm SEGMENT

40±15   33±19.5   272±47   43±15.7

FIG. 7. Accumulation of immunoreactive somatostatin proximal to a ligature on the sciatic nerve of the rat. (From Rasool et al., 1981).

TABLE 1. *Transport of somatostatin in rat sciatic nerve*

Apparent transport rate	79.8 mm/24 hr
Mobile fraction	19.2%
Calculated absolute transport rate	415.6 mm/24 hr
Amount transported distally	2.7 ng/24 hr
Sciatic ganglia content	0.3 ng
Calculated turnover in ganglia[a]	9 times/24 hr

[a]Based on peripheral contribution only.

This result is typical of standard neurotransmitter regulation and is illustrative of a number of observations showing that somatostatin in nerve cells is released by the same kinds of membrane-active factors that regulate release of classic neurotransmitters. The second experiment, shown in Fig. 10, involves the effect of γ-aminobutyric acid (GABA), which was found to inhibit

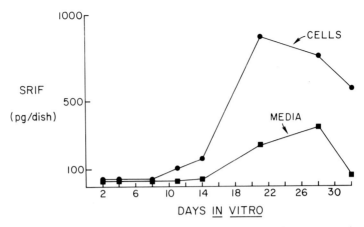

FIG. 8. Accumulation of somatostatin in cultured cerebral cortical cells of the rat grown in dispersed culture. (From Delfs et al., 1980, with permission.)

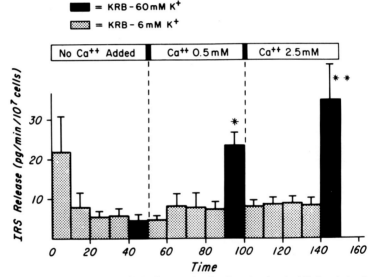

FIG. 9. Exposure of cerebral cortical cells in dispersed cell culture to elevated K+ levels leads to an increase in somatostatin release to the media, an effect not expressed in the absence of Ca^{2+}, and dependent on the concentration of Ca^{2+} at 0.5 and 2.5 mM (Robbins et al., 1982a).

somatostatin release from these cultures. This effect was blocked by its antagonist, picrotoxin. It has recently been reported by Gamse and coworkers (1980) that GABA suppresses somatostatin release from hypothalamic cells in culture.

These results indicate that somatostatin release by cerebral cortical neurons can be affected by neurotransmitters present in the cortex. What is not known, and should be studied, is the regulation of somatostatin-containing neurons in other parts of the brain. Whether the results observed with cerebral cortical mantle are valid for spinal cord and for the hypophysiotropic hormonal system is not known. It is clear that somatostatin is made by many different kinds of cells, and probably for each cell's purpose. The fact that a cell makes somatostatin does not characterize it completely as a functional entity.

BIOSYNTHESIS

The biosynthesis of somatostatin has been studied in several ways. Pulse-chase labeling experiments have been performed in both hypothalamic fragment incubates (Ensinck et al.,

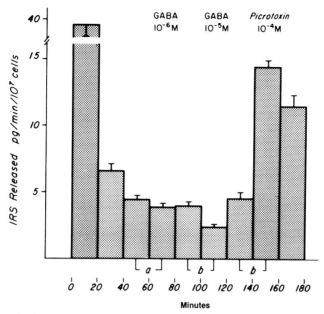

FIG. 10. Exposure of cultured cerebral cortical cells to GABA leads to suppression of somatostatin release. The first two epochs showing falling values are the "wash out" phase observed when cells are transferred to a serum-free medium (Robbins et al., 1982b).

1978) and pancreatic islets (Noe et al., 1978). The results suggest that a high molecular weight precursor of somatostatin is initially synthesized. In addition, gene cloning studies have demonstrated the existence of genes coding for a 119 amino acid (13.5 kDal) precursor to anglerfish pancreatic islet somatostatin (Goodman et al., 1980; Habener et al., *this volume*).

The molecular weights of substances that are made by the cultured rat cerebral cortical cells and react with somatostatin antibody are consistent with the notion that there are precursors of somatostatin. Chromatographic analysis reveals a major peak of immunoreactive somatostatin corresponding to somatostatin-14, as shown in Fig. 11A. Peaks corresponding to an 11.5-kD immunoreactive somatostatin and to somatostatin with 25 or 28 amino acids (unresolved in this separation) can also be seen. The nature of the small peak immediately preceding the large somatostatin-14 peak is unknown. Figure 11B (Scanlon et al., 1981) shows the pattern of somatostatin released from cultured rat cells. To obtain this information, a flow-through capillary culture system was established so that moment-by-moment output of somatostatin could be measured. Although somatostatin-14 is the predominant hormone, there is a considerable amount of somatostatin-25/28 and the 11.5-kD precursor hormone secreted from the cells. In general, the same proportions of these forms are secreted as are found within the cells.

Figure 11C shows a pattern of somatostatin found in human spinal fluid. Peaks corresponding to somatostatin-14, to the somatostatin-25/28, and to the large somatostatin precursor can be seen. Since spinal fluid reflects the secretion of neurons, human cerebral cortex appears to exhibit the same pattern of secretion as do rat cerebral cortical cells in dispersed culture.

These observations raise two questions regarding biological function. First, is each of these forms a hormone, and second, do the different forms of the peptide, as secreted, have different biological effects? Rorstad and coworkers (1979b) showed that each of these forms, when extracted from brain, was effective in inhibiting growth hormone secretion. Meyers and coworkers (1980) showed that purified somatostatin-28 is actually more potent than somatostatin-14 in eliciting some biological effects, notably the suppression of pancreatic insulin secretion. Konturek and coworkers (1981) showed that somatostatin-28 suppresses release of serum gas-

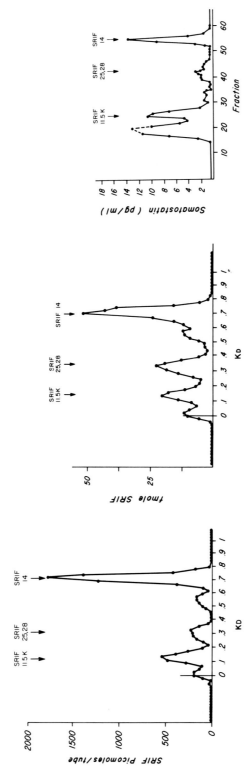

FIG. 11. A: Multiple molecular forms of immunoreactive somatostatin in extracts of cultured cerebral cortical cells (Biogel P-10): Somatostatin (SRIF) 14 corresponding to the 14-amino-acid peptide; SRIF 25, 28 to the N-terminal extended molecules (the chromatographic system does not distinguish between the two); and a still larger form corresponding to prosomatostatin. B: Pattern of molecular sizes of immunoreactive somatostatin in the secretory effluent of cultured cerebral cortical cells. Note that substantial amounts of the larger forms are secreted as such into the effluent (Scanlon et al., 1981). C: Pattern of molecular sizes of immunoreactive somatostatin in human cerebrospinal fluid (Biogel P-10). These probably represent the secretion of neurons, and suggest that all forms are released *in situ* as they are in neuronal effluent.

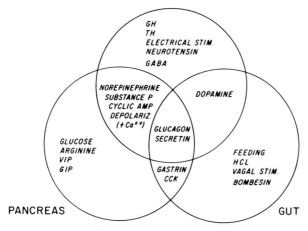

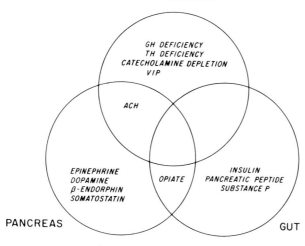

FIG. 12. Venn diagrams summarizing most of the reported studies in the literature that deal with regulation of somatostatin secretion, from its major sites of secretion. Some of the data are contradictory and many gaps in knowledge exist, but the important point is that somatostatin release from different cell types may be regulated in different ways. A particular class of cells, including neurons, cannot be classified by the nature of their secretion alone. (From Reichlin, 1981.)

trin, insulin, and pancreatic polypeptide, just as does somatostatin-14. However, unlike somatostatin-14, the larger form does not affect mesenteric circulation, oxygen uptake or intestinal motility. Thus, all of the forms exhibit hormonal activity, but they do not have identical biological effects. Brown and collaborators (1981) have shown recently that when administered intraventricularly, somatostatin-28 is more potent than somatostatin-14 in bringing about inhibition of bombesin-induced hyperglycemia. This centrally mediated effect indicates the 28-membered peptide has greater central nervous system effect than the smaller peptide. The available evidence indicates that all forms of somatostatin are hormones, and that the larger forms are prohormones as well. In this regard the situation with somatostatin is identical to that for gastrin. There is a large gastrin with 34 amino acids and a small gastrin with 17 amino acids. Both forms occur in gastrin-secreting cells and both have biological activity (Dockray, 1978). J. McKelvy et al. *(this volume)* offer additional examples illustrating the association of different biological activities with specific peptides and their fragments.

A key problem for the classic neurobiologist is to understand how a single regulatory molecule can appear in many places in the brain and presumably have a variety of functions; a larger question is how synthesis and release of somatostatin itself is regulated. Figure 12 is a Venn diagram summarizing the factors that have been implicated in regulation of the three principal sites in the body where somatostatin is made (Reichlin, 1981). This is a compilation of all of the published data on the biological effects of somatostatin, and it includes several contradictory claims. Glucagon and secretin can be viewed as universal somatostatin secretagogues. These two peptides have been shown to release somatostatin from hypothalamus, from the gastrointestinal tract, and from the pancreas (see Reichlin, 1981, for references). In general, however, particular stimuli will release somatostatin from certain types of cells and not from others. Figure 12B shows a number of factors that have been reported to inhibit somatostatin release. The point illustrated in Fig. 10 is that the control of expression and release of somatostatin is dependent on the type of somatostatin-synthesizing cell being studied. The precise details remain to be worked out in each site.

This chapter summarizes some of the highlights of what is known about central somatostatinergic pathways. Classic immunohistochemical methods have defined the anatomy of the pathways, and classic physiological studies have begun to define factors that regulate biosynthesis, secretion, processing, and electrophysiological effects of the peptide. The advent of molecular biology has now revealed the structure of the preprohormone and a hint about posttranslational processing. Nevertheless, the roles of somatostatin in brain and peripheral neural function are still largely unknown. Integration of the classic approaches with the newer molecular biological techniques will continue to provide many stimulating new insights.

REFERENCES

Abe, H., Kato, Y., Chiba, T., Taminato, T., and Fujita, T. (1978): Secreted forms of somatostatin. *Life Sci.*, 23:1647.

Alpert, L. C., Brawer, J. R., Patel, Y. C., and Reichlin, S. (1976): Somatostatinergic neurons in anterior hypothalamus: Immunohistochemical localization. *Endocrinology*, 98:225–258.

Arimura, A., and Schally, A. V. (1976): Increases in basal and thyrotropin releasing hormone (TRH)-stimulated secretion of thyrotropin (TSH) by passive immunization with antiserum to somatostatin in rats. *Endocrinology*, 98:1069–1072.

Brazeau, P., Vale, W., Burgus, R., Ling, N., Butcher, M., Rivier, J., and Guillemin, R. (1973): Hypothalamic polypeptide that inhibits the secretion of immunoreactive pituitary growth hormone. *Science*, 179:77–79.

Brown, M., Speiss, J., Reubi, J. C., Perrin, M., and Vale, W. (1981): Somatostatin-28 (SS28), somatostatin-14 (SS-14) and their analogs: Comparative biological actions. *Program 63rd Ann. Meeting, The Endocrine Society*, Cincinnati, p. 164A.

Brownstein, M., Arimura, A., Sato, H., Schally, A. V., and Kizer, J. S. (1975): The regional distribution of somatostatin in the rat brain. *Endocrinology*, 96:1456–1461.

Davies, P., Katzman, R., and Terry, R. D. (1980): Reduced somatostatin-like immunoreactivity in cerebral cortex from cases of Alzheimer disease and Alzheimer senile dementia. *Nature*, 288:279–280.

Delfs, J., Robbins, R., Connolly, J. L., Dichter, M., and Reichlin, S. (1980): Somatostatin production by rat cerebral neurons in dissociated cell culture. *Nature*, 283:676–677.

Dockray, G. J. (1978): Gastrin overview. In: *Gut Hormones*, edited by S. R. Bloom, pp. 129–139. Churchill Livingstone, Edinburgh.

Ensinck, J. W., Laschansky, E. C., Kanter, R. A., and Fujimoto, W. Y. (1978): Somatostatin biosynthesis and release in the hypothalamus and pancreas of the rat. *Metabolism*, 27:1207–1210.

Epelbaum, J., Willoughby, J. O., Brazeau, P., and Martin, J. B. (1977): Effects of brain lesions and hypothalamic deafferentation on somatostatin distribution in the rat brain. *Endocrinology*, 101:1495–1502.

Forssmann, W. G. (1978): A new somatostatinergic system in the mammalian spinal cord. *Neurosci. Lett.*, 10:293–297.

Gamse, R., Vaccaro, D. E., Gamse, G., DiPace, M., Fox, T. O., and Leeman, S. E. (1980): Release of immunoreactive somatostatin from hypothalamic cells in culture: Inhibition by γ-aminobutyric acid. *Proc. Natl. Acad. Sci. USA*, 77:5552–5556.

Goodman, R. H., Jacobs, J. W., Chin, W. W., Lund, P. K., Dee, P. C., and Habener, J. E. (1980): Nucleotide sequence of a cloned structural gene coding for a precursor of pancreatic somatostatin. *Proc. Natl. Acad. Sci. USA*, 77:5869–5873.

Havlicek, V., and Friesen, H. G. (1979): Central nervous system effects of hypothalamic hormones and other peptides. In: *Comparison of Behavioral Effects of Somatostatin and B-endorphin in Animals*, edited by R. Collu, A. Barveau, J. R. Ducharme, and J. -G. Rochefort, pp. 381–402. Raven Press, New York.

Hökfelt, T., Efendic, S., Johansson, O., Luft, R., and Arimura, A. (1974): Immunohistochemical localization of somatostatin (growth hormone release-inhibiting actor) in the guinea pig brain. *Brain Res.*, 80:165–169.

Hökfelt, T., Elde, R., Johansson, O., Luft, R., and Arimura, A. (1975): Immunohistochemical evidence for the presence of somatostatin, a powerful inhibitory peptide, in some primary sensory neurons. *Neurosci. Lett.*, 1:231–235.

Hökfelt, T., Johansson, O., Ljungdahl, A., Lundberg, J. M., and Schultzbert, M. (1980): Peptidergic neurons. *Nature*, 284:515–521.

Konturek, S. J., Tasler, J., Jaworek, J., Pawlik, W., Walus, K. M., Schusdziarra, V., Meyers, C. A., Coy, D. H., and Schally, A. V. (1981): Gastrointestinal secretory, motor, circulatory, and metabolic effects of prosomatostatin. *Proc. Natl. Acad. Sci. USA*, 78:1967–1971.

Krisch, B. (1978): Hypothalamic and extrahypothalamic distribution of somatostatin-immunoreactive elements in the rat brain. *Cell Tissue Res.*, 195:499–513.

Krisch, B. (1980): Differing immunoreactivities of somatostatin in the cortex and the hypothalamus of the rat. (A light and electron microscopic study.) *Cell Tissue Res.*, 212:457–464.

Kronheim, S., Berelowitz, M., and Pimstone, B. L. (1976): A radioimmunoassay for growth hormone release-inhibiting hormone: Method and quantitative tissue distribution. *Clin. Endocrinol.*, 5:619–630.

Lechan, R. N., Nestler, J., Jacobson, S., and Reichlin, S. (1980): The hypothalamic "tuberoinfundibular" system of the rat as demonstrated by horseradish peroxidase (HRP) microiontophoresis. *Brain Res.*, 195:13–27.

Lundberg, J. M., Hökfelt, T., Nilsson, G., Terenius, L., Rehfeld, J., Elde, R., and Said, S. (1978): Peptide neurons in the vagus, splanchnic and sciatic nerves. *Acta Physiol. Scand.*, 104:499–501.

Marks, N., and Stern, F. (1975): Inactivation of somatostatin (GH-RIH) and its analogs by crude and partially purified rat brain extracts. *FEBS Lett.*, 55:220–224.

Meyers, C. A., Murphy, W. A., Redding, T. W., Coy, D. H., and Schally, A. V. (1980): Synthesis and biological actions of prosomatostatin. *Proc. Natl. Acad. Sci. USA*, 77:6171–6174.

Noe, B. D., Fletcher, D. J., Bauer, G. E., Weir, G. C., and Patel, Y. (1978): Somatostatin biosynthesis occurs in pancreatic islets. *Endocrinology*, 102:1675–1685.

Patel, Y. C., and Reichlin, S. (1978): Somatostatin in hypothalamus, extrahypothalamic brain, and peripheral tissues of the rat. *Endocrinology*, 102:523–530.

Pelletier, G. (1980): Immunohistochemical localization of somatostatin. *Prog. Histochem. Cytochem.*, 12:1–41.

Petrusz, P., Sar, M., Grossman, G. H., and Kizer, J. S. (1977): Synaptic terminals with somatostatin-like immunoreactivity in the rat brain. *Brain Res.*, 137:181–187.

Rasool, C. G., Schwartz, A. L., Bollinger, J. A., Reichlin, S., and Bradley, W. G. (1981): Immunoreactive somatostatin distribution and axoplasmic transport in rat peripheral nerve. *Endocrinology*, 108:996–1001.

Reichlin, S. (1981): Systems for the study of regulation of neuropeptide secretion. In: *Neurosecretion and Brain Peptides*, edited by J. B. Martin, S. Reichlin, and K. L. Bick, pp. 573–597. Raven Press, New York.

Reichlin, S., Robbins, R. I., and Lechan, R. (1982): Somatostatin and the nervous system. *Metabolism (in press)*.

Renaud, L. P., Martin, J. B., and Brazeau, P. (1975): Depressant action of TRH, LH-RH and somatostatin on activity of central neurons. *Nature*, 255:233–235.

Robbins, R., Sutton, R., and Reichlin, S. (1982a): Sodium- and calcium-dependent somatostasin release from dissociated cerebral cortical cells in culture. *Endocrinology*, 110:496–499.

Robbins, R., Sutton, R., and Reichlin, S. (1982b): Effects of neurotransmitters and cyclic AMP on somatostatin release from cultured cerebral cortical cells. *Brain Res.*, 234:377–386.

Rorstad, O. P., Brownstein, M. J., and Martin, J. B. (1979a): Immunoreactive and biologically active somatostatin-like material in rat retina. *Proc. Natl. Acad. Sci. USA*, 76:3019–3023.

Rorstad, O. P., Epelbaum, J., Brazeau, P., and Martin, J. B. (1979b): Chromatographic and biological properties of immunoreactive somatostatin in hypothalamic and extrahypothalamic brain regions of the rat. *Endocrinology*, 105:1083–1092.

Scanlon, M. F., Robbins, R., Bolaffi, J. L., and Reichlin, S. (1981): Somatostatin and TRH secretion by hypothalamic and cortical neurons in a longterm maintenance capillary perfusion system. *Program 63rd Ann. Meeting, The Endocrine Society*, Cincinnati, p. 197.

Srikant, C. B., and Patel, Y. C. (1981): Somatostatin receptors: Identification and characterization in rat brain membranes. *Proc. Natl. Acad. Sci. USA*, 78:3930–3934.

Terry, L. C., Willoughby, J. O., Brazeau, P., Martin, J. B., and Patel, Y. (1976): Antiserum to somatostatin prevents stress-induced inhibition of growth hormone secretion in the rat. *Science*, 192:565.

Tessler, A., Goldberger, M., Murray, M., Himes, B. T., and Artymyshyn, R. (1980): Lack of plasticity in somatostatin systems in CAT lumbar spinal cord. *10th Ann. Meeting Soc. for Neurosciences*, Cincinnati, 59.11 p. 173.

Vale, W., Brazeau, P., Rivier, C., Brown, M., Ross, B., Rivier, J., Burgus, R., Ling, N., Guillemin, R. (1975): Somatostatin. *Recent Prog. Horm. Res.*, 31:365–397.

Vale, W., Rivier, C., and Brown, M. (1977): Regulatory peptides in the hypothalamus. *Ann. Rev. Physiol.*, 39:473–527.

Wilson, D. L., and Stone, G. C. (1979): Axoplasmic transport of proteins. *Ann. Rev. Biophys. Bioeng.*, 8:27–45.

Yamada, T., Marshak, D., Basinger, S., Walsh, J., Morley, J., and Stell, W. (1980): Somatostatin-like immunoreactivity in the retina. *Proc. Natl. Acad. Sci. USA*, 77:1691–1695.

Early Translational Events in the Synthesis of Acetylcholine Receptor

David J. Anderson, Peter Walter, and Günter Blobel

ABSTRACT

Early events in the biosynthesis of the *Torpedo* acetylcholine receptor (AChR) have been studied, using a cell-free protein synthesizing system supplemented with canine pancreas microsomal membranes. Using subunit-specific antisera, it has been demonstrated that each AChR subunit is translated from a separate mRNA, and is integrated into the endoplasmic reticulum membrane by a process that results in a transmembrane orientation for each polypeptide. The *in vitro* synthesized subunits are glycosylated, but are apparently not assembled into a functional quaternary complex. Purified signal recognition protein (SRP), previously shown to be obligatory for the translocation of nascent secretory proteins across the microsomal membrane, was shown to be required as well for the integration of AChR subunits into the lipid bilayer. Therefore, AChR subunits apparently undergo membrane integration by the same mechanism established in the case of the simple viral transmembrane protein, vesicular stomatitis virus (VSV) G protein.

There are a number of reasons for pursuing a study of the molecular biology of acetylcholine receptor (AChR) at the translational level. First, AChR is one of the best studied of all neurotransmitter receptors, making it an attractive model system for biosynthetic studies on this class of neuronal proteins. Second, a number of interesting regulatory phenomena that involve the AChR have been described in the literature. For example, denervation of mature skeletal muscle causes a specific increase in synthesis of AChR (Brockes and Hall, 1975), the basis for denervation supersensitivity in this tissue. In addition, factors have been extracted from brain that can increase the synthesis of AChRs in cultured skeletal muscle (Jessel et al., 1979). To investigate the mechanistic bases of such phenomena, it is important to understand the complete biosynthetic pathway of AChR, from transcription of the gene(s) to delivery of the assembled protein to the plasma membrane. One can then determine at which stages in this pathway receptor synthesis is regulated. Furthermore, from the standpoint of basic cell biology, AChR is interesting in that it represents the first multimeric integral plasma membrane protein to be synthesized *in vitro*. In this respect one would like to know whether each subunit is translated from a separate mRNA, or if all subunits are cleaved from a common precursor. Furthermore, one wonders whether the membrane integration of such a complex protein follows the same principles established for single-subunit transmembrane proteins. Finally, the development of efficient *in vitro* translation systems for AChR mRNA provides a first step toward identifying and isolating cloned AChR genes.

IN VITRO SYNTHESIS OF AChR SUBUNITS

Figure 1 illustrates a current model of protein translocation across, and integration into, the rough endoplasmic reticulum (RER) membrane. Several points should be clarified. First, the

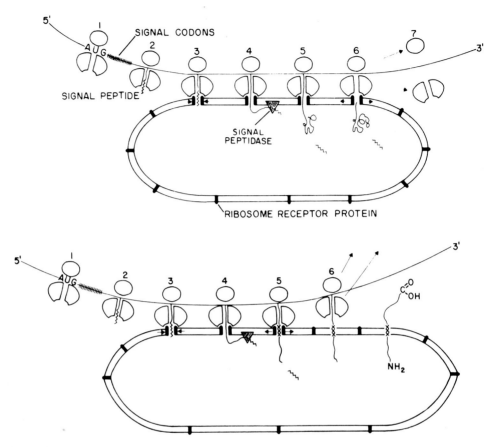

FIG. 1. Conceptualization of protein translocation across and integration into the RER. For both secretory and membrane proteins, the nascent polypeptide chain contains information at its NH_2-terminus, termed the signal sequence, which is recognized and decoded by specific receptors in the RER membrane. This receptor–ligand interaction is proposed to initiate the formation of a functional ribosome-membrane junction, permitting vectorial transfer of the growing polypeptide across the lipid bilayer. Concomitant with or shortly after this translocation, the signal peptide is cleaved by a specific protease. Cleavage is not necessary for translocation, as some membrane and secretory proteins are synthesized with uncleaved (Goldman and Blobel, 1981) or internal (Lingappa et al., 1979) signal sequences. In the case of a transmembrane protein *(bottom)*, translocation of the polypeptide across the lipid bilayer is incompletely effected, perhaps due to the presence of a specific "stop-transfer" sequence that abrogates the interaction between the ribosome and the membrane.

proposed mechanism for the translocation of soluble secretory proteins across membranes can be extended in a simple way to explain the integration of intrinsic membrane proteins into the lipid bilayer. Second, the concepts described in this figure have been developed from *in vitro* studies in which the processing events that occur early in biosynthesis were reconstituted by supplementing cell-free translation systems with heterologous rough microsomal membrane vesicles (Blobel and Dobberstein, 1975). A crucial control in these early studies was to demonstrate that these heterologous, reconstituted cell-free systems in fact seemed to reconstruct accurately the events that occur in the RER *in vivo*. In the case of transmembrane glycoproteins, this was first shown by Katz and coworkers (1977) for the G protein of vesicular stomatitis virus (VSV). Specifically, it was demonstrated that the protein not only underwent correct proteolytic processing and glycosylation, but also exhibited the same transmembrane orientation as material obtained from pulse-labeled cells infected with virus.

To apply such methods to a particular membrane protein, one needs an appropriate mRNA preparation that directs the synthesis of the protein, and an antibody with which to purify the translation products from the reaction mixtures by immunoprecipitation. In the case of the AChR, the electric organs of electric rays have long been known to contain copious quantities of nicotinic receptors and were therefore apparently the tissue of choice for extracting AChR mRNA. Unexpectedly, this tissue in fact contains very small amounts of mRNA. Apparently, the high steady-state levels of AChR in electric organ result from slow turnover of the protein. Despite the fact that the tissue yields only small amounts of mRNA, the receptor polypeptide mRNAs make up a fairly substantial fraction, amounting to almost 2% of the total message (Mendez et al., 1980.)

Several laboratories have contributed to the biochemical characterization of AChR, including Arthur Karlin (Weill et al., 1974), Michael Raftery (Vandlen et al., 1979), and Jean-Pierre Changeux (Sobel et al., 1977). In the electric organ, this protein is a pentameric structure composed of four polypeptide subunits, α, β, γ, and δ, of molecular weights 40, 50, 60, and 65 Kd, respectively, in a stoichiometric ratio of 2:1:1:1. Although peptide mapping studies (Froehner and Rafto, 1979; Lindstrom et al., 1979) originally indicated that these four subunits were biochemically distinct, direct amino acid sequencing (Raftery et al., 1980) has recently revealed 30 to 50% sequence homology between the chains in the first NH_2-terminal 15% of the polypeptides. These four chains form a quaternary, noncovalently associated oligomeric complex that can be dissociated only by denaturing detergents such as sodium dodecyl sulfate (SDS). The complex migrates as a 9S species on sucrose density gradients and binds 2 moles of acetylcholine per mole of receptor. There is also a dimeric form of the receptor produced by the formation of a disulfide bridge between two monomers (Chang and Bock, 1977.)

Fig. 2A, lane 1, shows the electrophoretic profile of an AChR preparation purified by affinity chromatography on cobratoxin-sepharose (Froehner and Rafto, 1979). The subunits were separated by preparative electrophoresis in SDS, and eluted as individual polypeptides. The homogeneity of these preparations is shown in lanes 2 to 5. Using methods established by Lindstrom and his coworkers (1978), these subunits were injected into rats and antisera were produced against each of the subunits. The specificities of these reagents for particular subunits are shown in Fig. 2B. To characterize these antisera, receptor protein was purified and iodinated under nondenaturing conditions. This material is shown in lane 1. The subunits were dissociated with SDS and different antisera were tested for their ability to immunoprecipitate these polypeptides. It is clear that each serum reacts primarily with the subunit to which it was originally raised (lanes 1 to 5). Interestingly, if this same assay is performed using AChR not dissociated in SDS, each serum precipitates all four subunits (data not shown). This result reflects the association of the subunits in the oligomeric complex. Hence, this assay provides an easy way to determine whether multisubunit assembly has occurred *in vitro*.

These reagents were used to isolate the primary translation products corresponding to each of the four AChR subunits. Messenger RNA from the electric organ was translated in a wheat germ cell-free protein synthesizing system in the absence of rough microsomal membranes. Translation products were denatured in SDS, immunoprecipitated with the subunit-specific sera, and displayed by electrophoresis on polyacrylamide gels. Figure 3 shows the results obtained for the γ and δ subunits. The following discussion will be restricted to these two chains. However, similar results were obtained for the two smaller subunits, α and β.

The results of this experiment are shown in lanes 2 of both panels. There is a separate band for each of the two subunits, and in both cases the primary translation product is smaller in molecular weight than its mature counterpart. This is to be expected, since all of the AChR subunits are known to be glycosylated (Vandlen et al., 1979), and the primary translation

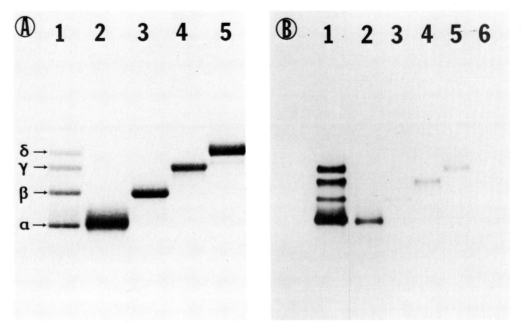

FIG. 2. **A:** Purification of *Torpedo* AChR subunits. Shown is a linear gradient 7.5 to 15% SDS polyacrylamide gel electrophoretogram. Protein was visualized by coomassie blue staining. Lane 1, AChR eluted from a cobratoxin-Sepharose affinity column. α, β, γ, and δ indicate the 40-, 50-, 60-, and 65-kD subunits, respectively. Lanes 2 to 5, the α, β, γ, and δ subunits rerun in the same gel system after preparative electrophoresis in SDS. Note the absence of significant cross-contamination in the subunit preparations. (From Anderson and Blobel, 1981.) **B:** Characterization of subunit specificities of anti-AChR antisera. AChR purified under nondenaturing conditions (10 ml Tris HCl, pH 7.4, 0.01% NaCholate) was radioiodinated using the IODOGEN reagent (New England Nuclear, Inc.). Parallel samples of [^{125}I]AChR were heated to 100°C in 1% SDS, 10 mM dithiothreitol (DTT) for 3 min, and then diluted with four volumes of buffer containing 1.25% Triton X-100, prior to addition of antisera. After an overnight incubation with the antibody, immunoprecipitates were adsorbed to Protein A-Sepharose, washed and subjected to electrophoresis in the same system described for **A**. Shown is an autoradiograph of the dried gel. Antisera used were: lanes 2 to 5, anti- α, β, γ, and δ, respectively; lane 6, nonimmune serum. Lane 1 shows the radioiodinated purified receptor. (Anderson and Blobel, 1981.)

products do not contain carbohydrate. The identity of these translation products was confirmed by the ability of the corresponding unlabeled *in vitro*-synthesized subunits to compete out the immunoprecipitation of their *in vitro*-synthesized counterparts. The molecular weights of the primary translation products obtained for the four chains correlate well with those published by Baxter and coworkers (Mendez et al., 1980) using a reticulocyte lysate system and an antiserum that reacted with all four subunits.

As mentioned earlier, one reason for carrying out this study was to determine whether there is a separate mRNA for each subunit, or whether all are derived from a common polyprotein precursor. R. Lerner *(this volume)* has indicated that such polyproteins are in fact synthesized in the case of viral membrane glycoproteins. The fact that distinct translation products corresponding to each subunit were observed, rather than a common 250,000-dalton species, strongly argues against a polyprotein precursor. However, because cell-free systems contain proteases that might rapidly cleave such a precursor, rigorous evidence was provided with the use of radioactive [^{35}S]formylated methionyl (fMet) tRNA labeling of the primary translation product (Mihara and Blobel, 1980).

The basis for the experiment is the fact that the first amino acid of a protein sequence, the translation initiation site, and only that site, codes for fMet in eukaryotic mRNAs. A polyprotein

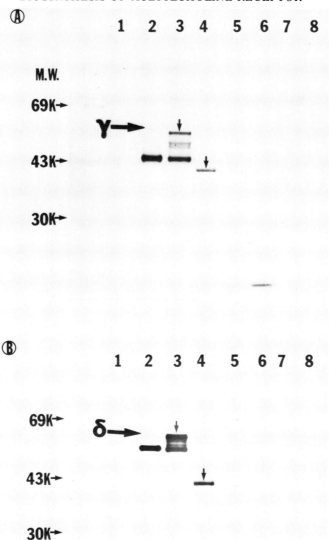

FIG. 3. *In vitro* synthesis and membrane integration of the γ and δ AChR subunits. After solubilization in 1% SDS and dilution with buffer containing a fivefold excess of Triton X-100, translation products were immunoprecipitated with either anti-γ **(A)** or anti-δ **(B)** antisera. Shown is a fluorograph of a 7.5 to 15% SDS polyacrylamide gel. M. W., molecular weight markers in kilodaltons. Lane 1, purified unlabeled γ subunit **(A)** or δ subunit **(B)** was run on the same gel and its position marked by aligning the autoradiographic film with the original gel; lanes 2 to 8 contain *in vitro* translation products labeled with [^{35}S]methionine. Translations were carried out in either the absence (lanes 2, 6, 7) or presence (lanes 3, 4, 5, 8) of two A_{280}/ml rough microsomes, and after translation was complete, the samples in lanes 2, 6, and 7 received two A_{280}/ml of microsomes, and were further incubated for 90 min at 26° C. In lanes 7 and 8, samples were immunoprecipitated after proteolysis with 300 μg/ml trypsin at 20° C for 30 min. In lane 5 the trypsinization was performed in the presence of 0.1% Nikkol, a nonionic detergent. Samples were reduced with 100 mM DTT and alkylated with 500 mM iodoacetamide prior to electrophoresis. In lanes 3, *arrows* indicate major glycosylated form; in lanes 4, major tryptic fragment. (Anderson and Blobel, 1981.)

precursor would yield, therefore, only one subunit labeled by the initiator methionine. The fact that all four subunits were labeled under these conditions shows that there is a separate mRNA for each chain (Anderson and Blobel, 1981). Given this analysis of the primary translation products, the next step was to examine the forms of the subunits that were synthesized when the translation system was supplemented with dog pancreas rough microsomal membranes.

Microsomes from this source are widely used because they function in a variety of cell-free systems without inhibiting protein synthesis (Shields and Blobel, 1978). Microsomal membranes from other tissues such as hen oviduct (Thibodeau and Walsh, 1980), bovine adrenal medulla (Boime et al., 1980), and Ehrlic ascites cells (Szczesna and Boime, 1976) have also been shown to effect proper processing of nascent presecretory proteins synthesized *in vitro*.

The results of this experiment are shown in lanes 3 of Fig. 3. A new major form of each chain that, in this gel system, migrates close to the position of the corresponding authentic subunit is obtained under these conditions. Some minor intermediate species are also seen. All of these new forms can be competed out of the immunoprecipitation by adding an excess of the corresponding unlabeled authentic subunit (lanes 8) indicating that they are immunologically indistinguishable from material synthesized in the absence of membranes (lanes 2). The increase in molecular weight on addition of microsomal membranes suggests that core glycosylation of the newly synthesized subunits may be occurring in this system.

TRANSMEMBRANE ORIENTATION OF AChR SUBUNITS

In mature AChR in *Torpedo* plasma membrane vesicles, Wennogle and Changeux (1980) showed that the α and δ subunits are insensitive to proteolytic attack at the extracytoplasmic surface of the membrane, but are readily cleaved once the vesicles are disrupted by sonication or alkaline extraction. This result suggested that both subunits had protease-sensitive sites at the cytoplasmic surface of the membrane. No evidence was available for the other two chains, however. The cell-free system used in the present study has the advantage that the rough microsomal membranes containing *in vitro*-synthesized AChR subunits are topologically inside-out with respect to AChR-rich *Torpedo* vesicles. Since the cytoplasmic surface of the microsomes is exposed to the aqueous medium, sensitivity of an AChR subunit(s) to proteolytic attack can be taken as an unambiguous indication that the polypeptide(s) communicates with the cytoplasm, without the complications of interpretation created by the use of membrane-disruptive procedures.

The results of this experiment are shown in lane 4 of Fig. 3. Treatment with high concentrations of trypsin caused both γ and δ to be reduced to a smaller fragment. In the case of the δ subunit, this fragment is almost 20 kD smaller than the uncleaved polypeptide. In contrast, lanes 6 show that the precursor synthesized in the absence of membranes is completely degraded by this procedure or is reduced to very small fragments. Thus, the bands in lanes 4 are not due to intrinsic insensitivity of the protein to proteolysis. Finally, if the trypsinization was performed on material synthesized with membranes, but with nonionic detergent present to disrupt the lipid bilayer, the subunits are again completely degraded, as seen in lanes 5. This result shows that the incomplete degradation of subunits in the presence of microsomes is due to protection of part of the chain by the protease-impermeable lipid bilayer. Thus, some part of each subunit has been translocated across or integrated into the membrane. However, the fact that substantial portions of each chain can be cleaved off by this procedure indicates that in fact all four subunits have regions exposed on the cytoplasmic surface of the membrane. Since labeling studies have shown that all four subunits are exposed on the extracellular surface of the membrane (Strader et al., 1979), it may be concluded that each of the four AChR subunits is a transmembrane protein.

Core glycosylation of the membrane-integrated forms of the AChR subunits was demonstrated by an affinity chromatography of the *in vitro* translation products on concanavalin A (Con A) Sepharose. The principle of the fractionation is that glycosylated polypeptides containing the mannose-rich core oligosaccharide will bind to the lectin-agarose, whereas nonglycosylated material is not retained. A representative result is shown for the δ subunit in Fig. 4.

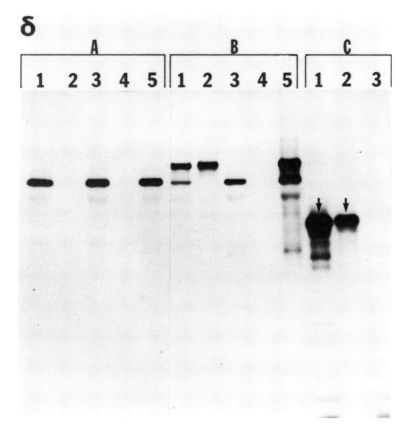

FIG. 4. Demonstration of *in vitro* glycosylation of the δ subunit by affinity chromatography on Con A-Sepharose. **A**, Products synthesized without membranes; **B** and **C**, products synthesized with membranes. In group **C**, the products were trypsinized immediately after translation (see Fig. 3, lane 4). Lanes 1, an aliquot immunoprecipitated without further fractionation; lanes 2, material that was adsorbed to the Con A-Sepharose; lanes 3, material not retained by the lectin. Groups **A** and **B**, lanes 4 and 5, are as lanes 2 and 3 but incubation with the Con A-Sepharose was in the presence of 0.4 M α-methylmannoside. Samples were eluted from the Con A by boiling in 1% SDS, and immunoprecipitated with anti-δ antiserum. The lower bands in lane **C**1 are contaminants that do not appear in lanes **C**2 or 3 due to loss of material incurred during the affinity chromatography procedure. *Arrows* indicate major tryptic fragment (see also Fig. 3B, lane 4). (Anderson and Blobel, 1981.)

The 60-kD "precursor" to δ synthesized in the absence of membranes is not glycosylated, and appears in the fraction not retained by the Con A (group A, lane 3). In contrast, material synthesized in the presence of microsomes (group B, lane 1) can be fractionated into a 65-kD form of δ, which is adsorbed by the lectin (lane 2), and residual unprocessed "pre-δ," which is not retained (lane 3). This binding could be abolished by adding excess α-methylmannoside as a "competitive inhibitor" of the lectin-glycoprotein interaction, i.e., none of the species was adsorbed (lane 4) and all appeared in the flow-through fraction (lane 5), indicating that the binding of the 65-kD form to the Con A was due to the presence of a specific carbohydrate on the protein.

Additionally, it was shown that the glycosylated residues of δ are contained in that part of the protein that is protected by the membrane from exogenously added protease. Thus, the 44-kD trypsin-resistant fragment of δ (group C, lane 1, arrow), when subjected to fractionation on Con A-Sepharose, was specifically adsorbed to the lectin (lane 2, arrow), indicating the presence of mannose residues. The presence of core oligosaccharide on the membrane-integrated forms of the AChR subunits has recently been confirmed by an independent method, using the enzyme Endoβ N-acetylglycosaminidase H (Anderson et al., 1982).

Figure 5 depicts some of the possible topological orientations of the AChR subunits in the membrane. The simplest one, on the left, shows the chain spanning the membrane only once. This is consistent with the fact that the subunits exhibit a single protease-resistant domain that is glycosylated. It is also in keeping with the precedent established for VSV G (Katz et al., 1977), and the HLA antigens (Ploegh et al., 1979). However, the present study is limited by the necessity of immunoprecipitation, and it is possible that there are some additional membrane-integrated and protease-protected segments that do not react with the antibody used. The more complex orientation implied by this possibility is suggested in the middle diagram. Finally, although by analogy to VSV G one might expect that the amino terminus of the polypeptide is on the inside of the vesicle, one can equally envision the orientation shown in the diagram on the right. Sequencing studies are currently being performed to investigate this issue.

As previously discussed, mature AChR is a pentameric complex of composition $\alpha_2\beta\gamma\delta$ which binds the cholinergic antagonist α-bungarotoxin with extremely high affinity ($k_d = 10^{-11}$M). When the *in vitro*-synthesized, membrane-integrated AChR subunits were assayed for toxin binding and oligomeric assembly by several independent methods, neither property could be demonstrated *(data not shown)*. Trivial explanations for these negative results have been ruled out. Thus, the cell-free system does not contain an inhibitor of α-bungarotoxin binding, since exogenously added mature AChR binds the ligand in a normal manner after incubation in a mock translation.

For what reasons might oligomeric assembly and functional maturation not occur *in vitro*? If the mechanism of assembly is simply lateral diffusion of the subunits and aggregation in the plane of the membrane, it is unlikely that such an assembly would occur with this system. In the cell, the RER is an extensive reticulum of membrane sheets. What is added to the cell-free system is a suspension of membranous vesicles, obtained from homogenized tissue. Since electron microscopy reveals that each vesicle can accommodate between 4 and 10 polysomes, given that each AChR subunit comprises 0.5% of total protein synthesis (and assuming this reflects mRNA abundance), it can be calculated that the probability of getting one polysome for each of the four subunits on a single vesicle is about one in a million. Another possible reason for the lack of functional assembly of AChR in this system is that under these conditions only those events that normally occur in the RER are reconstructed, whereas the AChR may require transport to another intracellular organelle, such as the Golgi apparatus, before the toxin binding site can form.

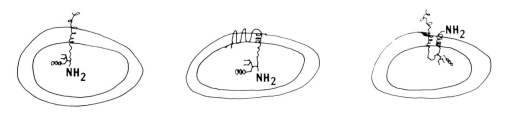

FIG. 5. Possible transmembrane orientations of the δ subunit in the RER membrane. The inside of the vesicles is topologically equivalent to the outside of the cell. "NH$_2$" indicates the amino terminus of the polypeptide chain. The branched structure represents a core oligosaccharide group, but does not imply that there is only one of these groups per chain, nor that its position in the protein is precisely as indicated.

MECHANISMS OF COTRANSLATIONAL INTEGRATION OF RECEPTOR SUBUNITS INTO RER MEMBRANE

A protein, essential for the vectorial translocation of nascent secretory proteins across the RER membrane, has been purified from pancreas rough microsomes (Walter and Blobel, 1980). This protein is a complex of six polypeptide chains that can be removed from the microsomes by extraction with buffers containing high monovalent salt concentrations. In subsequent studies, it was further determined that this complex functions in the initial binding of secretory polysomes to the rough microsomal membrane (Walter and Blobel, 1981). Moreover, the protein in solution displays a high affinity for polysomes synthesizing secretory (but not cytoplasmic) proteins. Since the only difference between these polysomes is that those containing nascent secretory proteins also express a consensus structure at the NH_2-terminus termed the signal sequence, the protein has been named signal recognition protein (SRP).

Figure 6 reiterates the point that transmembrane protein integration can be envisioned by a simple extension of the principles that operate in the translocation of secretory proteins. Recently, it was determined that SRP is obligatory for the integration of the AChR subunits into the membrane (D. J. Anderson, P. Walter, and G. Blobel, 1982). In other words, if SRP was removed from rough microsomes by salt extraction, these membranes were incapable of integrating or glycosylating the AChR subunits. However, addition of purified, homogeneous SRP to these inactive membranes restored their ability to integrate the transmembrane proteins concomitant with their capacity to translocate secretory proteins. Apparently, polysomes synthesizing secretory and membrane proteins destined for the RER share a common initial step in their interaction with the membrane (Lingappa et al., 1978). This step is mediated by a specific protein component, SRP, that functions as a signal sequence receptor. Radiosequencing studies are in progress to determine whether the four AChR subunits are each synthesized with signal sequences, as would be predicted from this result.

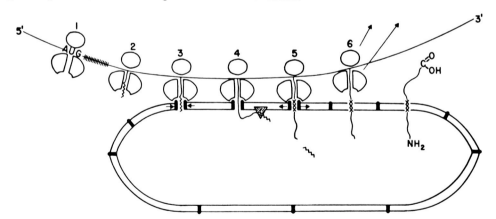

FIG. 6. Model established for the integration of VSV G, a single subunit transmembrane glycoprotein, into the RER membrane (Katz et al., 1977). A similar mechanism is postulated to operate in the case of each of the four AChR subunits, although in the latter cases the cytoplasmic domains of the polypeptides may be considerably larger than in that of the former. The dark particles in the membrane represent RER-associated proteins that may be required for one or more stages in the integration process. These include: (1) recognition of the signal sequence; (2) binding of the polysome to the membrane; (3) modification of the lipid bilayer permeability barrier to allow passage of the nascent chain through the membrane; and (4) dissociation of the ribosome-membrane junction and stabilization of membrane-integrated hydrophobic domains. At least one of these events (1) is in fact mediated by a protein, SRP, which does not distinguish between secretory and membrane proteins, but which does distinguish between these and cytoplasmic proteins.

It is thought that the signal sequence of secretory and membrane proteins interacts with SRP in a manner analogous to a classic receptor-ligand interaction (Walter and Blobel, 1981). If the leucine residues of preprolactin are substituted with the amino acid analog β-hydroxyleucine, translocation and processing of the polypeptide do not occur (Hortin and Boime, 1980). Furthermore, polysomes synthesizing such substituted preprolactin chains no longer bind SRP in solution at the high-affinity site (Walter and Blobel, 1981). Such experiments rule out the possibility that it is the secretory protein mRNA, and not the nascent polypeptide itself, that interacts with specific receptor sites on the membrane.

Regarding the role of the carboxy-terminal end of the protein in membrane insertion, it has been postulated that proteins that span the bilayer once, such as VSV G or HLA heavy chain, contain a so-called stop-transfer sequence near the carboxy-terminus (Lingappa et al., 1978). This sequence would actually stop the protein from being completely transferred across the RER membrane. It remains to be seen whether a separate protein is necessary to recognize such a stop-transfer sequence, or whether recognition is a function of the very complicated multifunctional SRP complex.

CONCLUSION

Figure 7 summarizes current understanding of the AChR in the context of its complete putative biosynthetic scheme, and suggests future approaches to solving this problem. Present data indicate that the four subunits of the AChR are translated on membrane-bound ribosomes as individual polypeptides, and integrated into the RER membrane by a cotranslational, SRP-mediated mechanism that results in a transmembrane orientation for each chain. At some point after synthesis is complete, the subunits assemble and the complex becomes competent to bind cholinergic agonists. Since the *in vitro* system seems unable to reproduce this assembly process, the problem must be approached in cell culture, by pulse-chase and immunoprecipitation experiments. Merlie and Sebbane (1981) have shown, and we have independently confirmed, that in cultured embryonic myotubes there is a 15- to 20-min delay between the time the AChR is synthesized and the time it can bind α-bungarotoxin. The work of Fambrough and his colleagues (Devreotes et al., 1977) has shown that muscle AChRs require 3 hr between synthesis and appearance on the plasma membrane. Taken together, these findings suggest the possibility that assembly of a multisubunit plasma membrane protein involves a unique posttranslational event that may be a rate-limiting step in its biosynthesis and hence a possible site of regulatory action. Finally, to extend our knowledge of AChR biosynthesis to the transcriptional level, it would be helpful to obtain cloned genes for each of the receptor subunits. Cell-free translation of receptor mRNA is certainly a first step toward this goal, and, in several laboratories, experiments have already been undertaken to produce electroplax cDNA libraries that can be screened for AChR-related clones.

GENERAL DISCUSSION

In eukaryotic cells mitochondria, chloroplasts, peroxisomes, and glyoxysomes possess integral membrane proteins that are synthesized on free ribosomes and posttranslationally inserted into the lipid bilayer. The mechanism of insertion of these proteins is presently not known. However, one could postulate, by analogy to the system described here, the existence of specific proteins in these organelles, which recognize signal sequences that are specific for a given organelle. These other signal recognition proteins might exist in an equilibrium between organelle membrane-associated forms and soluble cytoplasmic forms, as has been postulated in the case of RER SRP (Walter and Blobel, 1981). The soluble forms might interact with the

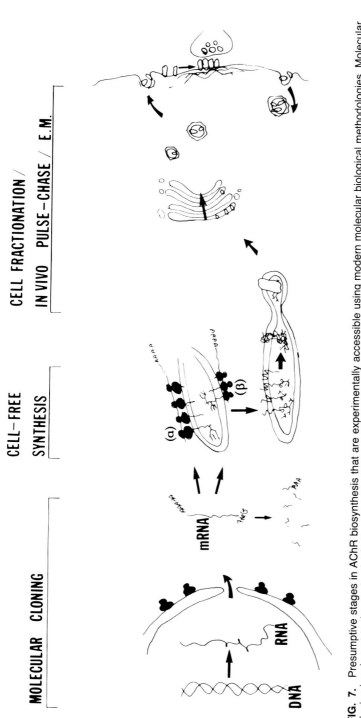

FIG. 7. Presumptive stages in AChR biosynthesis that are experimentally accessible using modern molecular biological methodologies. Molecular cloning is relevant to the study of AChR mRNA synthesis. Current evidence suggests that this process may require the transcription of at least four separate genes. Stages in posttranslational processing beyond the level of RER are not yet immediately accessible to in vitro manipulation and must be approached using classic cell biological methods. A current major problem is the lack of a cell culture system in which sufficient quantities of AChR could be synthesized to apply all of these approaches in concert.

nascent chains as they emerged from the free polysomes, but in contrast to the RER, no subsequent interaction between the polysome and the membrane would occur. Rather, the recognition protein, in a complex with the completed polypeptide, might "pilot" the protein through the cytoplasm to the appropriate site of integration. Implicit in this hypothesis is that the energy of protein synthesis is not needed to drive these integral membrane proteins into the lipid bilayer.

The classic example of multiple membrane insertions of a protein is bacteriorhodopsin. In that system there are seven separate transmembrane alpha-helical segments in a 26,000-dalton polypeptide. Lingappa and coworkers (1979) postulated that multiple internal signal sequences are used to stitch such a protein in and out of the membrane. However, the biochemistry of protein integration into prokaryotic membranes has not been clarified to the extent that it has in the RER system.

ACKNOWLEDGMENTS

This work was supported by a National Science Foundation Predoctoral Fellowship and a grant from the Muscular Dystrophy Foundation.

REFERENCES

Anderson, D. J., and Blobel, G. (1981): *In vitro* synthesis, glycosylation, and membrane insertion of the four subunits of *Torpedo* acetylcholine receptor. *Proc. Natl. Acad. Sci. USA*, 78:5598–5602.

Anderson, D. J., Walter, P., and Blobel, G. (1982): Signal recognition protein is required for the integration of acetylcholine receptor δ subunit, a transmembrane glycoprotein, into the endoplasmic reticulum membrane. *J. Cell Biol.*, 93:501–506.

Blobel, G., and Dobberstein, B. (1975): Transfer of proteins across membranes. II. Reconstitution of functional rough microsomes from heterologous components. *J. Cell Biol.*, 67:835–851.

Boime, I. Bielinska, M., Rogers, G., and Ruckinsky, T. (1980): *In vitro* processing of placental peptide hormones by smooth microsomes. *Ann. N. Y. Acad. Sci.*, 343:69–78.

Brockes, J. P., and Hall, Z. W. (1975): Synthesis of acetylcholine receptor by denervated rat diaphragm muscle. *Proc. Natl. Acad. Sci. USA.*, 72:1368–1372.

Chang, H. W., and Bock, E. (1977): Molecular forms of acetylcholine receptor. Effects of calcium ions and a sulfhydryl reagent on the occurrence of oligomers. *Biochemistry*, 16:4513–4520.

Devreotes, P. N., Gardner, J. M., and Fambrough, D. M. (1977): Kinetics of biosynthesis of acetylcholine receptor and subsequent incorporation into plasma membranes of cultured chick skeletal muscle. *Cell*, 10:365–373.

Froehner, S. C., and Rafto, S. (1979): Comparison of the subunits of *Torpedo Californica* acetylcholine receptor by peptide mapping. *Biochemistry*, 18:301–307.

Goldman, B. M., and Blobel, G. (1981): *In vitro* biosynthesis, core glycosylation and membrane integration of Opsin. *J. Cell Biol.*, 90:236–242.

Hortin, G., and Boime, I. (1980): Inhibition of pre-protein processing in ascites tumor lysates by incorporation of a leucine analog. *Proc. Natl. Acad. Sci. USA*, 77:1356–1360.

Jessel, T. M., Siegel, R. E., and Fishbach, G. D. (1979): Induction of acetylcholine receptors on cultured skeletal muscle by a factor extracted from brain and spinal cord. *Proc. Natl. Acad. Sci. USA*, 76:5397–5401.

Katz, F. M., Rothman, J. E., Lingappa, V. R., Blobel, G., and Lodish, H. F. (1977): Membrane assembly *in vitro*: Synthesis, glycosylation and asymmetric insertion of a transmembrane protein. *Proc. Natl. Acad. Sci. USA*, 74:3278–3282.

Lindstrom, J., Einarson, B., and Merlie, J. (1978): Immunization of rats with polypeptide chains from *Torpedo* acetylcholine receptors causes an autoimmune response to receptors in rat muscle. *Proc. Natl. Acad. Sci. USA*, 75:769–773.

Lindstrom, J., Merlie, J., and Yogeeswaran, G. (1979): Biochemical properties of acetylcholine receptor subunits from *Torpedo californica*. *Biochemistry*, 18:4465–4469.

Lingappa, V. R., Katz, F. N., Lodish, H. F., and Blobel, G. (1978): A signal sequence for the insertion of a transmembrane glycoprotein. *J. Biol. Chem.*, 253:8667–8670.

Lingappa, V. R., Lingappa, J. R., and Blobel, G. (1979): Chicken ovalbumin contains an internal signal sequence. *Nature*, 281:117–121.

Mendez, B., Valenzuela, P., Martial, J. A., and Baxter, J. D. (1980): Cell-free synthesis of acetylcholine receptor polypeptides. *Science*, 209:695–697.

Merlie, J. P., and Sebbane, R. (1981): Acetylcholine receptor subunits transit a precursor pool before acquiring α-bungarotoxin binding activity. *J. Biol. Chem.*, 256:3605–3608.

Mihara, K., and Blobel, G. (1980): The four cytoplasmically made subunits of yeast mitochondrial cytochrome c oxidase are synthesized individually and not as a polyprotein. *Proc. Natl. Acad. Sci. USA*, 77:4160–4164.

Ploegh, H. L., Cannon, L. E., and Strominger, J. L. (1979): Cell-free translation of the mRNAs for the heavy and light chains of HLA-A and HLA-B antigens. *Proc. Natl. Acad. Sci. USA*, 76:2273–2277.

Raftery, M. A., Hunkapiller, M. W., Streader, C. B. D., and Hood, L. E. (1980): Acetylcholine receptor: Complex of homologous subunits. *Science*, 208:1454–1457.

Shields, D., and Blobel, G. (1978): Efficient cleavage and segregation of nascent presecretory proteins in a reticulocyte lysate supplemented with microsomal membranes. *J. Biol. Chem.*, 253:3753–3756.

Sobel, A., Weber, M., and Changeux, J.-P. (1977): Large scale purification of the acetylcholine receptor protein in its membrane-bound and detergent extracted forms from *Torpedo marmorata* electric organ. *Eur. J. Biochem.*, 80:215–224.

Strader, C. B. D., Revel, J.-P., and Raftery, M. A. (1979): Demonstration of the transmembrane nature of the acetylcholine receptor by labeling with anti-receptor antibodies. *J. Cell Biol.*, 83:49–51.

Szczesna, E., and Boime, I. (1976): mRNA dependent synthesis of authentic precursor to human placental lactogen: Conversion to its mature form in ascites cell-free extracts. *Proc. Natl. Acad. Sci. USA*, 73:1179–1183.

Thibodeau, S. N., and Walsh, K. A. (1980): Processing of precursor proteins by preparations of oviduct microsomes. *Ann. N. Y. Acad. Sci.*, 343:180–191.

Vandlen, F. L., Wu, W. C.-S., Eisenach, J. C., and Raftery, M. A. (1979): Studies of the composition of purified *Torpedo californica* acetylcholine receptor and of its subunits. *Biochemistry*, 18:1845–1854.

Walter, P., and Blobel, G. (1980): Purification of a membrane-associated protein complex required for protein translocation across the endoplasmic reticulum. *Proc. Natl. Acad. Sci. USA*, 77:7112–7116.

Walter, P., and Blobel, G. (1981): Translocation of proteins across the endoplasmic reticulum. II, Signal recognition protein (SRP) mediates the selective binding to microsomal membranes of *in vitro* assembled polysomes. *J. Cell Biol.*, 91:551–556.

Weil, C. L., McNamara, M. G., and Karlin, A. (1974): Affinity labeling of purified acetylcholine receptor from *Torpedo californica*. *Biochem. Biophys. Res. Commun.*, 61:997–1003.

Wennogle, L. P., and Changeux, J.-P. (1980): Transmembrane orientation of proteins present in acetylcholine receptor-rich membranes from *Torpedo marmorata* studied by selective proteolysis. *Eur. J. Biochem.*, 106:381–393.

Section XI

Genetically Determined Disorders of the Nervous System

An important goal of the effort to increase knowledge of the structure and function of the genome is to improve techniques for investigating and alleviating genetically determined defects in humans. Genetic analysis of mutation in experimental animals also provides an opportunity to examine various aspects of normal development, and in so doing to improve our basic understanding of physiological function and organization.

Because various neuronal properties such as localization, morphology, synaptogenesis, and neurotransmitter selection are apparently genetically determined, genetic analysis is likely to illuminate the function, organization, and development of both the central and peripheral nervous systems. R. L. Sidman describes mouse mutations that provide especially appropriate material with which to illustrate both the promise and the problems that arise in the application of molecular genetic techniques to neurobiological and neuropathological issues. Sidman emphasizes the potential value of molecular genetic technology in the examination of genetic defects of the nervous system when coupled with an understanding of various aspects of the system under study.

At least half of the gene loci of human chromosomes may be involved in neural function. V. A. McKusick reviews progress in the delineation and localization of specific human gene loci, with special emphasis on disorders of the nervous system. This is an important initial step in the investigation of disorders for which the nature of the primary defect is unknown. An example of such a disorder is Huntington's disease, for which no effective treatment has yet been devised. It is a hereditary illness marked by degeneration of specific brain regions. J. B. Martin presents the clinical and neurological features of the disease in an historical framework and describes the various experimental strategies and findings that have evolved in the course of its investigation. Although the precise defect has yet to be identified, a very promising experimental approach for isolating and characterizing the gene or genes involved is outlined by D. E. Housman and J. F. Gusella. This involves the application of recombinant DNA technology to determine the precise chromosomal localization of the relevant gene and subsequent analysis of it and the surrounding genome as well as the products encoded by this region. These procedures may help provide new insights into the nature of this disease as well as others suspected of having a genetic basis.

Molecular Genetic Neuroscience, edited by
F. O. Schmitt, S. J. Bird, and F. E. Bloom.
Raven Press, New York © 1982.

Mutations Affecting the Central Nervous System in the Mouse

Richard L. Sidman

ABSTRACT

Three sets of mutations in mice are reviewed to illustrate prospects and problems in genetic analysis of nervous system organization and behavior. The shiverer locus affects myelin basic proteins in the central and peripheral nervous systems in a gene dose-dependent manner, acts intrinsically in cells of peripheral nerves other than the neuron and its axon, and is ripe for molecular genetic analysis. The weaver mutation reduces neurite formation, migration, and viability of cerebellar granule cell neurons *in vivo* and *in vitro*, also in a gene dose-dependent manner, but the range of intrinsically affected cells is uncertain and the molecular defect is unknown. Finally, several independent mutations affecting pigmentation of the retinal pigment epithelium also cause quantitative abnormalities of retinal ganglion cell neurons and of their axonal trajectories, with further reverberations on central visual organization and behavior; these manifold effects seem to be many steps removed from the molecular and genetic common denominator.

The approach of neurogeneticists to the study of the nervous system has usually been to identify a particular behavioral defect and then attempt to work backward to characterize the mutant gene responsible. Although this strategy has been applied to a wide variety of organisms, from protozoa and nematodes to fruit flies, mice, and human beings, it has not, in general, been completely satisfactory. In mammals, morphological and developmental analysis of behavioral mutants often does at least implicate a major affected cell type. Such data must be carefully interpreted, however, because a more detailed analysis frequently reveals developmental effects that occur prior to the one that at first seemed primary. A logical criterion for determining that the effect being measured is closely related to the primary defect rather than being an indirect result of the mutation is the observation of gene dosage effect on the phenotype of the affected cell type. Once the cell type of genetic interest has been identified, the next step for the neurogeneticist is to determine whether the gene action is normally intrinsic or extrinsic to this lineage. This can sometimes be done by designing an appropriate confrontation between wild-type and mutant cells using a graft, chimera, or tissue culture system.

Molecular biologists, on the other hand, have started with known gene defects, such as the human inherited aminoacidurias or lysosomal hydrolase deficiencies, and have tried to explain the complex nervous system effects of these mutations, including such disorders as mental retardation and epilepsy. In this endeavor, the molecular biologist has been no more successful than the neurogeneticist in achieving an understanding of how a specific behavioral defect can be caused by a specific defective gene product. The new techniques of molecular genetics should make it possible to bridge this gap. A potential starting point is the large existing collection of mouse neurological mutants, some of which have been extensively analyzed by neurogeneticists.[1]

Three cases, the shiverer mutation, the weaver mutation, and mutations affecting the visual system, will be considered in terms of the opportunities and difficulties they present for the molecular genetic approach.

SHIVERER MOUSE

The shiverer mutation is one of the best candidates in the mouse for eventually explaining a behavioral defect and its genetics at a molecular level. The shiverer mouse is characterized by a markedly decreased amount of myelin in the central nervous system (CNS) relative to control mice (Bird et al., 1978; Privat et al., 1979). This defect is expressed at the molecular level by a drastic and disproportionate reduction in the content of myelin basic protein (MBP) in the myelin membranes (Dupouey et al., 1979). Myelin basic protein is an extrinsic protein of the surface membrane of Schwann cells and oligodendroglial cells (Rumsby and Crang, 1977), as shown schematically in Fig. 1. The molecule has an axial ratio of about 10:1, and the polypeptide chain contains a hairpin turn at the site of a proline triplet (Martenson, 1981). In the mouse, several immunoreactive bands in the two-dimensional gels have been detected in the 14,000 to 19,000 dalton range, as well as a few very much higher molecular weight forms (Barbarese et al., 1978). It has been inferred that MBP resides at the cytoplasmic faces of mature myelin membranes, and since it is inserted at the time that myelin is becoming compact, it has been postulated that MBP plays a role in the developmental transition from loose to compact myelin (Braun, 1977). However, it is now clear that MBP is not an absolute requirement for zippering up the myelin into a compact structure (Kirschner and Ganser, 1980).

In shiverer homozygotes, the myelin of the CNS is not only decreased in amount but also displays an altered periodic structure relative to wild-type mice (Fig. 2) (Privat et al., 1979; Ganser and Kirschner, 1980). Unlike normal animals, in mutants virtually all myelin in the CNS has a loose structure; almost no compact myelin can be found. The amount of MBP in CNS myelin, as measured by radioimmunoassay, is 1/1,000 or less than the amount in normal mice (Privat et al., 1979; Ganser and Kirschner, 1980). Similarly, MBP in myelin of the peripheral nervous system (PNS) of the shiverer mutant has been reduced to undetectable levels (Kirschner and Ganser, 1980). Surprisingly, the myelin of the PNS, unlike that in the CNS, appears completely normal both in amount and in periodicity in the mutant animals (Kirschner and Ganser, 1980), as shown in Fig. 2.[2]

Shiverer heterozygotes display the gene dosage effect expected if the deficiency in MBP is a direct result of the mutation (Barbarese and Carson, 1981). Heterozygotes contain about half the normal levels of MBP in the CNS. The apparent absence of MBP in shiverer homozygotes has precluded the molecular analysis that might indicate whether the shiverer locus represents the structural gene for MBP. Polyacrylamide gel electrophoretic studies of membrane proteins reveal no polypeptides of less than 20,000 daltons in shiverer myelin, nor has any other evidence

---

[1]Virtually all the 100 or more mutations now available arose spontaneously and were detected by a caretaker at the Jackson Laboratories (Bar Harbor, Maine) or at comparable laboratories around the world. Although mice displaying abnormal locomotor behavior would be recognized in this situation, those with other classes of phenotypic abnormalities, i.e., defects, would not. No systematic effort has yet been attempted to obtain neurological mutants by deliberate use of mutagens.

[2]There is no evidence that the myelin of the PNS of the shiverer mutant does not operate properly. Thus, the role for MBP in peripheral myelin is unknown.

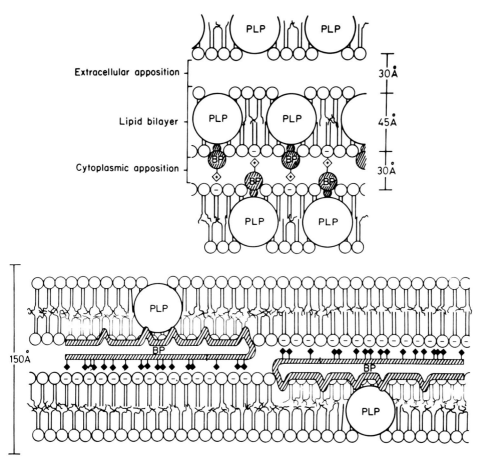

FIG. 1. Proposed organization of proteolipid (PLP)-basic protein (BP) units in compact myelin bilayers in cross-section **(top)** and (b) longitudinal section **(bottom)**. The latter shows the proposed hydrophobic penetration of the lipid bilayer by the "amino-half" of the basic protein and the ionic interactions across the cytoplasmic apposition by the basic residues on the "carboxyl-half" of the molecule. Basic amino acid residues are indicated by the diamond symbols. (From Rumsby and Crang, 1977.)

been obtained of increased MBP breakdown during the analytical procedures themselves (Kirschner and Ganser, 1980).

Another mutation, "myelin deficient," has been described (LaChapelle et al., 1980) and recently shown to be allelic with shiverer (LaChapelle et al., 1980; J. Cowen, 1980). In myelin-deficient mice, the CNS myelin contains reduced but detectable levels of MBP (3% of normal amounts in homozygotes, 60 to 70% of normal in heterozygotes) (Matthieu et al., 1980; J. R. Cohen, *unpublished observations*). Analysis of purified MBP from myelin-deficient homozygotes should prove helpful in establishing whether this locus represents the structural gene or a regulatory locus for MBP.

Despite the apparently normal amount and periodicity of myelin in the PNS of shiverer homozygotes, it is possible, by immunocytochemistry, to demonstrate the complete absence of MBP (R. L. Sidman, R. Altschuler, and B. Trapp, 1981, *unpublished observations*). In the upper panels of Fig. 3, control and shiverer PNS have been stained with antibody against P_0, the major glycoprotein of PNS myelin (Trapp et al., 1980); the lower panels illustrate the

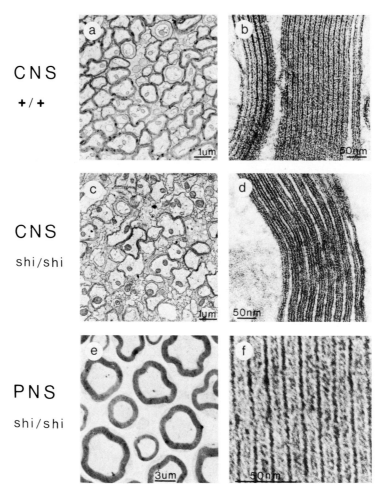

FIG. 2. Electron micrographs from normal (+/+) and shiverer (shi/shi) nervous tissue. (a), (b), and (c): Dorsal funiculi of cervical spinal cord, (d): optic nerve, (e) and (f): sciatic nerve. (From Ganser and Kirschner, 1980.)

reaction with anti-MBP antibody in normal and mutant, respectively. It is apparent that the defect in shiverer homozygotes is selective for MBP and does not affect levels of P_0.

In light of the evidence that (1) myelin is a component of the Schwann or oligodendroglial cell (Geren, 1954; Peters et al., 1970), (2) it forms only when the potentially myelinating cell has contacted an appropriate neuron (Webster, 1975), and (3) its formation may require, in the PNS, participation of a fibroblast or some extracellular collagen-associated product (Bunge, 1981), the question arises whether the shiverer locus acts extrinsically or intrinsically in the myelinating (MBP-synthesizing) cell itself. This question is approachable through the classic "confrontation" method of experimental embryology, elegantly applied to peripheral nerve in normal and mutant mice and humans by Aguayo and colleagues (1979). A segment of donor mouse or human nerve is grafted in end-to-end anastomosis with the proximal and distal stumps of the transected sciatic or other appropriately sized nerve of a host mouse. Donor axon segments, cut off from their original neuronal somas, will degenerate, but donor Schwann, fibroblast, and endothelial cells will survive, and may interact not only with one another but with host

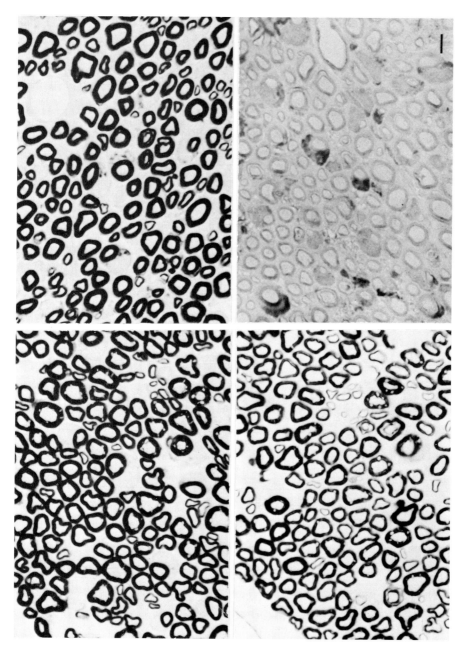

FIG. 3. Cross-sections of 19-day-old littermate mouse sciatic nerves, fixed in glutaraldehyde-paraformaldehyde, Epon-embedded, sectioned at 1 μm and stained by the PAP immunocytochemical method. Antisera were diluted 1:500. *Upper left:* normal, stained for P_o protein; *upper right:* shiverer, stained for P_o protein; *lower left:* normal, stained for MBP; *lower right:* shiverer,, stained for MBP. Bar = 5 μm. (From R. L. Sidman, R. Altshuler, and B. Trapp, *unpublished data.*)

axons that regenerate proximally to distally through the graft. Likewise, the segments of host axon distal to the original transection site will degenerate and will be replaced within a few months by regenerated host axons, now in contact, distal to the graft, with host Schwann cells.

Schwann cells in both the graft and distal host territory may proliferate, but most cells remain confined to their own territory, with only slight mixing close to the sites of anastamosis.

When shiverer sciatic nerve grafts were inserted into sciatic transection sites of wild-type host mice, myelin formed in abundance along regenerated axons both within the graft and distal to it (R. L. Sidman, J. Cowen, and T. Brushart, 1981, *unpublished data*). The tissue was examined at different levels immunocytochemically for MBP; several outcomes were possible. If MBP were found in myelin in the graft and distal to it, one could conclude that host axons control MBP expression. If MBP were found in neither site, one might suspect that regenerating axons were less effective than normal axons, or that the mutant graft elicits a systemic inhibitory effect on the host. If myelin within the graft lacked MBP whereas myelin more distally along the same regenerated axons contained MBP, one could conclude that the axon is not the determinant, and that the site of action of wild-type and mutant alleles at the shiverer locus is intrinsic to some cell type or types within the nerve itself. This last possibility is the one actually obtained (Fig. 4). The further choice among these putative target cell types is likely to come via cell culture, now that peripheral sensory neurons, Schwann cells, and nerve fibroblasts can be cultured as highly enriched, almost pure populations and can be recombined from different sources under conditions that allow myelination (Bunge, 1981).

The shiverer mouse, then, represents for the molecular geneticist a well-characterized mutant in which restricted cell types and a specific protein have been implicated as target and product of the locus in question. Although the system is complicated by the fact that Schwann cells usually fail to express MBP in pure cell culture (Mirsky et al., 1980), it should be possible to use oligodendroglial cells, which do express MBP for weeks *in vitro* in the absence of neurons (Mirsky et al., 1980), as a source of mRNA in order to obtain a cDNA clone for MBP. Such a cDNA clone will be invaluable whether the shiverer locus proves to be a regulatory locus or the structural gene for all or for certain subclasses of MBP.

WEAVER MOUSE

The weaver mutation is more complicated for molecular genetic analysis. In this case, no specific defective protein has been implicated, but the cell type in which the defect is expressed has now been tentatively identified, based on observation of a quantitative gene dosage effect. The cell that appears to be primarily affected is the granule cell neuron. These neurons are normally generated on the external surface of the cerebellum from a pool of proliferating cells. The postmitotic cells that initially lie at the base of the external granule cell layer display three easily recognizable characteristics: (1) they generate bipolar processes that extend parallel to the external surface in the transverse plane of the brain; (2) the cell bodies migrate (translocate) inward to reach a position in the granule layer deep in the Purkinje cell bodies and dendrites; and (3) granule cells persist throughout the life of the animal. In the weaver mutant, granule cell neurons are defective in all three properties: they do not make processes or migrate, and they die within 2 weeks after birth (Rakic and Sidman, 1973).

A gene dosage effect can be demonstrated in tissue culture preparations. Cell survival can be measured by culturing neurons from 7-day-old cerebellum of weaver homozygotes, weaver heterozygotes, and wild-type mice as shown in Fig. 5 (Willinger et al., 1981a). The neurons from weaver homozygotes die in such numbers that after 6 days in culture, fewer than 20% of the cells present at day 1 still survive. Wild-type cells survive almost quantitatively after day 1 *in vitro*, whereas neurons from heterozygotes display an intermediate level of survival. When neurons are prelabeled *in vivo* with [^{3}H]thymidine 6 hr before being removed and placed in culture, it can be seen that among cells from weaver homozygotes, the labeled cells die

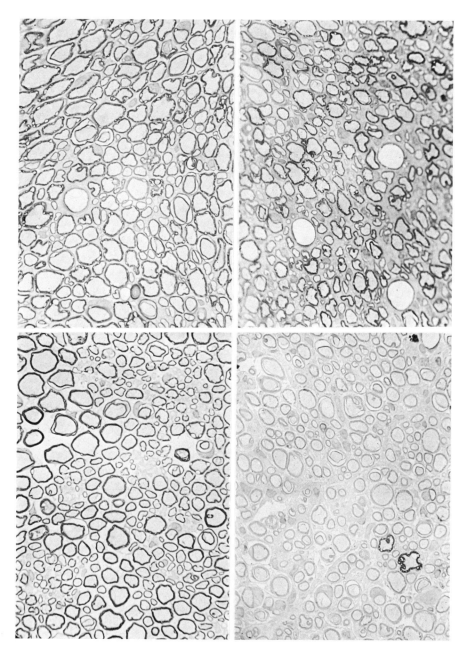

FIG. 4. Cross-sections of sciatic nerves from an adult normal mouse that had received a grafted segment of sciatic nerve from a shiverer mouse donor three months earlier. Tissue processed as for Fig. 3, and stained immunocytochemically for MBP. *Upper left:* control intact left sciatic nerve, *upper right:* proximal level of operated sciatic nerve, where host Schwann cells relate to intact host axons; *lower left:* distal level of the graft itself, where shiverer Schwann cells relate to regenerated host axons and almost all myelin sheaths are unstained for MBP; *lower right:* distal level of host sciatic nerve, where host Schwann cells relate to regenerated host axons. (From R. L. Sidman, J. Cowen, and T. M. Brushart, *in preparation*.)

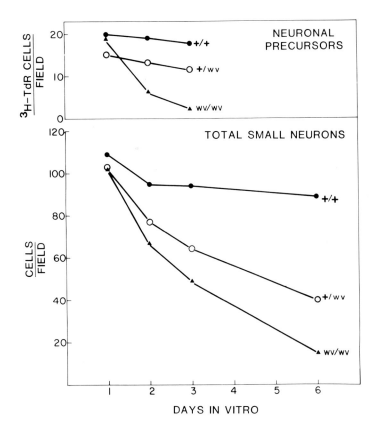

FIG. 5. Granule cell survival in cultures of weaver (WV) cerebellum. *Upper panel:* mice were injected with ³H-thymidine 6 hr prior to sacrifice on postnatal day 8. Dissociated cerebellar cells were plated at a density of $3 \times 10^4/mm^2$. The number of neurons with grains over their nuclei was counted in 5 to 10 random fields in duplicate cultures, each point representing an average of two independent experiments. *Lower panel:* Cells with granule cells nuclear morphology were counted in 6 contiguous fields in each of duplicate cultures of +/+, +/WV, and WV/WV postnatal day 7 cerebellum. The points represent the mean for each genotype. (From Willinger et al., 1981a.)

preferentially, whereas for heterozygote and wild-type neurons they do not (Fig. 5). This suggests that the weaver cells that die are those that were most recently generated, whereas the heterozygote cells that die represent a somewhat older population. This parallels the *in vivo* situation in which the weaver homozygote cells die soon after their genesis, whereas a smaller number of heterozygote neurons dies, and do so after a longer interval (Rezai and Yoon, 1972).

Neurite outgrowth can also be followed in tissue culture. The mean total length of processes generated per neuron in culture was measured (Fig. 6) and by including mean number of cells per field one can calculate mean total neurite production per microscopic field. In each case, a clear gene dosage effect is seen in the ability of weaver homozygotes and heterozygotes to generate neurites as compared with wild-type cells (Willinger et al., 1981a). Time-lapse microcinematography indicates that the defect in weaver cells is not in the inititation of neurites, but in neurite maintenance and elongation. In fact, weaver cells initiate many more neurites than do heterozygote or wild-type neurons *in vitro*, but in weaver cells these processes fail to persist. Semiquantitative analysis of the time-lapse films demonstrates that the rate of neurite

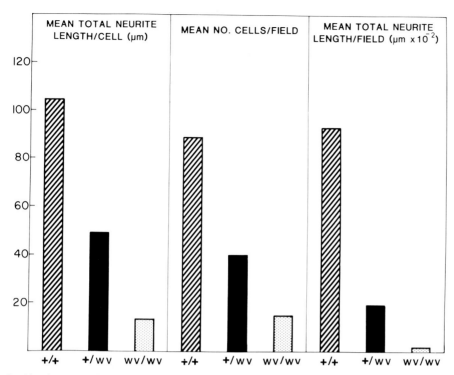

FIG. 6. Neurite growth by cells of postnatal day 7 mouse cerebellum maintained for 6 days *in vitro*. The mean total neurite length/cell was obtained by computer-assisted digitization of all fibers originating from a total of 20 neurons in triplicate cultures each of +/+, +/WV and WV/WV specimens. This value was multiplied by the number of cells/field (see Fig. 5) to obtain the mean total neurite length/field. (From Willinger et al., 1981a).

elongation is slowed by the weaver gene. The migratory behavior of mutant cells is also different from wild-type cells (Willinger et al., 1981b).

Despite these gene dosage effects, one cannot be absolutely sure that the granule cell neuron is the primary or only target of the weaver gene. The *in vitro* defects described above may reflect earlier events *in vivo*, some of them possibly extrinsic to the granule cell. Experiments with weaver—wild-type chimeras indicate that the granule cell neuron is intrinsically affected, but do not as yet contribute to the question of whether other cell types are also affected (Goldowitz and Mullen, 1980). Tissue culture recombination experiments in which wild-type neurons are plated on weaver glial cells, and vice versa, are incomplete as yet (M. Willinger, *personal communication*).

MUTATIONS AFFECTING THE VISUAL SYSTEM

Albino mutations occur in many mammalian species, and in all of them the visual system of the mutant is abnormal. The abnormality can be detected behaviorally, physiologically, and anatomically at all levels of the visual system from the retina to area 17 of the cortex and across the interhemispheric corpus callosum (Guillery, 1974; Creel and Giolli, 1976). The primary defect appears to be in the retinal neuron (Guillery, 1969) or its axon, which crosses in the chiasm to innervate the diencephalon and midbrain (Guillery, 1974).

The albino locus is generally believed to represent the structural gene for the enzyme tyrosinase and, at first glance, this might seem like an ideal case for applying the techniques of

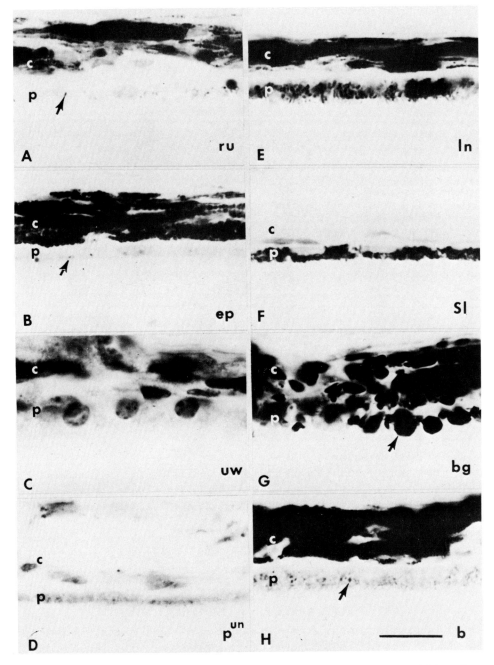

FIG. 7. Photomicrographs to show pigmentation in choroid (c) and pigment epitheliuim (p) from the eyes of various adult mutant mice. Pigment granules are indicated by arrows. Symbols for the mutant genes are b, brown; bg, beige; ep, pale ears; ln, leaden; p^{un}, pink-eyed dilution-unstable; ru, ruby; sl, steel; and uw, underwhite. Bar = 10 μm. (From LaVail et al., 1978.)

molecular genetics to explain a specific behavioral defect.[3] However, when one examines other mutants affecting the pigmentation system, it becomes clear quite quickly that adequate neurogenetic groundwork must be done before molecular genetic analysis can be effectively applied. Specifically, the examination of other mutants demonstrates that one would be misled to study tyrosinase in an attempt to gain a specific understanding of the routing of optic nerve fibers. The visual defect relates more generally to the problem of hypopigmentation (LaVail et al., 1978). Figure 7 displays sections of retina from a number of nonalbino mouse mutations. Regardless of the site of the mutation, the organizational problem in the visual system of these mutants relates, in a quantitatively graded fashion, to the degree of pigmentation; less pigmentation corresponds to greater neurological abnormality.

These mutants can also be used to identify the cell type primarily involved in this phenomenon. There are two pigmented layers of cells in the retina, the pigment epithelium, which is of neuroepidermal origin, and the choroidal layer, which derives from melanocytes of neural crest origin. The mutants shown in Fig. 7 display different phenotypes with respect to the effect of the mutation on each pigment layer (LaVail et al., 1978). The ruby mutation, for example, affects the pigment epithelium but not the choroidal zone; the steel mutation has the opposite phenotype. The degree of neurological abnormality in these various mutants correlates inversely with the degree of pigmentation of the retinal epithelium. This would seem to be an unusually intriguing problem in developmental biology, and a very difficult case for the application of molecular genetics.

The time appears to be at hand when the molecular geneticist will be able to contribute greatly to our understanding of mutations affecting behavior. This has been made possible not only by recent advances in general genetic concepts and technology, but also by the strategies of phenotypic analysis, especially the search for gene dosage effects, and of confrontation analysis,[4] as used by developmental geneticists and neurogeneticists. The application of recombinant DNA techniques to the investigation of mutations affecting the nervous system will be most successful if systems chosen for study are carefully examined based on knowledge of the affected cell type and of the developmental time of action of the mutation.

ACKNOWLEDGMENTS

This research has been supported in part by NIH grants No. 5-RO1-EY03225 and 2 RO1 NS11237.

REFERENCES

Aguayo, A. J., Bray, G. M., and Perkins, C. S. (1979): Axon-Schwann cell relationships in neuropathies of mutant mice. *Ann. N. Y. Acad. Sci.*, 317:512–531.
Barbarese, E., and Carson, J. H. (1981): Level and polymorphism of myelin basic protein in Shiverer mouse. *Am. Soc. Neurochem. Abstr.*, 12:103.
Barbarese, E., Carson, J. H., and Braun, P. E. (1978): Accumulation of the four myelin basic proteins in mouse brain during development. *J. Neurochem.*, 31:779–782.
Bird, T. D., Farrell, D. F., and Sumi, S. M. (1978): Brain lipid composition of the shiverer mouse (genetic defect in myelin development). *J. Neurochem.*, 31:387–391.

---

[3]Data in the literature are unclear as to whether the albino locus actually does represent the structural gene for tyrosinase, although the locus is commonly stated to be that gene. This case was chosen to illustrate the potential pitfalls in relying on a small amount of evidence, however attractive that evidence might be.

[4]The term confrontation analysis is suggested to indicate any of the many biological experimental designs that juxtapose cells with new neighbors to observe their interactions, e.g., experimental chimeras *in vivo* or a mixture of cells of different genetic sources in culture.

Braun, P. E. (1977): Molecular architecture of myelin. In: *Myelin*, edited by P. Morell, pp. 201–231. Plenum Press, New York.

Bunge, R. P. (1981): The development of myelin and myelin-related cells. *Trends Neurosci.*, 4:175–177.

Cowen, J. (1980): Shiverer and myelin deficient allelism. (Personal communication.) *Mouse News Letter*, 63:14.

Creel, D., and Giolli, R. A. (1976): Retinogeniculate projections in albino and ocularly hypopigmented rats. *J. Comp. Neurol.*, 166:445–455.

Dupouey, P., Jaque, C., Baurre, J. M., Cesselin, F., Privat, A., and Baumann, N. (1979): Immunochemical studies of myelin basic protein in Shiverer mouse devoid of major dense line of myelin. *Neurosci. Lett.*, 12:113–118.

Ganser, A. L., and Kirschner, D. A. (1980): Myelin structure in the absence of basic protein in the Shiverer mouse. In: *Neurological Mutations Affecting Myelination*, edited by N. Baumann, pp. 171–176. Elsevier/North-Holland Biomedical, Amsterdam.

Geren, B. B. (1954): The formation from the Schwann cell surface of myelin in the peripheral nerves of chick embryos. *Exp. Cell. Res.*, 7:558–562.

Goldowitz, D., and Mullen, R. J. (1980): Weaver mutant granule cell defect expressed in chimeric mice. *Soc. Neurosci. Abstr.*, 6:743.

Guillery, R. W. (1969): An abnormal retinogeniculate projection in Siamese cats. *Brain Res.*, 14:739–741.

Guillery, R. W. (1974): Visual pathways in albinos. *Sci. Am.*, 230:44–54.

Kirschner, D. A., and Ganser, A. L. (1980): Compact myelin exists in the absence of basic protein in the Shiverer mutant mouse. *Nature*, 283:207–210.

Lachapelle, F., DeBaecque, C., Jacque, C., Bourre, J. M., Delassalle, A., Doolittle, D., Hauw, J. J., and Baumann, N. (1980): Comparison of morphological and biochemical defects of two probably allelic mutations of the mouse myelin deficient (mld) and Shiverer (shi). In: *Neurological Mutations Affecting Myelination*, edited by N. Baumann, pp. 27–32. Elsevier/North-Holland Biomedical, Amsterdam.

LaVail, J. H., Nixon, R. A., and Sidman, R. L. (1978): Genetic control of retinal ganglion cell projections. *J. Comp. Neurol.*, 182:399–421.

Martenson, R. E. (1981): Prediction of the secondary structure of myelin basic protein. *J. Neurochem.*, 36:1543–1560.

Matthieu, J. M., Ginalski, H., Friede, R. L., Cohen, S. R., and Doolittle, D. P. (1980): Absence of myelin basic protein and major dense line in CNS myelin of the mld mutant mouse. *Brain Res.*, 191:278–283.

Mirsky, R., Winter, J., Abney, E. R., Pruss, R. M., Gavrilovic, J., and Raff, M. C. (1980): Myelin-specific proteins and glycolipids in rat Schwann cells and oligodendrocytes in culture. *J. Cell Biol.*, 84:483–494.

Peters, A., Palay, S. L., and Webster, H. deF. (1970): *The Fine Structure of the Nervous System: The Neurons and Supporting Cells*, 2nd ed. Harper and Row, New York.

Privat, A., Jacque, C., Bourre, J. M., Dupouey, P., and Baumann, N. (1979): Absence of the major dense line in myelin of the mutant mouse "Shiverer." *Neurosci. Lett.*, 12:107–112.

Rakic, P., and Sidman, R. L. (1973): Sequence of developmental abnormalities leading to granule cell deficit in cerebellar cortex of weaver mutant mice. *J. Comp. Neurol.*, 152:103–132.

Rezai, Z., and Yoon, C. H. (1972): Abnormal rate of granule cell migration in the cerebellum of weaver mutant mice. *Dev. Biol.*, 29:17–26.

Rumsby, M. G., and Crang, A. J. (1977): The myelin sheath—a structural examination. In: *The Synthesis, Assembly and Turnover of Cell Surface Components*, edited by G. Poste and G. L. Nicolson, pp. 247–362. Elsevier/North-Holland Biomedical, Amsterdam.

Trapp, B. D., Sternberger, N. H., Quarles, R. H., Itoyama, Y., and Webster, H. deF. (1980): Immunocytochemical localization of peripheral nervous system myelin proteins in epon sections. *J. Neuropathol. Exp. Neurol.*, 39:392.

Webster, H. deF. (1975): Development of peripheral myelin sheaths during their formation and growth in rat sciatic nerves. *J. Cell Biol.*, 48:348–367.

Willinger, M., Margolis, D., and Sidman, R. L. (1981a): Neuronal differentiation in cultures of weaver *(wv)* mutant mouse cerebellum. *J. Supramol. Struc. & Cell. Biochem.*, 17:79–86.

Willinger, M., Margolis, D., and Sidman, R. L. (1981b): Granule cell behavior in dissociated cultures of weaver *(wv)* mutant cerebellum. *Soc. Neurosci. Abstr.*, 7:346.

Genetic Disorders of the Human Nervous System: A Commentary

Victor A. McKusick

ABSTRACT

Delineation of specific gene loci and mapping of these loci to specific chromosomes and regions of chromosomes are central to human biology, including neurobiology. This is a rapidly advancing field, scarcely more than a decade old. Progress in this field is reviewed with particular reference to disorders of the nervous system.

To establish how genetic mutations could affect the nervous system, it is important to know the number and chromosomal location of genes critical to nerve development and function. Extensive genetic analysis of the mouse indicates that there are on average about 20 gene loci per centimorgan, a recombinational unit (T. H. Roderick, 1976, *personal communication*). The human genome, estimated to be some 33,000 centimorgans in "length," could therefore contain some 50,000 to 100,000 structural gene loci. Another estimate comes from considering the 65,000-base-pair (65-Kb) segment of DNA containing the β-globulin gene locus and 4 related loci on chromosome 11 (see T. Maniatis et al., *this volume*). Given a total of 3 million Kb of DNA, the human haploid genome could contain 200,000 structural gene loci. Studies of mouse brain DNA-mRNA complexity suggest that out of 150,000 structural gene loci, about 80,000 are expressed in brain (see W. Hahn, *this volume*). Thus, if there were 100,000 to 200,000 gene loci in humans, at least half might be involved in neural function.

Another number critical to this consideration is the number of inherited diseases in humans. Over the years a catalog has been maintained of separately identified Mendelizing phenotypes in humans (McKusick, 1978). These phenotypes have been subdivided according to whether their mode of inheritance is autosomal dominant, autosomal recessive, or X-linked (Table 1). This catalog represents the number of gene loci for which allelic variations are compatible with life and those that have been demonstrated experimentally, e.g., by somatic cell hybridization. The most recent figure for established Mendelizing loci in humans is 1,611, and the total number of established and tentative is 3,303. At least 500 of these traits may be manifest predominantly in the nervous system. However, it is clear that even when the nervous system is the most dramatically affected organ, the primary site of gene action may not be in nerve cells. For example, in phenylketonuria the defect is in phenylalanine hydroxylase, an enzyme located in the liver, yet mental retardation and poor myelination are hallmarks of the mutation. Thus even in a disease such as Huntington's disease, where the pathology apparently involves only central nervous tissue, the primary lesion may not be in a gene expressed in the brain.

TABLE 1. Number of loci identified mainly by mendelizing phenotypes[a]

Phenotype	Verschuer 1958[b]	Mendelian inheritance in humans					
		1966 (1st ed.)	1968 (2nd ed.)	1971 (3rd ed.)	1975 (4th ed.)	1978 (5th ed.)	Sept. 10, 1981 (unpublished)
Autosomal dominant	208	269 (+568)	344 (+449)	415 (+528)	583 (+635)	736 (+753)	904 (+880)
Autosomal recessive	89	237 (+294)	280 (+349)	365 (+418)	466 (+481)	521 (+596)	591 (+692)
X-linked	38	68 (+ 51)	68 (+ 55)	86 (+ 64)	93 (+ 78)	107 (+ 98)	116 (+120)
Total	412	574 (+913)	692 (+853)	866 (+1,010)	1,142 (+1,194)	1,364 (+1,447)	1,611 (+1,692)
Combined total	412	1,487	1,545	1,876	2,336	2,811	3,303

[a]Numbers in parentheses refer to loci not yet fully identified or confirmed.
[b]Lehrbuch der Humangenetik (Munich: Urban and Schwartzenberg, 1958).
Modified from McKusick, 1978.

Information about the location of genes on human chromosomes is also available (Fig. 1) and expanding at a rapid rate. Of the total of almost 1,500 known loci on autosomes, mapping information is available for over 20%. In some cases, assignments have been made to specific regions of the chromosome as identified by Giemsa staining of metaphase chromosomes. Over 115 genetic loci have been assigned to the human X chromosome. The Y chromosome contains only a few genes, which appear to be involved in sex determination, histocompatibility antigens, and height.

A number of procedures have been used in gene mapping. Classically, such studies involved family studies to establish cosegregation (linkage) of traits (the so-called family method). Linked genes tend to be inherited together. Genes on the same chromosome are said to be syntenic, but may not be linked in the strict genetic sense, depending on how far apart they are on the chromosome. However, substantial mapping has been carried out using the parasexual approaches made possible by somatic cell genetic techniques. These involved somatic cell hybrids and gene transfer. More recently, cDNA probes for genes of interest have been used in combination with somatic cell techniques. This approach will serve to fill in rapidly the position of most single-copy DNA sequences, even when their gene product is not yet identified (see F. Ruddle, *this volume*).

So far only a few genes known to be important in neural function have been mapped. The first gene mapped was a "neuronal gene," that for color blindness, located on the X chromosome (Wilson, 1911), subsequently found to be closely linked to the gene for glucose-6-phosphate dehydrogenase (G6PD) (Porter et al., 1962) and to the gene responsible for classic hemophilia (Haldane and Smith, 1947). The close linkage of G6PD and hemophilia was demonstrated by Boyer and Graham (1965). Recently, the gene responsible for a neurological disease, adrenoleukodystrophy, has also been assigned to this gene cluster (Moser et al., 1980). This cluster has been assigned to the terminal segment of the X chromosome (see Fig. 1) by somatic cell hybridization analysis of the segregation of X chromosome rearrangements (Pai et al., 1980); G6PD was the marker used in making that assignment. Other genes mapped to specific locations on the X chromosome that have critical roles in neural function include the gene coding for hypoxanthine phosphoribosyltransferase, a deficiency of which causes the Lesch-Nyhan syndrome; the gene responsible for Duchenne's muscular dystrophy (Jacobs et al., 1981); and the gene coding for the flavin-containing subunit of monoamine oxidase (Pintar et al., 1981). In addition, the inheritance of a "fragile site" on the X chromosome is correlated with X-linked mental retardation in family pedigrees.

To date about 450 specific gene loci have been mapped. An interesting finding is that there is a great deal of linkage homology between mice and humans. Considering the evolutionary divergence of these species, this finding suggests that selective pressure is exerted on gene position. The application of molecular biology approaches to problems of gene number and position will soon make possible the identification of a large number of new structural genes, including those that are critical to the development and function of the nervous system.

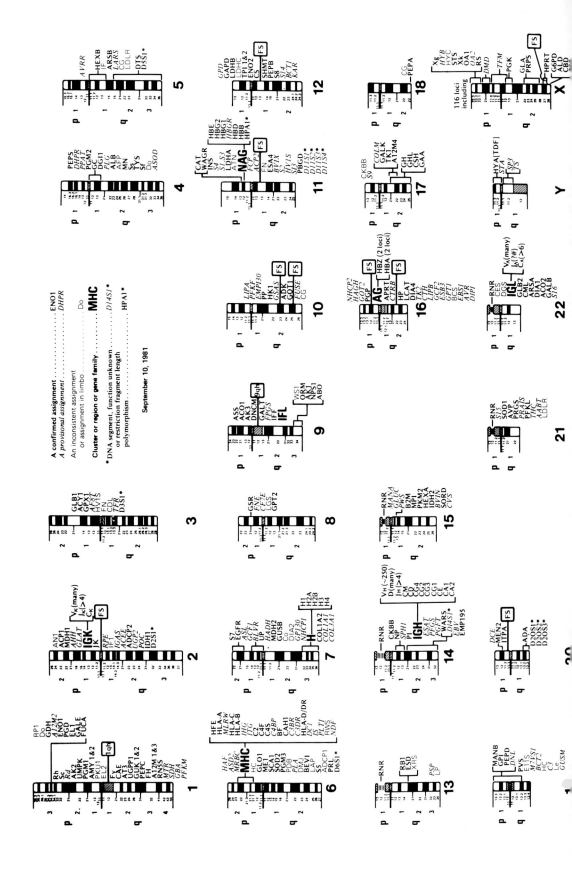

REFERENCES

Boyer, S. H., and Graham, J. B. (1965): Linkage between the X chromosome loci for glucose-6-phosphate dehydrogenase electrophoretic variation and hemophilia A. *Am. J. Hum. Genet.*, 17:320–324.
Haldane, J. B. S., and Smith, C. A. B. (1947): A new estimate of the linkage between the genes for colour-blindness and hemophilia in man. *Ann. Eugen.*, 14:10–31.
Jacobs, P. A., Hunt, P. A., Mayer, M., and Bart, R. D. (1981): Duchenne muscular dystrophy (DMD) in a female with an X/autosome translocation: Further evidence that the DMD locus is at Xp21. *Am. J. Hum. Genet.*, 33:513–515.
McKusick, V. A. (1978): *Mendelian Inheritance in Man: Catalogs of Autosomal Dominant, Autosomal Recessive and X-linked Phenotypes*, 5th Ed. Johns Hopkins University Press, Baltimore.
Moser, H. W., Moser, A. B., Kawamura, N., Migeon, B., O'Neill, B. P., Fenselau, C., and Kishimoto, Y. (1980): Adrenoleukodystrophy: Studies of the phenotype, genetics and biochemistry. *Johns Hopkins Med. J.*, 147:217–224.
Pai, G. S., Sprenkle, J. A., Do, T. T., Mareni, C. E., and Migeon, B. R. (1980): Localization of loci for hypoanthine phosphoribosyltransferase and glucose-6-phosphate dehydrogenase and biochemical evidence of nonrandom X-chromosome expression from studies of human X-autosome translocation. *Proc. Natl. Acad. Sci. USA*, 77:2810–2813.
Pintar, J. E., Barbosa, J., Francke, U., Castiglione, C. M., Hawkins, M., Jr., and Breakefield, X. O. (1981): Gene for monamine oxidase type A assigned to the human X chromosome. *J. Neurosci.*, 1:166–175.
Porter, I. H., Schulze, J., and McKusick, V. A. (1962): Genetic linkage between the loci for glucose-6-phosphate dehydrogenase deficiency and colour blindness in American Negroes. *Ann. Hum. Genet.*, 26:107–122.
Wilson, E. B. (1911): The sex chromosomes. *Arch. Mikrosk. Anat. Entwicklungsmech.*, 72:249–271.

FIG. 1. A diagrammatic synopsis of the human gene map. p, short arm; q, long arm. (For definition of symbol, see McKusick, 1978.)

Huntington's Disease: Genetically Determined Cell Death in the Mature Human Nervous System

Joseph B. Martin

ABSTRACT

Huntington's disease is an autosomal dominant hereditary disorder of the human central nervous system. Progressive, uncontrolled choreiform movements and dementia appear in middle life after normal development. The neuropathologic changes consist of premature nerve cell death in the caudate and putamen (neostriatum) and in the cerebral cortex. The mechanism of this cell death is unknown.

Huntington's disease (HD) is an autosomal dominant hereditary disorder of the human central nervous system. The disease is characterized by uncontrolled choreic movements and dementia, both of which commence insidiously in middle age and progress in severity until the affected individual is completely incapacitated. Death usually occurs some 15 years after onset, often from pneumonitis caused by aspiration. There are two features of Huntington's disease that make it of particular interest to the study of the central nervous system. The first is that it displays a remarkably low mutation rate. In fact, no proven case of new mutation causing Huntington's disease has been discovered. The second unusual aspect of this genetic disease is that it does not appear to affect the development of the cell types of special importance to it. Rather, neuronal development and connectivity appear to occur normally; only later do certain cells begin to die prematurely in a regionally selective manner. Therefore, Huntington's disease presents an example of programmed cell death in the central nervous system. An understanding of the disorder might yield significant insight into the complexity and properties of the cell types involved not only in the disease but also in brain function, normal and abnormal.[1]

HISTORICAL FEATURES

Huntington's disease was first formally described by George Huntington (1872), a physician in private practice on Long Island. His classic paper, presented to the Medical Society of Ohio, grew out of the observations that he and his father, also a physician, had made of affected individuals in a number of families in the area.

---

[1]There are other human disorders that also show a late onset of effects on a specific cell type after apparently normal development. For example, in Duchenne muscular dystrophy, a sex-linked disorder, there is relatively normal development of muscle and nerves but rather early in life muscles degenerate. Similarly, Friedreich's ataxia and other cerebellar disorders affect a group of brainstem spinal cord neurons or peripheral nerves while sparing all other areas. Furthermore, other late-onset cerebellar ataxias also affect a very select group of cells after apparently normal development.

The history of the disorder in North America can actually be traced as far back as 1630 when John Winthrop sailed from England to become governor of the Massachusetts Bay Colony (Vessie, 1932). Among the group that landed at Salem, Massachusetts were several people who had a disorder whose description closely parallels that of Huntington's disease. The choreic family lines to which these individuals have given rise include families that later settled in Long Island. Two of the affected passengers were brothers, Nichols and Wilkie, and their progeny are particularly noteworthy. Wilkie had children by two wives. In the 1650s, some of his children migrated to Long Island where they probably gave rise to the choreic lines that were later studied by George Huntington. Another of Wilkie's children was the famous Groton Witch. She was accused and hanged as a witch because of her marked abnormality of movement which implies that she probably inherited the disease.

Nichols' wife and granddaughter were also accused of being witches, and it is possible that his wife had the disease, since many of the early settlers were closely related. In all likelihood, based on the family lines reported by Vessie, many of the unfortunate individuals executed for being witches in Salem were actually exhibiting symptoms of Huntington's disease.

CLINICAL FEATURES

Huntington's disease is currently found in most populations around the world at a frequency of 4 to 7 per 100,000 population. As an autosomal dominant disease it affects men and women with equal frequency, and each child of an affected parent has a 50% chance of inheriting the disease gene.

In most cases the onset of the disease occurs in the fourth or fifth decade of life, usually after an individual has married and had children. Since no presymptomatic test is available, the offspring of affected individuals often live for many decades in the fear that they too have inherited the gene. The first detectable signs of the disorder are subtle and may consist of clumsiness, fidgeting, sudden falls, absent-mindedness, irritability, or chronic depression. The uncontrolled movements gradually become more exaggerated until the patient must be confined to a wheelchair or bed. Speech at first becomes slurred, then incomprehensible, and finally ceases altogether as facial expressions become distorted and grotesque. Mental functions undergo similar deterioration, and the ability to reason eventually disappears.

The progression from onset to eventual death occurs over a period of approximately 15 years. There is no treatment capable even of slowing the course of the disease, much less of curing it. Once onset has occurred, the affected individual faces several years of gradually decreasing capacity leading to total disability and, ultimately, death. Occasionally the onset of the disease occurs in childhood and follows a more rapid course. One particularly baffling aspect of the genetics of this disorder is that in these juvenile cases, in which onset occurs before the age of 10 years, the disease gene has invariably been inherited from the father (Bird et al., 1974; Wallace, 1979).

NEUROPATHOLOGY

The selective localized loss of neurons in Huntington's disease can be demonstrated in living individuals by use of computer-assisted tomographic (CAT) analysis. In patients with advanced disease, there is an obvious enlargement of the ventricular system, especially of the frontal horns due to atrophy of the caudate and putamen (neostriatum). In normal individuals ventricles are small, whereas in the Huntington's patient up to one-third of the brain weight may be lost, which is reflected in tissue loss in the CAT scan.

The neuropathology can be more closely studied in postmortem sections of brains from affected individuals. The loss of tissue from the region of the anterior lateral ventricle is evident from the atrophy of the neostriatum, which may show more than 90% loss of tissue due to degeneration and neuronal cell loss. This loss is thought to be the cause of the choreic movements characteristic of the disease. The effects of the disease on cognitive functions that eventually include dementia are presumably due to the loss of cortical neurons.

The effects of loss of neurons from the caudate and putamen must be considered in light of their functional role in the normal brain. The basal ganglia normally function to monitor inputs from other areas of the nervous system and to direct them toward a reasoned output from the motor cortex to the voluntary muscles in the limbs. Figure 1 illustrates the four major inputs into the caudate and putamen: a massive input from all areas of the cerebral cortex, dopaminergic inputs from the substantia nigra, thalamic inputs, and serotonergic inputs of the raphe (Graybiel and Ragsdale, 1979).

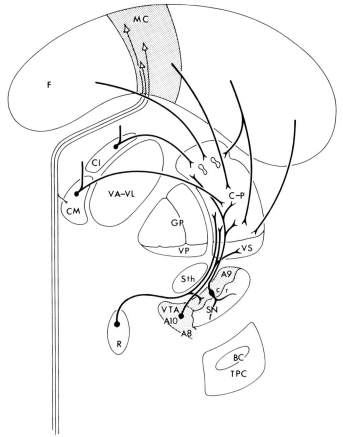

FIG. 1. Schematic diagram illustrating the main afferents of the striatum. Cerebral cortex is shown above (F, frontal; MC, motor cortex). Below are the basal ganglia and allied nuclei: C-P, caudate-putamen; GP, globus pallidus; VS and VP, ventral striatum and ventral pallidum; Sth, subthalamic nucleus; SN, substantia nigra and related cell groups (r, pars reticulata; c, pars compacta corresponding to dopamine cell group A9; VTA, ventral tegmental area, corresponding to dopamine cell group A10 and partly continuous with dopamine cell group A8). Thalamic correspondents of basal ganglia are VA-VL: nuclei ventralis anterior and ventralis lateralis; CL, centre lateralis; CM, centre médian. In the mesencephalon: dorsal raphe nucleus (R), and the nucleus tegmenti pedunculopontinus (TPC) near the brachium conjunctivum (BC). (From Graybiel and Ragsdale, 1979.)

The primary outflow from the caudate and putamen goes through the globus pallidus to the ventral anterior and ventral lateral thalamus and then to cerebral cortex (Fig. 2) (Graybiel and Ragsdale, 1979). Thus the caudate and putamen form the point of convergence of many inputs as well as the beginning of the output system that regulates motor function by the extrapyramidal system. If cell loss is investigated early in the course of the disease, the caudate and putamen appear maximally affected. Some cell loss also occurs in the cerebral cortex, globus pallidus, and hypothalamus, but other areas of the brain remain relatively unaffected. When cell loss does occur in other brain regions, it appears to be due to the death of specific neurons without any obvious local reactivity on the part of glial cells in the area.

The cause of neuronal loss in Huntington's disease is not understood. No evidence of inflammation, perivascular infiltration, or of any cellular immune response has been found. Neither has there been any significant association of a particular HLA haplotype. Investigations of viral interactions or immunological responses have all proved negative. No specific cellular pathology has been found that might explain the selective nature of the cell loss; all other organs and tissues are unaffected by the gene defect. The mechanism by which a dominant mutation could cause the premature death of specific neurons after half the normal lifespan has already elapsed remains a matter of conjecture.

NEUROTRANSMITTER CHANGES

There have been many attempts to correlate the observed neuropathology in Huntington's disease with concentrations of specific neurotransmitters in the affected areas. The results of

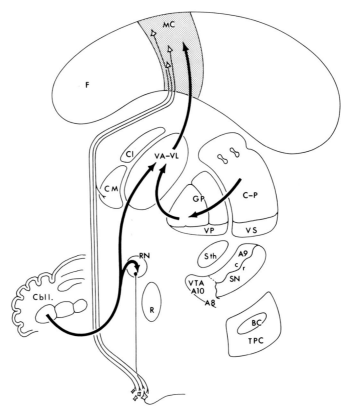

FIG. 2. Schematic diagram illustrating "pyramidal" and "extrapyramidal" motor systems. Cbll., cerebellum; RN, red nucleus. Other abbreviations as in Fig. 1. (From Graybiel and Ragsdale, 1979.)

these studies can be roughly divided into three groups based on the change in concentration observed. Decreased concentrations in affected areas have been reported for γ-aminobutyric acid (GABA) (Bird et al., 1973), acetylcholine (McGeer et al., 1973), substance P (Kanazawa et al., 1979; Buck et al., 1981), methionine enkephalin (Emson et al., 1980), and cholecystokinin (Krieger and Martin, 1981). Concentrations of norepinephrine, serotonin, and vasoactive intestinal polypeptide (Emson et al., 1979; Krieger and Martin, 1981) remain unchanged (Bird, 1980). Concentration increases have been seen in some studies for dopamine (Bird, 1980), thyrotropin releasing hormone (Spindel et al., 1980), and somatostatin (Cooper et al., 1981). Unfortunately, it does not appear that any of these changes represent the primary defect, since in each case concentrations of a given neurotransmitter remain normal in unaffected areas of the brain. Thus, it is likely that the changes observed are epiphenomena resulting from the selective cell death.

Examination of the structure of the caudate nucleus reveals how cell death could have such a multitude of effects. Figure 3 shows the mixture of local neurons interconnecting cells in the caudate and putamen and long projecting neurons that are contained in the caudate nucleus (Pasik et al., 1979). The two principal efferents that go both to the globus pallidus and to the substantia nigra are the GABA system and the substance P pathway. There are also similar enkephalin systems. Local neurons include various interneurons producing acetylcholine, GABA,

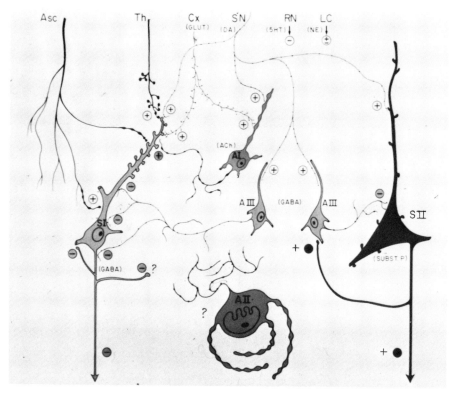

FIG. 3. Tentative neuronal circuits in the neostriatum. Afferents from brainstem or diencephalon (Asc), thalamus (Th), cerebral cortex (Cx), substantia nigra (SN), raphe nuclei (RN), locus coeruleus (LC). Interneurons: aspiny I (A I), aspiny II (A II), aspiny III (A III). Output neurons: spiny I (S I), spiny II (S II). Excitatory and inhibitory synapses indicated by + and −, respectively. Gabaminergic neurons are inhibitory, substance P, excitatory. Neurotransmitters are designated in parentheses near the element that contains them. ACh, acetylcholine; glut, glutamine; DA, dopamine; 5HT, serotonin; NE, norepinephrine; GABA, γ-aminobutyric acid; Subst P, substance P. (From Pasik et al., 1979, with permission.)

and a number of neuropeptides. It is obvious that the selective death of cells in the caudate nucleus could be accompanied by a variety of changes in the neurotransmitter concentrations and that these changes reveal little concerning the primary cause of neuronal loss.

RESEARCH STRATEGIES

A number of research strategies could potentially lead to a better understanding of the cause of Huntington's disease. Although many of these are being actively pursued, in other cases it is not yet possible to perform the desired experiments.

A search for neuronal markers characteristic of the population of cells that die in affected individuals would be fruitful. It is possible that the isolation of monoclonal antibodies directed against cell-surface antigens of neurons in the caudate and putamen could identify molecules characteristic of particular cell types.

A second line of investigation might be to discover more about the actual sequence of cell death. The first cells to die appear to be in the caudate and putamen where there are massive inputs from the cerebral cortex, which is the second most affected region of the brain. It is possible that, due to transsynaptic degeneration, the cortical loss is secondary to loss of cells in the basal ganglia. On the other hand, it is also possible that the primary defect affects the cerebral cortex, resulting in psychiatric and behavioral disorders that appear first in some patients; the movement disorders that appear later may be due to secondary cell loss in the basal ganglia.

In many developmental pathways, there is a role for programmed cell death. In the brain, perhaps 50% of the cells fail to survive the early developmental period. Furthermore, neuronal loss continues and brain size decreases as part of the normal aging process. An experimental strategy is greatly needed that, by examining the normal programming of the cells in the basal ganglia, would give clues as to why cells would develop and function normally for decades and then cease their function prematurely and die.

Another possible approach to understanding the mechanism of cell death might be to investigate the role of trophic factors, i.e., substances synthesized in one cell that are essential for the life of another cell. A number of such secreted molecules exist (E. Shooter, *this volume*), but none has yet been shown to have a clear role in the central nervous system.

Although measurements of neurotransmitter concentrations have not revealed the defect in Huntington's disease, it is possible that an alteration in the primary structure of a neurotransmitter could have the observed pathological effects. For example, kainic acid, a rigid analog of glutamate, does induce similar nerve cell death in rats (Coyle et al., 1979). Thus, an investigation of the primary structures of the various neurotransmitters in patients with Huntington's disease might be beneficial.

One explanation for the dominant nature of the HD gene is that it might affect a property of the cell membrane. Considerable effort has already gone into studies of the membranes of fibroblasts and red blood cells from affected individuals, but these have all been inconclusive (Comings et al., 1981; Lakowicz and Sheppard, 1981). Similar investigations of the actual affected cell type might yield more definitive results.

Studies of the central nervous system can often be faciliated by the existence of mutant animals (see R. Sidman, *this volume*). Unfortunately, there is as yet no animal model for this disease. Furthermore, there appears to be no animal mutant in which normal development occurs but is followed by premature cell death of a particular cell type. A systematic attempt using controlled mutagenesis to isolate autosomal dominant mutants affecting the nervous system of experimental animals could be of tremendous value.

A final approach that holds great promise is the use of recombinant DNA-generated genetic markers in linkage studies to locate and ultimately clone the HD gene (D. Housman and J. Gusella, *this volume*).

Huntington's disease clearly represents a unique model system for the study of human autosomal dominant mutations. Its extremely low mutation rate combined with the fact that it provides a clear case of programmed cell death in the nervous system make it a considerable challenge for the biologist. It is, however, a challenge that should be met. The disease causes an enormous degree of prolonged human suffering, not only affecting those carrying the gene but also a host of relatives and friends who must stand by and helplessly witness the slow degeneration of a loved one. It is to be hoped that research into its cause will not only increase our understanding of the nervous system, but also alleviate the anguish of those who are afflicted.

ACKNOWLEDGMENTS

Supported by PHS grants AM26252 and NS16367 (Huntington's Disease Center without Walls).

REFERENCES

Bird, E. D. (1980): Chemical pathology of Huntington's disease. *Ann. Rev. Pharmacol. Toxicol.*, 20:533–551.
Bird, E. D., Caro, D. J., and Pilling, J. B. (1974): A sex related factor in the inheritance of Huntington's chorea. *Am. J. Hum. Genet.*, 37:255.
Bird, E. D., MacKay, A. V. P., Rayner, C. N., and Iversen, L. L. (1973): Reduced glutamic acid-decarboxylase activity of post-mortem brain in Huntington's chorea. *Lancet*, 1:1090–1092.
Buck, S. H., Burks, T. F., Brown, M. R., and Yamamura, H. I. (1981): Reduction in basal ganglia and substantia nigra substance P levels in Huntington's chorea. *Brain Res.*, 209:464–469.
Comings, D. E., Pekkala, A., Schuh, J. R., Kuo, P. C., and Chan, S. I. (1981): Huntington's disease and tourette syndrome. *Am. J. Hum. Genet.*, 33:166–174.
Cooper, P. E., Aronin, N., Bird, E. D., Leeman, S. E., and Martin, J. B. (1981): Increased somatostatin in basal ganglia of Huntington's disease. *Neurology*, 31:64.
Coyle, J. T., Condon, E. D., Biziere, K., and Zaczek, R. (1979): Kainic acid neurotoxicity: Insights into the pathophysiology of Huntington's disease. In: *Advances in Neurology: Huntington's Disease, vol. 23*, edited by T. N. Chase, N. S. Wexler, and A. Barbeau, pp. 593–608. Raven Press, New York.
Emson, P. C., Arregui, A., Clement-Jones, V., Sandberg, B. E. B., and Rossor, M. (1980): Regional distribution of methionine-enkephalin and substance P-like immunoreactivity in normal human brain and in Huntington's disease. *Brain Res.*, 199:147–160.
Emson, P. C., Fahrenkrug, J., and Spokes, E. G. S. (1979): Vasoactive intestinal polypeptide (VIP): Distribution in normal human brain and in Huntington's disease. *Brain Res.*, 173:174–178.
Graybiel, A. M., and Ragsdale, C. W., Jr. (1979): Fiber connections of the basal ganglia. *Prog. Brain Res.*, 51:239–283.
Huntington, G. (1872): On chorea. *Med. Surg. Reporter*, 26:317–321.
Kanazawa, I., Bird, E. D., Gale, J. S., Iversen, L. L., Jessell, T. M., Muramoto, O., Spokes, E. G., and Sutoo, D. (1979): Substance P: Decrease in substantia nigra and globus pallidus in Huntington's disease. In: *Advances in Neurology: Huntington's Disease, vol. 23*, edited by T. N. Chase, N. S. Wexler, and A. Barbeau, pp. 495–504. Raven Press, New York.
Krieger, D. T., and Martin, J. B. (1981): Brain peptides. *New Engl. J. Med.*, 304:876–885; 944–951.
Lakowicz, J. R., and Sheppard, J. R. (1981): Fluorescence spectroscopic studies of Huntington fibroblast membranes. *Am. J. Hum. Genet.*, 33:155–165.
McGeer, P. L., McGeer, E. G., and Fibiger, H. C. (1973): Choline acetylase and glutamic acid decarboxylase in Huntington's chorea. *Neurology*, 23:912–917.
Pasik, P., Pasik, T., and Difiglia, M. (1979): The internal organization of the neostriatum in mammals. In: *The Neostriatum*, edited by I. Divac and R. G. Oberg, pp. 5–36. Pergamon Press, New York.
Spindel, E. R., Wurtman, R. J., and Bird, E. D. (1980): Increased TRH content of the basal ganglia in Huntington's disease. *N. Engl. J. Med.*, 303:1235–1236.
Vessie, P. R. (1932): On the transmission of Huntington's chorea for 300 years: The Bures family group. *J. Nerv. Ment. Dis.*, 76:553–573.
Wallace, D. C. (1979): Distortion of mendelian segregation in Huntington's disease. In: *Advances in Neurology: Huntington's Disease, vol. 23*, edited by T. N. Chase et al., pp. 73–81. Raven Press, New York.

Molecular Genetic Neuroscience, edited by
F.O. Schmitt, S.J. Bird, and F.E. Bloom.
Raven Press, New York © 1982.

Molecular Genetic Approaches to Neural Degenerative Disorders

David Housman and James Gusella

ABSTRACT

There are a large number of human genetic disorders in which no defective gene product has been identified. The advent of recombinant DNA technology has opened an avenue to locating the gene product responsible for any genetic disorder that shows an unequivocal single-gene pattern of inheritance. The approach relies on the use of variations in DNA sequence as markers to identify the exact chromosomal location of the gene by genetic linkage analysis. The entire region surrounding the defective gene can then be isolated and analyzed by combining somatic cell genetics and DNA cloning techniques. The application of this method to the study of Huntington's disease is outlined here.

In recent years, molecular genetic techniques have been successfully applied in many systems to the isolation of specific genes where an easily identifiable gene product is present. The DNA segments coding for globin, immunoglobulin, and insulin have all been cloned using recombinant DNA technology. These are only a few examples of genes encoding for well-characterized proteins that have been isolated in this manner. Unfortunately, in most genetic diseases no specific protein defect has yet been identified. Therefore, a method by which these mutant genes could be cloned and characterized without any prior knowledge of the defective gene product would be of immense benefit. Huntington's disease (HD), an autosomal dominant disorder of the human nervous system (J. Martin, *this volume*), is used here as a model system for the application of such an approach, but since the approach is general, it could potentially be applied to any genetic disease, human or animal, that shows a clear pattern of inheritance.

LOCATING THE HD GENE

Huntington's disease is a human genetic disorder that has its onset in middle age and that affects the central nervous system. There is no clue to the gene product involved in this disease, despite the fact that a significant effort has been made by many investigators to find one. Current knowledge of the disease includes two undisputed facts: the mutant gene acts in an autosomal dominant fashion with high penetrance, and it causes premature loss of a specific set of neurons in the brain. An approach whereby this information can be used to identify and characterize the gene product responsible for this disease is described here.

The first step is to use the approach of the classic geneticist to determine the chromosomal localization of the HD gene. The clear pattern of inheritance of this disorder makes it possible to trace the gene among members of affected pedigrees. The classic genetic approach of linkage analysis could, therefore, be used to identify a polymorphic marker located on the same chromosome in close proximity to the HD gene. Using standard polymorphic protein markers, this strategy has already been applied but with only limited success; the gene has been excluded from some 15 to 20% of the human genome (Pericak-Vance et al., 1979; Went and Volkers, 1979). Unfortunately, there are not enough genetic markers currently available to test for linkage in the remaining 80 to 85% of the genome.

How many new genetic markers are needed to make such a study feasible? The human genome can be divided into some 3,300 centimorgans by using frequency of recombination as a measure. There is a 1% chance of recombination occurring in any given meiosis between any two genetic markers separated in the DNA by 1 centimorgan. This distance represents on the order of 10^6 base pairs of DNA. Positive linkage can usually be readily detected at a distance of 10 centimorgans because markers linked this closely are inherited in concert 90% of the time. Since each marker can be used to detect linkage for a distance of 10 centimorgans on either side of it, some 165 (3,300/2 × 10) equally spaced genetic markers are needed to ensure the ability to detect linkage anywhere in the genome. If these markers were found, the mutant gene causing any disease with a clear pattern of inheritance could be localized without fail within the genome.

GENERATION OF POLYMORPHIC MARKERS

The major difficulty in the linkage approach is clearly the need for the discovery of new polymorphic markers. Thus far, discoveries of this sort have proceeded at a relatively modest rate. Fortunately, the advent of recombinant DNA technology has made the generation of large numbers of polymorphic markers an easier task. It is clear that genetic markers, whether they are expressed as individual differences in visible phenotypes, differences in electrophoretic mobility of proteins, or differences in antigenic determinants, are all ultimately due to individual differences in the primary sequence of the DNA. Thus, differences in DNA sequence between individuals could potentially be used directly as genetic markers without the need for these differences to be expressed as proteins.

The techniques of molecular genetics are well suited to the visualization of differences in the primary sequence of DNA. For example, Fig. 1 shows schematically two alleles of an arbitrary genetic marker. The arrows indicate the cleavage sites in this DNA segment for a particular restriction endonuclease. The normal sequence contains four recognition sites for this enzyme, but in the variant allele, a change in base sequence has occurred that has altered one recognition site. The fragments resulting from digestion of the two segments with the restriction enzyme will, therefore, differ significantly in size. Differences could also occur if insertion or deletion of DNA segments takes place within these fragments. These size differences can be detected by resolving digested DNA by agarose gel electrophoresis, transferring it to a nitrocellulose filter by the Southern blotting procedure, and then visualizing the fragments by hybridizing with a probe DNA sequence cloned from this region.

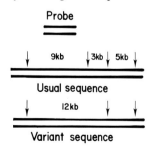

FIG. 1. A hypothetical restriction fragment length polymorphism. *Arrows* indicate cleavage sites in this DNA segment for a particular restriction endonuclease.

An example of this process is shown in Fig. 2 where Pst I-digested DNA from 8 individuals has been digested with restriction endonuclease Pst I and hybridized with a cloned β-globin probe. The first 6 individuals (lanes 1 to 6) show a normal pattern of hybridization and are homozygous for a 4.4-Kb fragment and a 2.2-Kb fragment. The other 3 people (7, 8, and 9) have thalassemia and are heterozygous for a deletion that has reduced one allele of the 4.4-Kb fragment to 3.8 Kb. This difference, as expected from a pedigree analysis, is inherited as a mendelian codominant marker.

Although the DNA sequence used as a probe in Fig. 2 came from a known gene, the same procedure can be used for any single-copy DNA sequence regardless of whether it codes for a protein. The use of DNA markers, unlike that of standard polymorphic protein markers, is essentially an iterative process because the same set of resolved DNA fragments can be used to monitor several such markers by sequential hybridization with a number of probe sequences.

OBTAINING CHROMOSOMALLY LOCALIZED MARKERS

Ultimately it will be necessary to generate new markers that span the entire genome. One approach to this problem is to use clones of random DNA segments to identify polymorphisms and then to assign these segments to a particular chromosome by either *in situ* hybridization or the use of panels of somatic cell hybrids that have segregated human chromosomes. A more systematic approach would be to find markers on one chromosome at a time until enough markers are identified to cover the desired autosome. Once a possible linkage to the HD gene is detected, it will be desirable to generate a large number of markers from the region of the marker, both to confirm the linkage and to localize the HD gene more closely. These pursuits require a method to clone human DNA segments with knowledge of the chromosome or subchromosomal region in which they are located.

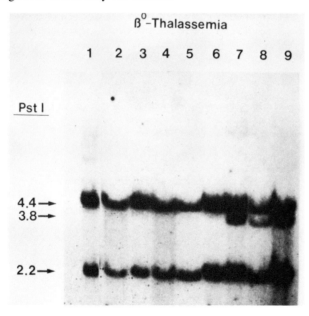

FIG. 2. A restriction fragment length polymorphism in the β-globin gene. DNA from six normal individuals (lanes 1–6) and from three patients with β-thalassemia (lanes 7–9) was digested with restriction endonuclease Pst I, fractionated by agarose gel electrophoresis, transferred to a nitrocellulose filter, and hybridized to a radioactive probe for the β-globin locus. Hybridized probe was detected by autoradiography. (From Orkin, 1980.)

Such a method has been devised by interfacing somatic cell genetics with recombinant DNA technology (Gusella et al., 1980). Over the past 15 years, a large body of work in somatic cell genetics has resulted in the isolation in hybrid cells of particular human chromosomes on a background of rodent chromosomes. It was reasoned that if recombinant DNA technology could be used to clone pieces of human DNA from such hybrid cells, then libraries of DNA segments from particular chromosomal regions could be obtained.

The method used to distinguish between segments of human DNA and rodent DNA relies on the presence in mammalian genomes of repetitive DNA segments interspersed among the segments of single-copy DNA. To determine whether human and rodent repetitive sequences differed enough to distinguish them easily, three recombinant DNA libraries were constructed in the vector Charon 4A using mouse, hamster, and human DNA. In Fig. 3, 100 phages from each library were arranged in a grid format. Each phage contained cloned DNA segments 15 to 20 consecutive Kb. When the phages containing human DNA were hybridized to radioactively labeled human repetitive DNA, greater than 90% of them were detected by the probe. In fact,

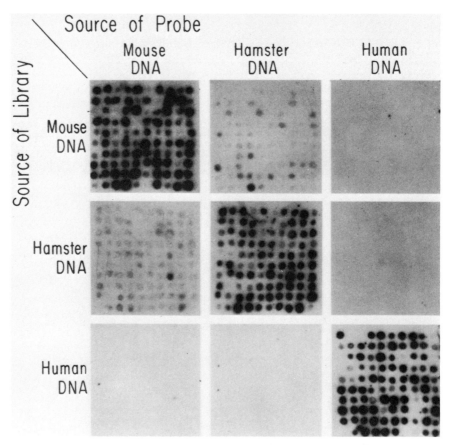

FIG. 3. Detection of species-specific reiterated DNA in recombinant libraries. DNA fragments from the indicated species were used to construct libraries in the bacteriophage vector Charon 4A. Individual recombinant clones from each library were spotted in an array in a bacterial culture dish, and three replicate nitrocellulose filters were prepared from each dish (Benton and Davis, 1977). The filters were hybridized to probe for repetitive DNA from the species indicated according to the procedure described by Gusella and coworkers (1980). Radioactivity bound to the filters was detected by autoradiography. (From Housman and Gusella, 1981).

in more extensive experiments, more than 99% of 15-Kb segments of human DNA contained repetitive sequences (J. Gusella and D. Houseman, *unpublished results*). When the same phages were probed with radioactive mouse or hamster repetitive DNA, none of them displayed detectable hybridization. The technique can in fact be pushed much harder, since even when up to 15,000 phages per plate are screened under these conditions, there appears to be no detectable cross-homology between human and rodent DNA (D. Oates, J. Gusella, and D. Houseman, *unpublished results*). Apparently, interspersed repetitive DNA sequences are species-specific and can be used as a tag to indicate the source of DNA fragments. Thus hybridization with human repetitive DNA is an effective method of identifying human DNA segments.

This technique should clearly be applicable to the cloning of chromosomally localized DNA segments by using appropriate somatic cell hybrids. The procedure was tested by attempting to clone DNA segments from the short arm of human chromosome 11. A library of 15- to 20-Kb DNA segments was constructed from the hybrid cell J1–11, which contains only the short arm of human chromosome 11 and a full complement of hamster chromosomes. Phages from the hamster library, the human library, and the J1–11 library appear in Fig. 4 in a grid format. The phage HβG1, containing the cloned human β-globin gene that resides on chromosome 11, was included as a positive control. When human DNA was used as a probe against these phages, most of the human DNA-containing plaques, including the HβG1 plaque, showed intense hybridization, presumably due to the presence of interspersed repetitive sequences. The phages from the hamster library displayed little hybridization. Of the phages from the J1–11 library, one showed intense hybridization and, therefore, presumably contained a cloned DNA segment from the short arm of human chromosome 11 (Gusella et al., 1980).

Using a series of somatic cell hybrids, it was possible not only to confirm the presence of human chromosome 11 DNA, but also to localize the DNA segment to a particular region of

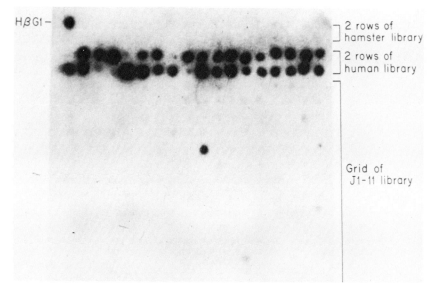

FIG. 4. Isolation of recombinants containing human DNA from a library of human-Chinese hamster hybrid cell DNA. Recombinant Charon 4A plaques from a hamster library, a human library, and a hybrid cell library (constructed from DNA of the hybrid cell line J1–11, which contains the human chromosome 11 short arm on a Chinese hamster background) were spotted into an array as indicated. The phage HβG1 containing the human and globin genes was also included as a positive control. A nitrocellulose replica was made and hybridized to 10^6 cpm of [^{32}P] human DNA. One phage plaque from the J1–11 library showed intense hybridization. (From Gusella et al., 1980.)

the chromosome. The procedure outlined here has now been used to localize more than 10 such cloned DNA segments. Kao and coworkers (1977) have used cytotoxic antisera directed against human cell-surface antigens encoded on chromosome 11 to generate a panel of hybrids with increasing terminal deletions of the short arm of human chromosome 11. If the DNA segment contained in a human recombinant from the J1–11 library were used directly as a probe against DNA from these hybrid cells, the repetitive sequences present in the clone would hybridize to all the DNAs. Thus, it is first necessary to isolate a subsegment of the clone that contains no repetitive DNA. An example of this procedure is shown in Fig. 5 in which the left-hand pair of lanes show the ethidium bromide-stained DNA fragments resulting from digestion of a clone with two different restriction endonucleases. The right-hand pair of lanes display the pattern of hybridization of these fragments to human repetitive DNA. In this case, fragment g was chosen as a marker for chromosome 11 since it contained no repetitive sequences.

As shown in Fig. 6, fragments identified in this fashion were used as probes against DNAs from the panel of hybrids that contain different regions of chromosome 11. For the case shown in Fig. 6, the probe hybridized as expected to DNA from the J1–11 hybrid from which it was cloned, but not to DNA from the parent hamster cell. Furthermore, by analyzing the pattern of hybridization of the probe to the other hybrids in the panel, it was possible to obtain a regional assignment on the short arm. When this procedure is applied for many clones, it should be possible to obtain a fine-structure map of the short arm of human chromosome 11.

The methods that can be used to generalize this strategy to other chromosomes will not be discussed here. Suffice it to say that using a number of approaches it is now possible to obtain chromosomally localized DNA from virtually anywhere in the human genome.

BENEFITS OF MAPPING THE HD GENE

It is clear that by applying the various procedures outlined here, it should be possible to first detect linkage of a marker with the HD gene and then to generate enough closely spaced markers

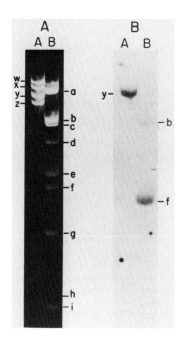

FIG. 5. Detection of fragments containing human repetitive DNA in a recombinant phage. Lane A, EcoRI; lane B, BamH1. **Left:** Ethidium bromide-stained restriction fragments from clone H11–3, which contains a piece of human DNA from chromosome 11. Restriction fragments were resolved on a 1% agarose gel and visualized by staining with ethidium bromide at 0.5 μg/ml. Fragment y is the single 15-Kd EcoRI fragment of human DNA inserted in this clone. **Right:** Fragments shown at left were transferred to nitrocellulose and hybridized to radioactively labeled human DNA: b and f show some hybridization ot the probe and therefore contain some repetitive DNA; previous studies have shown that c, d, e, and h consist entirely of Charon 4A DNA; a, g, and i show no hybridization to human repetitive DNA. Fragment g was chosen to be a single-copy probe. (From Gusella et al., 1980.)

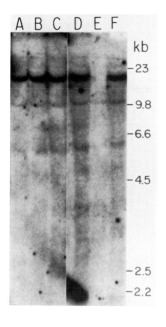

FIG. 6. Localization of clone H11–3 on human chromosome 11. DNA from a number of human-Chinese hamster hybrid cells containing different regions of human chromosome 11 were digested with restriction endonuclease EcoRI. The resulting fragments were resolved by agarose gel electrophoresis, transferred to nitrocellulose, and hybridized to radioactively labeled fragment g from Fig. 5. The probe detects a 15-Kb band in all hybrid cell lines (lanes A, B, C, D, and F) but fails to hybridize to hamster DNA (lane E). Because the only region of human chromosome 11 common to all the hybrid cell lines is the centromere region, the insert in clone H11–3 has been localized to this region. (From Gusella et al., 1980.)

to localize it within 1 or 2 centimorgans.[1] Obtaining the map position of the HD gene will have two important consequences. First, it will have the practical effect of allowing a prenatal or presymptomatic test to be designed to aid in genetic counseling efforts. Perhaps of more importance, however, it will provide a handle to clone the gene itself and determine the genetic basis of the disease. As pointed out by J. Martin *(this volume)*, we do have a sense of where the HD gene operates because it causes the loss of an identified set of neurons. Once the map location has been determined, it should be possible to clone the entire 10^6 base pairs of DNA in the stretch surrounding the HD gene. This amount of DNA has the coding capacity for perhaps 50 functions, and the cloned DNA itself can be used to determine which mRNAs are represented in this region. Once this has been done, the analysis of this restricted set of proteins in the affected cell type could well lead to the discovery of the primary defect in Huntington's disease and an identification of the gene sequence itself.

DISCUSSION

L. E. Hood pointed out that once the location of the HD gene has been narrowed down to a region of 10^6 base pairs, the problem of identifying the actual gene is still not trivial. The coding capacity of such a region is theoretically up to 2,000 polypeptides of 200 amino acids each, and although the existence of large intervening and flanking sequences suggests that the actual number of genes present will probably be lower, this is the size of the major histocompatibility locus (MHC) of the mouse, which codes for a large number of different gene products. The localization of a single gene in a region of this size would be very difficult unless new techniques become available, particularly if virtually every gene including the HD gene belongs

---

[1] D. Housman agreed that chromosome sorting, suggested by W. E. Hahn, would provide a good complement to hybrid technology in isolating chromosome-specific probes, but it has the disadvantage that it is not absolutely clean (especially for smaller chromosomes) and the level of contamination with fragments of other chromosomes can be significant. Chromosome sorting circumvents the problem that some chromosomal regions are commonly lost in hybrid cells.

to a multigene family. However, it would be possible to find a specific gene in such a region if it belonged to a relatively simple family.

F. H. C. Crick indicated that there was reason to be optimistic since one might quickly discover a polymorphism that represents the actual HD gene. If not, there will certainly be other methods, possibly involving selection of mRNAs, that will facilitate the localization of the gene. In any case, there is clearly a great need for a large number of markers to be placed on the human gene map for many purposes, including linkage analysis. A comparison of the gene for Huntington's disease to the MHC, which contains a large multigene family of histocompatibility antigens, may be misleading since the quest for the HD locus is more likely to represent a search for a single gene. Furthermore, the concept of a multigene family differs with its application. Thus, although it might be very difficult to locate any single gene known to be part of the MHC, it should be quite possible to locate the homoglobin genes if in a region of 1 centimorgan, using a knowledge of hemoglobin genetics.

D. Housman pointed out that human dominant disorders are often caused by deletion of DNA. If this were also true of Huntington's disease, the gene should be relatively easy to find. Furthermore, Huntington's disease is a good candidate for applying molecular genetic techniques precisely because there is no other biological clue to approaching the disorder.

ACKNOWLEDGMENTS

We thank Dr. Stuart Orkin for use of his data on the β-globin polymorphism. This work is supported by PHS grant NS16367 (Huntington's Disease Center without Walls).

REFERENCES

Benton, W., and Davis, R. (1977): Screening gt recombinant clones by hybridization to single plaques in situ. *Science*, 196:180–182.
Gusella, J., Keys, C., Varsangi-Breiner, A., Kao, F. T., Jones, C., Puck, T. T., and Housman, D. (1980): Isolation and localization of DNA segments from specific human chromosomes. *Proc. Natl. Acad. Sci. USA*, 77:2829–2833.
Housman, D., and Gusella, J. (1981): Use of recombinant DNA techniques for linkage studies in genetically based neurological diseases. In: *Genetic Research Strategies for Psychobiology and Psychiatry*, edited by E. S. Gershan, S. Matthysse, X. O. Breakefield and R. D. Ciaranello, pp. 17–24. Boxwood Press, Pacific Grove, Calif.
Kao, F. T., Jones, C., and Puck, T. T. (1977): Genetics of cell-surface antigens: Regional mapping of three components of the human cell-surface antigen complex, A_L, on chromosome 11. *Somatic Cell Genet.*, 3:421–429.
Orkin, S. (1980): Specific abnormalities of globin gene organization in the thalassemia syndromes. *Ann. N. Y. Acad. Sci.*, 344:48–61.
Pericak-Vance, M. A., Conneally, P. M., Merritt, A. D., Roos, R. P., Vance, J. M., Yu, P. L., Norton, J. A., and Antel, J. P. (1979): Genetic linkage in Huntington's disease. *Adv. Neurol.*, 23:59–71.
Went, L. N., and Volkers, W. S. (1979): Genetic linkage. *Adv. Neurol.*, 23:37–42.

Section XII

Trophic and Instructive Factors

More than half a century ago it was discovered that certain substances which an organism could not synthesize itself were required to maintain life or particular aspects of metabolism. These substances were called "vitamins" (because the earliest one discovered was both vital and an amine). Soon after the discovery of the first three vitamins many more were discovered, at which time they became known by their appropriate chemical names.

So today, it has been discovered that neuronal development and survival require the continuous supply of such "trophic" substances from a target cell (e.g., another neuron or a muscle cell) with which any particular neuron makes synaptic contact. The first such substance, discovered and subsequently well characterized by Levi-Montalcini and Hamburger (1951), was called a "factor," i.e., nerve growth factor (NGF).

As is described by E. M. Shooter and colleagues, the chemical composition of the paired polypeptide chains of NGF, like those of a similar factor, epidermal growth factor (EGF), has been fully determined. The survival of certain sympathetic and sensory neurons depends on the continuous supply of NGF synthesized by target cells; after traversing the junctional structure, the NGF binds to high-affinity and/or to low-affinity receptors, and is internalized and transported within vesicles to the neuronal soma and nucleus. It is believed that NGF does not directly regulate genes but may act through the receptors to which it is bound or, as suggested by Thoenen and Barde (1981), through a yet unidentified second messenger (not a cyclic nucleotide).

Nerve growth factor is also able to influence neuronal neurotransmitter type through regulation of the expression of genes for enzymes (e.g., tyrosine hydroxylase and dopamine β-hydroxylase) necessary for the synthesis of adrenergic transmitters. However, Shooter points out that it may be the complex of NGF with both high- and low-affinity receptors that induces the synthesis of enzymes determining transmitter type.

P. H. Patterson found that a factor diffusing from heart muscle in culture is capable of transforming neonatal rat adrenergic neurons to cholinergic type. Furthermore, adrenergic neurons release into intercellular fluid a number of glycoproteins, certain of which mediate specification of transmitter type. This would seem to be an example of a paracrine effect that regulates expression of one or more genes depending not only on the neuroactive ligand, but also on the receptor protein.

Another factor produced by glial cells is immunologically and functionally different from NGF and is required for the survival of sensory neurons (Barde et al., 1980). Yet another factor is produced, by C-6 glioma cells, that enhances the survival of chick sensory neurons and stimulates neurite outgrowth from them (Barde et al., 1978). It would seem that "factors" are now established forms of neuroactive ligands that may induce specific gene expression. Perhaps, like neuropeptides, their numbers will grow greatly in the months and years to come.

According to A. N. Martonosi, Ca^{2+} may serve as a second messenger that induces gene expression by combining with DNA-bound protein that binds Ca^{2+} with high affinity and that, by so doing, activates one or more genes. Gene control is thus influenced by cytosolic $[Ca^{2+}]$ and is intimately related to alterations of calcium channel permeability. Thus, molecular genetic processes may also be entrained in bioelectrical phenomena.

Modulation of Neuronal Function by Nerve Growth Factor

Eric M. Shooter, Peter Frey, Peter W. Gunning, Gary E. Landreth, Arne Sutter, and Bruce A. Yankner

ABSTRACT

Nerve growth factor is a peptide hormone involved in the whole life cycle of sympathetic and some sensory neurons from the survival of the embryonic neuron to the maintenance of the fully differentiated cell. Its chemistry is well understood, and the mechanism of its biosynthesis in the adult submaxillary gland has been partially characterized. The interaction of nerve growth factor with specific receptors on the responsive cells leads to neurite outgrowth, some control of the direction of growth, and selective induction of certain neurotransmitter-synthesizing enzymes. The nature of the intracellular signals that result from the nerve growth factor receptor interaction are not yet known but probably result from internalization of the factor and its receptor.

The peripheral nervous system has been widely used to explore the development of nerve cells at a molecular level. Two key problems in this area of eukaryotic cell biology are how the morphology of the cell is related to gene expression, and how genes are turned on or off or up and down in particular cell types. Because nerve growth factor (NGF) profoundly affects the morphology of neurons and induces the synthesis of certain enzymes required for neurotransmitter biosynthesis, it is a valuable tool for exploration of these problems.

Nerve growth factor is essential for the survival and development of sympathetic and of some sensory neurons (Fig. 1) (Levi-Montalcini and Angeletti, 1968). Both *in vivo* and in culture these neurons grow and develop normally in the presence, but not in the absence, of NGF. Antibodies against NGF injected into newborn rats or mice significantly inhibit the development of the sympathetic nervous system, consistent with the finding that at this stage, sympathetic neurons are dependent on NGF for survival (Levi-Montalcini and Angeletti, 1966). These antibodies act by neutralizing circulating NGF in the newborn animals (Goedert et al., 1980). It is also possible to analyze the effects of NGF prior to birth by injecting female rats with murine NGF (Gorin and Johnson, 1979). Antibodies produced against this heterologous NGF cross-react with endogenous NGF in the animal; they also cross the placenta and block the action of NGF in developing embryos. The growth of sensory neurons in the embryo is inhibited, indicating that they are sensitive to NGF at an earlier stage in development than sympathetic neurons. Direct injection of embryos *in utero* with antibodies to NGF also results in a reduction in the size of the sensory neuron-containing dorsal root ganglia (Aloe et al., 1981a). These effects of NGF can be demonstrated in culture as well as *in vivo*. Neurons in newborn rat sympathetic ganglia in explant cultures degenerate and die in the absence of NGF, but with NGF these neurons survive and send out extensive neurites (Fig. 2). The same

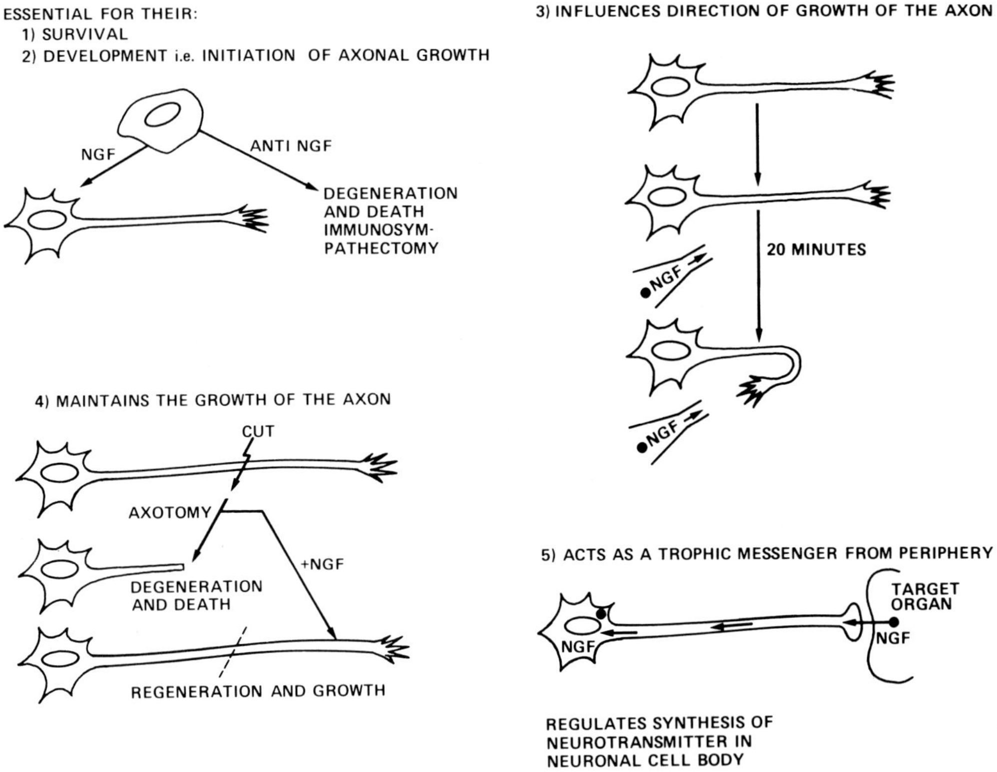

FIG. 1. The roles of nerve growth factor in the development and maintenance of sympathetic and sensory neurons.

phenomenon is observed with cultures of dissociated neurons from embryonic chick sensory ganglia and forms the basis of the original bioassay (Levi-Montalcini et al., 1954). The effect of NGF on the direction of neurite extension can be shown by applying high concentrations of NGF to one side of the line of growth of a neurite from sympathetic neurons (Gundersen and Barrett, 1979). In a relatively short time the growth cone turns toward the source of NGF. Although NGF is not the only substance that influences the directionality of neurite growth *in vitro* (Gundersen and Barrett, 1980), striking effects on sympathetic neurite growth are observed when NGF is injected into unusual locations in neonatal rats (Menesini-Chen et al., 1978). Interestingly, the first evidence that NGF could orient as well as stimulate neurite outgrowth was obtained in the original experiments that led to the discovery of NGF (Levi-Montalcini and Hamburger, 1951, 1953).

It is now also clear that the growing neurite depends on continuous exposure to NGF. If the retrograde transport of NGF in the axon of a sympathetic neuron is blocked, e.g., by axotomy or removal of the target organ, the developing neuron degenerates and dies (Fig. 1). However, if an extra supply of NGF is provided, at the same time the injury or excision is made the neurons regenerate neurites and make functional synapses (Thoenen and Barde, 1980). These results have led to the hypothesis that the target organs of sympathetic and sensory neurons synthesize the NGF required for the growth and maintenance of the axon. Nerve growth factor from the target traverses the synapse and is carried by retrograde transport up the axon to the cell body (Fig. 1). The existence of this highly specific pathway for NGF translocation and its

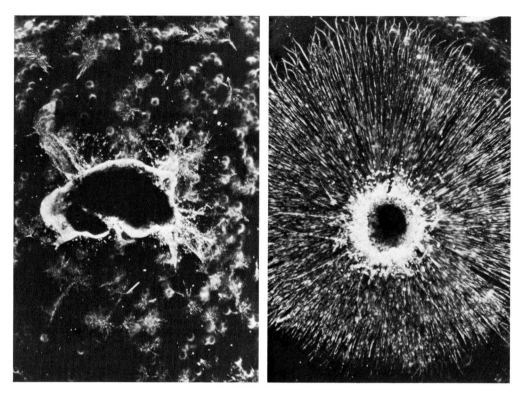

FIG. 2. The stimulation of neurite outgrowth from a newborn rat sympathetic ganglion. **Left:** Control without NGF. **Right:** Ganglia exposed to 1 μg/ml NGF for 2 days. (Courtesy of Dr. Robert Stickgold.)

persistence for the lifetime of sympathetic and sensory neurons have been amply documented (Thoenen and Barde, 1980). One further action of the translocated NGF is to regulate the synthesis of the neurotransmitter, norepinephrine, used by sympathetic neurons. Nerve growth factor arriving in the interior of the cell body by retrograde transport selectively induces two key enzymes in catecholamine biosynthesis, tyrosine hydroxylase (TH) and dopamine-β-hydroxylase (DBH) (Thoenen et al., 1971). Impairment of the retrograde transport of NGF during development or in the adult causes significant decreases in TH and DBH levels. Although it is attractive to consider that sympathetic and sensory targets synthesize NGF, much remains to be done to substantiate this hypothesis. It is known that one sympathetically innervated tissue, the adult male mouse submaxilllary gland, does synthesize NGF (Berger and Shooter, 1978), although the rate is unusually high and regulation is achieved by testosterone. Ebendal and coworkers (1980) have shown that the rat iris contains NGF when initially transplanted into the anterior chamber of the eye but not after reinnervation is established. Denervation of the iris leads again to NGF accumulation in the iris, suggesting that NGF is synthesized in this effector organ and that its synthesis may be regulated by neuronal contacts.

NGF PROTEIN AND ITS BIOSYNTHESIS

The NGF peptide has been purified from the submaxillary gland of the adult male mouse and characterized in detail (Varon et al., 1967, 1968). Another tissue that contains significant levels of NGF is the prostate gland of the guinea pig, rabbit, or bull (Harper and Thoenen, 1980). The submaxillary glands of both male and female *Mastomys natalensis* (an African rat) are also rich sources of NGF (Aloe et al., 1981b). It is reasonable to assume that the NGFs from these sources are closely related peptides. The βNGF protein (referred to as NGF) consists of two identical peptide chains each containing 118 amino acids and three internal disulfide bridges (Fig. 3) (Angeletti and Bradshaw, 1971; Greene et al., 1971). The amino acid sequence of these chains shows some homology with the amino acid sequence of proinsulin (Frazier et al., 1972), and there are similarities as well in the mode of action of insulin and NGF. The NGF in the submaxillary gland is stored in a 7S complex with two α peptides, two trypsin-like γ peptides, and two zinc ions (Fig. 4). Interestingly, insulin is also stored in the secretory granules of the pancreatic cells as a complex with zinc; in this case, each zinc ion is joined to three insulin monomers (Blundell et al., 1972). The kinetics of removal of zinc ions from 7S NGF parallels the kinetics of removal of zinc ions from the zinc-insulin hexamer (Dunn et al., 1980). In analogy with the biosynthesis of insulin, NGF is also produced by proteolytic cleavage of a precursor polypeptide by the γ trypsin-like enzyme. However, in the case of NGF, the proteolytic trypsin-like enzyme remains attached to its product, the NGF chain, initiating the formation of the 7S complex (Berger and Shooter, 1977). Pro-NGF, the immediate precursor of NGF, is about 9,000 daltons greater in molecular weight than the 13,000-dalton NGF chain. The absolute specificity of the processing of pro-NGF by the γ peptide has not yet been demonstrated. This has been shown, however, for another growth factor, epidermal growth factor (EGF), also found in the adult male mouse submaxillary gland (Fig. 5). Epidermal growth factor is also a dimer of two identical peptide chains ending in arginine residues. The factor is isolated from the gland complexed with two copies of a trypsin-like proteolytic enzyme (EGF-BP) and two zinc ions (Taylor et al., 1970). The pro-EGF precursor is cleaved at a specific Arg-containing peptide bond to produce the EGF peptide, and the only trypsin-like enzyme that is known to carry out this cleavage is EGF-BP, associated with the EGF complex (Frey et al., 1979). Thus for these two growth factors, NGF and EGF, a mechanism for the last stage of biosynthesis has evolved that ensures that they are packaged separately in stable complexes ready for export on stimulation of the submaxillary gland. Indeed, when the adult

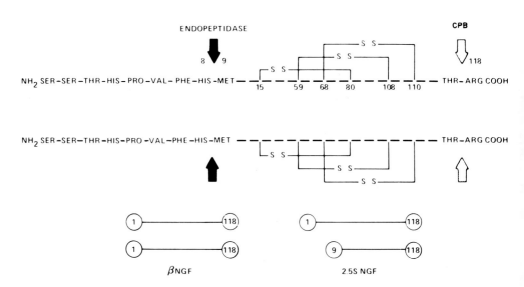

FIG. 3. The primary structure of mouse NGF. CPB, cleavage site for carboxypeptidase B.

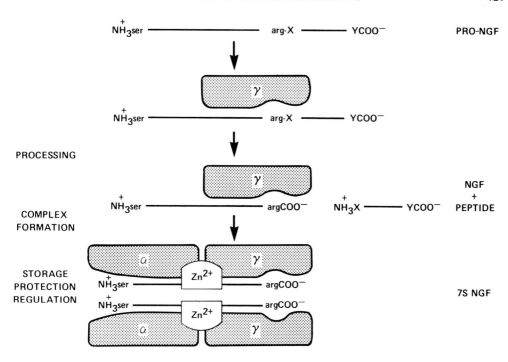

FIG. 4. The 7S NGF complex is formed during the last stage of the biosynthesis of NGF.

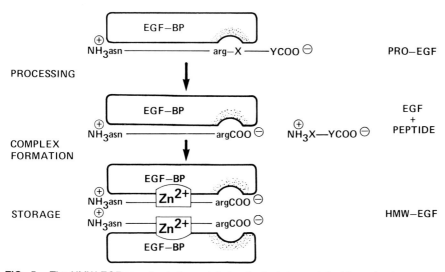

FIG. 5. The HMW-EGF complex is formed during the last stage of the biosynthesis of EGF.

male mouse is stimulated to salivate, both high molecular weight complexes of EGF and NGF are secreted in saliva (Burton et al., 1978).

The present methodology for investigating the precursors of NGF is not sensitive enough to examine its biosynthesis in target organs. For this reason, a number of laboratories have made attempts to isolate and translate the message for NGF *in vitro*. By analogy with the insulin system, one would expect to find a prepro-NGF as the primary translation product of NGF mRNA. However, attempts to identify this translation product in cell-free translation systems have so far been unsuccessful. Interestingly, it has been possible to translate the mRNA for the α-subunit of 7S NGF in this system. The problem may result from the inability of antibodies prepared against NGF to recognize the initial translation product (Thoenen and Barde, 1980). Antibodies that have been used include rabbit antibodies against native NGF, reduced and denatured NGF, and a monoclonal antibody of rather high affinity for NGF (Zimmermann et al., 1981). Further analysis of the biosynthesis of NGF therefore awaits the availability of a cDNA probe for the NGF mRNA. Such a probe could also be used to screen for biosynthesis of NGF in target organs and to explore in more detail the regulation of NGF by innervating neurons.

MECHANISM OF ACTION OF NGF

The mechanism of action of NGF begins with its interaction with specific cell surface receptors on responsive cells. Receptors with two binding affinities have been described on sensory and sympathetic neurons from the chick (Sutter et al., 1979a), as well as the clonal cell line PC12 of a rat pheochromocytoma (Landreth and Shooter, 1980). The equilibrium dissociation constants of these two sites are 10^{-11} and 10^{-9} M, and they are referred to as high- and low-affinity receptors, respectively. High-affinity receptors are only found on neurons, whereas low-affinity receptors have been found on other cell types as well (Sutter et al., 1979b). The expression of receptor type also changes during development. In sensory neurons high-affinity receptors appear at the embryonic age at which NGF induces neurite outgrowth. The number of NGF receptors decreases markedly at the embryonic age when the cells cease responding to NGF by neurite outgrowth. On the other hand, on sympathetic neurons, receptor numbers and the ratio of high- to low-affinity receptors on sympathetic neurons remains nearly constant throughout embryonic life, in keeping with their responsiveness to NGF at all ages.

The actions of NGF can apparently be categorized with respect to the types of receptors involved. The concentration of NGF required to initiate neurite outgrowth on all the above cells corresponds to a fractional occupancy of the high-affinity receptors, whereas that for the induction of TH and DBH corresponds essentially to complete occupancy of these receptors and significant occupancy of low-affinity receptors. Data on the retrograde transport of NGF in sympathetic neurons also lead to the conclusion that transport is mediated by two types of receptors, one of high affinity and low capacity, and the other of low affinity and high capacity (Dumas et al., 1979).

The two different NGF receptor affinities are reflected almost entirely by differences in the rate of dissociation of NGF from the receptors (Sutter et al., 1979a). Dissociation is slow from the high-affinity receptors and fast from the low-affinity receptor (Fig. 6). When sensory neurons, preloaded with a low concentration of [^{125}I]NGF to establish occupancy of only the high-affinity receptors, are diluted into a large excess of cold NGF, the loss of bound [^{125}I]NGF is monophasic with a half-life of about 10 min. When the cells are loaded at a much higher concentration of [^{125}I]NGF, so that all receptors are occupied, the dissociation curve is biphasic: there is a very rapid phase with a half-life of a few seconds corresponding to dissociation from the low-affinity receptors, followed by a slower phase corresponding to dissociation from the high-affinity receptors. The biphasic dissociation curve shows that the low-affinity receptors are not generated from high-affinity receptors by negative cooperativity. The same phenomena is seen on PC12 cells: dissociation at an intermediate preloading concentration shows both a fast component (the low-affinity receptor) and a slow component (the high-affinity receptor). If the temperature is lowered from 37° to 0°C, then release from the high-affinity sites is reduced

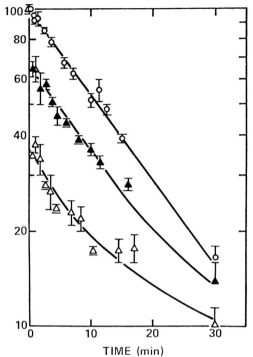

FIG. 6. Dissociation of [^{125}I]βNGF after preequilibration with different [^{125}I]βNGF concentrations at 37°C. Cells (3 × 10^6/ml) were preincubated for 30 min with 0.6 × 10^{-11}M (○), 1.6 × 10^{-10}M (▲) and 1.36 × 10^{-9}M [^{125}I]βNGF. The dissociation of [^{125}I]βNGF was induced by the addition of 3.8 × 10^{-7} M of unlabeled βNGF. The data are corrected for nonspecific binding. (Reproduced from Sutter et al., 1979a.)

even further and becomes essentially insignificant. It is, therefore, possible to wash NGF off the low-affinity receptors, while leaving it bound to the high-affinity receptors. When this is done, it is apparent that naive PC12 cells that have never seen NGF have only low-affinity receptors (Landreth and Shooter, 1980). Low-affinity NGF binding is converted to high-affinity binding in the absence of free NGF, suggesting a conversion from low- to high-affinity binding induced by the ligand NGF (Fig. 7).

LINKING STEPS IN THE MECHANISMS

There is good evidence that NGF, after binding to its specific receptors, is internalized by the cell. Indeed, one of the earliest examples of the cellular internalization of peptide hormones was the demonstration of the internalization of NGF at synapses of sympathetic neurons and its transport by retrograde flow to the cell body (Hendry et al., 1974). Is this internalization part of the mechanism of action of NGF or is the interaction of NGF with the surface receptors sufficient by itself to generate the appropriate intracellular signals? A definitive answer is not yet available, as the following discussion indicates.

Since the induction of the catecholamine-synthesizing enzymes by NGF does not require ongoing RNA synthesis, it probably occurs at the translational level (Thoenen and Barde, 1980). The NGF-induced stimulation (regeneration) of neurite outgrowth from explanted ganglia is also independent of RNA synthesis. However, NGF-mediated neurite extension from naive PC12 cells does require transcription, although regeneration of neurites from PC12 cells previously exposed to NGF is again independent of RNA synthesis (Greene and Shooter, 1980).

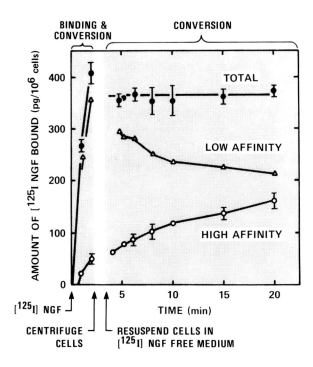

FIG. 7. The conversion of low-affinity NGF binding to high-affinity binding in the absence of free NGF. PC12 cells were incubated with 35 pM [^{125}I]NGF at 37°C for 2 min. The cells were centrifuged for 1 min at 1,000 × g, the [^{125}I]NGF-containing medium was aspirated, and the cells were resuspended in the same volume of fresh medium *(arrow)*. Total binding (●), high-affinity binding (○), and low-affinity binding (△) were determined. Each point represents the mean ± SD of triplicate determinations. (Reproduced from Landreth and Shooter, 1980.)

Thus, it appears that NGF can act both by transcription-dependent and transcription-independent processes.

It seems unlikely from available evidence that NGF acts solely through a second messenger like cyclic AMP, cyclic GMP, or calcium because, like ecdysone, steroid hormones, and EGF, neuron cell bodies must be treated with NGF for 3 to 4 hr before the cells become committed to neurite outgrowth or enzyme induction. Although some investigators have reported increases in cyclic AMP in sensory cells in response to NGF (Schubert and Whitlock, 1977), others have presented contrary data (Otten et al., 1978). A possible action of cyclic AMP in initiation of neurite outgrowth has been suggested because dibutyryl cyclic AMP can produce this effect on PC12 cells (Schubert et al., 1978). In fact, NGF and cyclic AMP act synergistically with respect to neurite outgrowth, producing very rapid and extensive growth (Gunning et al., 1981). However, whereas NGF alone requires RNA transcription to produce neurites from naive PC12 cells, the combination of these two agents does not; therefore, cyclic AMP appears to override a transcription-dependent stage in the NGF-induced neurite outgrowth mechanism. It would be of considerable interest to identify the substrates for the cyclic-AMP-dependent protein kinases in these cells.

In sympathetic neurons, NGF interacting with receptors in growth cones or synapses is internalized in vesicles and carried by retrograde transport into the interior of the cell body. There it is found in vesicular bodies and lysosomal-like structures (Schwab et al., 1979). A significant proportion of this NGF is relatively stable since it can be isolated and shown, both by size and by immunological reactivity, to be intact (Dumas et al., 1979). Nerve growth factor taken up in this way is presumably responsible for enzyme induction (Paravicini et al., 1975), but the nature of messengers involved in these reactions is unknown.

Internalization of NGF by the cell bodies of sensory and sympathetic neurons is not so obvious, at least as judged by the extent of lysosomal degradation (Sutter et al., 1979a), but it is a very evident feature of the interaction of NGF with the PC12 cells (Marchisio et al., 1980; P. Layer, *unpublished data*). A pronounced NGF receptor downregulation occurs, and the internalized NGF is degraded in the lysosomes. This process does not lead to the generation of intracellular signals for neurite outgrowth because the latter still occurs even when lysosomal degradation of NGF is substantially inhibited by antipain and leupeptin. A similar conclusion stems from the finding that the introduction of anti-NGF antibody into the cytoplasm of the PC12 cell does not inhibit NGF-induced neurite outgrowth (Heuman et al., 1981). Biochemical but not morphological evidence suggests that internalized NGF is eventually translocated to the nucleus where it binds to NGF receptors on the inner nuclear membrane. Given that free NGF introduced into the cytoplasm does not accumulate in the nucleus or induce neurite outgrowth (Heuman et al., 1981), the translocation would have to occur in vesicles. Correlations are found between the extent of nuclear-bound NGF and the commitment of the PC12 cell to initiate neurite outgrowth. The nuclear accumulation of NGF in PC12 cells has the same characteristic lag of 3 or 4 hr observed for initiating neurite outgrowth. The capacity to regenerate neurites a second time from PC12 cells also correlates with the amount of nuclear-bound NGF. It is also clear that NGF brings about the phosphorylation of a number of specific proteins in PC12 cells (Halegoua and Patrick, 1980), although again the significance of these events in terms of the mechanism is not known.

Nerve growth factor offers an especially interesting challenge for the use of molecular genetic techniques in neurobiology (see H. Thoenen, *this volume*). Because the amino acid sequence of NGF is known, it should in principle be possible to use nucleotide primers to obtain the cDNA probe for the NGF polypeptide and eventually to isolate the gene itself. Application of such techniques will help to establish the processing of the mRNA and precursor polypeptide for NGF, the cellular sites of synthesis, and the regulation of NGF biosynthesis during neuronal development and regeneration. In addition, the probes may be helpful in discovering new neurotrophic polypeptides that are related to NGF and that are also essential in neuronal development.

ACKNOWLEDGMENTS

Original research described in this paper was supported by grants from NIH (NS 04270), NSF (BNS 7914088), and the American Cancer Society (BC 325).

REFERENCES

Aloe, L., Cozzari, C., Calissano, P., and Levi-Montalcini, R. (1981a): Somatic and behavioral postnatal effects of fetal injections of nerve growth factor antibodies in the rat. *Nature*, 291:413–415.

Aloe, L., Cozzari, C., and Levi-Montalcini, R. (1981b): The submaxillary salivary glands of the african rodent *Praomys (Mastomys) natalensis* as the richest available source of the nerve growth factor. *Exp. Cell Res.*, 133:475–480.

Angeletti, R. A., and Bradshaw, R. A. (1971): Nerve growth factor from mouse submaxillary gland: Amino acid sequence. *Proc. Natl. Acad. Sci. USA*, 68:2417–2420.

Berger, E. A., and Shooter, E. M. (1977): Evidence for pro-βNGF, a biosynthetic precursor to β nerve growth factor. *Proc. Natl. Acad. Sci. USA*, 74:3647–3651.

Berger, E. A., and Shooter, E. M. (1978): The biosynthesis of β nerve growth factor on mouse submaxillary gland. *J. Biol. Chem.*, 243:804–810.

Blundell, T., Dodson, G., Hodgkin, D., and Meriola, D. (1972): Insulin: The structure in the crystal and its reflection in chemistry and biology. *Adv. Protein Chem.*, 26:279–402.

Burton, L. E., Wilson, W. H., and Shooter, E. M. (1978): Nerve growth factor in mouse saliva—rapid isolation procedures for and characterization of 7S nerve growth factor. *J. Biol. Chem.*, 253:7807–7812.

Dumas, M., Schwab, M. E., and Thoenen, H. (1979): Retrograde axonal transport of specific macromolecules as a tool for characterizing nerve terminal membranes. *J. Neurobiol.*, 10:179–197.

Dunn, M. F., Pattison, S. E., Storm, M. C., and Quiel, E. (1980): Comparison of the zinc binding domains in the 7S nerve growth factor and the zinc insulin hexaner. *Biochemistry*, 19:718–725.

Ebendal, T., Olson, L., Seiger, A., and Hedlund, K. O. (1980): Nerve growth factors in the rat iris. *Nature*, 286:25–28.

Frazier, W. A., Angeletti, R. H., and Bradshaw, R. A. (1972): Nerve growth factor and insulin. *Science*, 176:482–488.

Frey, P., Forand, R., Maciag, T., and Shooter, E. M. (1979): The biosynthetic precursor of epidermal growth factor and the mechanism of its processing. *Proc. Natl. Acad. Sci. USA*, 76:6294–6298.

Goedert, M., Otten, U., Schafer, T., Schwab, M., and Thoenen, H. (1980): Immunosympathectomy—lack of evidence for a complement-mediated cytotoxic mechanism. *Brain Res.*, 201:399–409.

Gorin, P., and Johnson, E. M. (1979): Experimental autoimmune model of nerve growth factor deprivation: Effects on developing peripheral sympathetic and sensory neurons. *Proc. Natl. Acad. Sci. USA*, 76:5382–5386.

Greene, L. A., and Shooter, E. M. (1980): Nerve growth factor—biochemistry and synthesis and mechanism of action. *Annu. Rev. Neurosci.*, 3:353–402.

Greene, L. A., Varon, S., Piltch, A., and Shooter, E. M. (1971): Substructure of the β subunit of mouse 7S nerve growth factor. *Neurobiology*, 1:37–48.

Gundersen, R. W., and Barrett, J. N. (1979): Neuronal chemotaxis: Chick dorsal-root axons turn toward high concentrations of nerve growth factor. *Science*, 206:1079–1080.

Gundersen, R. W., and Barrett, J. N. (1980): Characterization of the turning response of dorsal root neurites toward nerve growth factor. *J. Cell Biol.*, 87:546–554.

Gunning, P. W., Landreth, G. E., Bothwell, M. A., and Shooter, E. M. (1981): Differential and synergistic actions of nerve growth factor and cyclic AMP in PC12 cells. *J. Cell Biol.*, 89:240–245.

Halegoua, S., and Patrick, J. (1980a): Nerve growth factor mediates phosphorylation of specific proteins. *Cell*, 22:571–581.

Harper, G. P., and Thoenen, H. (1980): Nerve growth factor: Biological significance, measurement and distribution. *Neurochem.*, 34:5–16.

Hendry, I. A., Stöckel, K., Thoenen, H., and Iversen, L. L. (1974): The retrograde axonal transport of nerve growth factor. *Brain Res.*, 68:103–121.

Heuman, R., Schwab, M., and Thoenen, H. (1981): A second messenger required for nerve growth factor biological activity? *Nature*, 292:838–840.

Landreth, G. E., and Shooter, E. M. (1980): Nerve growth factor reception in PC12 cells: Ligand induced conversion from low to high affinity states. *Proc. Natl. Acad. Sci. USA*, 77:4751–4755.

Levi-Montalcini, R., and Angeletti, P. U. (1966): Immunosympathectomy. *Pharmacol. Rev.*, 18:619–628.

Levi-Montalcini, R., and Angeletti, P. U. (1968): Nerve growth factor. *Physiol. Rev.*, 48:534–569.

Levi-Montalcini, R., and Hamburger, V. (1951): Selective growth-stimulating effects of mouse sarcoma on the sensory and sympathetic nervous system of the chick embryo. *J. Exp. Zool.*, 116:321–362.

Levi-Montalcini, R., and Hamburger, V. (1953): A diffusable agent of mouse sarcoma producing hyperplasia of sympathetic ganglia and hyperneurotization of viscera in the chick embryo. *J. Exp. Zool.*, 123:233–278.

Levi-Montalcini, R., Meyer, H., and Hamburger, V. (1954): In vitro experiments on the effect of mouse sarcoma 180 and 37 on the spinal and sympathetic ganglia of the chick embryo. *Cancer Res.*, 14:49–57.

Marchisio, P. C., Noldini, L., and Calissano, P. (1980): Intracellular distribution of nerve growth factor in rat pheochromocytoma PC12 cells. Evidence for a perinuclear and intranuclear location. *Proc. Natl. Acad. Sci. USA*, 77:1656–1660.

Menesini-Chen, M. G., Chen, J. S., and Levi-Montalcini, R. (1978): Sympathetic nerve fibers ingrowth in the central nervous system of neonatal rodents upon intracerebral NGF injections. *Arch. Ital. Biol.*, 116:53–84.

Otten, U., Hatanaka, H., and Thoenen, H. (1978): Role of cyclic nucleotides in NGF-mediated induction of tyrosine hydroxylase in rat sympathetic ganglia and adrenal medulla. *Brain Res.*, 140:385–389.

Paravicini, U., Stoeckel, K., and Thoenen, H. (1975): Biological importance of retrograde axonal transport of nerve growth factor in adrenergic neurons. *Brain Res.*, 84:279–291.

Schubert, D., LaCorbiere, M., Whitlock, C., and Stallcup, W. (1978): Alterations in the surface properties of cells responsive to nerve growth factor. *Nature*, 273:718–723.

Schubert, D., and Whitlock, C. (1977): Alteration of cellular adhesion by nerve growth factor. *Proc. Natl. Acad. Sci. USA*, 74:4055–4058.

Schwab, M., Suda, K., and Thoenen, H. (1979): Selective retrograde transynaptic transfer of a protein, tetanus toxin, subsequent to its retrograde axonal transport. *J. Cell Biol.*, 82:798–810.

Sutter, A., Riopelle, R. J., Harris-Warrick, R. M., and Shooter, E. M. (1979a): Nerve growth factor receptors—characterization of two distinct classes of binding sites on chick embryo sensory ganglia cells. *J. Biol. Chem.*, 254:5972–5982.

Sutter, A., Riopelle, R. J., Harris-Warrick, R. M., and Shooter, E. M. (1979b): The heterogeneity of nerve growth factor receptors. In: *Transmembrane Signaling*, edited by M. Bitensky, R. J. Collier, D. F. Steiner, and C. F. Fox, pp. 659–677. Alan R. Liss, New York.

Taylor, J. M., Cohen, S., and Mitchell, W. M. (1970): Epidermal growth factor—high and low molecular weight forms. *Proc. Natl. Acad. Sci. USA*, 67:164–171.

Thoenen, H., Angeletti, P. U., Levi-Montalcini, R., and Kettler, R. (1971): Selective induction by nerve growth factor of tyrosine hydroxylase and dopamine-β-hydroxylase in the rat superior cervical ganglia. *Proc. Natl. Acad. Sci. USA*, 68:1598–1602.

Thoenen, H., and Barde, Y. A. (1980): Physiology of nerve growth factor. *Physiol. Rev.*, 60:1284–1335.

Varon, S., Nomura, J., and Shooter, E. M. (1967): The isolation of the mouse nerve growth factor protein in a high molecular weight form. *Biochemistry*, 6:2202–2209.

Varon, S., Nomura, J., and Shooter, E. M. (1968): Reversible dissociation of the mouse nerve growth factor protein into different subunits. *Biochemistry*, 7:1296–1303.

Zimmermann, A., Sutter, A., and Shooter, E. M. (1981): Monoclonal antibodies against β nerve growth factor and their effects on receptor binding and biological activity. *Proc. Natl. Acad. Sci. USA*, 78:4611–4615.

Cellular and Hormonal Interactions in the Development of Sympathetic Neurons

Paul H. Patterson

ABSTRACT

The type of neurotransmitter produced by postmitotic sympathetic neurons in culture can be controlled by several types of developmental signals: hormones, electrical activity, and macromolecular factors from heart cells. When the neurotransmitter phenotype is altered, specific changes also occur in the surfaces of the neurons and in the glycoproteins they spontaneously secrete into the medium. A shift in transmitter production similar to that seen in culture also appears to occur during the normal development of these neurons *in vivo*.

The techniques of molecular genetics should prove especially useful in analyzing the control of gene expression during neuronal development. One of the systems to which these methods could be applied is the transmitter decision made by sympathetic neurons. These neurons can become adrenergic (using catecholamines as their transmitter) or cholinergic (using acetylcholine as their transmitter), depending on the environment in which they develop, and a variety of molecular markers are available for each phenotype.

Sympathetic neurons arise from a transient embryonic structure called the neural crest, a group of cells located initially on the dorsal margin of the neural tube (Fig. 1). Neural crest cells migrate to different places in the embryo and give rise to a wide variety of derivatives, such as melanocytes, sensory neurons, and autonomic neurons. The phenotype adopted by the crest cells appears to be influenced by the environment through which they migrate, as well as the environment at their final location after migration (LeDouarin, 1980). These conclusions come from transplantation studies; crest cells transplanted to novel locations assume the phenotype typical of that location. In these *in vivo* studies, it is not clear whether a particular phenotype is induced in naive individual cells, or if a subpopulation of cells already committed to expressing that phenotype is selected for survival. In the studies of cultured postmitotic sympathetic neurons discussed below, the selection hypothesis has been ruled out.

During their migration from the neural crest, sympathetic neuron precursors apparently receive a signal to begin adrenergic differentiation (LeDouarin, 1980) and do so, even while still dividing (Rothman et al., 1978). Virtually all of the neurons in the neonatal rat superior cervical ganglion display catecholamine fluorescence (Eränkö, 1972), and when these cells are dissociated and placed in cell culture all of the varicosities initially formed contain adrenergic synaptic vesicles (Johnson et al., 1976; Landis, 1980). The neurons synthesize and accumulate catecholamines but little or no acetylcholine (ACh) (Mains and Patterson, 1973; Patterson and Chun, 1977b). This adrenergic differentiation is greatly enhanced, or stabilized, by growing the neurons under depolarizing conditions, or stimulating them with action potentials (Walicke

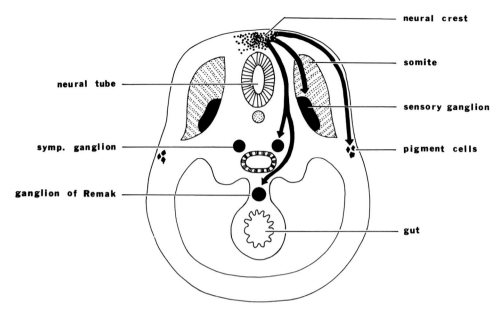

FIG. 1. Cross section of the developing chick embryo illustrating the migration pathways of some neural crest derivatives. The Remak ganglion is an example of a parasympathetic ganglion. (From Patterson, 1979.)

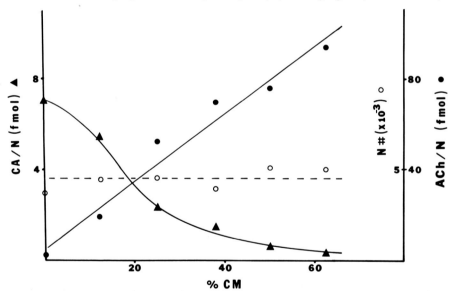

FIG. 2. The effects of conditioned medium on neuronal survival and transmitter production. Conditioned medium (CM) from heart cell cultures was mixed with fresh medium and added to neuronal cultures at the concentrations shown. After 3 weeks, neuron somas (○) were counted and the production of radioactive acetylcholine (ACh, ●) and catecholamines (CA, ▲) from [^{3}H]choline and [^{3}H]tyrosine was determined. (From Patterson and Chun, 1977a.)

et al., 1977; Landis, 1980). The effect of depolarization is mediated by calcium ions (Walicke et al., 1977; Walicke and Patterson, 1981).

On the other hand, these same sympathetic neurons can be influenced to become cholinergic by both diffusible (Patterson and Chun, 1977a; Landis, 1980) and surface-bound (Hawrot, 1980) factors from certain types of nonneuronal cells (Fig. 2). The neurons develop the ability

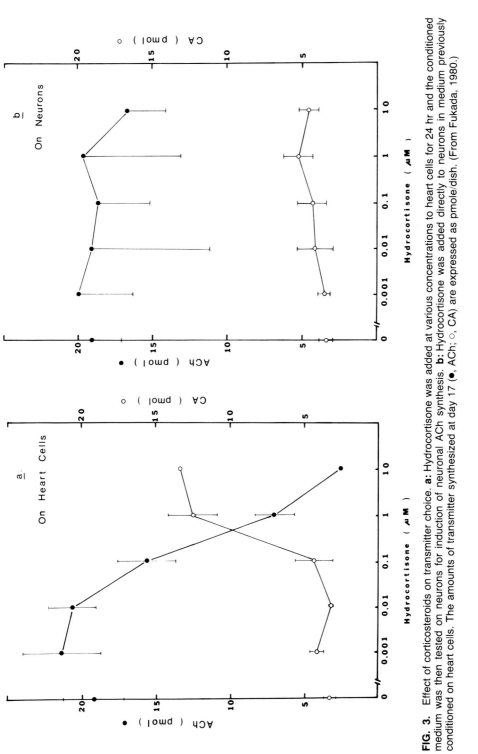

FIG. 3. Effect of corticosteroids on transmitter choice. **a:** Hydrocortisone was added at various concentrations to heart cells for 24 hr and the conditioned medium was then tested on neurons for induction of neuronal ACh synthesis. **b:** Hydrocortisone was added directly to neurons in medium previously conditioned on heart cells. The amounts of transmitter synthesized at day 17 (●, ACh; ○, CA) are expressed as pmole/dish. (From Fukada, 1980.)

to produce ACh and form functional cholinergic synapses with each other and with skeletal and heart muscle cells (Nurse and O'Lague, 1975; Furshpan et al., 1976; Ko et al., 1976; O'Lague et al., 1978). Since the neurons are initially adrenergic, and most can be influenced to become cholinergic (Reichardt and Patterson, 1977), it is reasonable to predict a transitional state. The most compelling demonstration of the transition from adrenergic to cholinergic has come from repeated electrophysiological assays of the transmitter status of single neurons as they develop individually in microcultures (Potter et al., 1981). Many partial transitions have been observed in these microcultures, always in the direction of adrenergic to dual-function (both types of synapses on the heart myocytes) or dual-function to cholinergic. One complete transition from purely adrenergic to dual-function to purely cholinergic has been observed thus far.

The diffusible cholinergic factor is especially interesting because of its specificity; although it can control the type of transmitter produced and the type of synapse formed, it has no effect on the survival or growth of these neurons (Patterson and Chun, 1977a). The factor chromatographs on Sephadex at a molecular weight of approximately 45,000 and has the properties of a glycoprotein (Weber, 1981). The production (or release) of the factor by heart cells is under hormonal control. For instance, when used at the concentration present in neonatal rat serum, glucocorticoids can prevent heart cells from producing or releasing the conditioned medium factor without alteration of the growth or health of the myocytes (Fig. 3) (Fukada, 1980). This indirect influence of hormones on the neuronal transmitter decision is an example of the subtlety of the controls acting on this crucial phenotypic choice.

How many of these culture findings apply directly to the normal development of sympathetic cholinergic neurons *in vivo*? The development of the cholinergic sympathetic innervation of the exocrine sweat glands of the rat footpad appears to display two unique properties in common with the cultured cholinergic neurons: they go through an adrenergic phase before becoming cholinergic, and they retain their ability to take up and store catecholamine even after becoming cholinergic (Landis and Keefe, 1980; Landis, 1981). A similar transmitter transition has been suggested for parasympathetic neurons in the embryonic mouse gut (Cochard et al., 1978; Teitelman et al., 1978; Jonakait et al., 1979). In addition, as with the sympathetic system, the adrenergic status of the gut neurons is prolonged by glucocorticoid injections (Jonakait et al., 1980). Thus the sympathetic and parasympathetic lineages may be more similar than previously suspected (Landis and Patterson, 1981), and the culture findings could be applicable to both systems *in vivo*.

What other changes occur in a cultured sympathetic neuron when it alters its transmitter status and forms a (functionally) different kind of synapse? Several lines of evidence have demonstrated that the surface membranes of the cholinergic neurons are different from those of the adrenergic neurons: (1) Two-dimensional gel analysis after metabolic or external labeling reveals a striking increase in the labeling of a 55,000-molecular-weight glycoprotein on the surface of cholinergic neurons and a decrease in the labeling of a 145,000-molecular-weight glycoprotein in adrenergic neurons (Braun et al., 1981). (2) Colloidal gold-conjugated soybean agglutinin binds to adrenergic axons with a fivefold greater density than to cholinergic axons (Schwab and Landis, 1981). Recent evidence suggests that a glycolipid (globoside) is the soybean lectin receptor on the neuron surfaces (Zurn and Patterson, 1981). (3) A monoclonal antibody (N10) has been produced, using cultured adrenergic neurons as the immunogen, which shows greater binding to the adrenergic neurons than to the cholinergic neurons (Chun et al., 1980).

Another striking difference between the two types of cultured neurons is observed in the family of glycoproteins spontaneously released into the medium (Sweadner, 1981). The neurons

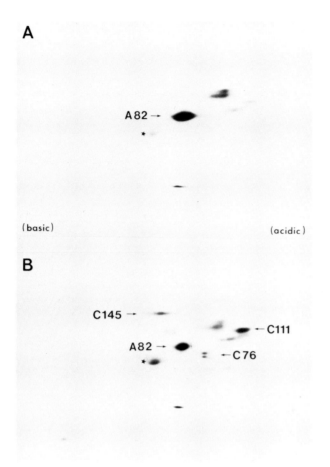

FIG. 4. Proteins spontaneously released from adrenergic and cholinergic neurons. Two-dimensional gel electrophoresis (isoelectric focusing followed by electrophoresis in SDS) of spontaneously released [³H]leucine-labeled protein from adrenergic cultures **(A)** and cholinergic cultures **(B)** was run and autoradiography was performed. The proteins marked with *arrows* are identified on the basis of their molecular weight in the second dimension. Faint labeling with serum albumin *(asterisk)* is visible in both gels. (From Sweadner, 1981.)

release 16 to 18 soluble proteins that are distinct from the cell membrane glycoproteins by two-dimensional gel analysis. Their release is spontaneous and is not stimulated by depolarization or inhibited by Co^{2+}, Mg^{2+}, or the divalent cation chelator EGTA. The expression of 5 of the 18 released proteins is correlated with the transmitter status of the neurons (Fig. 4). These transmitter-specific proteins are possible mediators of intercellular communication specific for different kinds of neurons. Antibodies directed against these glycoproteins as well as the transmitter-specific membrane molecules will be useful in the investigation of their functions.

In summary, the transmitter phenotype of a postmitotic neuron can be altered even after it has begun to express one particular set of differentiated properties. At least three qualitatively different exogenous factors can influence the choice of transmitter: (1) a macromolecular factor produced by heart cells can instruct the neurons as to transmitter without affecting their survival or growth; (2) depolarization or activity stabilizes the initial adrenergic choice and makes the neurons resistant to the cholinergic factor; and (3) corticosteroids act indirectly, via the heart cells, by inhibiting their production or release of the cholinergic factor. The change in transmitter and synapse phenotype is accompanied by specific alterations in the glycoproteins and glycolipids in the neuronal surface membranes and in the glycoproteins spontaneously released into the culture medium. Finally, the adrenergic-to-cholinergic transmitter shift seen in culture appears to also occur during the normal development of cholinergic sympathetic neurons *in vivo*.

ACKNOWLEDGMENTS

Many thanks go to my colleagues whose work is discussed here, and to the NINCDS, the Rita Allen Foundation, the Wellington Fund, the Muscular Dystrophy Association, the Medical Foundation, Charles A. King Trust, the Dysautonomia Foundation, and the American and Massachusetts Heart Associations for supporting this work.

REFERENCES

Braun, S. J., Sweadner, K. J., and Patterson, P. H. (1981): Neuronal cell surfaces: Distinctive glycoproteins of cultured adrenergic and cholinergic sympathetic neurons. *J. Neurosci.*, 1:1397–1406.
Chun, L. L. Y., Patterson, P. H., and Cantor, H. (1980): Preliminary studies on the use of monoclonal antibodies as probes for sympathetic development. *J. Exp. Biol.*, 89:73–83.
Cochard, P., Goldstein, M., and Black, I. B. (1978): Ontogenetic appearance and disappearance of tyrosine hydroxylase and catecholamines in the rat embryo. *Proc. Natl. Acad. Sci. USA*, 75:2986–2990.
Eränkö, L. (1972): Postnatal development of histochemically demonstrable catecholamines in the superior cervical ganglion of the rat. *Histochem. J.*, 4:225–236.
Fukada, K. (1980): Hormonal control of neurotransmitter choice in sympathetic neuron cultures. *Nature*, 287:553–555.
Furshpan, E. J., MacLeish, P. R., O'Lague, P. H., and Potter, D. D. (1976): Chemical transmission between rat sympathetic neurons and cardiac myocytes developing in microcultures: Evidence for cholinergic, adrenergic and dual-function neurons. *Proc. Natl. Acad. Sci. USA*, 73:4225–4229.
Hawrot, E. (1980): Cultured sympathetic neurons: Effects of cell-derived and synthetic substrata on survival and development. *Dev. Biol.*, 74:136–151.
Johnson, M., Ross, D., Meyers, M., Rees, R., Bunge, R., Wakshull, E., and Burton, H. (1976): Synaptic vesicle cytochemistry changes when cultured sympathetic neurons develop cholinergic interactions. *Nature*, 262:308–310.
Jonakait, G. M., Bohn, M. C., and Black, I. B. (1980): Maternal glucocorticoid hormones influence neurotransmitter phenotypic expression in embryos. *Science*, 210:551–553.
Jonakait, G. M., Wolfe, J., Cochard, P., Goldstein, M., and Black, I. B. (1979): Selective loss of noradrenergic phenotypic characters in neuroblasts of the rat embryo. *Proc. Natl. Acad. Sci. USA*, 76:4683–4686.
Ko, C.-P., Burton, H., Johnson, M. I., and Bunge, R. P. (1976): Synaptic transmission between rat superior cervical ganglion neurons in dissociated cell cultures. *Brain Res.*, 117:461–485.
Landis, S. C. (1980): Developmental changes in the neurotransmitter properties of dissociated sympathetic neurons: A cytochemical study of the effects of medium. *Dev. Biol.*, 77:349–361.
Landis, S. C. (1981): Environmental influences on the postnatal development of rat sympathetic neurons. In: *Development in the Nervous System*, edited by D. R. Garrod and J. D. Feldman, pp. 147–160. Cambridge University Press, Cambridge.
Landis, S. C., and Keefe, D. (1980): Development of cholinergic sympathetic innervation of eccrine sweat glands in rat footpad. *Soc. Neurosci. Abstr.*, 6:379.
Landis, S. C., and Patterson, P. H. (1981): Neural crest cell lineages. *Trends Neurosci.*, 4:172–175.
LeDouarin, N. M. (1980): The ontogeny of the neural crest in avian embryo chimeras. *Nature*, 286:663–669.
Mains, R. E., and Patterson, P. H. (1973): Primary cultures of dissociated sympathetic neurons. I. Establishment of long term growth in culture and studies of differentiated properties. *J. Cell Biol.*, 59:329–345.
Nurse, C. A., and O'Lague, P. H. (1975): Formation of cholinergic synapses between dissociated sympathetic neurons and skeletal myotubes of the rat in cell culture. *Proc. Natl. Acad. Sci. USA*, 72:1955–1959.
O'Lague, P. H., Potter, D. D., and Furshpan, E. J. (1978): Studies on rat sympathetic neurons developing in cell culture. III. Cholinergic transmission. *Dev. Biol.*, 67:424–443.
Patterson, P. H. (1979): Epigenetic influences in neuronal development. In: *The Neurosciences: Fourth Study Program*, edited by F. O. Schmitt and F. G. Worden, pp. 929–936. MIT Press, Cambridge, Mass.
Patterson, P. H., and Chun, L. L. Y. (1977a): The induction of acetylcholine synthesis in primary cultures of dissociated rat sympathetic neurons. I. Effects of conditioned medium. *Dev. Biol.*, 56:263–280.
Patterson, P. H., and Chun, L. L. Y. (1977b): The induction of acetylcholine synthesis in primary cultures of dissociated rat sympathetic neurons. II. Developmental aspects. *Dev. Biol.*, 60:473–481.
Potter, D., Landis, S., and Furshpan, E. (1981): Adrenergic-cholinergic dual function in cultured sympathetic neurons of the rat. *Ciba Found. Symp.*, 83:123–150.

Reichardt, L. F., and Patterson, P. H. (1977): Neurotransmitter synthesis and uptake by isolated rat sympathetic neurons developing in microcultures. *Nature*, 270:147–151.

Rothman, T., Gershon, M., and Holtzer, H. (1978): Cell division and the acquisition of adrenergic characteristics by developing sympathetic ganglion cells. *Dev. Biol.*, 65:322–341.

Schwab, M., and Landis, S. (1981): Membrane properties of cultured rat sympathetic neurons: Morphological differentiation. *Dev. Biol.*, 84:67–78.

Sweadner, K. J. (1981): Environmentally regulated expression of soluble extracellular proteins of sympathetic neurons. *J. Biol. Chem.*, 256:4063–4070.

Teitelman, G., Joh, T. H., and Reis, D. J. (1978): Transient expression of a noradrenergic phenotype in cells of the rat embryonic gut. *Brain Res.*, 158:229–234.

Walicke, P. A., Campenot, R. B., and Patterson, P. H. (1977): Determination of transmitter function by neuronal activity. *Proc. Natl. Acad. Sci. USA*, 74:5767–5771.

Walicke, P. A., and Patterson, P. H. (1981): On the role of Ca^{++} in the transmitter choice made by cultured sympathetic neurons. *J. Neurosci.*, 1:343–350.

Weber, M. J. (1981): A diffusible factor responsible for the determination of cholinergic functions in cultured sympathetic neurons: Partial purification and characterization. *J. Biol. Chem.*, 256:3447–3453.

Zurn, A., and Patterson, P. (1981): Globoside as a possible surface membrane receptor for the lectin soybean agglutinin in cultured sympathetic neurons. *Soc. Neurosci. Abstr.*, 7:847.

Molecular Genetic Neuroscience, edited by
F. O. Schmitt, S. J. Bird, and F. E. Bloom.
Raven Press, New York © 1982.

A Possible Role for Cytoplasmic Calcium in the Regulation of Gene Expression

Anthony N. Martonosi

ABSTRACT

The Ca^{2+} ionophores ionomycin and A23187 selectively promote the synthesis of an 80,000- and a 100,000-dalton membrane protein in cultured muscle cells, fibroblasts, HeLa, CV1, and LSP cells. Increased synthesis of the 80,000-dalton protein was also observed upon cell-free translation of poly(A) RNA isolated from ionomycin-treated as compared with control muscle cultures. The effect of ionophores is probably mediated through an increase in cytoplasmic $[Ca^{2+}]$, and is consistent with the proposed regulation of gene transcription through Ca^{2+}-binding chromosomal proteins.

Detailed information is available concerning calcium activation of enzymes in many systems (Kretsinger, 1976; Scarpa and Carafoli, 1978; Cheung, 1980). By contrast, virtually nothing is known about the effect of calcium at the level of gene regulation. Most studies in this area have involved monitoring of cell growth or total RNA or protein synthesis following abrupt changes in the cytoplasmic calcium concentration. As a result, elevated cytoplasmic calcium concentrations have been correlated with the burst of synthetic activity observed after fertilization of eggs (Epel, 1980; Jaffe, 1980), transformation of lymphocytes (Freedman et al., 1975; Greene et al., 1976), initiation of collagen synthesis in chondrocytes (Deshmukh and Sawyer, 1977), DNA synthesis in hemopoietic stem cells (Gallien-Lartigue, 1976), embryonic induction in *Rana pipiens* (Barth and Barth, 1974), and in many other processes (Leffert, 1980). In none of these cases, however, has calcium been shown to affect the expression of a specific gene product by direct action. Because of the important role played by calcium in muscle contraction and the accurate regulation of the cytoplasmic levels of this cation by the calcium transport mechanism of the sarcoplasmic reticulum, muscle would appear to be an ideal system in which to search for an effect of calcium on gene expression. The sarcoplasmic reticulum membrane contains a considerable amount of Ca^{2+} adenosine triphosphatase (ATPase), an enzyme involved in the regulation of cytoplasmic calcium levels, the synthesis of which might therefore be expected to be subject to regulation by calcium.

CA^{2+} ATPase AND REGULATION OF CYTOPLASMIC CALCIUM

Ca^{2+} ATPase is an intrinsic protein of the sarcoplasmic reticulum membrane (Martonosi, 1972). It has a mass of approximately 100,000 daltons (100 kD) and may represent as much as 60 to 80% of the protein of the sarcoplasmic reticulum of adult muscle (MacLennan and Holland, 1976; Hasselbach, 1979). The Ca^{2+} APTase acts to pump Ca^{2+} ions from the cytoplasm of the muscle cell into the tubules of the sarcoplasmic reticulum, with the expenditure of cytoplasmic adenosine triphosphate (ATP) (de Meis and Vianna, 1979).

The overall scheme of calcium regulation in the muscle cell is shown in Fig. 1. This complex system consists of Ca^{2+}-binding proteins in the cytoplasm, mitochondria, nucleus, and sarcoplasmic reticulum, as well as Ca^{2+} pumps in the surface and mitochondrial membranes, in addition to the Ca^{2+} ATPase of the sarcoplasmic reticulum. Calcium channels are also in the surface membrane and in the membrane of the transverse tubules. Since the cytoplasmic calcium concentration is maintained by the coordinated activity of all these components, it is likely that some integrative regulation could be involved in the synthesis of each individual component. The cytoplasmic calcium concentration is so tightly controlled by this system that during activation of the muscle, only enough calcium is released from the sarcoplasmic reticulum to saturate the troponin C component of the contractile myofilaments, whereas the cytoplasmic calcium concentration rises from 10^{-7} M to slightly above 10^{-6} M (Endo, 1977; Schneider, 1981). During relaxation, the Ca^{2+} ATPase pumps the calcium back into the sarcoplasmic reticulum.

EMBRYONIC DEVELOPMENT OF THE SARCOPLASMIC RETICULUM

The amount of Ca^{2+} ATPase in the sarcoplasmic reticulum undergoes drastic changes during embryonic development (Martonosi, 1975; Martonosi et al., 1980). In chicken pectoralis muscle before 10 days of development, the enzyme concentration is low, as indicated in Fig. 2 (Martonosi et al., 1977). After 10 days, there is a sharp 30- to 40-fold increase in the Ca^{2+} ATPase activity; maximum levels are reached about one month after hatching. Much of the increase in enzyme concentration occurs at the time of hatching, when vigorous muscle activity begins, triggered perhaps by the elevated levels of cytoplasmic calcium associated with muscle contraction.

An increase in Ca^{2+} ATPase can also be demonstrated by monitoring the protein composition of isolated sarcoplasmic reticulum using polyacrylamide gel electrophoresis (Boland et al.,

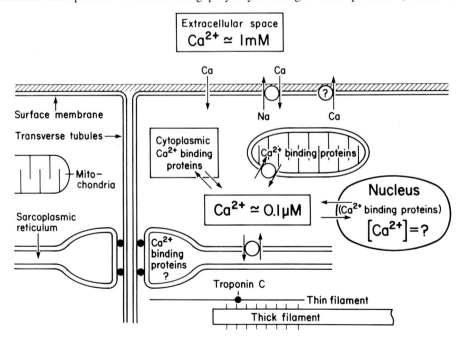

FIG. 1. Scheme of the systems that participate in the regulation of cytoplasmic $[Ca^{2+}]$ in skeletal muscle.

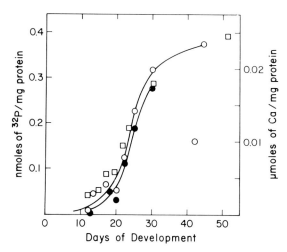

FIG. 2. Phosphoprotein formation and calcium uptake in muscle homogenates at various stages of development. The phosphoenzyme intermediate of the Ca^{2+} transport ATPase was measured by phosphorylation of the enzyme with [^{32}P]ATP. The phosphoprotein concentrations represent the difference between values obtained in the presence and absence of calcium and correlate with the concentration of Ca^{2+} ATPase. The calcium uptake was measured in the presence of 5 mM sodium azide and 5 mM ATP. ○, phosphoprotein concentration in 0.1 M KCl; ●, phosphoprotein concentration in 0.1 M NaCl; □, calcium uptake. (From Martonosi et al., 1977, with permission.)

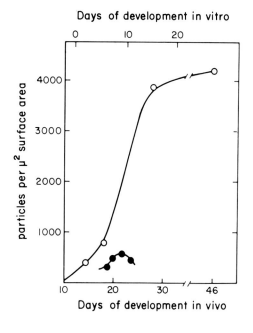

FIG. 3. Changes in the intramembranous particle density of sarcoplasmic reticulum of intact chick skeletal muscle during development, compared to the density of particles in tissue culture. ○, average number of 85 Å particles on the fracture faces of embryonic and posthatched chick skeletal muscle; ●, the average number of particles on the fracture faces of sarcoplasmic reticulum and T-system membranes of cultured chick skeletal muscle cells. The particles probably represent clusters of four Ca^{2+} ATPase molecules. (From Martonosi et al., 1977, with permission.)

1974). In addition, freeze-fracture electron microscopy has revealed an increase in the density of 85 Å intramembrane particles that parallels the increase in Ca^{2+} ATPase activity (Tillack et al., 1974; Martonosi et al., 1977). As shown in Fig. 3, the density of particles on the cytoplasmic fracture face of the membrane rises from an initial level of 300/μm^2 to more than 4,000/μm^2 by 46 days of development. It is interesting that this level far exceeds the approximately 1,000 particle/μm^2 reached by muscle cells developing *in vitro*. This is probably due, at least in part, to lack of innervation in culture since the amount of Ca^{2+}ATPase decreases to similar levels *in vitro*, following denervation.

Ca^{2+} ATPase IN TISSUE CULTURE

In primary cultures of chicken pectoralis muscle, the rate of incorporation of [^{35}S]methionine into the Ca^{2+}ATPase begins to increase after 2 days, and a maximum activity is reached at 7 days (Fig. 4). The myoblasts in such cultures fuse at the second or third day to form multinucleate myotubes, and the initial increase in calcium transport activity appears to coincide with cell fusion (Martonosi et al., 1977; Ha et al., 1979).

Fusion of myoblasts can be prevented in culture without affecting cell proliferation by reducing the calcium concentration of the medium to 0.1 mM from the usual 1.8 mM. Under these conditions, no increase in Ca^{2+}ATPase synthesis is observed (Fig. 4) (Ha et al., 1979). Returning the calcium concentration to 1.8 mM at any time initiates the accumulation of Ca^{2+}ATPase. Since the reduced calcium concentration in the culture medium is likely to be reflected in a reduced concentration of cytoplasmic calcium, it is possible that the latter exerts a direct effect on synthesis of the Ca^{2+}ATPase.

The effect of calcium on the Ca^{2+} ATPase content of sarcoplasmic reticulum can also be demonstrated by using the Ca^{2+} ionophores A23187 and ionomycin to promote transfer of calcium from the medium and from intracellular stores into the cytoplasm. In myoblasts cultured in the presence of either of these ionophores, an increase is observed in the steady-state level of Ca^{2+}ATPase (Martonosi et al., 1977). This elevation is probably due to increased synthesis of the protein, as shown in Fig. 5. In lane 1 the [^{35}S]methionine-labeled proteins of cells

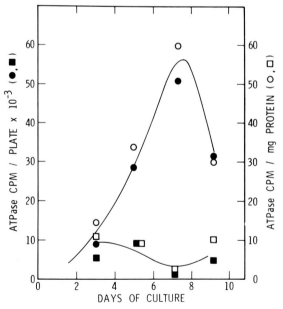

FIG. 4. Incorporation of [^{35}S]methionine into Ca^{2+} ATPase of cultured chicken pectoralis muscle cells at various stages of development. Cells were grown in a medium containing 1.8 mM calcium for 2 days and then in a medium containing either 1.8 mM (○, ●) or 0.15 mM calcium (□, ■). After an appropriate period, the cells were labeled with 25 μCi/ml [^{35}S]methionine for 24 hr, harvested, sonicated, and the ATPase was immunoprecipitated. After electrophoresis on SDS-polyacrylamide slab gel, the ATPase band was cut out and the radioactivity incorporated into the ATPase was determined. ○, □, radioactivity (cpm/mg total cell protein × 10^{-3}); ●, ■, radioactivity (cpm/plate × 10^{-3}). (From Ha et al., 1979, with permission.)

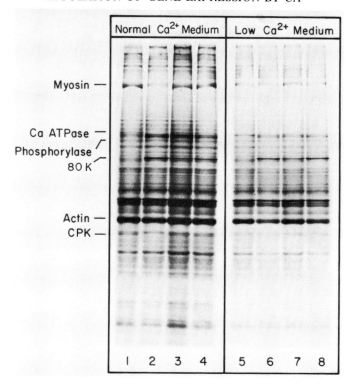

FIG. 5. Effect of Ca^{2+} ionophores on rate of synthesis of muscle proteins in tissue culture (autoradiogram). The cells were grown for 4 days in normal Ca^{2+} (1 to 4) or low Ca^{2+} (5 to 8) medium. On the fourth day, the cells were exposed to 10^{-5} M ionomycin (2,6), 10^{-6} M A23187 (3,7), or 10^{-6} M ionomycin (4,8) for 6 hr and labeled with [^{35}S]methionine (25 μCi/ml) for 2 hr. The cells were rinsed and prepared for gel electrophoresis and autoradiography. Samples 1 and 5 served as controls. (From Roufa et al., 1981.)

cultured in normal Ca^{2+} medium are displayed. Both the Ca^{2+} ATPase at 100 kD and another protein at about 80 kD are easily detectable. In lanes 2, 3, and 4 are the proteins from cells treated with the ionophores. In all three cases an increased expression of the 80 and 100 kD proteins appears to be the principal change resulting from the treatment (Roufa et al., 1981). A similar effect can also be demonstrated in low Ca^{2+} medium (lanes 5 to 8). The 100 kD protein is probably the Ca^{2+} ATPase since it reacts with a specific antiserum; however, the 80 kD protein has not yet been functionally identified.

INCREASED LEVELS OF mRNA FOR 80 kD PROTEIN

The increased synthesis of 80 kD protein is caused by a Ca^{2+}-induced increase in the level of its mRNA. This can be demonstrated by preparing cytoplasmic poly(A)-containing RNA from control and ionophore-treated cells and translating the RNA *in vitro* in a reticulocyte lysate system (Wu et al., 1981). As shown in Fig. 6, the ionomycin treatment induces a significant increase in translatable mRNA for the 80 kD protein in both high calcium and low calcium media.

It could be argued that the ionophores cause increased synthesis of these proteins by a specific effect not involving changes in Ca^{2+} concentration, but this possibility can be eliminated by modulating cytoplasmic Ca^{2+} without the use of ionophores. It is known that in low calcium medium, the calcium permeability of the muscle cell surface membrane increases (Winegrad, 1971). Thus, when these cells are transferred to high Ca^{2+} medium, there is an influx of Ca^{2+}

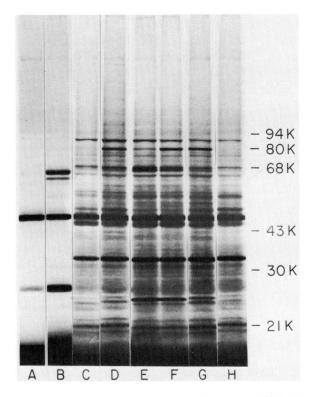

FIG. 6. Autoradiogram of *in vitro* translation products of poly(A)-enriched RNA isolated from muscle cell cultures. Myoblasts isolated from 12-day-old chicken embryos were plated in normal Ca^{2+} medium for 24 hr followed by growth in either normal calcium (lanes C and D) or low calcium medium (lanes E to H). After 64 hr of incubation, cultures D and F were exposed to 4×10^{-6} M ionomycin for 3 hr; cultures C and E served as control. Cultures G and H grown in low calcium medium were transferred into normal calcium medium after 64 hr of incubation and incubated for 3 hr (lane G) or 22 hr (lane H) in normal medium prior to RNA extraction. At the end of incubation, RNA was isolated from all cultures and fractionated into poly(A)-enriched RNA and non-poly(A) RNA on an oligo-(dT)-cellulose column. Poly(A)-enriched RNA (0.87 μg) from each sample was translated in rabbit reticulocyte lysate system *in vitro*. The polypeptide products were analyzed by electrophoresis on 6 to 12% polyacrylamide gradient gel slabs, dried, and autoradiographed. Note the intense labeling of the 80-kD band in lanes D, F, and G. Poly(A) RNA isolated from low Ca^{2+} cultures (E, F) is active in the synthesis of a 26- to 28-kD protein, irrespective of ionomycin treatment, but forms less myosin compared with cultures grown in normal medium (C, D). Less pronounced differences were observed in other protein bands. Lane A, reticulocyte lysate without further addition; lane B, reticulocyte lysate with globin mRNA added. (From Wu et al., 1981, with permission.)

that causes a temporary elevation of the cytoplasmic calcium concentration. Under these conditions, increased levels of mRNA for the 80 kD protein can be detected 3 hr after the transfer, but this increase has disappeared by 22 hr (Wu et al., 1981). Thus, the influx of calcium promotes the synthesis of mRNA coding for the 80 Kd protein, but this mRNA is degraded within 22 hr.

INCREASED LEVELS OF mRNA FOR THE 100 kD PROTEIN

Although the reticulocyte lysate translation system can be used to demonstrate the increase in mRNA for the 80 kD protein, it is less adequate for detecting the 100 kD protein unless it is followed by immunoprecipitation of the products with an antibody prepared against the Ca^{2+} ATPase of sarcoplasmic reticulum (Chyn et al., 1979). If this procedure is applied as seen in

Fig. 7, increased levels of mRNA can be visualized in ionomycin (lane 2) or A23187-treated cells (lane 3) as compared to control cells (lane 1). Thus, increased cytoplasmic Ca^{2+} in myoblasts acts to increase the level of mRNA coding for the 100 kD, as well as for the 80 kD protein. Further work is required to definitely identify the 100 kD protein with the Ca^{2+} transport ATPase.

EFFECT OF IONOMYCIN ON THE RATE OF SYNTHESIS OF 80-kD PROTEIN IN OTHER CELL TYPES

The rate of incorporation of [^{35}S]methionine into the 80 kD protein is also increased by ionomycin treatment of fibroblasts, LSP, HeLa, and CV1 cells (Fig. 8), indicating that the effect is not specific to muscle tissue. The widespread occurrence of this effect among mammalian cells argues for its significance in cell function. The expression of the 100 kD protein, on the other hand, is not increased to a similar extent by ionophores in all cell types tested (Fig. 8).

COMPARISON WITH HEAT-SHOCK PROTEINS

The specific stimulation of the 80 kD protein by Ca^{2+} ionophores raises the question of whether this protein is related to those known to be induced by heat shock, and whether in fact heat shock induces proteins by affecting calcium influx. When ionophore-treated cells are compared to cells exposed to 45° C to induce heat-shock proteins, it can be seen that there are no common proteins that show similarly increased levels of [^{35}S]methionine radioactivity among

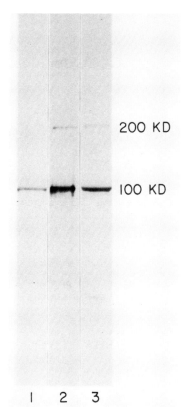

FIG. 7. Cell-free synthesis of the 100 kD protein. Muscle cells grown in normal Ca^{2+} medium for 3 days were treated with 3 μM ionomycin (lane 2) or A23187 (lane 3) for 3 hr followed by isolation and cell-free translation of total RNA in a reticulocyte lysate system. Untreated cell RNA (lane 1) served as control. The products of cell-free translation were treated with anti-ATPase antibodies (Chyn et al., 1979), and the immunoprecipitate was analyzed by SDS-gel electrophoresis and autoradiography.

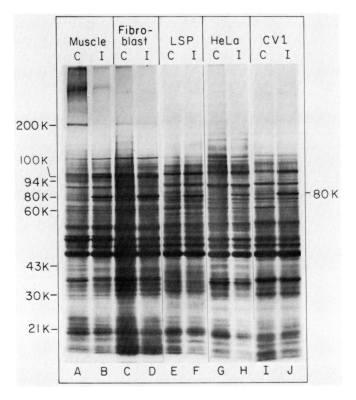

FIG. 8. The effect of ionomycin on the protein synthesis of different cell types (autoradiogram). The cells were grown in normal Ca^{2+} medium for 65 hr. One group of plates served as control (A, C, E, G, I), the other was treated with 4 μM ionomycin (B, D, F, H, J). After 3 hr both groups of cells were labeled with [^{35}S]methionine for 2 hr in a medium made from methionine-free minimum essential medium. The total cellular proteins were subjected to electrophoresis and autoradiography. Lanes A and B, primary cell cultures of chicken pectoralis muscles; lanes C and D, fibroblast culture of chicken embryos; lanes E and F, mouse LSP cells; lanes G and H, human HeLa cells; lanes I and J, CV1 cells of African green monkey kidneys. I, ionomycin treated; C, control (F. S. Wu and A. N. Martonosi, *unpublished data*).

the translation products of the two types of cells (Fig. 9) (F. S. Wu and A. N. Martonosi, *unpublished data*). Specifically, the increase in the 80 kD protein seen in heat-shocked cells is modest relative to that in ionophore-treated cells. It appears, therefore, that the effect of ionomycin treatment is rather selective. It can also be shown that this effect is not mimicked by glucose deprivation, or by amino acid analogs such as canavanine, ethionine, 1,2,4 triazo-3-alanine, and homoarginine, that have been reported to cause induction of specific proteins.[1]

POSSIBLE NUCLEAR EFFECTOR MOLECULES

The effect of elevated cytoplasmic calcium levels is to increase the cellular content of mRNA for the 80 and 100 kD proteins. It is likely that this action occurs at least in part at the level of transcription. If this is the case then the increased concentration of the cytoplasmic calcium

---

[1] S. Udenfriend suggested that the effects of ionophores in these experiments might not be due to a change in Ca^{2+} levels but to an effect on a different physiological parameter such as intracellular pH. Martonosi responded that while there is always that possibility, the only observable effect was that the treatment caused contraction of the muscle fibers. P. Mueller wondered whether the treatment did not also uncouple the mitochondria, but Martonosi indicated that treatment with dinitrophenol or cyanide does not mimic the effects of ionophore.

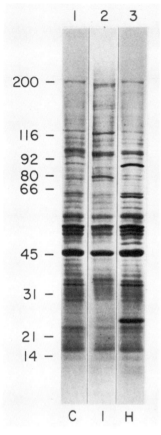

FIG. 9. Protein synthesis patterns in muscle cells after ionomycin treatment and heat shock. Three-day-old muscle cell cultures grown in normal media were exposed to 2 μM ionomycin (I), or heat shock (45°C) (H) for 5¼ hr and were labeled with 20 μCi/ml [^{35}S]methionine in the last 2 hr of incubation. Cells incubated at 37°C without further treatment served as control (C). The proteins were analyzed by SDS-polyacrylamide gel electrophoresis (F. S. Wu and A. N. Martonosi, *unpublished data*).

must somehow send a signal to the nucleus of the cell. Since the nuclear membrane is fully permeable to calcium, the nucleoplasmic calcium concentration is expected to follow the cytoplasmic concentration. How this change is reflected in gene transcription is not clear.

The nucleus does contain a number of calcium-binding proteins whose role might be to monitor calcium concentration (Schibeci and Martonosi, 1980a). The existence of these proteins can be demonstrated by resolving nonhistone chromosomal proteins on a nondenaturing polyacrylamide gel and by equilibrating the gel with radioactive $^{45}Ca^{2+}$ in the presence of 0.1 M KCl and 5 mM $MgCl_2$ (to eliminate nonspecific binding) (Schibeci and Martonosi, 1980b). Figure 10 shows the pattern of Ca^{2+} binding to the proteins from four fractions of muscle nuclei. One of the proteins peaks may correspond to calmodulin, but none of the others corresponds to any known calcium-binding protein such as troponin C, parvalbumin, or prothrombin. Thus there are several calcium-binding proteins in the soluble fraction of nuclear nonhistone chromosomal proteins that could mediate the effects of calcium on gene transcription. These effects might also be mediated by insoluble nuclear proteins not resolved in these studies. It is also possible, of course, that histones play a role since calcium has been shown to promote the phosphorylation of acetylated histone 3 in the nuclei of HeLa cells (Whitlock et al., 1980).

PLASTICITY OF SARCOPLASMIC RETICULUM

It is clear from the data reviewed here that sarcoplasmic reticulum does, in fact, represent a system that can be used to investigate the mechanism by which calcium regulates the expression of specific gene products.

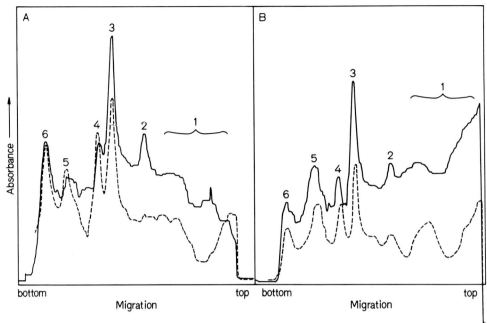

FIG. 10. Autoradiogram showing the influence of Ca^{2+} on the extraction of Ca^{2+}-binding proteins from nuclei. Nuclei were isolated from 15-day-old chick embryo skeletal muscle as described by Schibeci and Martonosi, 1980a. After incubation for 2 hr at 4°C in the presence of 100 mM KCl, 5 mM $MgCl_2$, 5 mM ATP, 10 mM Tris/HCl, pH 7.4, 50 μg/ml phenylmethylsulfonyl fluoride, 1 mM benzamidine, and either 5 mM EGTA or 1 mM $CaCl_2$, the suspension was briefly sonicated and centrifuged at 50,000 x g for 30 min. The sediment was washed with the same solution and centrifuged again. The sediment was frozen at −70°C until further use. The supernatants of each sample were combined, desalted by passage through a Sephadex G-25 column to remove most of the EGTA or Ca, and dialyzed first against 0.1 M KCl, 10 mM Tris/HCl, pH 7.5, and subsequently against water, followed by freeze-drying. The extracted proteins were dissolved in 10 mM Tris/HCl pH 6.8, 10 mM dithiothreitol, and 10% glycerol. The insoluble residue was dispersed in the same solution except that glycerol was replaced with 8 M urea. Electrophoresis was carried out on plain 10% polyacrylamide gels using 25 mMTris/192 mM glycine, pH 8.3, as buffer. The gels were exposed to ^{45}Ca followed by autoradiography and densitometry. **A:** *Solid line*, calcium extract of muscle nuclei; *dashed line*, EGTA extract of muscle nuclei. **B:** *Solid line*, calcium sediment of muscle nuclei; *dashed line*, EGTA sediment of muscle. Numbers indicate the principal radioactive bands. (From Schibeci and Martonosi, 1980a, with permission.)

The sarcoplasmic reticulum displays great plasticity both with respect to its amount in the cell and its protein composition, and it is possible that Ca^{2+} plays an even greater role in regulating plasticity than that already suggested. The extent of development of sarcoplasmic reticulum and its protein composition appears to be directly related to the functional pattern of the muscle in which it is found. In fast-acting skeletal muscle, the sarcoplasmic reticulum is extensively developed in comparison with slow-acting muscles such as slow-twitch skeletal or cardiac muscle (Jolesz and Sreter, 1981). In some fast-acting muscles, such as the lobster second-antenna motor muscle, which operates at high velocity at low temperatures, the total mass of sarcoplasmic reticulum tubules may be actually greater than the mass of myofibrils (Rosenbluth, 1969).

In addition to differences in the amount of sarcoplasmic reticulum, there are immunologically distinct isoenzymes of Ca^{2+} ATPase present within the same organism in the fast and slow skeletal and cardiac muscles (De Foor et al., 1980; Damiani et al., 1981). The fast muscles contain a catalytically fast enzyme, whereas the slow muscles possess a catalytically slow isozyme (Wang et al., 1979). The isozyme found in a particular muscle varies with the pattern of activity of that muscle. If a fast muscle is experimentally cross-innervated with a slow nerve, within a few months the rate at which Ca^{2+} is transported becomes comparable to that of a slow muscle (Mommaerts et al., 1969; Streter et al., 1975). Similarly, a slow muscle can be induced to transport Ca^{2+} slightly more rapidly if it is innervated by a fast nerve. Apparently, nerve activity affects the rate of calcium transport by determining the amount and the type of Ca^{2+} ATPase found in the muscle, although the immunological specificity of Ca^{2+} ATPase in cross-innervated muscle has not yet been established. The effect of cross-innervation may be due at least in part to the frequency of stimulation by the nerve, since electrical stimulation can be used as well as cross-innervation to alter the rate of Ca^{2+} transport by sarcoplasmic reticulum (Salmons and Sreter, 1976; Heilmann and Pette, 1979; Heilmann et al., 1981). The mechanism by which a muscle cell is able to translate frequency of stimulation into a metabolic signal that determines whether the fast or slow isozyme of Ca^{2+} ATPase is produced is not at all understood. Whether the frequency of change of cytoplasmic calcium concentration is the primary signal remains a matter of speculation.[2]

GENERAL DISCUSSION

F. Gros asked whether any proteins other than the Ca^{2+} ATPase and the 80 kD protein are subject to regulation by calcium. Martonosi answered that, although it had been his hope when he began his work that the entire battery of proteins that turn on during muscle differentiation would be regulated by calcium, this did not prove to be true. For example, troponin C, a calcium-binding protein, might be expected to be subject to such regulation, but apparently it is not.

ACKNOWLEDGMENTS

Work reported here was supported by research grants from NIH, NSF, and the Muscular Dystrophy Association. My thanks are due to Drs. Ricardo Boland, Doo Bong Ha, Takashi Morimoto, Shiro Ohnoki, Dikla Roufa, David D. Sabatini, Angelo Schibeci, Thomas W. Tillack, and Fang Sheng Wu for their cooperation in various phases of these studies, and to Ted Chyn, Eva Reyes, Judy Tsai, and David Cunningham for their assistance.

REFERENCES

Barth, L. G., and Barth, L. J. (1974): Ionic regulation of embryonic induction and cell differentiation in Rana pipiens. *Dev. Biol.*, 39:1–22.
Boland, R., Martonosi, A., and Tillack, T. W. (1974): Developmental changes in the composition and function of sarcoplasmic reticulum. *J. Biol. Chem.*, 249:612–623.
Cheung, W. Y., ed. (1980): *Calcium and Cell Function, Vol. 1. Calmodulin.* Academic Press, New York.
Chyn, T. L., Martonosi, A. N., Morimoto, T., and Sabatini, D. D. (1979): In vitro synthesis of the Ca^{2+} transport ATPase by ribosomes bound to sarcoplasmic reticulum membranes. *Proc. Natl. Acad. Sci. USA*, 76:1241–1245.
Damiani, E., Betto, R., Salvatori, S., Volpe, P., Salviati, G., and Margreth, A. (1981): Polymorphism of sarcoplasmic reticulum adenosine triphosphatase of rabbit skeletal muscle. *Biochem. J.*, 197:245–248.
De Foor, P. H., Levitsky, D., Biryukova, T., and Fleischer, S. (1980): Immunological dissimilarity of the calcium pump protein of skeletal and cardiac muscle sarcoplasmic reticulum. *Arch. Biochem. Biophys.*, 200:196–205.
de Meis, L., and Vianna, A. L. (1979): Energy interconversion by the Ca^{2+}-dependent ATPase of the sarcoplasmic reticulum. *Ann. Rev. Biochem.*, 48:275–292.
Deshmukh, K., and Sawyer, B. D. (1977): Synthesis of collagen by chondrocytes in suspension culture. Modulation by calcium, 3':5'-cyclic AMP and prostaglandins. *Proc. Natl. Acad. Sci. USA*, 74:3864–3868.

---

[2]L. L. Iversen pointed out that the underlying mechanism causing the changes in cross-innervated muscle is a matter of controversy and he wondered whether Martonosi had ruled out a role for trophic factors. Martonosi stated that cross-innervation did have effects on the levels of several muscle proteins other than the two regulated by calcium and that certainly trophic factors could be involved in these effects.

Endo, M. (1977): Calcium release from the sarcoplasmic reticulum. *Physiol. Rev.*, 57:71–108.
Epel, D. (1980): Ionic triggers in the fertilization of sea urchin eggs. *Ann. N. Y. Acad. Sci.*, 339:74–85.
Freedman, M. H., Raff, M. C., and Gomperts, B. (1975): Induction of increased calcium uptake in mouse T lymphocytes by concanavalin A and its modulation by cyclic nucleotides. *Nature*, 255:378–382.
Gallien-Lartigue, O. (1976): Calcium and ionophore A-23187 as initiators of DNA replication in the pluripotent haemopoietic stem cell. *Cell Tissue Kinet.*, 9:533–540.
Greene, W. C., Parker, C. M., and Parker, C. W. (1976): Calcium and lymphocyte activation. *Cell. Immunol.*, 25:74–89.
Ha, D. B., Boland, R., and Martonosi, A. (1979): Synthesis of the calcium transport ATPase of sarcoplasmic reticulum and other muscle proteins during development of muscle cells in vivo and in vitro. *Biochim. Biophys. Acta*, 585:165–187.
Hasselbach, W. (1979): The sarcoplasmic calcium pump. A model of energy transduction in biological membranes. *Top. Curr. Chem.*, 78:1–56.
Heilmann, C., Müller, W., and Pette, D. (1981): Correlation between ultrastructural and functional changes in sarcoplasmic reticulum during chronic stimulation of fast muscle. *J. Membr. Biol.*, 59:143–149.
Heilmann, C., and Pette, D. (1979): Molecular transformations in sarcoplasmic reticulum of fast-twitch muscle by electro-stimulation. *Eur. J. Biochem.*, 93:437–446.
Jaffe, L. F. (1980): Calcium explosions as triggers of development. *Ann. N. Y. Acad. Sci.*, 339:86–101.
Jolesz, F., and Sreter, F. A. (1981): Development, innervation, and activity-pattern induced changes in skeletal muscle. *Ann. Rev. Physiol.*, 43:531–552.
Kretsinger, R. H. (1976): Calcium binding proteins. *Ann. Rev. Biochem.*, 45:239–266.
Leffert, H. L. (1980): Growth regulation by ion fluxes. *Ann. N. Y. Acad. Sci.*, 339:1–335.
MacLennan, D. H., and Holland, P. C. (1976): The calcium transport ATPase of sarcoplasmic reticulum. In: *The Enzymes of Biological Membranes*, Vol. 3, edited by A. Martonosi, pp. 221–259. Plenum Press, New York.
Martonosi, A. (1972): Biochemical and clinical aspects of sarcoplasmic reticulum function. In: *Current Topics in Membranes and Transport*, Vol. 3, edited by F. Bronner and A. Kleinzeller, pp. 83–197. Academic Press, New York.
Martonosi, A. (1975): Membrane transport during development in animals. *Biochim. Biophys. Acta*, 415:311–333.
Martonosi, A., Roufa, D., Boland, R., Reyes, E., and Tillack, T. W. (1977): Development of sarcoplasmic reticulum in cultured chicken muscle. *J. Biol. Chem.*, 252:318–332.
Martonosi, A., Roufa, D., Ha, D. B., and Boland, R. (1980): The biosynthesis of sarcoplasmic reticulum. *Fed. Proc.*, 39:2415–2421.
Mommaerts, W. F. H. M., Buller, A. J., and Seraydarian, K. (1969): The modification of some biochemical properties of muscle by cross-innervation. *Proc. Natl. Acad. Sci. USA*, 64:128–133.
Rosenbluth, J. (1969): Sarcoplasmic reticulum of an unusually fast-acting crustacean muscle. *J. Cell Biol.*, 42:534–547.
Roufa, D., Wu, F. S., and Martonosi, A. (1981): The effect of Ca^{2+} ionophores upon the synthesis of proteins in cultured skeletal muscle. *Biochim. Biophys. Acta*, 674:225–237.
Salmons, S., and Sreter, F. A. (1976): Significance of impulse activity in the transformation of skeletal muscle type. *Nature*, 263:30–34.
Scarpa, A., and Carafoli, E. (1978): Calcium transport and cell function. *Ann. N. Y. Acad. Sci.*, 307:1–655.
Schibeci, A., and Martonosi, A. (1980a): Ca^{2+}-binding proteins in nuclei. *Eur. J. Biochem.*, 113:5–14.
Schibeci, A., and Martonosi, A. (1980b): Detection of Ca^{2+} binding proteins in polyacrylamide gels by $^{45}Ca^{2+}$ autoradiography. *Anal. Biochem.*, 104:335–342.
Schneider, M. F. (1981): Membrane charge movement and depolarization-contraction coupling. *Ann. Rev. Physiol.*, 43:507–518.
Sreter, F. A., Luff, A. R., and Gergely, J. (1975): Effect of cross-reinnervation on physiological parameters and on properties of myosin and sarcoplasmic reticulum of fast and slow muscles of the rabbit. *J. Gen. Physiol.*, 66:811–821.
Tillack, T. W., Boland, R., and Martonosi, A. (1974): The ultrastructure of developing sarcoplasmic reticulum. *J. Biol. Chem.*, 249:624–633.
Wang, T., Grassi de Gende, A. O., and Schwartz, A. (1979): Kinetic properties of calcium adenosine triphosphatase of sarcoplasmic reticulum isolated from cat skeletal muscles. A comparison of caudofemoralis (fast), tibialis (mixed), and soleus (slow). *J. Biol. Chem.*, 254:10675–10678.
Whitlock, J. P., Jr., Augustine, R., and Schulman, H. (1980): Calcium-dependent phosphorylation of histone H3 in butyrate-treated HeLa cells. *Nature*, 287:74–76.
Winegrad, S. (1971): Studies of cardiac muscle with a high permeability to calcium produced by treatment with ethylenediaminetetraacetic acid. *J. Gen. Physiol.*, 58:71–93.
Wu, F. S., Park, Y. C., Roufa, D., and Martonosi, A. (1981): Selective stimulation of the synthesis of an 80,000 dalton protein by calcium ionophores. *J. Biol. Chem.*, 256:5309–5312.

Section XIII

Molecular Genetic Approach to Neurobiological Problems

The sections and chapters of this volume thus far have been written by molecular geneticists whose conceptual and technological concerns have been primarily biochemical in nature, and by biologists, including endocrinologists and neurobiologists, already acquainted with the application of molecular genetic methods to their biological problems. Neuroscientists who have had no direct contact until now with molecular genetics, but who perceive the great opportunities awaiting the application of recombinant DNA technology to their field, may wish to have the basic procedures clarified. For this purpose, a chapter was prepared by H. Thoenen. The chapter obviously is not meant to be a *vade mecum* containing sufficient detail to permit moving directly to bench research in the field. Rather, Thoenen has outlined the "mainline" procedures likely to be needed in most biological investigations.

Thoenen also draws on other chapters in the book to illustrate, illuminate, and, in particular cases, to supplement, items dealt with in this chapter. Authors whose chapters fall in this category include D. J. Anderson and coworkers, H. M. Goodman and coworkers, P. H. Patterson, and E. M. Shooter.

The chapter's discussion is valuable not only for its elaboration of items presented in other chapters, but also for the important additional contributions made by the participants, including J. D. Baxter and coworkers, W. E. Hahn, E. Herbert and coworkers, and L. E. Hood (whose discussion of the great technical value of automatic analytic and synthetic "gene machines" further elaborates on his own chapter).

Strategies for Application of Molecular Genetics to Neurobiology

Hans Thoenen

ABSTRACT

For the benefit of neurobiologists who wish to use molecular genetic technology, a short, simplified version of methods that might be employed is presented.

Suggested applications to neurobiological problems include the comparative molecular analysis of the phenotypes of different populations of neurons and glial cells. Synthetic oligodeoxynucleotides (constructed according to partial amino acid sequences and employed as probes and primers on mRNA for reacting with reverse transcriptase) may be used for the production of cDNA of macromolecules that can be isolated by conventional methods only in quantities so small as to make the preparation of the corresponding mRNA impossible. The same procedure can be used for the production of the cDNA of macromolecules; antibodies against their end product may not recognize precursor molecules. From the cDNAs thus obtained, complete (indirect) determination of amino acid sequences may be obtained. Such macromolecules could be synthesized in large quantities by insertion of the corresponding cDNA into appropriate vectors of prokaryotic or eukaryotic cells, allowing their transcription and translation in a manner already shown for insulin, growth hormone, and interferon.

Neurobiologists and molecular geneticists do not speak a common language. A major purpose of this volume is to assist molecular geneticists and neuroscientists to become bilingual in each other's fields. As a neurobiologist, I wish to make a contribution in this direction, in particular by making suggestions concerning the use of the tools of molecular genetics to resolve problems in neurobiology. These suggestions will, it is hoped, provide a basis for interaction with those skilled in molecular genetics that may lead to the elucidation of rewarding approaches to problems of neurobiology.

This discussion will focus on the more heuristic concerns of biochemical neuroscience rather than on the more global problems such as analogies between the immune system and the nervous system.

Molecular genetics has already been applied to a degree to neuroscience problems. Thus, in the field of neuroendocrinology, well-characterized end products (protein hormones) have been used to isolate the corresponding mRNAs from which they were translated (see D. Krieger; H. Goodman et al.; and S. Udenfriend, *this volume*). This has led to a better understanding of the size of the precursor proteins, and the arrangement of active peptides within these multiproduct precursors. Tracing back further from mRNAs to the corresponding DNAs has provided information on the organization of these particular genes and has led to a more realistic approach to understanding the regulation of their expression.

Presented here are a few examples of neurobiological problems that lend themselves to molecular genetic investigation. Thus far, comparisons between different populations of neuronal or glial cells have been achieved by the use of "conventional" methods, including two-dimensional gel electrophoresis, surface characterization with ^{125}I-peroxidase labeling, affinity labeling with specific ligands, surface mapping with lectins, and surface characterization using monoclonal antibodies (see Table 1). For instance, P. Patterson *(this volume)* has demonstrated that *in vitro* peripheral sympathetic neurons adopt predominantly adrenergic or cholinergic properties depending on the culture conditions. This was possible because of the relatively large amount of information available regarding biochemical and physiological properties of adrenergic and cholinergic neurons (see Table 2). Included is the characterization of enzymes involved in the synthesis of each transmitter, descriptions of the high affinity uptake mechanisms of choline and of catecholamines, and the identification of specific storage vesicles for the transmitters of both types of neurons.

Gene technology could be used in the comparison of two populations of both glial cells and neurons (Fig. 1). First, the mRNAs of each population are isolated, then the corresponding cDNAs are produced, inserted into plasmids, cloned, and amplified in bacterial vectors. The amplified cDNA could then be excised, denatured, and used for hybridization with the mRNAs of the neurons to be compared. The comparison would include hybrid-arrested translation. If a complete set of cDNA clones derived from the mRNAs of a given population of neurons is added to an *in vitro* translation system to which the mRNAs of these neurons have been supplied, the translation of all the mRNAs will be arrested. If the same cDNAs are added to an *in vitro* translation system of mRNA from a population of neurons to be compared (e.g., cholinergic with adrenergic), only the translation of mRNAs common to both populations of neurons would be blocked by hybridizational inactivation. This selective comparison of the differences between

TABLE 1. *Comparison of conventional methods and gene technological approaches to determining differences between given populations of neurons or glial cells[a]*

Molecular genetics	Conventional approach
Hybrid-arrested translation to determine cell-specific proteins	Comparative TD electrophoresis
Selection of DNA clones that do not cross-hybridize	Comparative [^{125}I]-peroxidase labeling of surface molecules
Determination of sequence, synthesis of characteristic peptides, coupling to macromolecules, production of antibodies	Surface characterization by lectins
Insertion of DNA into plasmid (prokaryotic host) or virus (eukaryotic host) in a manner allowing transcription and translation in host as already shown for insulin and interferon	Monoclonal antibodies against specific surface markers, identification of specific antigens in neurons and glia

[a]The approach of molecular genetics is based on the production of cDNAs from mRNAs of two different populations of neurons and glial cells.

TABLE 2. *Summary of established differences between adrenergic and cholinergic neurons[a]*

Adrenergic		Cholinergic	
TH DBH DDC	Enzymes involved in catecholamine synthesis	CAT	Enzyme involved in acetylcholine synthesis
High affinity uptake for catecholamines		High affinity uptake for choline	
Specific storage vesicles for catecholamines		Specific storage vesicles for acetylcholine	

[a]TH, tyrosine hydroxylase; DBH, dopamine β-hydroxylase; DDC, dopadecarboxylase; CAT, choline acetyltransferase.

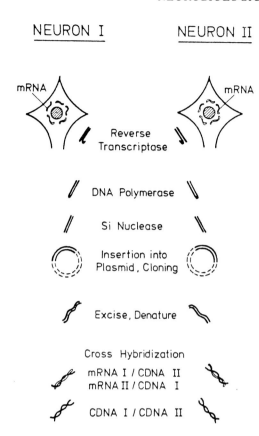

FIG. 1. An approach for comparing the phenotypes of two different populations of neurons (or glial cells) using molecular genetic methods: cDNAs produced from mRNAs of the two populations of cells are compared by cross-hybridization.

the mRNAs of the two populations of neurons or glial cells provides a more precise picture than the much more complex comparison of two-dimensional gels of proteins from the two populations. Moreover, this approach would select the cDNA clones that do not cross-hybridize. These would represent mRNAs, and their corresponding proteins, that are unique to the population of cells being studied. The base sequence of this cloned DNA could then be determined, and the corresponding amino acid sequence deduced. From this amino acid sequence, particular unique regions could be selected (after consulting a polypeptide library to avoid sequences homologous to other known polypeptides). In addition, the characteristic sequences could then be synthesized by solid-phase methods, the resulting peptide linked to a macromolecule (e.g., bovine serum albumin or thyroglobulin) and used for the production of antibodies (see R. Lerner, *this volume*). This would allow isolation of larger quantities of the corresponding protein molecule (end product or precursor) by affinity chromatography. Such antibodies could also be used for the cytological localization of the protein by immunocytochemical and electron microscopic methods.

Extending this technique further by analogy, the neuron-specific cDNAs themselves might be used as *in situ* hybridization probes to investigate the expression of specific genes at the transcriptional level. This could be done at different stages of nervous system development. For example, as stem cells migrate from the neural crest in the developmental system described by P. Patterson *(this volume)*, they differentiate under the influence of epigenetic factors to assume the characteristics of cholinergic or adrenergic neurons. Specific cDNA hybridization probes might prove useful in the analysis of neuron-specific gene expression during this developmental progression. Such probes would provide more direct information on the level of expression of specific genetic information than do immunocytochemical techniques that require the presence of the end product. Increased sensitivity of *in situ* hybridization would allow the detection of a few copies of mRNA and the development of electron microscopic procedures using ferritin or horseradish peroxidase (HRP) coupled to cDNA probes. Moreover, this procedure would avoid the problem that antibodies produced against end products may not recognize the precursor.

A further application of molecular genetic technology in the comparison of two populations of different neurons or glial cells would be the insertion of the cDNA of interest into a plasmid (prokaryotic host) or virus (eukaryotic host), e.g., SV40, in a manner that allows transcription and translation in the host, as has been shown for insulin, interferon, growth hormone, and so on (see H. Goodman et al., *this volume*). This would permit the production of these polypeptides in a nonneuronal system in relatively large quantities, thus greatly expediting their further study.

These approaches are used exclusively for peptides; glycoproteins are not yet included. However, recent experiments have provided evidence that the glycosylation of polypeptides is not cell-specific (B. Dobberstein, *personal communication*), i.e., the site and nature of glycosylation are determined by the polypeptide amino acid sequence. This is in agreement with the experiments of D. Anderson et al., *(this volume)* on the translation of mRNAs of the subunits of the acetylcholine receptor in a cell-free system in combination with a heterologous microsomal preparation. Glycosylation of the proteins synthesized *in vitro* occurs similarly in *in vivo* glycosylation.

Another potentially useful technique is the synthesis of "primer" or "probe" DNA corresponding to known amino acid sequences (Noyes et al., 1979; Agarwal et al., 1981; Rehfeld, 1981). This technique is especially important because, frequently, the quantities of macromolecules of high physiological significance in the function of neurons and glia can be isolated only in extremely small quantities. This generally precludes the production of antibodies against these macromolecules and consequently the use of a conventional approach to the isolation of mRNA by immunoprecipitation of polysomes, producing the nascent chains of the polypeptide in question. Moreover, even if the production of antibodies against a purified end product is possible, these antibodies may not recognize the corresponding precursors (Noyes et al., 1979; Thoenen et al., 1981). Synthetic primer or probe DNA constructed according to known amino acid sequences can be used for the selection of specific mRNAs and the production of the corresponding cDNA. This cDNA can be cloned and inserted into vectors in a manner that allows the transcription and translation of the product of interest. One of the main applications of these methods would be in cases in which the antibody against a protein end product does not cross-react with the precursor. This approach has been successfully used for the isolation of the precursor of gastrin and cholecystokinin (Noyes et al., 1979; Rehfeld, 1981), and will most probably also prove important for the isolation of the mRNA of nerve growth factor (NGF). Cell-free translation of mRNA isolated from the male mouse submandibular gland, a rich source of NGF, did not result in the production of a translation product that was recognized by antibodies produced against NGF (E. Shooter, *personal communication*; H. Thibault and H. Thoenen, *unpublished results*).

The failure of the antibody produced against the end product to recognize the precursor can also be deduced from electron microscopic immunological investigations of Schwab and coworkers (1976) and of Thoenen and coworkers (1981), who demonstrated that, in secretory cells of the salivary gland that produce NGF, binding of HRP-like antibodies occurred only in the lumen and in the granules of apical regions adjacent to the lumen, whereas in the basal part of the cell, where biosynthesis is thought to occur, there was no binding of antibody with the contents of the secretory granules. Because NGF is not a glycoprotein, the most probable explanation for this change in the immunoreactivity is cleavage of the end product from a precursor that is not recognized by the antibodies against the final product.

Techniques for DNA probe synthesis may find important application in the study of macromolecules present as a very small proportion of the total protein isolated from neuronal tissue. For example, choline acetyltransferase exists in such minute amounts that isolation of a pure (> 95%) preparation from total brain requires a purification factor of over one million (Eckenstein et al., 1981). It is apparent that the isolation of the corresponding mRNA by conventional methods, such as immunoprecipitation or affinity chromatography of polyribosome preparations containing the mRNA and a nascent polypeptide recognized by the antibody, will most probably be unsuccessful. The methodology developed by L. Hood *(this volume)* for sequencing picomolar quantities of proteins may advance the study of such rare proteins. It expands the scale of application of molecules of unknown function and cellular localization that heretofore have been characterized exclusively by two-dimensional gel electrophoresis. Both the localization and function of these proteins might then be traced by the combined use of amino acid microsequencing and rapid synthesis of oligodeoxynucleotides corresponding to characteristic polypeptide sequences.

The synthesis of oligodeoxynucleotides as probes and primers for the production of cDNA, corresponding to proteins with known, or at least partially known, amino acid sequences, has been hampered by the fact that only two amino acids, tryptophan and methionine, have unambiguous codons. Agarwal and coworkers (1981) demonstrated that the normal RNA-DNA base pairs G-dC and C-dG can be replaced without loss of hybridization efficiency by G-dT and U-dG. This increases the number of "useful" (unambiguous) amino acids from 2 to 11 (see Table 3). Together with the methods for rapid and efficient synthesis of oligodeoxynucleotides described by Hood, this discovery gives the molecular genetic approach greater power and usefulness for problems in neurobiology.

Many molecules of great interest or even potential therapeutic importance, such as trophic factors, cannot be extracted from neuronal tissue in sufficient quantities to allow detailed study or clinical application. Thus, even at this early juncture, the techniques of recombinant DNA technology, already used for the large-scale production of insulin, growth hormone, and interferon, will have an important place in future neurobiological research. These techniques will

TABLE 3. *Summary of the RNA/DNA codons of amino acids that can be used for the synthesis of "useful" oligonucleotide probes*[a]

Amino acid	RNA codon	Deduced DNA codon	Amino acid	RNA codon	Deduced DNA codon
Phenylalanine	UUU UUC	AAG	Glutamine	CAA CAG	GTT
Tyramine	UAU UAC	ATG	Lysine	AAA AAG	TTT
Histidine	CAU CAC	GTG	Glutamic acid	GAA GAG	CTT
Asparagine	AAU AAC	TTG			
Aspartic acid	GAU GAC	CTG	Tryptophan	UGG	ACC
Cysteine	UGU UGC	ACG	Methionine	AUG	TAC

[a]The extension from tryptophan and methionine (the only two amino acids with unambiguous, nonwobbling codons) to nine additional amino acids is based on the fact that the usual RNA/DNA base pairs G/cC and dG/C can be replaced by G/dT and dG/U without impairing hybridization.

permit the production of large quantities of rare molecules that are of interest to neurobiologists, which, in turn, will permit more extensive investigation leading to a more complete understanding of their biological and clinical actions. With respect to therapeutic applications, this approach should facilitate the selective production of molecules such as trophic factors that are human-specific and thus avoid immunological complications in their potential therapeutic use.

GENERAL DISCUSSION: RATIONALE AND MODIFICATION OF APPROACH

L. Hood agreed that Thoenen's general approach will prove very exciting. In the future, microgram quantities (or less) of an interesting polypeptide will be adequate to determine the amino acid sequence of the molecule. With that information, protein fragments could be synthesized and used to make specific antibodies. In addition, the amino acid sequence could be examined for regions of minimal degeneracy as well as for synthesis of DNA probes that would bind to cloned genes. Thus, progress is possible in both directions: the amino acid sequence can be used in the production of antibodies and in the synthesis of its gene.

The power of the new machines that synthesize DNA probes lies in their ability to overcome base sequence ambiguity problems. Perfect matches are required in these very short sequences to get hybridization that allows the sequences to serve either as primers or as direct probes for analysis. Thus in 14-mers, even one centrally placed base pair mismatch prevents hybridization. With a machine cycle time of ½ hr, two 11 to 14 base oligonucleotides can be synthesized a day. Therefore, even if the sequence shows four- or eightfold degeneracy, all the combinations may be synthesized in a reasonable period of time. These probes can then be used as a mixture; only the appropriate probe will bind and serve as a primer.

Hood also described experiments performed by K. Itakura *(unpublished data)* in the City of Hope laboratories that will avoid the mRNA level completely. A mixed 15-mer oligodeoxynucleotide containing all the alternative degenerate positions for the human gene, β-2-microglobulin, was synthesized. This mixture of probes was used to examine a cDNA library and pick out the corresponding cDNA sequence. Thus, determination of protein sequence circumvents the need to determine the cell type that actually expresses that gene product; it is possible to go directly from a protein sequence to a cDNA library. Alternatively, Hood suggested that two similar 15-mer probes could be synthesized to examine a genomic library directly: The first probe would yield a set of clones, 2 to 3% of which contain the desired sequence, and the second probe would select among those clones to determine exactly which clones actually contain the sequence of interest.

Although neurobiologists are often more interested in the phenotype of the cell than its genotype, Hood explained that by going directly to the genome one could get nucleic acid probes that would then permit going back and doing the type of *in situ* hybridizations Thoenen had proposed to localize the gene products of interest. The greatest difficulty in doing such experiments may be that often the proteins of interest will prove to be products of multigene families. Initial examination of the genome will permit isolation of multigene members and, following determination of their structure, probes unique to individual gene products can be synthesized at either the protein or DNA level. This will obviate the necessity of determining which cells expressed particular members of this multigene family, when the process occurred, and other items of interest. This level of refinement is required for defining developmentally and spatially where this kind of information is expressed.

E. Herbert pointed out that some types of experiments mentioned by Thoenen and Hood had actually been carried out with a number of different genes. One example is the work of S.

Udenfriend *(this volume)* on enkephalin genes. In addition, J. Herbert and coworkers *(this volume)* have demonstrated that short DNA pieces can be effectively used in the search for genes that are expressed at low levels. Thus, these techniques have general applicability, and work with gastrin and cholecystokinin described by Thoenen do not represent special cases.

D. Anderson suggested that conventional methods combined with the technical advances of molecular genetics could be applied to a functionally important set of molecules in the nervous system that are present in very small amounts, namely the receptors for various neurotransmitters. Although these proteins are present in small quantities, the ligands they bind can often be obtained in large quantities. Antisera to these ligands can be prepared and, in conjunction with covalent photoaffinity derivatives of the ligands, the antibodies can be used to purify the receptors in a covalent complex with their ligand. Then, by using microsequencing techniques, genes for neurotransmitter receptors could be cloned.

Thoenen raised a question concerning the number of copies of a particular mRNA needed to produce the corresponding cDNA as a basis to pursue the approaches offered by molecular genetics. Hood suggested that rather than mRNA, a sample of the protein is necessary, because everything can be done by working back from the amino acid sequence. However, F. H. C. Crick pointed out that, if one is trying to clone a cDNA copy of the mRNA, then only a minute amount of mRNA is required. On the other hand, the level of RNA required to localize mRNA in the cell with the probe is another matter. One could probably determine whether a particular message was present in the putamen, which has a large number of cells, but in neurobiology it is frequently desirable to localize the mRNA to one particular kind of cell. This would be done on tissue sections. Although it may be feasible for a common message, it may not be for a rare message. It is important for neurobiologists to learn the limitations as well as the possibilities of molecular genetic methods.

W. E. Hahn agreed that the problem is not in obtaining probes, but rather in using them effectively in complex cell populations. For example, in the sea urchin, which undergoes a division of the cells in micromeres and macromeres in the very early stages of development, one can clone some of the cDNAs that represent, on a global scale, the relatively infrequent messages. This can be followed by *in situ* hybridizations with a cDNA probe representing a rare-class message, and that message can be identified in a specific cell or in a cluster of cells from the micromere. However, it may prove difficult to identify such messages, even though the hybridization kinetics can be adjusted to optimize the sensitivity. A specific message in a given cluster of micromere cells has yet to be identified by this approach, although it is possible to distinguish messages that are in the micromere but not in the macromere.

Thoenen referred to a promising procedure, presented in a recent paper, that described *in situ* hybridization with prolactin cDNA in the pituitary (Pochet et al., 1981). J. D. Baxter observed that the prolactin mRNA represented about 30% of the messages in that system. Similar experiments on a histological section of kidney revealed the distribution of kallikrein mRNA, which, in this case, is less than 1% of the total mRNA (although the actual abundance has not been accurately determined) (Hudson et al., 1981). Baxter indicated that this technique was useful for RNAs of medium abundance.

Herbert pointed out that peptides or proteins that are exported by cells would be turned over in time and therefore their mRNA levels are apt to be fairly high. Thus, the mRNA for some transmitters or peptide hormones might be detectable by autoradiographic methods (*in situ* hybridization). However, mRNAs for proteins that are made during the process of development or in small numbers may be present in less than one copy of RNA per cell. In this case, detection by such autoradiographic methods may be a problem, and techniques developed for the peptide hormones might not be applicable.

Hood mentioned attempts to develop fluorescent hybridization probes. Probes using biotin linkage or fluorescence that are as sensitive as HRP-linked probes may soon be developed. Such probes would increase by manyfold the ability to detect products of single-copy genes.

REFERENCES

Agarwal, K. L., Brunstedt, J., and Boyes, B. E. (1981): A general method for detection and characterization of an mRNA using an oligonucleotide probe. *J. Biol. Chem.*, 256:1023–1028.

Eckenstein, F., Barde, Y.-A., and Thoenen, H. (1981): Production of specific antibodies to choline acetyl-transferase purified from pig brain. *Neuroscience*, 6:993–1000.

Hudson, P., Penschow, J., Shine, J., Ryan, G., Niall, H., and Coghlar, J. (1981): Hybridization histochemistry: Use of recombinant DNA as a homing probe for tissue localization of specific mRNA populations. *Endocrinology*, 108:353–356.

Noyes, B. E., Mevarech, M., Stein, R., and Agarwal, K. L. (1979): Detection and partial sequence analysis of gastrin mRNA by using an oligodeoxynucleotide probe. *Proc. Natl. Acad. Sci. USA*, 76:1770–1774.

Pochet, R., Brocas, H., Vassart, G., Toubeau, G., Seo, H., Refetoff, S., Dumont, J. E., and Pasteels, J. L. (1981): Radioautographic localization of prolactin messenger RNA on histological sections by in situ hybridization. *Brain Res.*, 211:433–438.

Rehfeld, J. F. (1981): General molecular aspects of the gastrin-cholecystokinin system. In: *Hormones and Cell Regulation*, Vol. 5, edited by J. E. Dumont and J. Nunez, pp. 169–183. Elsevier/North Holland Biomedical Press, Amsterdam.

Schwab, M. E., Stöckel, K., and Thoenen, H. (1976): Immunocytochemical localization of nerve growth factor (NGF) in the submandibular gland of adult mice by light and electron microscopy. *Cell Tissue Res.*, 169:289–299.

Thoenen, H., Schäfer, Th., Heumann, R., and Schwab, M. (1981): Nerve growth factor as a retrograde macromolecular messenger between effector cells and innervating neurons. In: *Hormones and Cell Regulation*, vol. 5, edited by J. E. Dumont and J. Nunez, pp. 15–34. Elsevier/North Holland Biomedical Press, Amsterdam.

Prospects from Retrospect

Floyd E. Bloom

ABSTRACT

The recently developed methods and concepts of molecular genetics research surveyed in this volume offer technical strategies of immense value to research in the neurosciences. This powerful armamentarium of molecular methodology seems especially pertinent to important unsolved questions in neuronal development and organization, heritable disorders of the nervous system, and in the discovery process by which new neurotransmitters and their functionally related protein entourages (e.g., synthetic and catabolic enzymes, and receptors) are discovered and characterized. Some of the more obvious prospects for fruitful applications of these approaches are highlighted in this chapter with special attention devoted to specific examples of neuropeptide systems.

My fellow neuroscientists may wonder at the wisdom of gaping wistfully at a new complicated field like molecular genetics with its recombinant DNA technology when there are obviously many critical and equally complex questions still unanswered in neurosciences. However, as an instigator of the enforced courtship between our disciplines, I fully believe that molecular genetics and neurosciences have much to offer each other. In this retrospective highlighting of selected topics covered, I intend to indicate some of what I see as the more immediate virtues of this emerging frontier area. On first encounter, neuroscientists will be dismayed by the new jargon to be learned to reflect on the relevance of molecular genetics to our own pet projects, but "their" jargon is no less formidable than our circuitry diagrams, channel kinetics, or synaptic transmitter litanies. Having been tutored incessantly in my newly formed collaborations, I can attest that at least this old dog has acquired some new approaches without any real discomfort. At the outset, all that is really needed is the realization that the powerful methods of modern molecular genetics research are based on the ability to create novel combinations of DNA in bacterial plasmids or other live vectors, and to colonize (or clone) each of these recombinant DNA molecules in bacteria that have been "transformed" (i.e., infected more-or-less permanently with this special vector). The bacterial clones can be induced to express large amounts of the products of the synthetic genes either in the bacterial clone—more frequently—or in another step by some other cellular or cell-free system. An impressive array of innovative techniques has been devised to purify abundant mRNAs, to enrich for low abundance mRNAs, and to use the nucleic acid components of cloned plasmids as probes to analyze a variety of problems from genomic organization to *in vitro* expression of medically important peptide hormones. As the data summarized in Table 1 indicate, there have already been many serious applications of molecular genetics to the problems of neuroscience. This body of experimental work can actually be traced back to 1972 (see DeLarco and Guroff, 1972; Zomzely-Neurath et al., 1972). However, in the past four years it has been greatly intensified, and important contributions have been made at three levels: (1) organization of the genes coding for pituitary

TABLE 1. *Recombinant DNA strategies applied to neurobiological problems*

Strategy	Reference
Genome	
Hormone gene located on chromosome maps	Owerbach et al., 1980
Structural organization of hormone gene (pro-mRNA)	Chang et al., 1980; Nakanishi et al., 1980; Roberts et al., 1979
mRNA	
Cellular localization of mRNAs	Hudson et al., 1981
Characterize abundance or variety	Hahn et al., 1978; Morrison et al., 1979
Changes with development	Morrison et al., 1981
Changes with aging	Colman et al., 1980
Changes with drug, hormone modulation	Baetge et al., 1981
Injected mRNA transforms cell properties	Dahl et al., 1981
Gene products: Expression (E) or sequence (S) by cloning	
Acetylcholine receptor (E)	Mendez et al., 1980; Anderson and Blobel, 1980; Sumikawa et al., 1981
Tubulins (S)	Gozes et al., 1975, 1980; Valenzuela et al., 1981
Actin (E)	Minty et al., 1981
Somatostatin (E)	Itakura et al., 1977
Prosomatostatin (S)	Goodwin et al., 1981
Pro-opiomelanocortin, "γ-MSH" (S)	Nakanishi et al., 1979
Gastrin (E)	Mevarech et al., 1979
Vasopressin-neurophysin common precursor (E)	Schmale and Richter, 1981
Polyenkephalins (S)	Kilpatrick et al., 1981

and brain hormones; (2) exploitations of partially purified mRNAs for a variety of normative measurements and for effects of experimental perturbations; and (3) *in vitro* expression or sequence determinations of the ultimate gene products.

Molecular cloning of a different sort is also obtained through the hybridoma technology, which allows clones of antibody-producing cells to be raised under conditions that yield identical immunoglobulins directed at the same immunogen. These monoclonal antibodies have already demonstrated their usefulness in detecting unique cellular markers in the nervous system (see E. Haber; C. Barnstable; *this volume*). The present momentum in the field is likely to make this table and my prospective predictions obsolete in short order. Nevertheless, it may be useful to consider some of these possible applications as an indicator of future developments. In particular, the continuing saga of the pro-opiomelanocortin system exemplifies vividly how recombinant genetic techniques may yield important advances in neurotransmitter research, especially discovery of novel peptides, their precursors and related cleavage products, and their posttranslational processing.

NEW PEPTIDES FROM OLD

Pro-opiocortin was the name given to the apparent common precursor for both corticotropin and β-lipotropin-derived pituitary peptides (Rubinstein et al., 1978), which are distinct from possible precursors for the enkephalin pentapeptides (see Bloom et al., 1978; Watson et al., 1978; also see D. Krieger; E. Herbert et al., *this volume*). Recently, the N-terminal region of bovine pituitary pro-opiocortin was determined by nucleotide sequencing of a cDNA clone prepared from purified intermediate lobe mRNA (Nakanishi et al., 1979). This sequence indicated the existence of a zone of high amino acid sequence homology with the heptapeptide sequence of corticotropin 4–10, which also, therefore, is found in α-melanocyte-stimulating hormone (α-MSH) and in the β-MSH fragment of γ-lipotropin (Nakanishi et al., 1979). The newly discovered homologous sequence was given the tentative name (γ-MSH) based only on the sequence homology without knowledge of the independent existence of this putative hormone or its precise length. The entire precursor has since been renamed pro-opiomelanocortin (POMC) (Chretien et al., 1979; Chang et al., 1980).

Synthetic replicates of γ-MSH with varying C-terminal extensions (termed γ_1-, γ_2-, or γ_3-MSH) were found to possess very low melanotropic activity and no adenohypophyseal releasing activity *in vitro* (Table 2) (Ling et al., 1979). Most recently, specific antisera were raised to γ_3-MSH, which do not cross-react with corticotropin, α, β-MSH or β-endorphin (see Shibasaki et al., 1980, 1981). In collaboration with T. Shibasaki, N. Ling, and R. Guillemin, sites immunoreactive to γ_3-MSH have been localized within rat pituitary and brain and these sites have been compared with those previously described (Bloom et al., 1978) that exhibit immunoreactivity to β-endorphin (Bloom et al., 1980). The general locations of preterminal and terminal fiber immunoreactivity for either γ-MSH or β-endorphin staining were very similar; this was particularly striking for the similar morphology of the axons and their extremely large varicosities alternating with long arrays of fine beaded fibers, as well as less frequent thick fascicles of fibers of passage through zones of white matter. What is intriguing is that numerous instances of consistent differences in relative fiber staining density were observed, in the γ_3-MSH-immunoreactive fibers and apparent nerve terminals were more pronounced than endorphin-reactive fibers and boutons in several of the target zones of presumed axon termination. Within the hypothalamus, the neuronal perikarya reactive for γ_3-MSH were fewer in number and smaller in size than those stained for β-endorphin, yet there were more immunoreactive fibers in this medial arcuate area for γ_3-MSH than for β-endorphin.

Methods developed for simultaneous detection of two tissue antigens in the same section have been applied to the distribution of POMC component peptides in the basal hypothalamic cell body area of the rat (McGinty and Bloom, 1980). This analysis indicated that endorphin- and corticotropin-reactive cell bodies were probably a single homologous class, but that endorphin-positive cell bodies were not all reactive for γ_3-MSH immunoreactivity (i.e., there were fewer γ-MSH reactive cells). Since the cell body region showed many more γ-MSH-reactive fibers than endorphin-reactive fibers, the discrepancies in costaining of the cell bodies cannot be attributed entirely to species differences between rat and bovine peptides. These three sets of results on individual immunocytochemical comparisons, dual simultaneous immunocytochemical comparisons, and regional radioimmunoassay comparisons all imply that cells expressing the gene for synthesis of the POMC common precursor process this precursor peptide differentially.

Studies of the posttranscriptional processing of the corticotropin- and endorphin-containing fragments of the C-terminal half of the common precursor have indicated that the precise mode of cleavage is a cell-specific variable (Zakarian and Smythe, 1979). As presently conceived,

TABLE 2. *Amino acid sequences of MSH-like peptides*

α-MSH	Ac-Ser-Tyr-Ser-Met-Glu-His-Phe-Arg-Trp-Gly-Lys-Pro-Val-NH$_2$
β-MSH	——Met-Glu-His-Phe-Arg-Trp-Gly-Ser-Pro——
γ_1-MSH	——Met-Gly*-His-Phe-Arg-Trp-Asp*-Arg-Phe-NH$_2$——
γ_2-MSH	——Met-Gly-His-Phe-Arg-Trp-Asp-Arg-Phe-Gly——
γ_3-MSH	——Met-Gly-His-Phe-Arg-Trp-Asp-Arg + 17 amino acids——
	Glu = GA$\genfrac{}{}{0pt}{}{G}{A}$ Gly = GGN Asp = GA$\genfrac{}{}{0pt}{}{C}{U}$

Nakanishi et al., 1979; Shibasaki et al., 1980, 1981.

corticotrophs process the precursor mainly to corticotropin and β-endorphin, whereas cells of the intermediate lobe and of the hypothalamus process the precursor to α-MSH and β-endorphin. The results on MSH-related peptides suggest that the N-terminal half of the common precursor may also be differentially processed according to cell-specific variables. The differences in regional fiber density or radioimmunoassay content observed for β-endorphin or γ_3-MSH could simply reflect the greater extent of processing to γ_3-MSH in some β-endorphinergic neurons than in others, possibly as a result of differences in local synaptic traffic or biochemical regulation.

However, with an eye toward the future, it may also be of value to consider two other possibilities: (1) in cells that all express the same precursor molecule, the terminal steps of posttranscriptional processing may vary within different branches of the same neuron, proceeding to β-endorphin, α-MSH, or γ_3-MSH in some cases, and to other secretory products for other fiber branches; and (2) within different neurons of the same general cell cluster location, a given subpopulation of cells and their distal nerve fiber terminals preferentially process the common precursor to the same single final secretory product, e.g., either β-endorphin or γ_3-MSH, at every secretory locus of those cells. Should future data support the idea that certain neurons or their processes within an anatomically defined population of chemically homologous neurons can process common precursor molecules to different final secretory products, then a wide range of functional regulation for such processing must be envisioned. Clearly, monoclonal antisera directed at different regions of a precursor could be critical tools in following such processing phenomena.

The physiological properties of γ-MSH are unlike those of comparable doses of β-endorphin after intraventricular injection to rats; γ_3-MSH is much more potent in reducing body temperature than β-endorphin, although the mechanisms responsible in both actions appear to be depression of heat production (Henriksen et al., 1980). However, γ-MSH does not produce analgesia or epileptiform effects on the limbic system, and does not interfere with the latter actions of β-endorphin. Studies are under way to evaluate the cellular actions of these peptides at target cell sites likely to receive fibers immunoreactive for both peptides.

One of the major unexpected general developments of molecular genetics research has been the recognition that all DNA is not directly transcribed into mRNA (see J. Darnell and M. Wilson; P. Sharp and M. Wilson; *this volume*), but rather that considerable editing of the pro-mRNA transcript is done, to remove noncoding zones (introns) and to splice together the RNA that will be expressed in the final message (exons). Although the meaning of this extra major phase is as yet unknown, it has already proved fruitful to theoretical speculation (see Crick, 1979; Sharp, 1981). Given the several probable active products of the POMC precursor, one might have anticipated that each active zone is coded for separately in the gene and brought together as the pro-mRNA is processed. However, studies of the organization of the POMC gene have revealed that the entire actively expressed message resides within a single continuous nucleotide domain of the gene (Chang et al., 1980; Nakanishi et al., 1980). The amino acid sequence homologies between the three MSH-like segments are strong indications that the present POMC gene evolved at least in part by duplications of an ancestral gene with a single MSH gene primordium. Similar presumptive gene duplications have been detected in other peptide hormone families with similar amino acid sequences, such as vasotocin, oxytocin, and vasopressin, where the similarities were sufficiently pervasive to be recognized from the physiology and more classic peptide isolation studies. However, as was the case for the MSH-related sequences, the homologies found between larger peptide hormones such as prolactin, growth hormone, and placental lactogen were undetected until nucleotide sequencing revealed structural similarities. The duplications differ from the POMC case not only by having introns in the DNA segments, but in actually residing on different chromosomes (see J. Baxter et al., *this volume*). Thus, the retention of the POMC gene as a single continuous genetic unit could be construed as prima facie evidence that the multiple independent gene products serve some interactive functional role, perhaps as self-modifiers. Figure 1 shows that even the minor amino acid discrepancies between the three heptapeptide segments can be accounted for by single nucleotide substitutions (Gly for Glu or Asp). Obviously, only direct testing of physiological properties can determine whether these minor peptide changes, in a sequence that is otherwise highly preserved, are "silent" substitutions (Jukes, 1980) or have placed important functional changes in the nature of the agonist properties of the cleaved peptide product. It should perhaps be noted that none of these conceptual advances would have occurred if only "known" products of the gene were being sought, and that the chances of finding a γ-MSH by conventional peptide purification methods would be not only time-consuming but probably futile.

DO THESE RESULTS SUGGEST A GENERAL STRATEGY?

The discovery of an apparently novel neuropeptide wholly through the nucleotide sequence of the common precursor represents an entirely new strategy for the discovery of such regulatory molecules. Rather than processing tissue extracts for conventional purifications based on bioassay potencies and the now routine methods of protein purification, isolation, and eventual sequencing, the process here began with trial peptide syntheses based on the sequence inferred from the nucleotide sequences resulting from analysis of the cloned genetic insert derived from the mRNA (see Ling et al., 1979). From these presumptive peptides, immune sera lead to assays capable of establishing the independent existence of the new peptide *in vivo*. This part of the new discovery process has general utility (see R. Lerner, *this volume*; Sutcliffe et al., 1980; Walter et al., 1980) as a means of proceeding rapidly from a previously unknown DNA sequence to studies of structure and function of the cells where this gene is expressed. In this case, the new peptide shows biological effects different from the previously known processed parts of the precursor molecule.

If this process were to be considered as a general strategy, two levels of prediction can be conceived. First, the recently recognized larger forms of somatostatin (Schally et al., 1980; Brazeau et al., 1981; Mandarino et al., 1981; also see Habener et al.; S. Reichlin, *this volume*), vasopressin (Schmale and Richter, 1981; H. Gainer, *this volume*), and of the adrenal medullary-brain polyenkephalins (Kilpatrick et al., 1981; also see S. Udenfriend; E. Herbert et al.; D. Krieger, *this volume*) may also be viewed as possible sources for other novel transmitter substances with different physiological profiles of action. Second, mRNAs of brain may be viewed as a fertile but enigmatic source of new transmitter substances and their entourages (such as synthetic or catabolic enzymes, or receptors). The enormous difficulties of identifying such "needles-in-haystacks" are obvious, yet the potential yield may make this effort valuable.

To evaluate the feasibility of this approach, clones derived from an abundant mRNA in the embryonic chicken brain (Milner et al., 1980) and the adult rat brain were studied. The embryonic chicken brain contains several predominant mRNA species, of which the most abundant code for tubulin and actin. Studies have focused on the third most abundant species, which codes for a previously unknown protein with molecular weight around 36 Kd. Using recombinant DNA techniques, as well as DNA sequencing methods, a partial amino acid sequence of this protein has been obtained, sufficient to determine its identity and bypass the standard biochemical methods of protein purification and classic amino acid sequencing. The cloned plasmids (generously provided by D. W. Cleveland, University of California, San Francisco) used in these studies consisted of double-stranded DNA produced from the mRNA coding for the 36,000-dalton protein and inserted into the Hind III site of the plasmid PBR322. The complete nucleotide sequence of the cDNA insert of one such plasmid was determined by partial chemical degradation procedure (Maxam and Gilbert, 1977). Translation of the nucleotide sequence into its appropriate amino acid sequence using the genetic code showed that one strand of the insert gave a continuous reading frame of 675 bases, equivalent to 225 residues of amino acid sequence. Comparison of this amino acid sequence with known sequences indicated that there was a very high homology with the enzyme glyceraldehyde-3-phosphate dehydrogenase (GAPDH). The nucleic acid sequence contained in the chick brain plasmid contains coding information equivalent to amino acid residues 2 to 226 of pig GAPDH. The predominant mRNA species in chick brain from which this clone was derived can therefore be unequivocally identified as coding for the glycolytic enzyme GAPDH. Experiments are now in progress, using the cloned plasmid and antibodies against the purified protein, to determine the localization and possible functional role of this enzyme in the developing chick.

OPPORTUNITIES ABOUND

Before closing, two other applications of molecular genetics methodologies seem so close to being realized that I will offer them as additional examples of the benefits of our interdisciplinary discourse. Readers of this volume should study the chapters by D. Housman and by J. Gusella and J. Martin: the latter to realize the profoundly serious nature of inheritable degenerative disorders of the nervous system for which there are yet no treatments, and the former to envision how chromosomal mapping studies that derive directly from the techniques of somatic cell genetics (see F. Ruddle; V. McKusick; *this volume*) may in fact lead to methods by which prospective inheritors of these degenerative disorders may be able to learn whether a particular embryo has in fact a high likelihood of expressing the gene. These methods of chromosomal mapping have so far revealed very few locations for proteins specific to neuronal function. However, there may be enough mapped cases of peptide hormone genes to predict that, when related hormones such as growth hormone, prolactin, and placental lactogen (see J. Baxter et al., *this volume*) occur on separated chromosomes, this separation may imply independent cell regulatory functions for the hormones that have been segregated to different gene loci after duplication to secure independent expression controls.

These considerations lead me to one final admittedly speculative thought, which comes from trying to keep in my head at the same time the enormously complex bodies of evidence in the chapters by L. Hood, T. Maniatis et al., and P. Patterson. Together, these bodies of data suggest some possible general scenarios for which we may want to probe the developmental history of neurons. The work described by Maniatis elegantly defines the relationship within sequential regions of the globin genes, between segments coding for the globin molecules i.e., those that are needed at the three major phases of mammalian development for the proper expression of the right hemoglobin at the right time. In my simplistic view, it is as though the maturing organism is able to cause the expression of different forms of a roughly similar functional system when the organism most needs them. Could this progressive interchange of genetic readout have any bearing on the data reviewed by Patterson, in which the presumptive "growth factor" signals released by different neuronal target cells (e.g., smooth muscles innervated by sympathetic or parasympathetic nerves *in vivo*) are able to induce a "switch" in the phenotypic expression of a sympathetic neuron from adrenergic (making all the enzymes and expressing all the reuptake systems appropriate for an adrenergic neurons) to cholinergic (and then expressing instead the neuronal enzymes and uptake systems appropriate to this chemical class of cell)? Is it conceivable that the switch is a result of the progressive genomic readout of adjacent gene segments, such that maturation signals from the environment can activate one series of enzymes, markers, and other proteins and suppress the other? Perhaps these related "one-or-the-other" systems may be coupled like those described by M. Ptashne *(this volume)* in phage λ.

Could we begin to suppose that the families of peptide hormones that operate in brain, gut, and endocrine systems may share specific, functionally related genomic locations, in which a differentiating cell is able to have the transmitter (or hormone, if you prefer) selected from a family set (e.g., dopamine, norepinephrine, and epinephrine with or without a coexisting peptide), which is able best to elicit responses from the target cells to which it is about to find itself connected? If sequential genome readout could conceivably function in selecting transmitter phenotype, then perhaps similar controls could be envisioned for some other major definable properties of neurons that are also considered to be regulated genetically: cell location, orientation, size, cytoskeletal shape, cell connections and their stabilization, ionic properties, receptive response mechanisms, and surface markers. Of course the sequential readout would not necessarily require adjacent genomic sequences. Progressive modifications could occur through switches, through a shuffling reorganization of DNA segments, through processing of RNA transcripts, or through posttranslational processing of specific proteins or their enzymatic products. The lessons of modern molecular genetics so far suggest all these mechanisms are likely to be operative. I personally am very eager to follow the data that must be close to realization, in which these methodologies may yield more rapid insights into the functional regulation of the brain.

ACKNOWLEDGMENTS

Supported by USPHS grants DA 01785 and AA 03504, and grants from the Sun Company and McNeil Laboratories. I thank Dr. Robert Milner for advice and Mrs. Nancy Callahan for manuscript preparation.

REFERENCES

Anderson, D. J., and Blobel, G. (1980): In vitro synthesis and membrane integration of the subunits of torpedo acetylcholine receptor. *Soc. Neurosci. Abstr.*, 6:209.
Baetge, E. E., Kaplan, B. B., Reis, D. J., and Joh, T. H. (1981): Translation of tyrosine hydroxylase from poly (A) mRNA in phenochromocytoma cells is enhanced by dexamethasone. *Proc. Natl. Acad. Sci. USA*, 78:1269–1273.
Bloom, F. E., Battenberg, E. L. F., Rossier, J., Ling, N., and Guillemin, R. (1978): Neurons containing β-endorphin in rat brain exist separately from those containing enkephalin: Immunocytochemical studies. *Proc. Natl. Acad. Sci. USA*, 75:1591–1595.
Bloom, F. E., Battenberg, E. L. F., Shibasaki, T., Benoit, R., Ling, N., and Guillemin, R. (1980): Localization of γ-melanocyte stimulating hormone (γMSH) immunoreactivity in rat brain and pituitary. *Reg. Peptides*, 1:205–222.
Brazeau, P., Ling, N., Esch, F., Bohlen, P., Benoit, R., and Guillemin, R. (1981): High biological activity of the synthetic replicates of somatostatin-28 and somatostatin-25. *Reg. Peptides*, 1:255–264.
Chang, A. C. Y., Cochet, M., and Cohen, S. N. (1980): Structural organization of human genomic DNA encoding the peptide. *Proc. Natl. Acad. Sci. USA*, 77:4890–4894.
Chretien, M., Benjannet, S., Gossard, F., Gianoulakis, C., Crine, P., Lis, M., and Seidah, N. G. (1979): From β-lipotropin to β-endorphin and 'proopiomelanocortin.' *Can. J. Biochem.*, 57:1111–1121.
Colman, P. D., Kaplan, B. B., Osterburg, H. H., and Finch, C. E. (1980): Brain poly (A) RNA during aging: Stability of yield and sequence complexity in two rat strains. *J. Neurochem.*, 34:335–345.
Crick, F. (1979): Split genes and RNA splicing. *Science*, 204:264–271.
Dahl, G., Azarnia, R., and Werner, R. (1981): Induction of cell-cell channel formation by mRNA. *Nature*, 289:683–685.
DeLarco, J., and Guroff, G. (1972): The synthesis of RNA which binds to oligo(dT)-cellulose by brain *in vivo* and *in vitro*. *Biochem. Biophys. Res. Commun.*, 49:1233–1237.

Goodwin, R. H., Lund, P. K., Barnett, F. H., and Habener, J. F. (1981): Intestinal pre-prosomatostatin: Identification of mRNA coding for a precursor by cell-free translational and hybridization with a cloned islet cDNA. *J. Biol. Chem.*, 256:1499–1501.

Gozes, I., de Baetselier, A., and Littauer, U. Z. (1980): Translation *in vitro* of rat brain mRNA coding for a variety of tubulin forms. *Eur. J. Biochem.*, 103:13–20.

Gozes, I., Schmitt, H., and Littauer, U. Z. (1975): Translation *in vitro* of rat brain messenger RNA coding for tubulin and actin. *Proc. Natl. Acad. Sci. USA*, 72:701–705.

Hahn, W. E., Van Ness, J., and Maxwell, J. H. (1978): Complex population of non-polyadenylated messenger RNA in mouse brain. *Cell*, 18:1341–1349.

Henriksen, S. J., Benabid, A. L., Madamba, S., Bloom, F. E., and Ling, N. (1980): γ3-melanocyte stimulating hormone (γ3-MSH): Electrographic, thermoregulatory and behavioral effects. *Soc. Neurosci. Abstr.*, 6:681.

Hudson, P., Penschow, J., Shien, J., Ryan, G., Niall, H., and Coghlan, J. (1981): Hybridization histochemistry: Use of recombinant DNA as a "homing probe" for tissue localization of specific mRNA populations. *Endocrinology*, 108:353–359.

Itakura, K., Hirose, T., Crea, R., Riggs, A. D., Heyneker, H. L., Bolivar, F., and Boyar, H. W. (1977): Expression in *E. coli* of a chemically synthesized gene for the hormone somatostatin. *Science*, 198:1056–1063.

Jukes, T. H. (1980): Silent nucleotide substitutions and the molecular evolutionary clock. *Science*, 210:973–978.

Kilpatrick, D. L., Taniguchi, T., Jones, B. N., Stern, A. S., Shively, J. E., Hullihan, J., Kimura, S., Stein, S., and Udenfriend, S. (1981): A highly potent 3200 Dalton adrenal opioid peptide that contains both a [Met] and [Leu] enkephalin sequence. *Proc. Natl. Acad. Sci. USA*, 78:3265–3268.

Ling, N., Ying, S., Minick, S., and Guillemin, R. (1979): Synthesis and biological activity of four γ-melanotropins derived from cryptic region of the adrenocorticotropin/β-lipotropin precursor. *Life Sci.*, 25:1773–1780.

Mandarino, L., Stenner, D., Blanchard, W., Nissen, S., Gerich, J., Ling, N., Brazeau, P., Bohlen, P., Esch, F., and Guillemin, R. (1981): Selective effects of somatostatin-14, -25, and -28 on in vitro insulin and glucagon secretion. *Nature*, 291:76–77.

Maxam, A. M., and Gilbert, W. (1977): A new method for sequencing DNA. *Proc. Natl. Acad. Sci. USA*, 74:560–564.

McGinty, J. F., and Bloom, F. E. (1980): Double immunocytochemical labeling demonstrates distinctions among opioid peptidergic neurons. *Soc. Neurosci. Abstr.*, 6:354.

Mendez, B., Vanenzuela, P., Martial, J. A., and Baxter, J. D. (1980): Cell free synthesis of acetylcholine receptor polypeptides. *Science*, 209:695–697.

Mevarech, M., Noyes, B. E., and Agarwal, K. L. (1979): Detection of gastrin-specific mRNA using oligodeoxynucleotide probes of defined sequence. *J. Biol. Chem.*, 254:7472–7475.

Milner, R. J., Sutcliffe, J. G., Shinnick, T., Ray, J., Lerner, R. A., and Bloom, F. E. (1980): Nucleic acid sequence of a predominant mRNA from chick embryo brain. *Soc. Neurosci. Abstr.*, 6:136.

Minty, A. J., Caravatti, M., Benoit, R., Cohen, A., Daugas, P., Weydest, A., Grag, F., and Buckingham, M. E. (1981): Mouse actin mRNAs construction and characterization of a recombinant plasmid molecule containing a complementary DNA transcript of mouse α-actin mRNA. *J. Biol. Chem.*, 256:1008–1014.

Morrison, M. R., Brodeur, R., Pardue, S., Baskin, F., Hall, C. L., and Rosenberg, R. N. (1979): Differences in the distribution of Poly (A) size classes in individual messenger RNA's from neuroblastoma cells. *J. Biol. Chem.*, 254:7675–7683.

Morrison, M. R., Pardue, S., and Griffin, W. S. T. (1981): Developmental alterations in the levels of translationally active messenger RNA's in the postnatal rat cerebellum. *J. Biol. Chem.*, 256:3550–3556.

Nakanishi, S., Inoue, A., Kita, T., Nakamura, M., Chang, A. C. Y., Cohen, S. N., and Numa, S. (1979): Nucleotide sequence of cloned cDNA for bovine corticotropin β-lipotropin precursor. *Nature*, 278:423–428.

Nakanishi, S., Teranishi, Y., Noda, M., Natake, M., Watanabe, Y., Kakidani, H., Jingami, H., and Numa, S. (1980): Construction of bacterial plasmids that contain the nucleotide sequence for bovine corticotropin-β-lipotropin precursor. *Nature*, 287:752–755.

Owerbach, D., Rutter, W. J., Martial, J. A., Baxter, J. D., and Shows, T. B. (1980): Genes for growth hormone, chronic somatomammotropin, and growth hormone-like gene on chromosome 17 in humans. *Science*, 209:289–292.

Roberts, J. L., Seeburg, P. H.,, Shine, J., Herbert, E., Baxter, J. D., and Goodman, H. M. (1979): Corticotropin and β-endorphin: Construction and analysis of recombinant DNA complementary to mRNA for the common precursor. *Proc. Natl. Acad. Sci. USA,* 76:2153–2157.

Rubinstein, M., Stein, S., and Udenfriend, E. (1978): Characterization of pro-opiocortin, a precursor to opioid peptides and corticotropin. *Proc. Natl. Acad. Sci. USA*, 75:669–671.

Schally, A. V., Huang, W.-Y., Chang, R. C. C., Arimura, A., Redding, T. W., Miller, R. P., Hunkapiller, M. W., and Hood, L. E. (1980): Isolation and structure of prosomatostatin: A putative somatostatin precursor from pig hypothalamus. *Proc. Natl. Acad. Sci. USA*, 77:4485–4493.

Schmale, H., and Richter, D. (1981): Immunological identification of a common precursor to arginine vasopressin and neurophysin II synthesized by in vitro translation of bovine hypothalamic mRNA. *Proc. Natl. Acad. Sci. USA*, 78:766–769.

Sharp, P. A. (1981): Speculations on RNA splicing. *Cell*, 23:643–646.

Shibasaki, T., Ling, N., and Guillemin, R. (1980): Pituitary immunoreactive γ-melanotropins are glycosylated oligopeptides. *Nature*, 285:416–417.
Shibasaki, T., Ling, N., Guillemin, R., Silver, M., and Bloom, F. E. (1981): The regional distribution of γ3-melanotropin-like peptides in bovine brain is correlated with adrenocorticotropin immunoreactivity but not with β-endorphin. *Reg. Peptides*, 2:43–52.
Sumikawa, K., Houghton, M., Emtage, J. S., Richards, B. M., and Barnard, E. A. (1981): Active multi-subunit ACh receptor assembled by translation of heterologous mRNA in Xenopus oocytes. *Nature*, 292:862–864.
Sutcliffe, J. G., Shinnick, T. M., Green, N., Liu, F.-T., Niman, H. L., and Lerner, R. A. (1980): Chemical synthesis of a poly-peptide predicted from nucleotide sequence allows detection of a new retroviral gene product. *Nature*, 287:801–805.
Valenzuela, P., Quiroga, M., Zalchiver, J., Rutter, W. J., Kirschner, M. W., and Cleveland, D. W. (1981): Nucleotide and corresponding amino acid sequences encoded by α and β-tubulin on RNA's. *Nature*, 289:650–655.
Walter, G., Scheidtmann, K.-H., Carbone, A., Laudano, A. P., and Doolittle, F. (1980): Antibodies specific for the carboxy- and amino-terminal regions of simian virus 40 large tumor antigen. *Proc. Natl. Acad. Sci. USA*, 77:5197–5200.
Watson, S. J., Richard, C. W., III, and Barchas, J. D. (1978): Adrenocorticotropin in rat brain: Immunocytochemical localization in cells and axons. *Science*, 200:1180–1182.
Zakarian, S., and Smythe, D. (1979): Distribution of active and inactive forms of endorphins in rat pituitary and brain. *Proc. Natl. Acad. Sci. USA*, 76:5972–5976.
Zomzely-Neurath, C., York, C., and Moore, B. W. (1972): Synthesis of a brain-specific protein (S100 protein) in a homologous cell-free system programmed with cerebral polysomal messenger RNA. *Proc. Natl. Acad. Sci. USA*, 69:2326–2330.

Abbreviations

A	adenine
AAf	acetoxyacetylaminofluorine
ACh	acetylcholine
AChR	acetylcholine receptor
ACTH	adrenocorticotrophic hormone, adrenocorticotropin, corticotropin
ad 2	adenovirus type 2
AMP	adenosine monophosphate
cAMP	adenosine 3′,5′-cyclic monophosphate
ATP	adenosine triphosphate
ATPase	adenosine triphosphatase
AVP	arginine vasopressin
Bt_2-cAMP	dibutyryl cAMP
C	cytosine
CA	catecholamine
CCA	1-methylcyclohexane carboxylic acid
CD	circular dichroism
CLIP	corticotropin-like intermediate lobe peptide
CNA	chlornaltrexamine
CNS	central nervous system
Con A	concanavalin A
CRF	corticotropin-releasing factor
d	2′-deoxyribo
DA	dopamine
DADLE	D-Ala^2-D-Leu^5-enkephalin
DBH	dopamine-β-hydroxylase
DDC	diethyl dithiocarbamate
DEAE	diethylaminoethyl
DFP	diisopropyl fluorophosphate
DHT	dihydrotestosterone
DMSO	dimethyl sulfoxide
DNA	deoxyribonucleic acid
cDNA	complementary DNA
gDNA	genomic DNA
DPA	sodium dipropylacetate
ECP	enkephalin-containing peptide
EDTA	ethylenediaminetetraacetic acid
EGF	epidermal growth factor
EGTA	ethyleneglycol-bis-(β-aminoethyl ether)-N, N′-tetraacetic acid
EIP	ecdysteroid-inducible polypeptide
E_2	estradiol
Fab	antigen-binding fragment
Fc	crystallizable fragment
FCCP	carboxycyanide p-tri-fluoromethoxyphenylhydrazone
fMET	formylmethionine
G	guanine
GABA	γ-amino-n-butyric acid
GAPDH	glyceraldehyde-3-phosphate dehydrogenase
GH	growth hormone
Glc	glucose
GlcNAc	N-acetylglucosamine
GTP	guanosine triphosphate
G6PD	glucose-6-phosphate dehydrogenase

HAT	hypoxanthine, aminopterin, and thymidine
20-HE	20-hydroxyecdysone
HPLC	high-performance or high-pressure liquid chromatography
HRP	horseradish peroxidase
5-HT	serotonin, 5-hydroxytryptamine
IC_{50}	concentration for 50% inhibition
LHRH	leuteinizing hormone-releasing hormone
LPH	lipotropin
LVP	lysine vasopressin
Man	mannose
MBP	myelin basic protein
MIF	MSH release-inhibiting factor
MSH	melanocyte-stimulating hormone; melanotropin
NAD	nicotinamide adenine dinucleotide
NAS	neuroactive substance
NE	norepinephrine
NGF	nerve growth factor
NMR	nuclear magnetic resonance
Np	neurophysin
OT	oxytocin
PCMB	p-chloromercuribenzoate
pI	isoelectric point
PL	placental lactogen
POMC	pro-opiomelanocortin
Prl	prolactin
PTH	phenylthiohydantoin
RER	rough endoplasmic reticulum
RNA	ribonucleic acid
hnRNA	heterogeneous nuclear RNA
mRNA	messenger RNA
tRNA	transfer RNA
Rot	conc. of RNA in moles of nucleotide per liter X time (sec)
SDS	sodium dodecyl sulfate
SME	stalk median eminence
SRP	signal recognition protein
T	testosterone or thymine
T_3	triiodothyronine
TAT	tyrosine aminotransferase
TH	tyrosine hydroxylase
TK	thymidine kinase
TLCK	Nα-p-tosyl-L-lysine chloromethyl ketone
TPCK	L-1-tosylamide-2-phenylethyl chloromethyl ketone
TRH	thyrotropin-releasing hormone
	thyroid stimulating hormone-releasing hormone
U	uracil
UV	ultraviolet
VSV	vesicular stomatitis virus

Amino acid symbols

Amino acid	Three letter symbol	One letter symbol
Alanine	Ala	A
Arginine	Arg	R
Asparagine	Asn	N
Aspartic acid	Asp	D
Asn + Asp	Asx	B
Cysteine	Cys	C
Glutamine	Gln	Q
Glutamic acid	Glu	E
Gln + Glu	Glx	Z
Glycine	Gly	G
Histidine	His	H
Isoleucine	Ile	I
Leucine	Leu	L
Lysine	Lys	K
Methionine	Met	M
Phenylalanine	Phe	F
Proline	Pro	P
Serine	Ser	S
Threonine	Thr	T
Tryptophan	Trp	W
Tyrosine	Tyr	Y
Valine	Val	V

The genetic code

First position (5' end)	Second position				Third position (3' end)
	U	C	A	G	
U	Phe	Ser	Tyr	Cys	U
	Phe	Ser	Tyr	Cys	C
	Leu	Ser	Stop	Stop	A
	Leu	Ser	Stop	Trp	G
C	Leu	Pro	His	Arg	U
	Leu	Pro	His	Arg	C
	Leu	Pro	Gln	Arg	A
	Leu	Pro	Gln	Arg	G
A	Ile	Thr	Asn	Ser	U
	Ile	Thr	Asn	Ser	C
	Ile	Thr	Lys	Arg	A
	Met	Thr	Lys	Arg	G
G	Val	Ala	Asp	Gly	U
	Val	Ala	Asp	Gly	C
	Val	Ala	Glu	Gly	A
	Val	Ala	Glu	Gly	G

Given the position of the bases in a codon, it is possible to find the corresponding amino acid. For example, the codon 5' AUG 3' on mRNA specifies methionine, whereas CAU specifies histidine, UAA, UAG, and UGA are termination signals. AUG is part of the initiation signal, in addition to coding for internal methionines.

Subject Index

Abbreviations, list of, 477–478
Abundance class, RNA, 336–341, 343–346
Acetoxyacetylaminofluorine (AAF), 17
Acetylcholine, 2
　changes in HD, 411–412
　coexistent peptides, 3, 4
　and opiates, 251–253
Acetylcholine receptor (AChR)
　cotranslational integration into RER, 381–382
　glycosylation, 378–380
　in vitro, 373–378
　precursor, 376
　mRNAs, 376–377
　and SRP, 381–382
　subunits, 375
　synthesis, 373–385
　translation products, 375–376
　transmembrane orientation, 378–380
　trypsinization, 378
Acetylcholinesterase (AChE), 279
N-Acetylgalactosamine, 166
N-Acetylglucosamine, 162, 164
ACTH
　antiserum, 222
　in brain and pituitary, 208
　inhibition by glucocorticoids, 232
　location on POMC gene, 231
　precursor. *See* Pro-opiomelanocortin
　processing, 213, 220
　products, 213
ACTH-producing tumors, 213
Actin, 343, 344
Actinomycin D, 269
Adenosine triphosphatase (ATPase), 181
Adenosine triphosphate (ATP), 181
Adenovirus type 2 (ad-2)
　genome, 25
　in vitro transcription, 51–52
　polyadenylation, 30
　mRNAs, 49
　splicing, 28–30, 56–58
　transcription, 26–28, 48–49, 53–55
　transcriptional control, 30–34
Adrenalectomy, 204
Adrenal medulla, 209
　opioid peptides, 239, 241–246
Adrenergic neurons, 437–442, 460
Adrenocorticotropic hormone. *See* ACTH
Adrenoleukodystrophy gene, 403
Alanine aminotransferase, 313
D-Alanine, 257
β-Alanyl histidine, 189
Albino mutations, 397
Albumin, 125

Aldosterone secretion, 232
Alpha cell, 149, 156
Alternate pairwise cooperativity, 41, 44
Alu gene family, 94
Alzheimer's disease, 361
Amacrine cells, 363
Amino acids; *See also* Peptides; Proteins
　for oligonucleotide synthesis, 463
　sequencing, 154
　symbols, 479
α-Aminoisobutyric acid, 310
Aminopeptidase, 190–193
Amino terminal domains. *See* N-terminal domain
Amygdala, 226, 362
Analgesia, 470
Androgen, 272, 289, 290, 297–299
Androgen receptors
　in androgen-resistant mutants, 297–299
　in brain, 266, 289–306
　and estrogen receptors, 294, 301–302
Androgen-resistant mutants, 290, 297–299
Anglerfish, 177, 350–355
Anisomycin, 269, 271
Anterior pituitary lobe, 225–227. *See also* Pituitary Pituitary
Antibodies; *See also* Monoclonal antibodies
　antipeptides, 119–125
　binding, and cellobiose, 128
　cell-specific, analysis of retina with, 137–146
　to chemically synthesized peptides, 119–125
　diversity, 79–82
　domains, 75–76
　to Z-DNA, 19
　to dynorphin, 258–260
　to enkephalin, 239
　genes, 75–85
　to NGF, 425
　to POMC domains, 221
　specificity, enhancement of, 131–133
　structure of, 75–76
　synthesis and use as tools, 117–146
Anti conformation, DNA, 14–15
Antigens, 75, 76, 82
Antimycin, 151
Antipain, 433
Antipeptide antibodies, 119–125
Antisera, limitations, 127
Antitoxin, 133
Area-code gene families, 82–83
Arginosuccinate synthetase, 66
Aromatization, 290
Asparagine, 162
Aspartic acid, 162
Assays, for globin gene expression, 95–97

ATA box, 90-91
Atropine, 253
AtT-20 cells, 210-213, 221-225, 232
Autonomic nervous system, 362
Auxotrophic donor cells, 63
Axonal transport
  and precursor processing, 171
  somatostatin, 360, 363-364
p-Azophenylarsonate, 129

Bacteriophage; See also specific types
  in construction of gene libraries, 89-90
  molecular switch, 37-45
Bacteriorhodopsin, 384
BAM peptides, 239-240, 246
Base stacking, DNA, 15-16, 18
B cells, 79-82, 129
Behavioral effects, brain steroid hormones, 265-275
Beta cell, 149, 150, 156
Biochemical reagents, synthetic antibodies as, 124
Biotin, 466
Birds, brain sexual differentiation, 266
Blood pressure, 204-205
Body temperature, 470
Bone marrow, 89
Bovine brain, 191, 214-215, 239-240, 246
Brain
  amount of genetic information used by, 324
  androgen and estrogen receptors in, 289-306
  circuitry, 1-2
  disorders of, and precursor processing, 158
  DNA methylation in, 21
  mouse, 323-334
  neuroactive substances in, 2-4
  neuropeptides in, 208-209, 349-358
  number of polypeptides in, 330
  number of structural gene loci expressed in, 401
  opioid peptides in, 207-209, 213, 238, 242, 249, 258-260
  poly(A)⁻ mRNA, 329-331
  mRNA in, 323-334
  mRNA complexity in, 342-343
  sequence complexity, 329
  sexual differentiation in, 266, 289
  somatostatin in, 359-372
  steroid hormone action on, 265-275
Brattleboro rats, 173
α-Bungarotoxin, 380, 382

Calcitonin, 349-358
Calcium
  ionophores, 448-449, 451-452
  nuclear, 453
  regulation of Ca^{2+}-ATPase, 445-446, 448
  regulation of gene expression, 445-456
  and somatostatin release, 364
  and steroid receptor action, 291
  transport, 455
Calcium-phosphate-precipitated DNA, 67
Calmodulin, 453
cAMP
  dibutyryl, 336

induction of TAT, 310
and neurite outgrowth, 432
phosphodiesterase, 310
and progesterone, 291
Cap site
  consensus sequences around, 56-58
  in globin genes, 96
  in insulin genes, 105
  splicing and structure, 27-28
Carbohydrate metabolism, 235-236
Carboxyl terminal domain. See C-terminal domain
Carboxypeptidases, 157, 158, 190, 239
Carboxypeptidase-B-like enzymes, 177
Carcinogens, 37, 44
Cardiac myosin, 130-133
Carnosine, 189
Carrier-mediated transport, 189
Catecholamines, 253
Catepsin-B, 177, 185
Caudate nucleus, in HD, 408-412
CCA, 336, 343-344, 346
C domains, 75-79
Cell adhesion, 344
Cell death, 407-413
Cell differentiation, and abundance classes, 342
Cell division, 344
Cell fusion techniques, 64-66
Cell membrane
  and HD, 412
  regional components, 141-142
Cellobiose, 128
Cell-specific antibodies, analysis of retina with, 137-146
Cell-surface glycoproteins, 161-169
Cellular effects, brain steroid hormones, 265-275
Cellular transcription assays, 96-97
Cerebral cortex. See Cortex
Cerebellum, 189, 294, 407
C gene segments, 76-79
Charon vectors, 89
Chemical circuitry, 1-2
Chicken
  brain, 471-472
  insulin genes, 104
  muscle, 446-449
Chinese hamster ovary cells, 162, 167, 168
Chlornaltrexamine (CNA), 257
Chloroquine, 185
Cholecystokinin, 3, 357, 411
Choline acetyltransferase, 463
Cholinergic factor, 439
Cholinergic neurons, 438-442, 460
Chymotrypsin, 158
Cholinergic transmission, 272-273
Chorionic somatommotropin gene, 232-237
Chromaffin granules, 181, 239-242
Chromatin, 48, 307-308
Chromosome, *Drosophila*, 277, 279
Chromosome mapping, 403, 417-420
  HD gene, 420-421
  insulin genes, 107
cI gene, 37-39
Circular dichroism (CD) spectra, 16-17

CI-628, 267
Class switch, 80, 82
Clathrin-coated vesicles, 165
Cleavage sites, 123–124
CLIP, 213, 220, 231
Cloned genes, 95–96, 98, 108–109, 145
CNS mutations, mouse, 389–400
Colchicine, 360
Color blindness gene, 403
Combinatorial strategies, antibodies, 81
Concanavalin A, 356, 379
Concerted evolution, 94
Confrontation analysis, 399
Constant domain, 75–79
Cooperativity, alternate pairwise, 41, 44
Cordycepin, 30
Corpus striatum, 190, 191
Cortex, 1–8
  monoclonal antibodies to, 144
  mRNA, 343
  somatostatin in, 362, 364–366
Corticotrophs, 469–470
Corticotropin. See ACTH
Corticotropin-like intermediate lobe peptide (CLIP), 213, 220, 231
Corticotropin-releasing factor (CRF), 220, 224
Cos-7 cells, 96–97
C-peptide, 150, 157
CRF, 220, 224
Critical period, 265
  sexual differentiation, 266, 289, 294–296
Cro gene, 37–39
C-terminal domain
  binding, 40, 42
  glucagon, 357
  peptide removal from, 123
  preprocalcitonin, 357
  role in membrane insertion, 302
CVl cells, 451
Cyanogen bromide, 175, 249
Cycloheximide, 31, 164, 285
Cytoplasmic calcium, 445–457
Cytoplasmic stability, 32
Cytosine, 20–21

DADLE, 257
D cells, 149, 156
DDC, 196
Defeminization, 289–290
Deletion sites; *See also* Introns; Splicing
  globin genes, 91
Deoxyguanosine, 14, 15
Development
  genetic control of, 63–72
  globin expression during, 88–89
Dexamethasone, 204, 237, 310, 315
Dextrorphan, 251, 253
DFP, 185
D gene segment, 76–79, 81
DHT. *See* Dihydrotestosterone
Diabetes, 107, 158
Diethyl dithiocarbamate (DDC), 196
Digitalin, 133
Digitoxin, 133

Digoxin, 133
Dihydrotestosterone (DHT), 268
  and androgens, 289, 294, 301
  and neurite outgrowth, 267
  and sexual differentiation, 266, 294
Dimethyl sulfoxide (DMSO), 336, 343, 344, 345
Diversity gene segments, 76–79, 81
DNA
  base stacking, 15–16, 18
  B-form of, 14–18, 20
  binding, 293, 294, 297
  Charon vectors, 89
  conformations, 14–15
  distinguishing rodent from human, 418–420
  gene transfer, 67–70
  markers, 417
  methylation, 17, 20–21, 105
  microinjection, 70
  molecular organization of, in relation to gene function, 11–22
  probes
    phage, 40–41, 44
    POMC, 228–229
    synthesis, 462–463
  nuclease-sensitive, 48
  repetitive, 236, 418–420
  replication of viral, and transcription, 49
  repressor binding, 38–41
  sequences
    antibodies to chemically synthesized peptides from, 119–125
    chromatin, 48
    differences in, and polymorphic markers, 416
    recombination in gene families, 323–324
  single copy, 325
  steroid receptor interactions, 292–293
  synthesizer, 83
  wild-type, 41
  Z-form of, 13–22
cDNA, 68–70
  to assay transcription rates, 31–34
  to bovine pituitary intermediate lobe precursor, 211
  calcitonin, 356
  EIP, 282–285
  gene mapping with, 65–66
  glucagon, 357
  growth hormone, 313
  hormone genes, 232–234
  hybridization, 324, 325–326, 461–462
  immunoglobulin, 76
  insulin genes, 103
  introns, 327
  neuropeptide, 350
  POMC, 214–216, 225–229, 231–232
  proenkephalin, 246–247
  proinsulin, 154–155
  prolactin, 236–237, 313
  rare-message genes, 84
  mRNA, 111, 330, 336
  S and P cells, 337–342
  single base changes, 97
  somatostatin, 355
  SV40-insulin gene recombinant, 112–113
gDNA

gDNA (contd.)
  hormone genes, 232–235
  introns, 327
  POMC, 215
DNA-cellulose chromatography, 294
DNase, 40
DNA-RNA hybrids, 111
Dolichol linkage, 162
Dominance-recessiveness tests, 313
Dopamine, 2, 3, 5, 6, 189, 411
Dopamine-β-hydroxylase (DBH), 427, 430
Drosophila, 52
  ecdysteroid effects on, 279
  polytene chromosome, 19, 21, 277
  steroid control of gene expression in, 277
Duchenne's muscular dystrophy, 403, 407
Dynorphin, 209
  amino acid sequence, 249
  biological activity, 244, 246, 249, 253
  implications of gene duplication to, 249–262
  localization, 249, 258–260
  molecular model, 251
  naloxone reversal, 257
  potency, 250, 257
  role of Arg-Arg pair, 256
  selective tolerance to, 257–258
  structure activity studies, 256–257
Dynorphin receptor, 249–262

Early puff, 278, 285
Ecdysteroids
  and cycloheximide, 285
  effect on Drosophila, 277–285
  effect on puffing, 277
Ecdysteroid-induced polypeptides (EIPs), 280–282, 285
Ecdysteroid receptors, 279
Eco RI, 89–90
Edman sequenator, 154
EDTA, 185
EGF-BP, 428
Embryonic development, hemoglobin, 88
Embryonic microinjection, 70, 71
Endorphin, 167
  in brain and pituitary, 208
  action of enkephalinase on, 191–192
  historical aspects, 203–218
  potencies, 207
  precursor, 209
  structure, 204
α-neo-Endorphin, 254, 260
β-Endorphin, 185, 232
  in adrenal medulla, 241
  biological activity, 204, 244
  location on POMC gene, 231
  physiological properties, 470
  potency, 250, 253
  precursor. See Pro-opiomelanocortin
  processing, 220
Enkephalin; See also Leu-enkephalin; Metenkephalin
  in bovine adrenal medulla, 239–240
  in brain and pituitary, 208
  historical aspects, 203–218
  and norepinephrine, 3
  precursor, 209
  structure, 204
Enkephalinase, 190
Enkephalin-containing polypeptides (ECPs), 239, 241–242, 247
Enkephalin endopeptidase, 190–193
Enolase, 335, 346
env gene, 120, 121, 123
Enzyme induction, 433
Epidermal growth factor (EGF), 428
Epileptiform effects, 470
Epitope, 127, 133
Erythropoiesis, 88–89
Escherichia coli, 23, 50, 292, 307
Estradiol (E_2)
  and androgens, 289, 294
  and female sexual behavior, 266, 268–271
  and neurite outgrowth, 266–268
  and prolactin, 315
  and serotonin, 273
  and sexual differentiation, 290
Estradiol receptor, 291, 301
Estrogen, 269, 271, 272–273, 307
Estrogen receptor, 301
  in brain, 266, 289–306
  control of cholinergic pathways, 273
  differentiated from androgen, 294
  and neurite outgrowth, 267–268
  and Tfm mutants, 299
Ethanol, 17
Eukaryotes
  DNA, 17, 19, 20–21
  gene families, 82–83
  genetic control of development, 63–72
  glycosylation, 162
  regulation of gene expression in, 25–36, 47–59
  RNA, 29–30, 48–56
  transcription, 52–56
Evolution
  androgen-estrogen receptor system, 301–302
  concerted, 94
  hormones, 234–236, 349
  human and mice linkage homology, 403
  insulin genes, 104–107
  neurotransmitters, 349
  POMC gene, 470–471
  somatostatin, 352
Exocytosis, 277
Exonuclease VII technique, 111

Fatty acylation, 167
Fetal development, hemoglobin, 88
α-Fetoprotein, 294
Fibroblasts, 66, 451
5' terminal, 28, 56–58. See also Cap site
Fluorescence-activated cell sorter, 156
Fluorescent hybridization probes, 466
14-3-2 protein, 335
Framework residues, 76
Frederick's ataxia, 407
Fucose residues, 165

GABA, 2, 365–366, 411–412
Galactose, 165

SUBJECT INDEX

β-Galactosidase gene, 232
Ganglion cells, 144–145
Gas-phase solid-state microprotein sequenator, 83
Gastric inhibitory peptide, 357
Gastrin, 368–369, 370
Gastrointestinal tract, 209. See also Intestine
Gene control
   of development, and somatic cell genetics and gene transfer techniques, 63–72
   in bacteriophage, 37–45
Gene conversion, 94
Gene deletions, 95
Gene diversification, 349
Gene dosage effects, 390, 394, 396–397
Gene duplication, 91–94, 105, 249–262, 349, 470–471
Gene expression
   and antibodies to chemically synthesized proteins, 119–125
   control of, 23–59
   eukaryote, 25–36
   growth hormone, 232–237
   hormone, 231–238
   human globin, 87–101
   insulin, 103–115
   neuroactive substances, 7
   POMC, 219–230, 231–232
   regulation of, 25–36, 47–59
   regulation during terminal neurogenesis, 335–347
   role of calcium in, 445–456
   steroid control of, 238–320, 265, 272–273, 277–287
   in variant and hybrid cells, 307–320
Gene families
   Alu, 94
   area-code, 82–83
   coding for antibodies, 76
   DNA sequence recombination in, 323–324
   human globin, 87
   split gene-multigene, 82–83
Gene function, molecular organization of DNA and, 11–22
Gene libraries, 89–90, 97
Gene mapping, 64–66, 403
Gene marker, prototrophic, 67–70
Gene matching, 93
Gene position, selective pressure on, 403
Gene regulation
   growth hormones, 237
   and DNA, 20–21
Genes; See also Structural genes; *specific genes*
   ecdysteroid-responsive, 279–285
   homology, 234–235
   neuronal, 403
   overlapping, 120
   primary responsive, 278, 279, 285
   rare-message, 82, 83–84, 465
   recombination selection for cloned, 98
   repetitive DNA in, 236
   split, 327
Gene segments, 76–79
Genetic Cell Depository, 107
Genetic code, 119, 479
Genetic disorders, 94–95, 401–405
Gene transfer techniques, 63–72, 403
Gene transformation frequencies, 70
GH_3 cells, 315–318
Giant polytene chromosome, 19, 21, 277
GL hybrids, 315–318
Glial cells, 139, 198, 459–462
Globin genes
   expression, 88–89, 94–97
   gene clusters, 91, 93–94
   human, mouse, and rabbit, 90
   mapping and isolation, 89
   mutant, 97–99
   pseudogenes, 91–93, 105
   in SV40 vectors, 96
   structure, 90–91
α-globin, 87–95
β-globin, 21, 29, 87–95, 98, 417
γ-globin, 88–89, 94, 95
δ-globin, 88, 89
Globoside, 441
Glucagon, 149, 155, 349–358, 370
Glucocorticoid
   and cholinergic factor, 440
   control in variants and hybrids, 309–318
   effects, 272, 308, 315
   inhibition by, 220, 224, 232
   variants resistant to, 308
Glucocorticoid receptor, 293, 308–309, 312
Glucose, 162, 164–165, 345
Glucose-6-phosphate dehydrogenase, 403
Glucosidases, 164, 165
Glue polypeptides, 277, 278, 285
Glutamic acid, 2
Glutamine synthetase, 308, 315
Glyceraldehyde-3-phosphate dehydrogenase (GAPDH), 472
Glycerol, 294
Glycerol phosphate dehydrogenase, 313
Glycine, 2
Glycoproteins, 161–169, 171, 439, 441–442
   dependence of glycosylation on, 164
   transmembrane, 374
   VSV, 165
Glycosidase, 356
Glycosylation, 162–166, 167, 462
   AChR subunits, 378–380
   POMC, 212, 224
Glycosyl transferase, 165, 166
Goldberg-Hogness box. See TATA box
Golgi apparatus, 165–166
Gonadal steroids, 266
Gonadotropin, 196
G protein, 374
gp 70 protein, 121
Granule cell, 394, 397
Growth factors, 427–430
Growth hormone, 232–237, 312–318, 359, 367
Guanosine triphosphate (GTP), 27
Guinea pig myenteric plexus, 249, 250, 251–253

HD gene, 415–421
Heart cells, 439
Heat shock proteins, 451–452
Heavy chain, 76, 81
HeLa cells, 25, 26, 96, 451, 453

Hemagglutinin, 124–125
Heme, 87
Hemoglobin, 87–88
Hemophilia gene, 403
Hereditary persistence of fetal hemoglobin (HPFH), 95
Herpes simplex virus, 53, 67–88
High performance liquid chromatography (HPLC), 241
Hippocampus, 272, 362
Histidyl-proline diketopiperazine, 198
Histofluorescence technique, 137
Histones, 48, 53, 453
HLA antigens, 380, 382
Hormones; See also Steroid hormones
 control of cholinergic factor, 439
 evolution, 234–236, 349
 localization of action, 265–266
 and neurite outgrowth, 266–268
 structure and expression of, 231–238
Horse hemoglobin, 87–88
HTC cell variants, 309–312
Human
 brain, 1–2, 107, 401–405, 407–413
 chromosome mapping, 403
 genetic disorders, 401–405, 407–413
 genome, 401, 416
 globin genes, 87–101, 107
 hormone genes, 234–238
 insulin genes, 104, 107
 retina, 138
 somatostatin, 367
Human-mouse hybrid cells, 64–65
Huntington's disease, 401, 407–421
Hybrid-arrested translation, 460–461
Hybrid cells, 64–65, 307–320
Hybridization, 225–229, 350, 355–357
 techniques, 127–135, 324, 335, 466
2O-Hydroxyecdysone (2O-HE), 277. See also Ecdysteroids
Hyperglycemia, 369–370
Hypervariable region, 81
Hypophysectomy, 204
Hypophysiotropic system, 359
Hypothalamus; See also Ventromedial hypothalamus
 hormone receptors in, 193–195, 269, 294–296
 and lordosis behavior, 268, 271
 neurite outgrowth in, 267
 POMC in, 226, 470
 and sexual differentiation, 266
 somatostatin in, 360–361, 369
 Tfm, 297
 transport of neurosecretory vesicles, 171–172
Hypoxanthin phosphoribosyltransferase, 403

Imaginal disk cells, 279
Immune response, restricted, 127
Immune system, 137
Immunogens, 119–125
Immunoglobulins
 class switch, 79–81
 combinatorial strategies, 81
 cDNA, 76

gene families, 76, 323–324
heavy chain, 28, 56–58
mouse, 66
mRNA, 76–77
somatic diversification, 81–82
structure, 75
variability, 76, 82
Imprinting behavior, 332
Influenza virus, 119–120, 124–125
Insulin, 369, 428
Insulin gene
 expression, 108–115
 mammalian, 103–115
 mapping, 107–108
 polymorphism, 107
 processing, 105–107, 150–151, 157
Insulinoma, 105
Interferon, 66
Intermediate pituitary lobe. See Pituitary
Intermolt puff, 278
Interneurons, 362, 363
Intervening sequences. See Introns
Intestine
 opioids in, 239, 242, 249, 260
 somatostatin in, 355–356, 369
Introns, 28, 56–58, 327
 globin gene, 90
 hormone gene, 235–236
 insulin gene, 104–105
 POMC gene, 231
 structural genes, 328, 329
Ionomycin, 448–449, 451–452
Islets of Langerhans, 149
Isochizomers, 105
Isotubulin, 335, 343, 344

J gene segment, 76–81
Joining mechanism, 78–79, 81
Junctional diversification, 81

Kainic acid, 42, 301
κ chain, 66, 76, 81
κ receptor, 258
Kc cell line, 279
Kidney, 297, 329
Kinetic hybridization, 325–326

lac gene, 44–45, 91, 292–293
Lactogen, 232–237
λ bacteriophage, 37–45
λ chain gene family, 66, 76, 81
Late puffs, 278
Learning and memory, 207
Lesch-Nyhan syndrome, 403
Leu-enkephalin, 191, 207, 239, 241
 potency, 250
 receptor, 257, 258
Leukemia virus, 124
Leukocyte interferon, 60
Leupeptin, 185, 433
Levorphanol, 250, 253
LHRH, 193–197
Light chain, 81

Linkage analysis, 403, 415
β-Lipotropin. *See* β-LPH
Liver, 329
Local circuits, 1
Locus coeruleus, 5
Lordosis behavior, 268-272
Lordosis quotient (LQ), 268
β-LPH, 185, 192-193
   and ACTH precursor, 210-211
   and endorphin, 204
   location on POMC gene, 231
   precursor. *See* Pro-opiomelanocortin
   processing, 213, 220
   products, 213
   structure, 207
γ-LPH, 213, 220
LSP cells, 451
Lucia antiserum, 259, 260
Luteinizing hormone-releasing hormone. *See* LHRH
Lysogen, 37-45
Lysosyme, 125

Macrophage activation, 197
Magnesium, 291
Magnocellular hypothalamic nuclei, 208
   and neuropeptide processing, 171
Mammals
   brain sexual differentiation in, 266
   gene structure and diversification, 72-116
   genome, mRNA encoded in, 323
   globin pseudogenes, 91-93
   insulin gene structure, evolution, and expression, 103-115
Mannose, 162, 164, 165
α-Mannosidase, 165
Masculinization, 289-290
Mass spectrometer, 83
Mast cells, 197
Mastomys natalensis, 427
Median eminence, 193-195, 360
Melanotropin. *See* MSH
Membrane-interacting steroid receptors, 290-291, 308
Membrane permeability
   effect of CCA, 345-346
   and calcium, 449-450
Membrane potential, 2
Mendelian genetics, 63, 64
Mendelizing phenotype catalog, 401
Met-enkephalin
   changes in HD, 411
   concentrations, 207
   and cyanogen bromide, 249
   and enkephalinase, 191
   in proenkephalin, 239, 241
Met-enkephalin receptor, 258
Methylamine, 180
Methylation
   DNA, 17, 20-21
   rat insulin gene, 105-107
1-Methylcyclohexane carboxylic acid (CCA), 336, 343-344, 346
α-Methylmannoside, 379

Metrizamide, 179
Microchemical facility, 83
β-2-Microglobulin, 464
Microinjection gene transfer, 70, 318, 386-390
Microprotein sequenator, 83
Mitochondria, 346
Molarity of transcription, 26, 29
Molecular switch, 37-46
Moloney leukemia virus, 120
Molting hormone. *See* Ecdysteroid
Monkey cell line, 96
Monoamine oxidase, 403
Monoamines, 5
Monoclonal antibodies; *See also* Antibodies
   advantages, 137-138, 145
   in analysis of retina, 137-146
   applications, 127-136
   to calcitonin, 356
   to glia, 139-141
   developmental differences, 140-141, 143-144
   and HD, 412
   to LHRH peptidase, 196-197
   to photoreceptors, 141-144
   uses, 145
Morphine, 190, 250
Mouse
   androgen-resistant, 297-299
   antibodies, 76-77, 81
   BALB/c, 81, 130
   brain, 294-296, 323-334
   embryonic microinjection, 70
   genome, 77, 79, 325
   globin, 29, 58, 91-93
   and humans, linkage homology, 403
   immunoglobulins, 66
   L cells, 315-318
   mutants, 96, 297, 389-400
   and NGF, 427, 428
   sexual differentiation in, 289-290
   tail withdrawal, 190-191
Mouse-human hybrid cells, 64-65
MSH, 198, 231, 469
α-MSH, 197, 213, 220
β-MSH, 207, 213
γ-MSH, 468-469, 470
Müller cells, 139-140
Mullerian-inhibiting substance, 134-135
Multigene, 82-83, 464
μ receptor, 257, 258
Muscle, 446, 454-455
Mutations
   globin genes, 97-99
   glucocorticoid-resistant, 308-312
   hormonally controlled mRNA synthesis in, 307-320
   mouse, 389-400
Myelin, 390-394
Myeloma, 76, 129
Myenteric plexus, 260
Myocardial infarction, 130
Myoglobin, 87, 125

Naloxone, 204, 207, 249, 253
   dissociation constant, 257

SUBJECT INDEX

Naloxone (contd.)
  structure, 251
Nascent chain analysis, 26–27
Neostriatum, 408–412
Nerve growth factor (NGF)
  biosynthesis, 427–430
  mechanism of action, 430–433
  neuromodulation, 425–435
  translocation, 426–427
Nervous system; See also Brain
  genetic disorders, 387–422
  Mendelizing loci, 401
Neural cell lines, 330
Neural crest cells, 437
Neural degenerative disorders, molecular genetic approach to, 415–422
Neurite outgrowth, 197, 266–268, 343, 345, 396–397, 426, 432
Neuroactive substances, 1–6
Neurobiology
  molecular genetic approach to, 455–475
  significance of genetic mapping and gene transfer to, 70–71
  significance of rare-message genes to, 83–84
Neuroblastoma cell lines, 335–347
Neuroendocrinology, 459
Neurofilaments, 343
Neurogenesis, 335–347
Neurohormones, 359
Neuromodulators, 359, 361, 425–435
Neuronal differentiation, 335–347
Neuronal function, 66, 271–273
Neuronal genes, 403
Neuronal loss, 410
Neurons
  differentiating from glial cells, 459–462
  POMC, 231–232
  precursor processing in, 171
  receptor-steroid complex action in, 291
Neuron-specific markers, 335
Neuroparacrine interactions, 6–8
Neuropeptides
  biosynthesis, 349–357
  cleavage sites, 198
  degradation, 189–199
  and monoclonal antibodies, 137
  processing, 171–185
  uptake mechanisms, 189
Neurophysin, 173
Neuroscience, 1–2, 7–8, 21, 66
Neurosecretory vesicles,
  membrane potential, 181
  pH, 177–179, 180–181, 183–185
  and precursor processing, 171–187
Neurotensin, 197
Neurotransmitters, 2
  changes in HD, 410–412
  coexistence with peptides, 3–4
  and hormones, 349
  and NGF, 427
  and opiates, 207
  and somatostatin, 366
  transition, 439–440
Neurotransmitter receptors, 272–273, 465
NGF. See Nerve growth factor

Norepinephrine, 2, 3, 5, 189, 301, 411, 427
Normorphine, 253, 257
N-terminal domain, 39–40, 42, 44, 231–232
Nuclease, 48
Nucleotide sequences, defining eukaryotic mRNA promotors, 48–56
Nucleus, calcium content, 452–453

Obesity, 204
Oligodeoxynucleotide synthesis, 463–464
Oligosaccharide donor, 162–166
Operator-repressor binding, 38–41
Opiate receptors, 207, 210
Opiates, 204–207, 250–251
Opioid peptides
  binding studies, 251–253
  biological activities, 242–246
  evolution, 261
  intraventricular application, 258
  message and address, 253–254
  molecular genetics of, 201–238
Opioid receptors, 249–262
Ovalbumin, 31
Ovomucoid transcript, 58
Oxytocin, 173, 174, 175, 198, 208

Pain, 204
Pancreas, 114, 149, 367, 369
Pancreatic polypeptide, 3, 149, 369
P_0 antibody, 391–392
Paracrine activity, 6–8
Paraventricular system, 360–361
pBR322, 67–68
P cells, 336
PCMB, 185
PC12 cells, 430–433
Pekalosomes, 68
Pepstatin A, 185
Peptidases, 189–198
Peptide E, 246
Peptide F, 241, 244
Peptides; See also Neuropeptides
  chemically synthesized, antibodies to, 119–125
  coexistence with transmitters, 3–4
  cryptic, 354–355, 357
  relation of neural and endocrine, 349
  role of double basic residues in, 254–256
  synthesis, 349–358
Percoll, 179
Periaqueductal gray, 204
Periventricular system, 360–361
p15E, 120–122
Phenylketonuria, 401
Phosphodiesterase, 310, 312
Phosphorylcholine, 81
Photoreceptor cells, 139, 141–144
Picotoxin, 366
Pigmentation mutants, 397–399
Pituitary
  and brain, neuropeptides in, 208–209
  dynorphin in, 249, 260
  endorphin and enkephalin in, 208

gonadotropin secretion, 193-195
neurosecretory vesicle transport to, 171-172
and pain, 204
processing of POMC and its products, 213, 219-220, 224-227, 470
post-proline-cleaving enzyme in, 198
somatostatin, 360
pIX, 31, 33
Placenta, 209, 213
Placental lactogen, 232-237
Plasma cell tumors, 76
Plasmids, 282-285, 350. See also specific types
Plasticity, and steroids, 265
Pneumococcal polysaccharide, 127-128
pol gene, 120
Poly(A) RNA, 326-327
  brain, 326-329, 343-346
  EIP, 282, 286
  polysomal, 336-342
  site, 24, 98
  and splicing
Poly(A)⁻ RNA, 329-331
Polyethylene glycol, 64
Polymorphic markers, 415-417
Polyoma virus, 27, 28, 56-58, 70
Polypeptides, 330
Polyproteins, 354, 376
Polyribosomes, 326, 336-342
Polytene chromosomes, 19, 21, 277
POMC. See Pro-opiomelanocortin
Post-proline-cleaving enzyme, 195-198
Posttranscriptional control, 329
Posttranslational processing, 147-199
Precursor processing, 150-151, 171-187, 349-350
Precursor-product relationship, 153-154
Preoptic area, 266
Preprocalcitonin, 356-357
Preproinsulin, 103, 114, 150, 149
Prepro-NGF, 429
Preprosomatostatin, 351, 354-356
Presecretory proteins, 152
Primary transcript, 28-29, 90
Primate divergence, 94
Procalcitonin, 356
Prodynorphin, 260, 261
Proenkephalin, 239-247
Progesterone, 196, 266, 268, 270-271, 291
Progestin receptor, 266, 268-270
Proglucagon, 154, 156
Proglutamyl peptidase, 198
Proinsulin, 114, 150, 151, 157, 428
Prolactin, 232-237, 312-318, 465
Proline endopeptidase, 197-198
Prolyl-leucyl-glycinamide, 198
Promotor, 39, 48-56
Promotor proximal sequence, 27
Pro-NGF, 428
Pro-opiocortin, 181-185, 210-216. See also Pro-opiomelanocortin
Pro-opiomelanocortin (POMC), 209-211, 468-471
  DNA, 227-229
  evolution, 470-471
  expression, 219-232
  processing, 219-220, 223-224, 469-470

RNA, 225-227
structure, 219, 231-232
Pro-oxyphysin, 174, 178
Propressophysin, 173-175
Prorelaxin, 157
Prosomatostatin, 154, 156, 177, 354-355
Prostate gland, 427
Protease activity, 157-158
Proteins
  DNA binding, 38-41
  evolution, 324
  folding, 162
  heat shock, and calcium, 451-452
  joining, 78, 79
  glycosylation, 161-168
  MBP, 390-394
  -protein interactions, 44
  recognition signals, 47
  SRP, 381-382
  steroid control, 272
  translocation, 373-374
  transmembrane, 378
  unknown, 471-472
  Z and Y, 343, 344
Protein synthesis
  and CCA, 344
  DNA probes for, 462-463
  effect of ecdysteroids on, 279-282
  effect of ionomycin, 448, 449, 451-453
  and glycosylation, 162-164
  and hormone activation, 266
  and lordosis, 268-269, 271
Proteolytic processing, 149-159, 224
Prototrophic donor cells, 63, 67, 68
Pseudogenes, 91-93
PTH (phenylthiohydantoin), 83
P22 phage, 44
Puffs, 277-285
Pulse-chase studies, 151, 162
Putamen, 408-412

Radioimmunoassay, "sandwich" type, 131
Raphe neurons, 5
Rabbit globin genes, 91-93
Rare-message genes, 82-84, 465
Rat
  brain, 106-107, 259, 260, 294-296
  gDNA, 215, 227-228
  growth hormone, 234-235
  hippocampus, 272
  hormone action in, 266, 268
  hypothalamus, 360
  insulin, 104-115, 151
  medullary carcinoma of thyroid, 351-357
  NGF, 427
  retinal antibodies, 138-145
  sexual differentiation in, 289-290
  somatostatin, 350-356, 360
  tfm mutants, 297-299
Receptors, 261, 430. See also specific types
Recognition sequences, 77-79
Recombinant DNA technology, 64, 72, 145
  applications to neurobiology, 229, 463-464, 466-475

Recombinant DNA technology (contd.)
  and brain-specific protein, 332
  and HD gene cloning, 413
  and insulin gene, 103
  and neuropeptide biosynthesis, 350
  and polymorphic markers, 416
  and POMC, 214–216, 231–232
  and rare-message genes, 84
  rat insulin-SV40, 108–115
  and somatic cell genetics, 418
  and splicing, 58
  and transcription, 31–34, 50
Recombinatorial events, 323–324
Relaxin, 157–158
Repressor, 37–41
RER membrane, 162–166, 373–374
  and AChR subunits, 378–382
Restriction enzyme maps, 50, 65, 66, 97, 106, 228
Restriction fragment length polymorphism, 20–21, 66, 107, 416
RET-G, 140–141
RET-P, 141–144
Reticular formation, 4–8
Reticulocytes, 89
Retina, 137–146, 397–399
Retrograde transport, NGF, 430
Retrovirus, 120, 123–124, 167
Reuptake mechanisms, 189
Ribosome-membrane complex, 150
RNA
  abundance classes, 336–341
  cytoplasmic, relation to hnRNA, 25–26
  -DNA codons, 463
  polymerase, 39, 43–44, 50, 301, 308
    effect of steroids on, 307, 308
    and globin gene, 95–96
    and TATA region, 53, 55
  processing, globin genes, 91, 95
  sequence complexity, 324–325
  and splicing, 28–30, 56, 80–81
  synthesis, and promotors 39
  transcription, 26–27, 89
hnRNA, 25–26, 326–327
mRNA
  abundance classes, 25, 30–34, 336–341, 343–346
  AChR subunits, 375–377
  and antibody genes, 76–77
  brain, 323–334, 342–343
  calcium, 449–451
  cap site, 27, 50–51
  copy frequency classes, 324
  differential stability, 33–34
  EIP, 282–285
  formation, pathway of, 25–28
  growth hormone, 236, 237
  IgM, 80–81
  neuropeptide selection, 350
  polyadenylated, 282, 326–329
  POMC, 225–226, 232
  precursor-product relation to hnRNA, 26
  proenkephalin, 246–247
  primary transcripts, 56–58
  promotors, nucleotide sequences defining, 48–56
  protein fitting to, 119
  somatostatin, 355

  steroid effects on, 307–320
  transcription, globin genes, 90, 91, 96
  transcription, SV40-insulin gene, 109–114
  transcription map, adenovirus, 2, 49
  transcription unit, 25–28
tRNA suppressor, 97
Rodents, sexual dimorphism in brain, 266
Rod-specific antibodies, 142
Rot value, 336
Rough endoplasmic reticulum. See RER membrane
Ruby mutation, 399

Salivary gland puff sites, 272–279, 285
Salt concentration, 16–17
Saturation hybridization, 324–325
Sarcoplasmic reticulum, 445–448, 453–455
S cells, 336
Schizophrenia, 204
Schwann cells, 392–394
Sciatic nerve, 363–364, 392–394
Second messenger, 432
Secretin, 357, 370
Secretory cell, 149–151
Secretory proteins, 149–159
Seizure activity, 207
Sensory neurons, 425–427, 430
Sequence complexity, 324–330, 343–346
Sequence divergence, 91–93
Serine, 162, 166
Serotonin, 2, 3, 5, 189, 273, 411
Serum gastrin, 368–369, 370
Sexual differentiation, 266, 272, 289, 294–299
Shiverer mutant, 390–394
Sialic acid, 165
Signal hypothesis, 152
Signal pairs, 254
Signal peptide, 103, 114, 123–124, 211
Signal recognition protein (SRP), 373, 381–382
Signal sequence, 162, 351, 354, 357
Sindbis virus, 167
Sodium dipropylacetate (DPA), 343
Somatic cell hybridization, 63–72, 81–82, 129, 312–318, 403, 418
Somatostatin, 149
  anglerfish and rat, 350–355
  biosynthesis, 349–358, 366–370
  effect of GABA on, 365–366
  in HD, 411
  hypothalamic and extrahypothalamic, 360–366
  intestinal, 355–356
  localization and function, 359
  in nervous system, 359–372
  neuromodulation, 361
  and norepinephrine, 3
  precursors. See Preprosomatostatin; Prosomatostatin
  release, 363–366
S100, 50
S1 nuclease, 111
Songbirds, 266, 272
Southern blot technique, 66, 67, 70
Sperm whale, 87
Spinal cord
  dynorphin in, 249

SUBJECT INDEX

somatostatin in, 362, 363
Spleen-myeloma hybrids, 129, 138–139
Splicing, 28–29, 56–58
  alternate patterns, 56, 58, 80–81, 349, 356
  insulin genes, 105, 111–113
  and polyadenylation, 29–30
  random, 58
  in β-thalassemia, 98
Staphylococcus aureus, 124
Steel mutation, 399
Stereospecific binding, 251–253
Steroid hormones
  acceptor, 312
  action in brain, 265–275
  direct and indirect effects, 308
  in Drosophila, 277
  effects, 265–273
  and gene expression, 238–320
  genomic and nongenomic effects, 272
  and mRNA synthesis, 307–320
  and synaptic transmission, 271–273
  and transcription, 307–308
Steroid receptors
  DNA binding, 294
  genetics and development of, 289–306
  heterogeneity and control of, 312
  mechanisms of action, 290–293
  mutants, 293, 308
  phosphorylation, 308
  and steroid action, 307–309
Stop-transfer sequence, 382
Stria terminalis, 267–268
Structural genes, 323, 329, 401
Submaxillary gland, 427, 428
Substance P, 3, 5, 6, 197–198, 411–412
Substantia nigra, 5, 6
Superior colliculus, 144
Supraoptic nucleus, 171, 173, 175–176, 198
SV40, 27, 28, 53, 70, 96
  -rat insulin recombinant, 108–115
  splicing, 56–58
Switch regions, 50
Sympathetic nervous system, 425–427
Sympathetic neurons, 430, 437–443
Synaptic transmission, 271–272
Synaptogenesis, 346
Syn conformation, DNA, 14–15
Synkaryon, 64
Syntenic genes, 403

T_3, 237, 315
TAT. See Tyrosine aminotransferase
TATA box, 52–56, 90–91, 96, 105
T cells, 308
Testosterone
  androgen binding, 289, 294, 299–301
  effect on adult rat, 268
  and neurite outgrowth, 266–267
  regulation of NGF, 427
  and sexual differentiation, 266
Tetanus antitoxin, 133
Tetrahymena paraformis, 215
Tfm mutants, 297–299
Thalassemia, 94–95, 98

σ receptor, 258
Thiocyanate, 181
Thiorphan, 190
3' splicing site, 56–58. See also Poly(A) RNA
Threonine, 162, 166
Thymidine kinase, 53, 67–68
Thyroid-releasing hormone (TRH), 3, 149, 315
  cleavage, 197, 198
  in HD, 411
  and somatostatin, 359
  transport, 189
Tiger salamander, 142
Tissue specificity, 285, 308–309
Torpedo, 373, 375, 378
Toxin, 133–134
Transfection, 108
Transcript, primary, 28–29, 56–58
Transcription
  adenovirus 2, 48–49, 51–52
  and calcium, 452–453
  globin gene, 89, 90, 91, 95–96
  and NGF, 431–433
  poly(A)⁻ mRNAs, 331
  promotor, 39, 52–56
  puffing, 277
  and steroids, 292, 307–308
  SV40-insulin gene recombinant, 109–114
  TATA region, 52–56
  unit, 25–28
Transcriptional control, 30–34
Transformation frequencies, 70
Transgenomes, 68
Translation, AChR, 373–385
Transmembrane proteins, 378
Transmitter transition, 442
Transplantation antigens, 137
Transport mechanisms, 1–2. See also Axonal transport
Triiodothyronine. See T_3
Tripartite leader, 28
Trophic and instructive factors, 423–456
Trypsin, 249, 378
Trypsin-like peptidases, 157, 158, 177
Tryptophan oxygenase, 313
Tuberoinfundibular somatostatinergic system, 360–361
Tumor immunology, 137, 213
Tunicamycin, 213
Tyrosinase, 397, 399
Tyrosinase aminotransferase (TAT), 309–318
Tyrosine hydroxylase (TH), 335, 427, 430

Ultraviolet (UV) irradiation, 26

Vas deferens, 257–258
Vasoactive intestinal polypeptide (VIP), 3, 357, 411
Vasopressin, 173, 175, 198, 208, 232
Ventromedial hypothalamus, 266, 271, 273
Vesicular stomatitis virus (VSV), 165, 374, 380, 382
Vimentin, 343, 344
Virus replication, 26–28, 49, 56–58
Visual system, 397–399
V domain, 75–82

Weaver mutant, 394–397

X chromosome, 403
Xenopus laevis, 50
X-ray crystallography, 11, 87–88

Y chromosome, 403
Yeast, 71

Zinc, 428
Zucker rats, 204